INORGANIC PLANT NUTRITION

INORGANIC PLANT NUTRITION

Hugh G. Gauch

UNIVERSITY OF MARYLAND

DOWDEN, HUTCHINSON & ROSS, Inc. Stroudsburg, Pa.

Exclusive distributor outside the United States
and Canada: John Wiley & Sons, Inc.

To my wife, Martha, our son, Hugh, Jr.,
our daughter, Susan Coles,
and her husband, Richard Coles

Preface

Students, teachers, and researchers have long felt and expressed the need of a book on the mineral nutrition of plants. This book was written to fullfil that need. It provides a text for courses in mineral nutrition, and will serve as a companion book for courses in advanced plant physiology, soil fertility, and technologies of various horticultural and agronomic crops. It is a sourcebook for student, faculty, industrial, and governmental researchers in plant nutrition.

There is intensive coverage of the literature from 1960 on, but certain earlier, important researches are also included. There are over 2600 references complete with titles of the papers.

The book begins with a discussion of certain aspects of water, soils, and soil–plant relationships which are vital to an understanding of the other chapters. Owing to the vast number of papers which have appeared in recent years, the chapters on aspects of absorption and on roles of elements cover the many important contributions in these areas. The last two of the 18 chapters cover limiting factors in plant growth, with emphasis on mineral nutrition as a limiting factor.

Various members of the Department of Botany of the University of Maryland contributed to my endeavor. I am particularly indebted to Dr. Robert W. Krauss, Head of our Department, and to the University of Maryland for granting a 6-month sabbatical leave. Certain of my colleagues, Drs. Neal Barnett, Raymond Galloway, Edward Karlander, and Glenn Patterson, read portions of the manuscript and offered helpful suggestions, particularly on the chapters dealing with roles and with permeability and membranes. Dr. Everette Legett, formerly with the U. S. Department of Agriculture (now with the Department of Agronomy, University of Kentucky) read the chapters on absorption, and the one on space, organic acids, and source of nutrients. In acknowledging the help of these very capable individuals, I absolve them of any errors of commission or omission, and accept responsibility for the final form of the manuscript.

Special thanks are extended to Mrs. Robert Stewart, Librarian, National Agricultural Library, Beltsville, Maryland for the zeal with which she pursued and obtained publications, some of which are quite obscure.

I am very much indebted to Dr. Cecil H. Wadleigh, currently Science Advisor to the Secretary of Agriculture, with whom it was my pleasure and privilege to be associated in my early career at the U. S. Salinity Laboratory, Riverside, California, and to his wife, Clarice. From him I learned much about mineral nutrition, water relations, and soils. Special thanks are extended to my typist-daughter, Mrs. Richard Coles, for her superb typing and for catching errors along the way; also to her husband for his forebearance during the hectic typing ordeal. I should also like to refer to my friends, Donald Harvey, author of a very interesting book on the Civil Service Commission, and his wife, Winifred. Not only did Donald acknowledge me in his book but, at my request, he persistently and mercilessly prodded me concerning the progress of my book throughout its gestation.

Hugh G. Gauch

March 1972

Contents

CHAPTER 1

Historical Background

When considering the history of plant nutrition, most writers refer to Van Helmont (1577–1644) and quote his famous "experiment." Although his conclusions were incorrect, he at least initiated scientific enquiry into the nature of plant nutrition. He planted a willow branch, weighing 5 lb, in a tub containing 300 lb of dry soil and added nothing but water. A cover prevented dust from falling into the soil. At the end of 5 yr, the branch had grown into a tree weighing 164 lb, but the soil weighed only 2 oz less than at the beginning of the experiment. Inasmuch as Van Helmont did not calculate his results on a dry weight basis, he reasoned that the increase in weight had come from water and not from soil, and that the tree had made all of its substance from water. He was unaware that the willow had gained additional C (from CO_2 of the atmosphere) or essential elements from the soil. He had, however, established that the nutriment of plants is not obtained similarly to that of animals.

The history of plant nutrition, and the various investigators who contributed various new ideas, is much too long to present in detail. For excellent historical coverage, the reader should consult Reed's *A Short History of the Plant Sciences* (1942). Also recommended for this subject is Moulton's *Liebig and after Liebig—A Century of Progress in Agricultural Chemistry* (1942). In a very interesting presentation, Steward (1963b) discusses trends in the inorganic nutrition of plants.

It is interesting to note that the more meaningful developments in plant nutrition began in the 1860s with Sachs, Knop, Pfeffer, and others, who began the practice of growing plants in artificial, chemically defined media in water culture. Following their leads, and beginning in the early 1900s, plant physiologists used water and sand cultures to determine which elements were essential for plant growth. From these early investigations, it was concluded that higher plants generally require 10 elements, namely, C, H, O, P, K, Ca, Mg, N, S, and Fe. In 1910, B—a "minor" or "trace" element—was shown to be essential, and since that time Mn, Cu, Zn, Mo, and Cl have been added to the list. Thus the search for additional "essential elements" was an intense one from around 1910 until the 1950s.

For approximately 20 years, beginning around 1915, there was a concentrated search for "the best nutrient solution" for plants. Many investigators joined this effort, since it was the current fad. The pH of solutions and of cells and tissues was deemed all-important for many years, starting in the 1920s, and numerous investigators turned to that subject. As Miller (1943) noted in his article, "Forty Years of Plant Physiology," many investigators have followed, at a particular time, whatever was the fashionable topic of the day.

Although it stresses primarily the research by Hoagland and colleagues, *Lectures on the Inorganic Nutrition of Plants* (Hoagland, 1944) summarizes the chief interests in mineral nutrition prior to 1944. The latest, excellent edition of *Hunger Signs in Crops—A Symposium* (Sprague, 1964) covers literature on deficiencies and illustrates many deficiency symptoms. The *Bibliography of the Literature on the Minor Elements and Their Relation to Plant and Animal Nutrition* (Chilean Nitrate Educational Bureau, 1948a,b, 1951, 1953, 1955) contains abstracts or synopses of approximately 15,000 papers.

The student may best gain a historical perspective of the development of the field of mineral nutrition by referring to various review articles that have appeared over the years. In addition, a list of such articles serves the further purpose of indicating leaders in the field of plant nutrition. Starting in 1932, the *Annual Review of Biochemistry* published review articles on the mineral nutrition of plants or on some aspect of nutrition. Chronologically, the following authors contributed articles. Hoagland (1932, 1933), Lundegardh (1934), Steward (1935), Mozé (1936), Bennet-Clark (1937), Collander (1937), Gregory (1937), Shive and Robbins (1939), Lipman (1940), Sommer (1941), Petrie (1942), Arnon (1943), Richards (1944), Chapman (1945), Lundegardh (1947), Burstrom (1948), and Wadleigh (1949).

In 1950, the *Annual Review of Plant Physiology* was initiated, and various researchers have reviewed the mineral nutrition of plants or some aspect of the subject. Chronologically, the reviewers are: Mulder (1950), Stout and Overstreet (1950), Hewitt (1951), Robertson (1951), Leeper (1952), Overstreet and Jacobson (1952), Ulrich (1952), Burris (1953), Wood (1953), Boynton (1954), Ketchum (1954), McElroy and Nason (1954), Lundegardh (1955), Pirson (1955), Webster (1955), Williams (1955), Brown (1956), Epstein (1956), Briggs and Robertson (1957), Collander (1957), Esau et al. (1957),

Gauch (1957), Steward and Pollard (1957), Bernstein and Hayward (1958), Krauss (1958), Reuther et al. (1958), Steinberg and Tso (1958), Yemm and Folkes (1958), Broyer and Stout (1959), Burris (1959), Laties (1959), Wittwer and Teubner (1959), Bollard (1960), Yocum (1960), Zimmermann (1960), Fried and Shapiro (1961), Nicholas (1961), Virtanen (1961), Webster (1961), Dainty (1962), Smith (1962), Wilson (1962), Brown (1963), Carnahan and Castle (1963), Gerloff (1963), Russell (1963), Stahmann (1963), Stewart (1963), Epstein and Jefferies (1964), Hutner and Provasoli (1964), Kessler (1964), Brouwer (1965), Kretovich (1965), Bollard and Butler (1966), Burris (1966), Evans and Sorger (1966), Bond (1967), Fowden (1967), Franke (1967), Mans (1967), Thompson (1967), Barber (1968), Leggett (1968), O'Kelley (1968), Price (1968), Beevers and Hageman (1969), Laties (1969), Lewin and Reimann (1969), Milthorpe and Moorby (1969), Shrift (1969), Stadelmann (1969), Yoshida (1969), Aleem (1970), Boulter (1970), Eschrich (1970), Packer et al. (1970), Tukey (1970), Barrs (1971), Bergersen (1971), Canny (1971), Cook (1971), MacRobbie (1971), and Schwartz (1971).

In the foregoing list fragmentation of the general topic "mineral nutrition of plants" is indicated by the appearance of several reviews each year.

In a treatise edited by Steward (1959), there are chapters on aspects of nutrition by Collander (1959), Bennet-Clark (1959), Steward and Sutcliffe (1959), Swanson (1959), and Biddulph (1959). In a later volume (Steward, 1963a), additional aspects of nutrition are presented by Bould and Hewitt (1963), Hewitt (1963), Nicholas (1963), and Nason and McElroy (1963).

For information on the specific subjects covered by each of the above-mentioned authors, the reader should consult the literature cited at the end of this chapter.

LITERATURE CITED

Aleem, M. I. H. 1970. Oxidation of inorganic nitrogen compounds. Annu. Rev. Plant Physiol. 21:67–90.

Arnon, D. I. 1943. Mineral nutrition of plants. Annu. Rev. Biochem. 12:493–528.

Barber, D. A. 1968. Microorganisms and the inorganic nutrition of higher plants. Annu. Rev. Plant Physiol. 19:71–88.

Barrs, H. D. 1971. Cyclic variations in stomatal aperature, transpiration, and leaf water potential under constant environmental conditions. Annu. Rev. Plant Physiol. 22:223–236.

Beevers, L., and R. H. Hageman. 1969. Nitrate reduction in higher plants. Annu. Rev. Plant Physiol. 20:495–522.

Bennet-Clark, T. A. 1937. Organic acids of plants. Annu. Rev. Plant Physiol. 6:579–594.

Bennet-Clark, T. A. 1959. Water relations of cells, pp. 105–191. *In* F. C. Steward (ed.) Plant physiology – A treatise, Vol. II. Academic, New York.

Bergersen, F. J. 1971. Biochemistry of symbiotic nitrogen fixation in legumes. Annu. Rev. Plant Physiol. 22:121–140.

Bernstein, L., and H.E. Hayward. 1958. Physiology of salt tolerance. Annu. Rev. Plant Physiol. 9:25−46.

Biddulph, O. 1959. Translocation of inorganic solutes, pp. 553−603. *In* F. C. Steward (ed.) Plant physiology − A treatise, Vol. II. Academic, New York.

Bollard, E. G. 1960. Transport in the xylem. Annu. Rev. Plant Physiol. 11:141−166.

Bollard, E. G., and G. W. Butler, 1966. Mineral nutrition of plants. Annu. Rev. Plant Physiol. 17:77−112.

Bond, G. 1967. Fixation of nitrogen by higher plants other than legumes. Annu. Rev. Plant Physiol. 18:107−126.

Bould, C., and E. J. Hewitt. 1963. Mineral nutrition of plants in soils and in culture media, pp. 15−133. *In* Steward, F. C. (ed.) Plant physiology − A treatise, Vol. III, Academic New York.

Boulter, D. 1970. Protein synthesis in plants. Annu. Rev. Plant Physiol. 21:91−114.

Boynton, D. 1954. Nutrition by foliar application. Annu. Rev. Plant Physiol. 5:31−54.

Briggs, G. E., and R. N. Robertson. 1957. Apparent free space. Annu. Rev. Plant Physiol. 8:11−30.

Brouwer, R. 1965. Ion absorption and transport in plants. Annu. Rev. Plant Physiol. 16:241−266.

Brown, J. C, 1956. Iron chlorosis. Annu. Rev. Plant Physiol. 7:171−190.

Brown, J. C. 1963. Interactions involving nutrient elements. Annu. Rev. Plant Physiol. 14:93−106.

Broyer, T. C., and P. R. Stout. 1959. The macronutrient elements. Annu. Rev. Plant Physiol. 10:277−300.

Burris, R. H. 1953. Organic acids in plant metabolism. Annu. Rev. Plant Physiol. 4:91−114.

Burris, R. H. 1959. Nitrogen nutrition. Annu. Rev. Plant Physiol. 10:301−328.

Burris, R. H. 1966. Biological nitrogen fixation. Annu. Rev. Plant Physiol. 17:155−184.

Burström, H. 1948. Mineral nutrition of plants. Annu. Rev. Biochem. 17:579−600.

Canny, M. J. 1971. Translocation: Mechanics and kinetics. Annu. Rev. Plant Physiol. 22:237−260.

Carnahan, J. E., and J. E. Castle. 1963. Nitrogen fixation. Annu. Rev. Plant Physiol. 14:125−136.

Chapman, H. D. 1945. Mineral nutrition of plants. Annu. Rev. Biochem. 14:709−732.

Chilean Nitrate Educational Bureau. 1948a. Bibliography of the literature on the minor elements and their relation to plant and animal nutrition, Vol. I, 4th ed. Chilean Nitrate Educational Bur., New York. 1037 pp.

Chilean Nitrate Educational Bureau. 1948b. Bibliography of the literature on sodium and iodine in relation to plant and animal nutrition, Vol. I, 1st ed. Chilean Nitrate Educational Bur., New York. 123 pp.

Chilean Nitrate Educational Bureau. 1951. Bibliography of the literature on the minor elements and their relation to plant and animal nutrition, Vol. II, 4th ed. Chilean Nitrate Educational Bur., New York. 152 pp.

Chilean Nitrate Educational Bureau. 1953. Bibliography of the literature on the minor elements and their relation to plant and animal nutrition, Vol. III, 4th ed. Chilean Nitrate Educational Bur., New York. 117 pp.

Chilean Nitrate Educational Bureau. 1955. Bibliography of the literature on the minor elements and their relation to plant and animal nutrition, Vol. IV, 4th ed. Chilean Nitrate Educational Bur., New York. 150 pp.

Collander, R. 1937. Permeability. Annu. Rev. Biochem. 6:1−18.

Collander, R. 1957. Permeability of plant cells. Annu. Rev. Plant Physiol. 8:335−348.

Collander, R. 1959. Cell membranes: Their resistance to penetration and their capacity for transport, pp. 3–102. *In* F. C. Steward (ed.) Plant physiology – A treatise, Vol. II. Academic, New York.

Cook, G. M. W. 1971. Membrane structure and function. Annu. Rev. Plant Physiol. 22:97–120.

Dainty, J. 1962. Ion transport and electrical potentials in plant cells. Annu. Rev. Plant Physiol. 13:379–402.

Epstein, E. 1956. Mineral nutrition of plants: Mechanisms of uptake and transport. Annu. Rev. Plant Physiol. 7:1–24.

Epstein, E., and R. L. Jefferies. 1964. The genetic basis of selective ion transport in plants. Annu. Rev. Plant Physiol. 15:169–184.

Esau, K., H. B. Currier, and V. I. Cheadle. 1957. Physiology of phloem. Annu. Rev. Plant Physiol. 8:349–374.

Eschrich, W. 1970. Biochemistry and fine structure of phloem in relation to transport. Annu. Rev. Plant Physiol. 21:193–214.

Evans, H. J., and G. J. Sorger. 1966. Role of mineral elements with emphasis on the univalent cations. Annu. Rev. Plant Physiol. 17:47–76.

Fowden, L. 1967. Aspects of amino acid metabolism in plants. Annu. Rev. Plant Physiol. 18:85–106.

Franke, W. 1967. Mechanisms of foliar penetration of solutions. Annu. Rev. Plant Physiol. 18:281–300.

Fried, M., and R. E. Shapiro. 1961. Soil-plant relationships in ion uptake. Annu. Rev. Plant Physiol. 12:91–112.

Gauch, H. G. 1957. Mineral nutrition of plants. Annu. Rev. Plant Physiol. 8:31–64.

Gerloff, G. C. 1963. Comparative mineral nutrition of plants. Annu. Rev. Plant Physiol. 14:107–124.

Gregory, F. G. 1937. Mineral nutrition of plants. Annu. Rev. Biochem. 6:557–578.

Hewitt, E. J. 1951. The role of the mineral elements in plant nutrition. Annu. Rev. Plant Physiol. 2:25–52.

Hewitt, E. J. 1963. The essential nutrient elements: Requirements and interactions in plants, pp. 137–360. *In* F. C. Steward (ed.) Plant physiology – A treatise, Vol. III. Academic, New York.

Hoagland, D. R. 1932. Mineral nutrition of plants. Annu. Rev. Biochem. 1:618–636.

Hoagland, D. R. 1933. Mineral nutrition of plants. Annu. Rev. Biochem. 2:471–484.

Hoagland, D. R. 1944. Lectures on the inorganic nutrition of plants (Prather lectures at Harvard University). Chronica Botanica, Waltham, Mass., 226 pp.

Hutner, S. H., and L. Provasoli. 1964. Nutrition of algae. Annu. Rev. Plant Physiol. 15:37–56.

Kessler, E. 1964. Nitrate assimilation by plants. Annu. Rev. Plant Physiol. 15:57–72.

Ketchum, B. H. 1954. Mineral nutrition of phytoplankton. Annu. Rev. Plant Physiol. 5:55–74.

Krauss, R. W. 1958. Physiology of the fresh-water algae. Annu. Rev. Plant Physiol. 9:207–244.

Kretovich, W. L. 1965. Some problems of amino acid and amide biosynthesis in plants. Annu. Rev. Plant Physiol. 16:141–154.

Laties, G. G. 1959. Active transport of salt into plant tissue. Annu. Rev. Plant Physiol. 10:87–112.

Laties, G. G. 1969. Dual mechanisms of salt uptake in relation to compartmentation and long-distance transport. Annu. Rev. Plant Physiol. 20:89–116.

Leeper, G. W. 1952. Factors affecting availability of inorganic nutrients in soils with special reference to micronutrient metals. Annu. Rev. Plant Physiol. 3:1–16.

Leggett, J. E. 1968. Salt absorption by plants. Annu. Rev. Plant Physiol. 19:333–346.

Lewin, J., and B. E. F. Reimann. 1969. Silicon and plant growth. Annu. Rev. Plant Physiol. 20:289–304.

Lipman, C. B. 1940. Aspects of inorganic metabolism in plants. Annu. Rev. Biochem. 9:491–508.

Lundegårdh, H. 1934. Mineral nutrition of plants. Annu. Rev. Biochem. 3:485–500.

Lundegårdh, H. 1947. Mineral nutrition of plants. Annu. Rev. Biochem. 16:503–528.

Lundegårdh, H., 1955. Mechanisms of absorption, transport, accumulation, and secretion of ions. Annu. Rev. Plant Physiol. 6:1–24.

McElroy, W. D., and A. Nason. 1954. Mechanism of action of micronutrient elements in enzyme systems. Annu. Rev. Plant Physiol. 5:1–30.

MacRobbie, E. A. C. 1971. Fluxes and compartmentation in plant cells. Annu. Rev. Plant Physiol. 22:75–96.

Mans, R. J. 1967. Protein synthesis in higher plants. Annu. Rev. Plant Physiol. 18: 127–146.

Miller, E. C. 1943. Forty years of plant physiology. Science 97:315–319.

Milthorpe, F. L., and J. Moorby. 1969. Vascular transport and its significance in plant growth. Annu. Rev. Plant Physiol. 20:117–138.

Moulton, F. R. 1942. Liebig and after Liebig – A century of agricultural chemistry. Pub. 16. Amer. Ass. Advance. Sci., Washington, D.C. 111 pp.

Mozé, P. 1936. The role of special elements (boron, copper, zinc, manganese, etc.) in plant nutrition. Annu. Rev. Biochem. 5:525–538.

Mulder, E. G. 1950. Mineral nutrition of plants. Annu. Rev. Plant Physiol. 1:1–24.

Nason, A., and W. D. McElroy. 1963. Modes of action of the essential mineral elements, pp. 451–536. *In* F. C. Steward (ed.) Plant physiology – A treatise, Vol. II. Academic, New York.

Nicholas, D. J. D. 1961. Minor mineral elements. Annu. Rev. Plant Physiol. 12:63–90.

Nicholas, D. J. D. 1963. Inorganic nutrient nutrition of microorganisms, pp. 363–447. *In* F. C. Steward (ed.) Plant physiology – A treatise, Vol. III. Academic, New York.

O'Kelley, J. C. 1968. Mineral nutrition of algae. Annu. Rev. Plant Physiol. 19:89–112.

Overstreet, R., and L. Jacobson. 1952. Mechanisms of ion absorption by roots. Annu. Rev. Plant Physiol. 3:189–206.

Packer, L., S. Murakami, and C. W. Mehard. 1970. Ion transport in chloroplasts and plant mitochondria. Annu. Rev. Plant Physiol. 21:271–304.

Petrie, A. H. K. 1942. Mineral nutrition of plants. Annu. Rev. Biochem. 11:595–614.

Pirson, A. 1955. Functional aspects in mineral nutrition of green plants. Annu. Rev. Plant Physiol. 6:71–114.

Price, C. A. 1968. Iron compounds and plant nutrition. Annu. Rev. Plant Physiol. 19:239–248.

Reed, H. S. 1942. A short history of the plant sciences. Chronica Botanica, Waltham, Mass. 320 pp.

Reuther, W., T. W. Embleton, and W. W. Jones. 1958. Mineral nutrition of tree crops. Annu. Rev. Plant Physiol. 9:175–206.

Richards, F. J. 1944. Mineral nutrition of plants. Annu. Rev. Biochem. 13:611–630.

Robertson, R. N. 1951. Mechanism of absorption and transport of inorganic nutrients in plants. Annu. Rev. Plant Physiol. 2:1–24.

Russell, R. S. 1963. The extent and consequences of the uptake by plants of radioactive nuclides. Annu. Rev. Plant Physiol. 14:271–294.

Schwartz, M. 1971. The relation of ion transport to phosphorylation. Annu. Rev.

Plant Physiol. 22:469−484.

Shive, J. W., and W. R. Robbins. 1939. Mineral nutrition of plants. Annu. Rev. Biochem. 8:503−520.

Shrift, A. 1969. Aspects of selenium metabolism in higher plants. Annu. Rev. Plant Physiol. 20:475−494.

Smith, P. F. 1962. Mineral analysis of plant tissues. Annu. Rev. Plant Physiol. 13:81−108.

Sommer, A. L. 1941. Mineral nutrition of plants. Annu. Rev. Biochem. 10:471−490.

Sprague, H. B. (ed.). 1964. Hunger signs in crops − A symposium, 3rd ed. David McKay, New York. 461 pp.

Stadelmann, E. J. 1969. Permeability of the plant cell. Annu. Rev. Plant Physiol. 20:585−606.

Stahmann, M. A. 1963. Plant proteins. Annu. Rev. Plant Physiol. 14:137−158.

Steinberg, R. A., and T. C. Tso. 1958. Physiology of the tobacco plant. Annu. Rev. Plant Physiol. 9:151−174.

Steward, F. C. 1935. Mineral nutrition of plants. Annu. Rev. Biochem. 4:519−544.

Steward, F. C. (ed.). 1959. Plant physiology − A treatise. Vol. II. Plants in relation to water and solutes. Academic, New York. 758 pp.

Steward, F. C. (ed.). 1963a. Plant physiology − A treatise. Vol. III. Inorganic nutrition of plants. Academic, New York. 811 pp.

Steward, F. C. 1963b. Trends in the inorganic nutrition of plants, pp. 1−14. *In* F. C. Steward (ed.). Plant physiology − A treatise, Vol. III. Academic, New York.

Steward, F. C., and J. K. Pollard. 1957. Nitrogen metabolism in plants: Ten years in retrospect. Annu. Rev. Plant Physiol. 8:65−114.

Steward, F. C., and J. F. Sutcliffe. 1959. Plants in relation to inorganic salts, pp. 253−478. *In* F. C. Steward (ed.) Plant physiology − A treatise, Vol. II. Academic, New York.

Stewart, I. 1963. Chelation in the absorption and translocation of mineral elements. Annu. Rev. Plant Physiol. 14:295−310.

Stout, P.R., and R. Overstreet. 1950. Soil chemistry in relation to inorganic nutrition of plants. Annu. Rev. Plant Physiol. 1:305−342.

Swanson, C. A. 1959. Translocation of organic solutes, pp. 481−551. *In* F. C. Steward (ed.) Plant physiology − A treatise, Vol. II. Academic, New York.

Thompson, J. F. 1967. Sulfur metabolism in plants. Annu. Rev. Plant Physiol. 18:59−84.

Tukey, H. B., Jr. 1970. The leaching of substances from plants. Annu. Rev. Plant Physiol. 21:305−324.

Ulrich, A. 1952. Physiological bases for assessing the nutritional requirements of plants. Annu. Rev. Plant Physiol. 3:207−228.

Virtanen, A. I. 1961. Some aspects of amino acid synthesis in plants and related subjects. Annu. Rev. Plant Physiol. 12:1−12.

Wadleigh, C. H. 1949. Mineral nutrition of plants. Annu. Rev. Biochem. 18:655−678.

Webster, G. C. 1955. Nitrogen metabolism. Annu. Rev. Plant Physiol. 6:43−70.

Webster, G. C. 1961. Protein synthesis. Annu. Rev. Plant Physiol. 12:113−132.

Williams, R. F. 1955. Redistribution of mineral elements during development. Annu. Rev. Plant Physiol. 6:25−42.

Wilson, L. G. 1962. Metabolism of sulfate: Sulfate reduction. Annu. Rev. Plant Physiol. 13:201−224.

Wittwer, S. H., and F. G. Teubner. 1959. Foliar absorption of mineral nutrients. Annu. Rev. Plant Physiol. 10:13−30.

Wood, J. G. 1953. Nitrogen metabolism of higher plants. Annu. Rev. Plant Physiol. 4:1−22.

Yemm, E. W., and B. F. Folkes. 1958. The metabolism of amino acids and proteins in plants. Annu. Rev. Plant Physiol. 9:245–280.

Yocum, C. S. 1960. Nitrogen fixation. Annu. Rev. Plant Physiol. 11:25–36.

Yoshida, S. 1969. Biosynthesis and conversion of aromatic amino acids in plants. Annu. Rev. Plant Physiol. 20:41–62.

Zimmermann, M. H. 1960. Transport in the phloem. Annu. Rev. Plant Physiol. 11:167–190.

CHAPTER 2

Water and Soil—Plant Relations

WATER

Buswell and Rodebush (1956) provided an excellent review of the nature and properties of water. They noted that the formula for water is not simply H_2O; nor is water a single substance. The purest water that can be made contains six isotopes which can be combined in 18 different ways. If one includes the various kinds of ions into which the addition or removal of an electron may transform water's atoms, it is found that pure water contains no fewer than 33 entities.

Water is the medium in which life presumably originated and by means of which all organisms exist. It is a solvent par excellence. Water serves as the "plasticizer" for proteins and is responsible for their flexibility (Buswell and Rodebush, 1956).

WATER ABSORPTION

The absorption of water by plants is controlled by physical forces, osmotic forces, and combinations of these forces in soils. In order to appreciate the significance of these forces as they affect the absorption of water (and in turn affect the absorption of inorganic nutrients), it is necessary to understand certain terms and their derivations.

Physical Constants

Certain soil constants have been proposed, and some of them can be obtained or estimated in several ways. Some terms have been criticized, and some have been replaced by others. It is important to understand the terms and their significance with regard to the absorption of water and the growth of plants.

Wilting Coefficient (WC). Briggs and Shantz (1911a,b) introduced this term and defined it as "the water content of the soil at the time when the leaves of the plant under study first undergo permanent reduction in water supply. By a permanent reduction is meant a degree of wilting from which the leaves cannot recover in an approximately saturated atmosphere unless water is added to the soil." The value is expressed as a percentage of the dry weight of the soil.

A direct determination of the WC, or "observed WC," involved growing plants in sealed containers until the plants wilted and did not recover when placed in a saturated atmosphere unless water was added to the soil. When plants reached this condition, the soil was sampled and the percentage of moisture determined. The method was used mostly for seedling plants (generally sunflower) in small containers. It was difficult to obtain a sharp end point of wilting with larger plants, and in large containers the soil moisture often was not evenly depleted throughout the soil mass.

Briggs and Shantz (1912) claimed that it was possible to calculate the WC, without growing plants, by the use of physical, laboratory measurements on soils. Although later work questioned the reliability of these simpler, indirect measurements for determining WC or wilting percentage (WP), their approach is certainly of considerable historical significance and interest.

The moisture equivalent of a soil is the "percentage of water that it can retain in opposition to a centrifugal force 1000 times that of gravity." According to Briggs and Shantz (1912), the WC could be calculated by:

$$WC = \frac{\text{moisture equivalent}}{1.84 \pm 0.013}$$

Later work by Veihmeyer and Hendrickson (1927, 1928), Veihmeyer et al. (1927), Hendrickson and Veihmeyer (1929b), Wadsworth and Das (1930), and Furr and Reeve (1945) showed that the constant 1.84 did not apply to all soils; the value could be as low as 1.37 (Wadsworth and Das, 1930), or as high as 3.82 (Veihmeyer and Hendrickson, 1928; Hendrickson and Veihmeyer, 1929b).

The hygroscopic coefficient is the "percentage of water which a dry soil attains when placed in a saturated atmosphere." Its relationship to the WC is indicated by:

$$WC = \frac{\text{hygroscopic coefficient}}{0.68 \pm 0.012}$$

The moisture-holding capacity is defined (Briggs and Shantz, 1912) as "the percentage of water that a soil can retain in opposition to the force of gravity when free drainage is provided. Its relationship to the wilting coefficient is:

$$WC = \frac{\text{moisture-holding capacity} - 21}{2.9 \pm 0.06}$$

Permanent Wilting Percentage (PWP). Veihmeyer and Hendrickson (1927) found that the WC did not bear a constant relationship to the previously mentioned soil constants as determined by indirect, laboratory measurements. They (Hendrickson and Veihmeyer, 1929a) therefore proposed the term permanent wilting percentage (PWP) to replace WC in describing the residual soil water percentage at permanent wilting; this term has been generally adopted.

Slatyer (1957) reported that, instead of being a constant, PWP is dependent on the plant being studied, since the osmotic pressure (OP) of plant leaves is the key factor determining the point at which permanent wilting occurs. The OP of leaves may vary from 5 to as high as 202.5 atm (Harris, 1934) for *Atriplex confertifolia*. In addition, Slatyer (1957) noted that the PWP value depended on the technique used for determining it. He suggested that direct determinations of PWP should be abandoned and replaced by physical soil determinations of total soil moisture stress levels of, for example, 15 atm.

First Wilting Percentage. Furr and Reeve (1945) proposed the term first wilting percentage for "wilting of basal leaves and cessation of elongation of growth."

Ultimate Wilting Percentage. Furr and Reeve (1945) suggested the term ultimate wilting percentage for the stage characterized by "wilting of all leaves" of the plant.

15-Atm Value. Richards and Weaver (1944), using a pressure membrane apparatus (Richards, 1941) and studying the same soils for which Furr and Reeve (1945) had determined the values for PWP, reported soil moisture percentages similar to those obtained by Furr and Reeve when they subjected each of the soils to 15 atm of pressure. The 15-atm value is the "percentage of water remaining in the soil when it is subjected to a pressure of 15 atm." Comparing their results with those of Furr and Reeve, Richards and Weaver (1944) reported that at the first wilting percentage the total soil moisture stress ranged from 7 to 16 atm, and at the ultimate wilting percentage from 21 to 36 atm.

Whereas Richards and Weaver (1944) reported that the first permanent wilting percentage corresponded to a soil moisture stress from 7 to 16 atm,

Richards et al. (1949) found that 80% of the stress values in their later work were above 20 atm.

⅓-Atm Value. The ⅓-atm value is the percentage of water remaining in a soil when it is subjected to a tension or negative pressure of ⅓ atm. The ⅓-atm value has been found to correlate closely with field capacity. Haise et al. (1955) reported that field capacity corresponded more closely with $\frac{1}{10}$ rather than the ⅓-atm value.

Field Capacity (FC). Field capacity of a soil is identical with moisture-holding capacity—a term used by Briggs and Shantz (1912)—and pertains to "the percentage of water that a soil can retain in opposition to the force of gravity when free drainage is provided." FC can also be characterized as a water potential of approximately -0.1 to -0.2 atm (Milthorpe and Moorby, 1969).

Total Soil Moisture Stress (TSMS). Soil water may be held physically by soil particles and osmotically by salts dissolved in soil water. Each component can be expressed in terms of atmospheres, and their sum is the integrated moisture stress (Wadleigh and Ayers, 1945; Wadleigh, 1946) or the total soil moisture stress (TSMS) (Wadleigh et al., 1946; Wadleigh and Gauch, 1948).

Water Absorption from Soils

Physical Retention. The physical force with which water is held by a soil may vary from approximately ⅓ atm at FC to as high as 1000 atm (Shull, 1916) in an air-dry Oswego silt loam. A tensiometer (Richards, 1942) can be used to measure the physical retention of water by a soil from 0 to 1 atm of tension.

Furr and Reeve (1945) determined the WP of several soils by using sunflower plants. Richards and Weaver (1944) subjected these same soils to 15 atm of pressure in a pressure membrane apparatus (Fig. 2.1A and B) (Richards, 1941; Reitemeier and Richards, 1944) and found that the percentages of water that remained at that pressure correlated closely with WP values as determined by Furr and Reeve (1945). Richards and Weaver (1944) therefore, concluded that at the WP soil moisture is held with a force or negative pressure of 15 atm.

The relation between soil moisture percentage and moisture tension for a nonsaline soil is shown in Fig. 2.2 (Wadleigh and Ayers, 1945). This soil has a WP of 6.1 and an FC of 18.0%. From approximately 30 to 10% moisture, water is held by a very low tension or negative pressure by the soil. As the WP is approached, tension rises very steeply with only the slightest change in soil moisture percentage. This indicates strikingly why (1) plants in a nonsaline soil may appear normal one moment and very wilted the next, and (2) why *most plants* wilt at *roughly the same* soil moisture percentage in a given soil.

When soil is extremely dry, and the remaining moisture is held with considerable tension (approximately 1000 atm), water may move from the plant roots to the soil (Breazeale et al., 1950; 1951b; Breazeale and McGeorge,

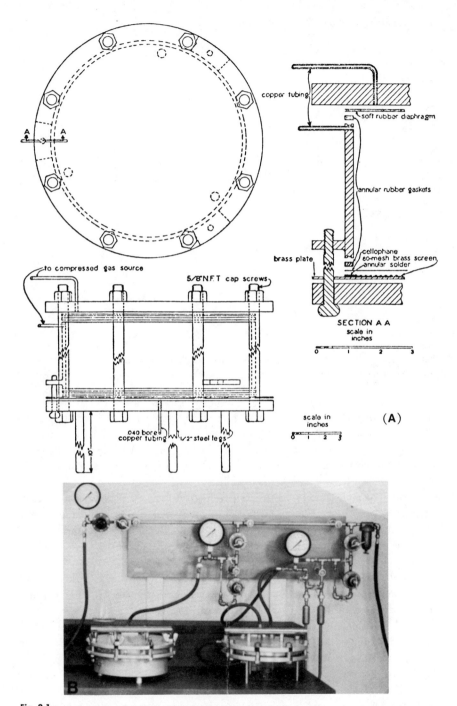

Fig. 2.1

(A) Pressure membrane apparatus for obtaining the soil solution. From L. A. Richards Soil Sci., 51:377–386. © 1941, The Williams and Wilkins Co., Baltimore, Md. (B) A pressure membrane apparatus in use for extraction of the soil solution. From R. F. Reitemeier and L. A. Richards (1944) Soil Sci., 57:119–135. © 1944, The Williams and Wilkins Co., Baltimore, Md.

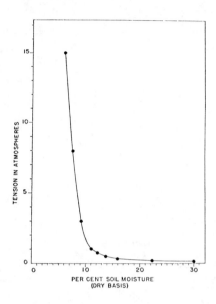

Fig. 2.2

Relationship between soil moisture percentage and moisture tension of a sample of Fallbrook loam. (Wadleigh and Ayers, 1945.)

1953). From an atmosphere of high humidity, tomato plants can absorb water through their leaves and exude water from the roots in considerable volume (Breazeale et al., 1950). If 90%-water-saturated air is passed around tomato roots, plants can reduce soil moisture well below the WP, in fact, to below the hygroscopic coefficient, without definite evidence of wilting (Breazeale et al., 1951a).

A thermocouple system has been developed to measure the relative vapor pressure of water in air near saturation with precision (Richards and Ogata, 1958). It can be used effectively in biological and soil systems at high humidity. The measurement depends on a temperature difference between wet and dry junctions of a sensitive thermocouple.

Osmotic Retention. The extent to which water is withheld "osmotically" from plants may vary from a very low value to an extremely high one depending on the concentration of soluble salts and the percentage of water in soil.

Total Soil Moisture Stress (TSMS). Plant growth is affected by a summation of forces holding water in soils—the physical and osmotic retentions. Each can be expressed in atmospheres, hence they are additive. The sum of these forces has been called the total soil moisture stress (Wadleigh and Ayers, 1945). It is possible to calculate the TSMS (Wadleigh, 1946) and to relate it to the growth of leaves (Wadleigh and Gauch, 1948), or to that of whole plants (Wadleigh et al., 1946).

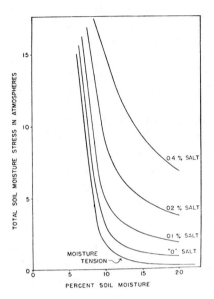

Fig. 2.3
Relation of moisture stress to moisture percentage at different levels of salinity. (Wadleigh and Ayers, 1945.)

When a soil is nonsaline, moisture is held increasingly strongly as the WP is approached (see curve marked "moisture tension" in Fig. 2.3). When salt is present, the slope of the TSMS curve does not show the sharp break characteristic of the moisture tension curve of a nonsaline soil (Wadleigh and Ayers, 1945). With each percentage reduction in soil moisture, the osmotic concentration of the soil solution increases. A reduction in soil moisture from 20 to 10% would, of course, double the osmotic concentration of dissolved salts. Over this range of moisture, in the absence of salt, there is very little change in moisture tension. Thus the more gentle slope of the TSMS curve for saline soil indicates why plants seldom show wilting when grown in the presence of salinity. Presumably, internal adjustments are made within the plant as the external moisture stress builds up.

Factors Affecting Water Absorption

Flooding. Flooding of the soil around plants results in yellowing of leaves (sometimes later even death), epinastic curvature of leaves, formation of callus usually at the water level and, in many species, development of adventitious roots from callus. Lack of water or nutrient absorption could not explain all of the usual symptoms of flooding (Kramer, 1951). It was suggested, therefore, that flooding might interfere with complete downward translocation of auxin, possibly resulting in its accumulation at the water level where callus and, later, adventitious roots are formed. It was also hypothesized that injury and death of leaves may be caused by upward movement of toxic substances from dead roots or the surrounding soil.

Tobacco is very sensitive to flooding and shows two types of reactions (Kramer and Jackson, 1954). If flooding is of short duration, there is temporary but severe wilting caused by a sudden decrease in the permeability of roots to water. Prolonged flooding causes injury and death of roots because of deficient aeration. The latter action is complicated by microorganisms which destroy roots and plug the xylem at the base of the stem.

In daytime flooded shrubs and trees of the Rio Negro in the Amazon reach zero turgor without any external sign of wilting (Scholander and Oliveira Perez, 1968). Sap pressures in these plants, standing in several meters of water, average between -15 and -20 atm in sunshine and above -10 in overcast.

Jackson (1956a) studied the reduced shoot growth and chlorosis of lower leaves of Marglobe tomato and Turkish tobacco plants following flooding. With approach-grafted and split root systems, one portion of the root system could be flooded while the other was not. Evidence indicated that growing shoots are dependent on a functional root system for growth factors (unidentified) other than water or nutrients. It was considered unlikely that toxins from the flooded environment were involved in the plant damage.

Working with Marglobe tomato plants, Jackson (1956b) studied the effects of (1) O_2 deficiency, (2) CO_2 excess, (3) microorganism and their by-products, and (4) debris from dead roots as factors in flooding injury. Of these factors only O_2 deficiency appeared to be clearly related to injury. Removal of CO_2, microorganisms, and their by-products, and dying roots from the root environment did not significantly improve the condition of the plants. Ordinarily, under field conditions, CO_2 does not accumulate to an extent to be toxic or to interfere with water absorption (Kramer, 1959).

Ammonia. Ammonia inhibited water uptake by sugar beet roots, but ATP at 0.2 mM partially or wholly prevented inhibition (Stuart and Haddock, 1968). It was reasoned that ATP might be involved in maintaining the structure of water pathways through the root. Ammonia has been shown to inhibit respiration of cotton seed (Altschul et al., 1946) and excised barley roots, garden beet discs, leaf discs of spinach and sugar beets, and garden beet mitochondria (Wedding and Vines, 1959; Vines and Wedding, 1960). Ammonia did not inhibit water uptake in roots lacking an epidermis, and therefore Stuart and Haddock (1968) concluded that the site of inhibition by ammonia resided in the epidermis.

Ammonia toxicity becomes increasingly acute above pH 7, when an ever-higher concentration of ammonia results from ammonium compounds. Ammonia diffuses very rapidly across cell membranes into cytoplasm, whereas the ammonium (NH_4^+) ion diffuses much more slowly (Hill, 1932; MacMillan, 1956; Warren and Nathan, 1958; Honert and Hooymans, 1959).

Water Absorption by Intact and Isolated Roots

Expressed as osmotic pressure, the range of concentrations of sucrose or KNO_3 that prevented water absorption by roots of *Allium cepa* varied from about 4.2–5.7 atm for attached roots to 1.8–3.3 atm for isolated roots and root segments (Rosene, 1941). There was no pronounced difference between the OP of solutions that stopped apical and basal entry of water into roots.

Water Absorption from the Atmosphere

Pinus ponderosa (Fowells and Kirk, 1945) and Coulter pine *(Pinus coulteri)* (Stone et al., 1950) can survive for considerable periods of time— several months—with the soil moisture at the permanent wilting point. It was demonstrated that Coulter pine can absorb moisture from the atmosphere under such conditions, and this has been termed negative transpiration (Stone et al., 1950).

WATER RELATIONS IN PLANTS AND SOIL

Relatively recent books or review articles deal with water relationships in plants and soils (Kramer, 1949, 1969; Philip, 1966; Slatyer, 1967; Preston, 1963, 1970), water deficits and plant growth (Kozlowski, 1968a,b), water absorption and movement within plants (Fogg, 1965; Briggs, 1967), and the nature of soil water (Day et al., 1967; DeWiest, 1969; Hillel, 1971). Philip (1966) discusses water relations in terms of the soil–plant–atmosphere continuum (SPAC). At all times he suggests that an investigator simultaneously consider water in the soil, plant, and atmosphere as a continuum.

Vaadia et al. (1961) noted that plant growth does not depend directly on soil moisture supply but rather on the balance between transpired and absorbed water. These investigators also discuss the effect of moisture stress on various physiological processes such as ion uptake, photosynthesis, and growth.

In pea plants, activity of indoleacetic acid (IAA) oxidase increased following a period of water stress (Darbyshire, 1971). With jack bean leaves (etiolated) under high relative humidity conditions, chlorophyll accumulation occurred rapidly in light, while low relative humidity conditions were associated with slow accumulation of chlorophyll and a much reduced incorporation of radioactive uracil into RNA (Bourque and Naylor, 1971).

Diffusion Pressure (DP)

As with a gas, water or any other liquid can be considered to have a diffusion pressure (DP). Pressure and temperature influence the diffusion pressure of pure water. In a solution the presence of solutes also affect the DP of the water, and the higher the concentration of solutes the lower the DP. The DP of water in a solution cannot readily be measured, but it is possible, for example, by means of an osmometer or other methods, to measure the extent to which the DP of water in a solution is *less than* that of pure water. This value represents the diffusion pressure deficit (DPD) of water in a solution. The concepts of DP and DPD have been explained in detail by Meyer (1945), Meyer and Anderson (1952), and Meyer et al. (1960).

Water Potential

Slatyer (1962) discusses the advantages and disadvantages of the DPD concept but favors water potential as a more appropriate term for describing the chemical potential of water in a plant. Price (1970) presents an excellent discussion of water potential.

Water potential (ψ_w) is defined in terms of the Gibbs free energy of water expressed as chemical potential (μ) in ergs mole^{-1}. Or, water potential may be considered as the difference between the partial specific Gibbs free energy of water in the system under consideration (μ_w) and of pure free water at the same temperature (μ_{w0})—the reference state (Slatyer 1962). As noted by Boyer (1969) in his recent review of the measurement of water status in plants, physiologists ordinarily use pressure concepts; water potential is generally given in pressure units by dividing the chemical potential by the partial molal volume of water ($\overline{V}$, in cm^3 mole $^{-1}$):

$$\psi_w = (\mu - \mu_0)/\overline{V}$$

Pressure units are then generally stated in bars (1 bar = 10^6 ergs cm^{-3}) or in atmospheres (1 atm = 0.987 bar). As a matter of convenience, water potential is often expressed in units of energy per unit volume rather than energy per unit mass of water. This gives the potential the dimensions of ergs cm^{-3}, which is dimensionally the same as dynes cm^{-2} and has the dimensions of pressure. This allows the bar (10^6 dynes cm^{-2} = 0.987 atm) to be used as a convenient working unit (Gardner, 1965). Gardner added that water movement through the plant is determined largely by osmotic and pressure potentials.

Water potential may be divided into various components:

$$\psi_w = \psi_s + \psi_m + \psi_p + \psi_g$$

where *s*, *m*, *p*, and *g* refer to the effects of solutes, matrix, pressure, and gravity, respectively. The matrix component represents the adsorptive and surface tension effects associated with solid surfaces. The gravity effect is important only for columns of water of considerable height, for example, very tall trees.

Hydrature (Walter, 1963; Walter and Stadelmann, 1968), which combines concepts both from the viewpoints of energy and the quantity, has not been widely adopted; it is similar to the matrix potential.

Vapor pressure is a sensitive indicator of the chemical potential of water. In turn, water potential is related to vapor pressure by the equation

$$\psi_w = -\frac{RT}{V}(\ln e/e_0)$$

where R is the gas constant (cm^3 bars mole^{-1} degree^{-1}), T is the Kelvin temperature, e is the vapor pressure of water in the tissue, and e_0 is the vapor pressure of pure water at atmospheric pressure.

Spanner (1951) showed that water potential could be measured with great accuracy by the use of thermocouples in the vapor phase of a system. In practice, a small amount of water is condensed on the thermocouple by passing an electric current through the junction for a short time in a direction that causes cooling (Peltier effect) and then allowing the condensate to evaporate into the air inside the chamber containing the plant tissue. The rate of evaporation is a function of the water potential of the system (e.g., plant tissue), and calibration is achieved with solutions of known water potential. It has been reported that the water potential of sliced leaf tissue may be higher than that of unsliced control tissue, since the remaining intact cells accumulate solutes from cut cells (Barrs and Kramer, 1969). This effect may cause spuriously high water potential values if very small pieces of tissue are used.

Various components of water potential can be distinguished or measured by modifying conditions, so that certain components become zero or are not sensed by the thermocouple. For example, pressure effects can be removed by freezing and thawing the tissue; readings are taken before and after these treatments (Ehlig, 1962). Sap can then be put on the thermocouple to remove the osmotic component of tissue, and then only the matrix component remains (Boyer, 1968).

A differential psychometer designed by Wylie (1963) was later modified for continuous measurements of transpiration, and the characteristics and performance of the instrument were reported (Slatyer and Bierhuizen, 1964). Boyer (1968) developed a technique for following rapid changes in leaf water potentials of intact leaves. A recently described psychrometer measures water potential of intact, transpiring leaves, since it is attached to only one side of the leaf with the other side free to transpire (Hofman and Splinter, 1968a, b).

Psychrometer measurements are subject to three sources of error: resistance of plant tissue to vapor transfer, heat produced by respiration, and adsorption of water by chamber walls. There can be systematic errors in the use of a thermocouple psychrometer in determining leaf water potential (Rawlins, 1964).

A second method for measuring water potential, the gravimetric vapor equilibration technique, involves placing tissue above solutions of known water potential and finding the solution that causes neither a gain nor a loss in weight of tissue (Slatyer, 1958).

A third technique, the pressure chamber, involves sealing a leaf in a chamber with the end of the petiole protruding from the chamber. Pressure is applied within the chamber until xylem sap appears on the end of the petiole (Scholander et al., 1964, 1965). Water potential is calculated from:

$$\psi_w = P + \psi_s \text{ xylem}$$

A pressure chamber can also be used to measure the osmotic potential (Scholander et al., 1964, 1965; Hammel, 1968) and the matrix potential (Boyer, 1967). The latter potential may also be measured by a pressure membrane apparatus (Wiebe, 1966).

A fourth measurement, relative water content (Weatherley, 1950), expresses water content as a percent of the turgid water content of tissue. This measurement is one of the most widely used ones for expressing the quantity of water in plants.

A fifth technique involves β-ray gauging, and it is based on the tendency for a stream of β rays to be attentuated as the mass of the particle pathway increases (Mederski, 1961). With a knowledge of the relationship between relative water content and ψ_w, water potential of intact leaves may be followed (Gardner and Nieman, 1964). Nakayama and Ehrler (1964) obtained estimates of relative water content with an accuracy of $\pm 0.25\%$.

O'Leary (1970) maintains that a positive water potential never develops in the xylem even during "root pressure." For example, he reasoned that if the solution in the xylem has 1.5 bars OP and a positive hydrostatic pressure of 1.0 bar, exudation or guttation could and would occur, but the water potential of the xylem fluid would be -0.5 bar. The apertures of stomata are controlled by water potential in the water-conducting system (Raschke, 1970). In nonirrigated plants in an arid environment, the density of stomata (stomata per square millimeter) and other epidermal cells is increased compared with that of plants with adequate moisture (Gindel, 1969).

For pepper plants in controlled-environment chambers, rate of transpiration was reduced by a decrease in osmotic potential, an increase in CO_2 concentration, and a decrease in light intensity (Janes, 1970). Movement of water from roots to leaves occurred in response to water potential gradient and not to the actual water potential in the leaves.

Soybean plants were grown with divided root systems (Michel and El-Sharkawi, 1970). By adding polyethylene glycol (Carbowax 6000) to reduce solute potential, or withholding water to reduce soil matric potential until water absorption from that root compartment stopped, the root xylem water potential could be measured. The experiment was successful when both compartments contained solutions or both contained soil, but not when one contained solution and the other soil. In soil the roots grew much less than did those in solution.

In maize and sugar beet plants, there was an inverse relationship between soil moisture and plant osmotic potential—especially in the lower one-third of the soil moisture range (Padurariu et al., 1969).

SOIL

Soils can be considered three-phase systems consisting of: (1) a solid phase composed of minerals and organic substances, (2) a liquid phase, and (3) a gaseous phase. In addition, one may consider the various soil organisms as a fourth phase—the living phase. In a typical silt loam, near optimum moisture content, the solid phase constitutes about 50% by volume, the liquid phase 25% by volume, and the gaseous phase 25% by volume. In order of decreasing numbers, bacteria, actinomyces, fungi, protozoa, earthworms, algae, and nematodes constitute about 5 tons of living tissue per acre (Truog, 1951). Organic matter may vary from less than 1% in sands to over 50% in peats. Soluble salts usually range from 100 to 1000 ppm. Carbon dioxide may be 10 or 20 times greater in soil air than in the atmosphere (Truog, 1951).

In a classic report, Edlefsen and Anderson (1943) discussed the thermodynamics of soil moisture. Movement of soil water has been reviewed by Childs (1964). The books by Russell (1915, 1926) are classics with regard to soil conditions as they affect crop production. Soil factors and the mineral nutrition of fruit crops have been covered in detail by Childers (1954).

Soil Moisture Availability

Veihmeyer and Hendrickson (1927, 1950) claimed that water is equally available from FC almost down to the PWP, but there is evidence that this may not always be the case (Kramer, 1952). Wadleigh and Richards (1951) noted that Veihmeyer and Hendrickson (1927) studied fruit trees in deep alluvial soils, and that under these conditions some of the roots may have had access to soil well above the WP. Schopmeyer (1939) reported that transpiration of pine seedlings decreased as soil moisture decreased—long before the PWP was reached. Kozlowski (1949) obtained similar results with oak and pine seedlings.

Soil Solution

Stout and Overstreet (1950) define the soil solution as "the displaceable soil moisture and its dissolved salts."

The soil solution may be obtained by displacement, involving the addition of water to soil in 1:1, 5:1, 10:1, or other proportions (Burd and Martin, 1923, 1924), or by use of a pressure membrane apparatus (Richards, 1941). The latter method even removes soil solution from soils with a moisture content below the PWP.

Comparing Burd and Martin's (1923) method with the pressure membrane technique, Reitemeier and Richards (1944) found that chemical analyses of the two types of soil solution indicated that the same solution was extracted by both methods. The most critical research on the nature of soil solutions was presented by Reitemeier (1946). He used a pressure membrane apparatus and studied soil solutions obtained from soils ranging from 500% water to somewhat less than FC. Soluble salts were quantitatively and qualitatively different for each soil moisture level of a given soil. NO_3^- and Cl^- increased more at low soil moisture percentages than could be accounted for on the basis of the percentage of the water in the soil, and thus these ions exhibited "negative adsorption" (Mattson and Wiklander, 1940). In other words, at low soil moisture percentages, not all the water is equally available for dissolving certain ions. A lower concentration of Cl^- and NO_3^- occurs in the water most tightly held by colloids than in water held less tightly.

For soils that contain an excess of crystalline $CaSO_4$, for example, a correction must be kept in mind when the displacement method is used and analyses are being converted from, for example, 500% moisture back to field moisture range. $CaSO_4$ has a solubility of approximately 25 meq/liter, and this concentration of $CaSO_4$ occurs at a field moisture percentage of 25 (not 500 meq $CaSO_4$/liter). When soil solution is obtained with the pressure membrane apparatus and with soils at moisture levels in the range from FC to the WP, extrapolation to field moisture levels is obviated. There is thus no danger of misinterpreting or miscalculating what the concentrations of various ions would be at typical field moisture percentages.

LITERATURE CITED

Altschul, M. M., M. L. Karon, L. Kyane, and C. M. Hall. 1946. Effect of inhibitors on the respiration and storage of cotton seed. Plant Physiol. 21:573–587.

Barrs, H. D., and P. J. Kramer. 1969. Water potential increases in sliced leaf tissue as a cause of error in vapor phase determination of water potential. Plant Physiol. 44:959–964.

Bourque, D. P., and A. W. Naylor. 1971. Large effects of small water deficits on chlorophyll accumulation and ribonucleic acid synthesis in etiolated leaves of jack bean (*Canavalia ensiformis* [L.] DC). Plant Physiol. 47:591—594.

Boyer, J. S. 1967. Leaf water potentials measured with a pressure chamber. Plant Physiol. 42:133—137.

Boyer, J. S. 1968. Relationship of water potential to growth of leaves. Plant Physiol. 43:1056—1062.

Boyer, J. S. 1969. Measurement of the water status of plants. Annu. Rev. Plant Physiol. 20:351—364.

Breazeale, E. L., W. T. McGeorge, and J. F. Breazeale. 1950. Moisture absorption by plants from an atmosphere of high humidity. Plant Physiol. 25:413—419.

Breazeale, E. L., W. T. McGeorge, and J. F. Breazeale. 1951a. Movement of water vapor in soils. Soil Sci. 71:181—185.

Breazeale, E. L., W. T. McGeorge, and J. F. Breazeale. 1951b. Water absorption and transpiration by leaves. Soil Sci. 72:239—244.

Breazeale, J. F., and W. T. McGeorge. 1953. Exudation pressure in roots of tomato plants under humid conditions. Soil Sci. 75:293—298.

Briggs, G. E. 1967. Movement of water in plants. Blackwell, Oxford. 142 pp.

Briggs, L. J., and H. L. Shantz. 1911a. A wax seal method for determining the lower limit of available soil moisture. Bot. Gaz. 51:210—219.

Briggs, L. J., and H. L. Shantz. 1911b. Application of wilting coefficient determinations in agronomic investigations. J. Amer. Soc. Agron. 3:250—260.

Briggs, L. J., and H. L. Shantz. 1912. The wilting coefficient and its indirect determination. Bot. Gaz. 53:20—37.

Burd, J. S., and J. C. Martin. 1923. Water displacement of soils and the soil solution. J. Agr. Sci. 13:265—295.

Burd, J. S., and J. C. Martin. 1924. Secular and seasonal changes in the soil solution. Soil Sci. 18:151—167.

Buswell, A. M., and W. H. Rodebush. 1956. Water. Sci. Amer. 194:77—89.

Childers, N. F. (ed.). 1954. Mineral nutrition of fruit crops. Hort. Pub. Rutgers Univ., New Brunswick, N.J. 907 pp.

Childs, E. C. 1964. The movement of soil water. Endeavour 23:81—84.

Darbyshire, B. 1971. The effect of water stress on indoleactic acid oxidase in pea plants. Plant Physiol. 47:65—67.

Day, P. R., G. H. Bolt, and D. M. Anderson. 1967. Nature of soil water, pp. 193—208. *In* R. M. Hagan et al. (eds.) Irrigation of agricultural lands. Amer. Soc. Agron., Madison, Wis.

DeWiest, R. J. M. 1969. Flow through porous media. Academic, New York. 530 pp.

Edlefsen, N. E., and A. B. C. Anderson. 1943. Thermodynamics of soil moisture. Hilgardia 15:31—298.

Ehlig, C. F. 1962. Measurement of energy status of water in plants with a thermocouple psychrometer. Plant Physiol. 37:288—290.

Fogg, G. (ed.). 1965. The state and movement of water in living organisms. Symp. Soc. Exp. Biol., no. 19. Cambridge Univ. Press, Cambridge. 432 pp.

Fowells, H. A., and B. M. Kirk. 1945. Availability of soil moisture to ponderosa pine. J. Forest. 43:601—604.

Furr, J. R., and J. O. Reeve. 1945. Range of soil-moisture percentages through which plants undergo permanent wilting in some soils from semiarid irrigated areas. J. Agr. Res. 71:149—170.

Gardner, W. R. 1965. Dynamic aspects of soil water availability to plants. Annu. Rev. Plant Physiol. 16:323—342.

Gardner, W. R., and R. H. Nieman. 1964. Lower limit of water availability to plants. Science 143:1460–1462.

Gindel, I. 1969. Stomata constellation in the leaves of cotton, maize, and wheat plants as a function of soil moisture and environment. Physiol. Plantarum 22:1143–1151.

Haise, H. R., H. J. Haas, and L. R. Jensen. 1955. Soil moisture studies of some Great Plains soils. II. Field capacity as related to 1/3 atmosphere percentage, and "minimum point" as related to 15- and 26-atmosphere percentages. Proc. Soil Sci. Soc. Amer. 19:20–25.

Hammel, H. T. 1968. Measurement of turgor pressure and its gradient in the phloem of oak. Plant Physiol. 43:1042–1048.

Harris, J. A. 1934. The physico-chemical properties of plant saps in relation to phyto-geography. Data on native vegetation in its natural environment. Univ. Minnesota Press, Minneapolis. 339 pp.

Hendrickson, A. H., and F. J. Veihmeyer. 1929a. I. Irrigation experiments with peaches in California. California Agr. Exp. Sta. Bull. 479.

Hendrickson, A. H., and F. J. Veihmeyer. 1929b. Some facts concerning soil moisture of interest to horticulturists. Proc. Amer. Soc. Hort. Sci. 26:105–108.

Hill, S. E. 1932. The effects of ammonia, of fatty acids, and their salts, on the luminescence of *Bacillus fisheri*. J. Cellular Comp. Physiol. 1:145–158.

Hillel, D. 1971. Soil and water — physical principles and processes. Academic, New York. 302 pp.

Hoffman, G. J., and W. E. Splinter. 1968a. Instrumentation for measuring water potentials of an intact plant-soil system. Trans. Amer. Soc. Agr. Eng. 11:38–40.

Hoffman, G. J., and W. E. Splinter. 1968b. Water potential measurements of an intact plant-soil system. Agron. J. 60:408–413.

Honert, T. H. van den, and J. M. Hooymans. 1959. Diffusion and absurption of ions in plant tissue. I. Observations on the absorption of ammonium by cut slices of potato tuber as compared to maize roots. Acta Bot. Neerl. 10:261–273.

Jackson, W. T. 1956a. Flooding injury studied by approach-graft and split root system techniques. Amer. J. Bot. 43:496–502.

Jackson, W. T. 1956b. The relative importance of factors causing injury to shoots of flooded tomato plants. Amer. J. Bot. 43:637–639.

Janes, B. E. 1970. Effect of carbon dioxide, osmotic potential of nutrient solution, and light intensity on transpiration and resistance to flow of water in pepper plants. Plant Physiol. 45:95–103.

Kozlowski, T. T. 1949. Light and water in relation to growth and competition of Piedmont forest tree species. Ecol. Monogr. 19:207–231.

Kozlowski, T. T. (ed.). 1968a. Water deficits and plant growth. Vol. I. Development, control and measurement. Academic, New York. 390 pp.

Kozlowski, T. T. (ed.). 1968b. Water deficits and plant growth. Vol. II. Plant water consumption and response. Academic, New York. 333 pp.

Kramer, P. J. 1949. Plant and soil water relationships, 1st ed. McGraw-Hill, New York. 347 pp.

Kramer, P. J. 1951. Causes of injury to plants resulting from flooding of the soil. Plant Physiol. 26:722–736.

Kramer, P. J. 1952. Plant and soil water relations on the watershed. J. Forest. 50:92–95.

Kramer, P. J. 1959. Transpiration and the water economy of plants, pp. 607–726. *In* F. C. Steward (ed.) Plant physiology — A treatise, Vol. II. Academic, New York. 758 pp.

Kramer, P. J. 1969. Plant and soil water relationships — A modern synthesis. McGraw-Hill, New York. 482 pp.

Kramer, P. J., and W. T. Jackson. 1954. Causes of injury to flooded tobacco plants. Plant Physiol. 29:241–245.

MacMillan, A. 1956. The entry of ammonia into fungal cells. J. Exp. Bot. 7:113–306.

Mattson, S., and L. Wiklander. 1940. The laws of soil colloidal behavior. XXI. A. The amphoteric points, the pH, and the Donnan equilibrium. Soil Sci. 49:109–134.

Mederski, H. J. 1961. Determination of internal water status of plants by beta ray gauging. Soil Sci. 92:143–146.

Meyer, B. S. 1945. A critical evaluation of the terminology of diffusion phenomena. Plant Physiol. 20:142–164.

Meyer, B. S., and D. B. Anderson. 1952. Plant physiology, 2nd ed. Van Nostrand, Reinhold, New York. 784 pp.

Meyer, B. S., D. B. Anderson, and R. H. Böhning. 1960. Introduction to plant physiology. Van Nostrand Reinhold, New York.

Michel, B. E., and H. M. ElSharkawi. 1970. Investigations of plant water relations with divided root systems of soybean. Plant Physiol. 46:728–731.

Milthorpe, F. L., and J. Moorby. 1969. Vascular transport and its significance in plant growth. Annu. Rev. Plant Physiol. 20:117–138.

Nakayama, F. S., and W. L. Ehrler. 1964. Beta ray gauging technique for measuring leaf water content changes and moisture status of plants. Plant Physiol. 39:95–97.

O'Leary, J. W. 1970. Can there be a positive water potential in plants? Bioscience 20:858–859.

Padurariu, A., C. T. Horovitz, R. Paltineanu, and V. Negomireanu. 1969. On the relationship between soil moisture and osmotic potential in maize and sugar beet plants. Physiol. Plantarum 22:850–860.

Philip, J. R. 1966. Plant water relations: Some physical aspects. Annu. Rev. Plant Physiol. 17:245–268.

Preston, R. D. (ed.). 1963. Advances in botanical research, Vol. I. Academic, New York. 396 pp.

Preston, R. D. (ed.). 1970. Advances in botanical research, Vol. III. Academic, New York. 322 pp.

Price, C. A. 1970. Molecular approaches to plant physiology. McGraw-Hill, New York. 398 pp.

Raschke, K. 1970. Stomatal responses to pressure changes and interruptions in the water supply of detached leaves of *Zea mays* L. Plant Physiol. 45:415–423.

Rawlins, S. L. 1964. Systematic error in leaf water potential measurements with a thermocouple psychrometer. Science 146(3644):644–646.

Reitemeier, R. F. 1946. Effect of moisture content on the dissolved and exchangeable ions of soils of arid regions. Soil Sci. 61:195–214.

Reitemeier, R. F., and L. A. Richards. 1944. Reliability of the pressure-membrane method for extraction of soil solution. Soil Sci. 57:119–135.

Richards, L. A. 1941. A pressure-membrane extraction apparatus for soil solution. Soil Sci. 51:377–386.

Richards, L. A. 1942. Soil moisture tensiometer materials and construction. Soil Sci. 53:241–248.

Richards, L. A., and G. Ogata. 1958. A thermocouple for vapor pressure measurement in biological and soil systems at high humidity. Science 128:1089–1090.

Richards, L. A., and L. R. Weaver. 1944. Moisture retention by some irrigated soils as related to soil-moisture tension. J. Agr. Res. 69:215–235.

Richards, L. A., R. B. Campbell, and L. H. Healton. 1949. Some freezing point depression measurements on cores of soil in which cotton and sunflower plants were wilted. Proc.

Soil Sci. Soc. Amer. 14:47—50.

Rosene, H. F. 1941. Control of water transport in local root regions of attached and isolated roots by means of the osmotic pressure of the external solution. Amer. J. Bot. 28:402—410.

Russell, E. J. 1915. Soil conditions and plant growth. Longmans, Green, London. 190 pp.

Russell, E. J. 1926. Plant nutrition and crop production. Univ. California Press, Berkeley.

Scholander, P. F., and M. de Oliveira Perez. 1968. Sap tension in flooded trees and bushes of the Amazon. Plant Physiol. 43:1870—1873.

Scholander, P. F., H. T. Hammel, E. A. Hemmingsen, and E. D. Bradstreet. 1964. Hydrostatic pressure and osmotic potential in leaves of mangroves and some other plants. Proc. Nat. Acad. Sci. 52:119—125.

Scholander, P. F., H. T. Hammel, E. D. Bradstreet, and E. A. Hemmingsen. 1965. Sap pressure in vascular plants. Science 148:339—346.

Schopmeyer, C. S. 1939. Transpiration and physico-chemical properties of leaves as related to drought resistance in loblolly pine and short-leaf pine. Plant Physiol. 14:447—462.

Shull, C. A. 1916. Measurement of the surface forces in soils. Bot. Gaz. 62:1—31.

Slatyer, R. O. 1957. The significance of the permanent wilting percentage in studies of plant and soil water relations. Bot. Rev. 23:585—636.

Slayter, R. O. 1958. The measurement of diffusion pressure deficit in plants by a method of vapour equilibration. Australian J. Biol. Sci. 11:349—365.

Slatyer, R. O. 1962. Internal water relations of higher plants. Annu. Rev. Plant Physiol. 13:351—378.

Slatyer, R. O., 1967. Plant-water relationships. Academic, New York. 366 pp.

Slatyer, R. O., and J. F. Bierhuizen. 1964. A differential psychrometer for continuous measurements of transpiration. Plant Physiol. 39:1051—1056.

Spanner, D. C. 1951. The Peltier effect and its use in the measurement of suction pressure. J. Exp. Bot. 2:145—168.

Stone, E. C., F. W. Went, and C. L. Young. 1950. Water absorption from the atmosphere by plants growing in dry soil. Science 111:546—548.

Stout, P. R., and R. Overstreet. 1950. Soil chemistry in relation to inorganic nutrition of plants. Annu. Rev. Plant Physiol. 1:305—342.

Stuart, D. M., and J. L. Haddock. 1968. Inhibition of water uptake in sugar beet roots by ammonia. Plant Physiol. 43:345—350.

Truog, E. 1951. Soil as a medium for plant growth, pp. 23—55. *In* E. Truog (ed). Mineral nutrition of plants. Univ. Wisconsin Press, Madison. 469 pp.

Vaadia, Y., F. C. Raney, and R. M. Hagan. 1961. Plant water deficits and physiological processes. Annu. Rev. Plant Physiol. 12:265—292.

Veihmeyer, F. J., and A. H. Hendrickson. 1927. Soil moisture conditions in relation to plant growth. Plant Physiol. 2:71—82.

Veihmeyer, F. J., and A. H. Hendrickson. 1928. Soil moisture at permanent wilting of plants. Plant Physiol. 3:355—357.

Veihmeyer, F. J., and A. H. Hendrickson. 1950. Soil moisture in relation to plant growth. Annu. Rev. Plant Physiol. 1:285—304.

Veihmeyer, F. J., J. Oserkowsky, and K. B. Tester. 1927. Some factors affecting the moisture equivalent of soils. Proc. 1st Int. Congr. Soil Sci., Ithaca, N.Y., 1926, 1:512—534.

Vines, H. M., and R. T. Wedding. 1960. Some effects of ammonia on plant metabolism and a possible mechanism for ammonia toxicity. Plant Physiol. 35:820—825.

Wadleigh, C. H. 1946. The integrated soil moisture stress upon a root system in a large container of saline soil. Soil Sci. 61:225–238.

Wadleigh, C. H., and A. D. Ayers. 1945. Growth and biochemical composition of bean plants as conditioned by soil moisture tension and salt concentration. Plant Physiol. 20:106–132.

Wadleigh, C. H., and H. G. Gauch. 1948. Rate of leaf elongation as affected by the intensity of the total soil moisture stress. Plant Physiol. 23:485–495.

Wadleigh, C. H., and L. A. Richards. 1951. Soil moisture and the mineral nutrition of plants, pp. 411–450. *In* E. Truog (ed.) Mineral Nutrition of plants. Univ. Wisconsin Press, Madison.

Wadleigh, C. H., H. G. Gauch, and O. C. Magistad. 1946. Growth and rubber accumulation in guayule as conditioned by soil salinity and irrigation regime. U. S. Dep. Agr. Tech. Bull. 925.

Wadsworth, H. A., and U. K. Das. 1930. Some observations on the wilting coefficient of a selected Waipio soil. Hawaiian Planters' Rec. 34:289–299.

Walter, H. 1963. Zur Klärung des spezifischen Wasserzustandes in Plasma und in der Zellwand bei höheren Pflanze und seine Bestimmung. Ber. Deut. Bot. Ges. 76:40–71.

Walter, H., and E. Stadelmann. 1968. The physiological prerequisites for the transition of autotrophic plants from water to terrestial life. Bioscience 18:694–701.

Warren, K. S., and D. G. Nathan. 1958. The passage of ammonia across the blood-brain barrier and its relations to blood pH. J. Clin. Invest. 37:1724–1728.

Weatherley, P. E. 1950. Studies in the water relations of the cotton plant. I. The field measurement of water deficits in leaves. New Phytol. 49:81–97.

Wedding, R. T., and H. M. Vines. 1959. Inhibition of reduced diphosphopyridinenucleotide oxidation by ammonia. Nature 184:1226–1227.

Wiebe, H. H. 1966. Matric potential of several plant tissues and biocolloids. Plant Physiol. 41:1439–1442.

Wylie, R. G. 1963. A new tool for transpiration experiments. C.S.I.R.O. Annu. Rep. 1962–1963:32–33.

CHAPTER 3

General Aspects of Nutrition

Although mineral nutrition is discussed throughout this volume, included in this chapter are certain general aspects not logically related to other sections.

DISEASES

Various workers have reported that the reactions of plants to infectious agents, including viruses (Spencer, 1935a, b), may be altered by varying their mineral nutrition (Schaffnit and Volk, 1927; Gassner and Hassebrauk, 1931; Volk, 1931; Fisher, 1935; Spencer and McNew, 1938; McNew and Spencer, 1939; Pryor, 1940; Wingard, 1941; Walker and Hooker, 1945a, b; Edgington and Walker, 1958; Corden, 1965).

Increasing the K^+ supply to tobacco plants decreased the severity of yellow mosaic (Spencer, 1935b).

Mineral nutrition exerted an influence on the host–pathogen complex of bacterial wilt of corn caused by *Phytomonas stewartii* (Spencer and McNew, 1938). N- or P-deficient seedlings were only slightly infected, whereas K-deficient plants were severely infected. Conversely, high-N and high-P plants showed intense wilting, and approximately half the high-N plants died within 2 weeks after inoculation. Inasmuch as the corn wilt bacterium multiplies almost exclusively in tracheal tubes during early stages of invasion, McNew and Spencer (1939) decided to evaluate the effect of N supply to the plants. They found that the external concentration of N determined the concentration of N in tracheal sap and, in turn, the suitability of this sap as a nutrient substrate for the pathogen. The higher the N supply, the greater the severity of wilting.

The severity of corn wilt was also affected by the K^+ supply to the plants (Shear and Wingard, 1944). When K^+ was deficient, NO_3^- concentration in tracheal sap was high. *Phytomonas stewartii* lives almost entirely in tracheal tubes during early stages of invasion, and apparently it is dependent on inorganic N for its parasitic existence. When K^+ was not deficient, there was less NO_3^- in tracheal sap, and thus conditions were less favorable for the organism.

The effects of mineral nutrition have been studied with regard to the development of clubroot of cabbage caused by *Plasmodiophora brassicae* Wor. Omission of K^+ (Pryor, 1940) and K^+ or P (Walker and Hooker, 1945b) from the nutrient solution decreased the severity of clubroot. Pryor (1940) reported that increasing the N supply in the nutrient solution increased the severity of disease, whereas Walker and Hooker (1945b) found that omission of N increased the disease rating.

Walker and Hooker (1945a) also studied the relation of nutrition to the development of cabbage yellows caused by *Fusarium oxysporum f. conglutinans* (Wr.) S. and H. In contrast with the effects of K^+ on clubroot, the omission of K^+ from the nutrient increased the disease rating for cabbage yellows; omission of either N or P reduced it. "Yellows" is a hypoplastic disease, while clubroot is a hyperplastic condition. This difference between the two diseases was not discussed with regard to the opposite effects of K^+ deficiency on the two diseases, but it is possible that the effect of K^+ may in some way be related to the nature of the two diseases.

In a study of mineral nutrition as affecting local lesion response of *Nicotiana glutinosa* to tobacco mosaic virus (TMV), Chessin and Scott (1955) noted that Ca^{2+} and Mg^{2+} deficiencies were without effect, but that Fe and S deficiencies resulted in lesions two to three times the size of those on control plants. They reasoned that if increased lesion size represented a tendency toward systematic infection, mineral nutrition of the host might affect spread of plant viruses in a manner not previously recognized.

Using foliar analysis of leaves, Thomas and Mack (1941) observed that tomato plants badly infected with "streak" (caused by a virus or mixture of viruses) had relatively high concentrations of N and low concentrations of K^+ in the leaves.

Neal (1928) reported that K^+ salts reduced cotton wilt in sand cultures, and that he obtained comparable results under field conditions in Mississippi. In Maryland, Stoddard (1942) reduced the severity of *Fusarium* wilt [*Fusarium oxysporum f. melonis* (L. and C.) S. and H.] of cantaloupe in the field by using a high ratio of K^+ to N in fertilizer.

Walker (1946) cautioned that one should proceed with caution in trying to apply to the field results obtained in the greenhouse and laboratory with regard to nutrition and disease development. He felt, however, that studies of host nutrition in relation to disease development might yield information on which future modification of soil practices might be based.

The severity of tomato wilt caused by *Fusarium oxysporum f. lycopersici* (Sacc.) Snyd. and Hans. is usually increased by low-Ca nutrition and decreased by excess Ca^{2+} (Fisher, 1935; Edgington and Walker, 1958; Corden, 1965). Growth of *Fusarium* in the water-conducting vessels of tomato stems was enhanced when Ca^{2+} concentration in the vascular sap was low (Corden, 1965). Higher, internal levels of Ca^{2+} inhibited *Fusarium* polygalacturonase, and it was therefore postulated that this Ca inhibition influenced the disease by interfering with decomposition of pectic substances within the host.

For citrus, high PO_4^{3-} concentrations in the root medium favored infection by *Thielavia basicola* (Chapman and Brown, 1942). Lowering the pH from 5.0 to 3.5 reduced infection by this organism even in the presence of a relatively high PO_4^{3-} concentration.

LIPID AND ORGANIC ACID METABOLISM

By modifying the level of organic P compounds, P could have a specific effect on lipid formation (Howell, 1954). P compounds are involved in the formation of fatty acids and their esterification to form lipids.

The levels of N and P supply can affect the concentrations of organic acids in the tops of soybean plants. Under a high N–low K regime, phosphoenolpyruvate (PEP) is apparently shunted into alternate pathways which results in a shortage of acetyl-CoA (Pattee and Teel, 1967). Oxaloacetate appears to be preferentially formed from PEP in K-deficient plants, inasmuch as organic acid derivatives accumulate. Increases in K^+ supply increase the concentration of citrate—irrespective of the level of N supply—and decrease the concentration of malate. On the contrary, Cooil (1948) reported that low K^+ supply increases citrate in guayule.

The decrease in malate as K^+ was increased under low N regimes (Pattee and Teel, 1967) could be explained by an increase in rate of conversion of PEP to pyruvate by pyruvic kinase (Evans, 1963). If this step were impeded by a K deficiency, one would expect an increase in the rate of conversion of PEP into alternative compounds. Formation of malate, by conversion of PEP to organic acids, and the conversion of organic acids to malate by malic dehydrogenase, should increase if PEP is shunted into alternate pathways (Pattee and Teel, 1967).

When NO_3- is absorbed and then converted to NH_3 for incorporation into amino acids, and so on, organic acid anions are formed which replace the lost NO_3^- anion (Briggs et al., 1961).

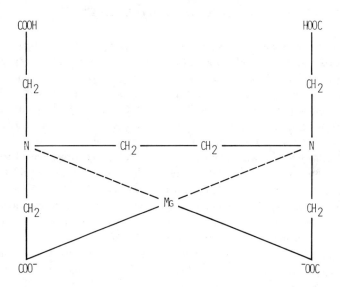

Fig. 3.1
Chemical structure of Mg–EDTA.

CHELATES

When chelates such as EDTA (Fig. 3.1) are used, plants may grow success-fully over a wider range of pH—as has been reported for desmids (Waris, 1953). By using ^{14}C-labeled EDTA, evidence was obtained that algae obtain metals by absorbing and decomposing the chelate (Krauss, 1957). In addition, Fe can become available from Fe•EDTA, in the medium or in algal cells, by the action of light which decomposes the complex (Jones and Long, 1952; Krugers and Agterdenbos, 1957).

Fe chelates as well as Fe_2SO_4 can be used to control Fe deficiency in yards and gardens.

Brown (1963) discusses chelates as metal-binding agents in biological sys-tems. Stewart (1963) covers naturally occurring chelates in soils and in plants, and the uptake of chelates by plants. Price (1970) describes chelates in con-siderable detail—including stability constants of certain metal–EDTA com-plexes.

SIGNIFICANCE OF EQUIVALENT SYSTEM

The equivalent weight of a substance is "that weight of a substance which is equivalent in chemical combining power with 8 g ($\frac{1}{2}$ g-atom) of oxygen." Or, equivalent weight can be defined as the "weight, in grams, of a substance divided by its hydrogen equivalent." For example, the equivalent weight of H_2SO_4 is one-half its molecular weight. For HCl or NaCl, the equivalent and molecular weights are the same. For $CaCl_2$, the equivalent weight is one-half the molecular weight, since the Ca ion is the chemical equivalent of two H ions. In other words, when solutions are made on an equivalent basis, they have the same chemical reactivity. One milliliter of a solution of NaOH, for example, is exactly neutralized by 1 ml of a solution of H_2SO_4. A milliequivalent is, of course, 1/1000 of an equivalent weight, and the former unit is a convenient one for expressing concentrations of cations and anions in a nutrient solution, in plant materials, irrigation waters, and so on.

With regard to a justification for using the equivalent system, let us consider several situations as examples. In making nutrient solutions, if the stock solutions contain 1 equiv wt/liter, it follows that each milliliter contains 1 meq. If, in the base nutrient solution to be made from the stock solutions, one wishes to have 3 meq of a given salt per liter, one needs only to take 3 ml of that stock solution in order to obtain 3 meq. Three milliequivalents of a salt, such as KCl, give 3 meq of K^+ and 3 meq of Cl^-. The total number of milliequivalents of cations obviously equals the total number of milliequivalents of anions. On the basis of parts per million, for example, the totals for cations and anions would *not* be equal. In Table 3.1, note how easily a nutrient solution can be conceived and formulated on this basis.

Table 3.1
Desired Composition of a Solution and Salts Required to Achieve It

Equivalent (or normal) solution	Milliliters/liter	Desired solution (meq/liter) Ca = 5, Mg = 3, K = 4, SO4 = 4, NO3 = 8, H2PO4 = 1						
		Ca	Mg	K	Na	SO4	NO3	H2PO4
$Ca(NO_3)_2$	5	5	—	—	—	—	5	—
KNO_3	3	—	—	3	—	—	3	—
KH_2PO_4	1	—	—	1	—	—	—	1
$MgSO_4$	3	—	3	—	—	3	—	—
Na_2SO_4	1	—	—	—	1	1	—	—
Totals:			(13)					(13)	

In addition, a chemical analysis of such a solution can be checked *instantly* for its accuracy *if* the analysis is stated in terms of *milliequivalents*. Milliequivalents of cations and milliequivalents of anions *must* equal each other if the analysis is accurate. Similarly, the accuracy of an analysis of well water or irrigation water can be checked instantly when results are on the basis of milliequivalents per liter. By contrast, the total parts per million of cations and total parts per million of anions would not be equal.

In soil chemistry, which involves the adsorption of bases on surfaces of clay or organic colloids, it follows that two K ions, for example, could occupy the space that one Ca ion might occupy. The base exchange capacity of a soil is expressed in terms of milliequivalents per 100 dry weight of soil. Substitution of ions from the soil solution for those on clays, or vice versa, occurs on the basis of chemical reactivity or chemical equality, and this is the essence of the equivalent system for expresssing concentration.

Later, as we shall see, the sum of cations in plant material equals the sum of anions. The sum of inorganic cations exceeds the sum of inorganic anions in most plant materials. Some of the anions in plant material are organic—that is, malate, citrate, and so on. When organic anions are also determined and *expressed on an equivalent basis*, total cations indeed equal total anions (inorganic and organic) when *all* are expressed on an equivalent basis.

The foregoing, then, serve as examples of the desirability of using the equivalent system rather than percent, parts per million, molarity, or other units for concentrations of ions.

COMPOSITION OF PLANTS

Concentrations Associated with Deficiency, Sufficiency, and Excess

There are numerous sources of information concerning the concentrations of essential elements associated with deficiency, sufficiency, and excess. Such information appears in individual papers and in review articles and various books. Beeson (1941), Goodall and Gregory (1947), Miller (1958), and Chapman (1966) provide numerous references to the composition of plants. Largely by use of these sources, the data in Table 3.2 were compiled. In the table data are given for each of the 16 elements currently recognized as essential for higher plants. The selection of species was based on various considerations including amount of research on a given species, importance of the element for a given species and, in some instances, the fact that concentrations from deficiency to excess were listed for a given element and species.

The concentration of an element in plants may not necessarily be an indica-

Table 3.2

Concentrations of Essential Elements in Representative Plants Associated with Deficiency, Normal Growth, and Toxicity

Element	Species and plant part	Stage	Concentration (dry wt basis) associated with:			Investigator
			Deficiency	Normal growth	Toxicity	
C	Many—entire plant	Maturity	—	44%	—	—
H	Many—entire plant	Maturity	—	6%	—	—
O	Many—entire plant	Maturity	—	44%	—	—
N	Apple leaves	Midseason	<1.48%	1.65–1.80%	—	Wander (1946)
	Apple leaves	Midseason	—	1.85–2.00%	—	Boynton and Compton (1945)
	Peach leaves	Midseason	—	3.50%	—	Smith and Taylor (1952)
P	Apple leaves	Midseason	<0.10%	0.12–0.39%	—	Walrath and Smith (1952)
	Peach leaves	Midseason	<0.11%	0.14–0.34%	—	Lilleland and Brown (1942); Cullinan et al. (1939)
	Alfalfa tops	Flowering	0.11%	0.35%	—	Hallock and Attoe (1954)
	Soybean leaves	Flowering	0.19%	0.27%	—	Kamprath and Miller (1958)
	Tomato leaves	Recently matured	0.10–0.18%	0.44–0.90%	—	Cannell et al. (1960)
K	Apple leaves	Midseason	0.45–0.93%	1.53–2.04%	—	Goodall (1945)
	Corn leaves	—	0.58–0.78%	0.74–1.49%	—	Lucas and Scarseth (1947)
	Peach leaves	5-yr-old tree	0.59–1.35%	1.58–2.65%	—	Cullinan and Batjer (1943)
	Tomato leaves	Apr.–May	0.28–1.44%	1.40–2.40%	—	Wall (1940a, b)
	Tobacco leaves	Sept.	2.79–3.72%	4.37–5.29%	—	Lagatu and Maume (1935)
	Ryegrass leaves	—	—	3.50%	—	Hylton et al. (1967)
Ca	Alfalfa tops	7 wk old	0.58%	1.55%	—	Bower and Turk (1946)
	Apple leaves	—	0.56%	1.10%	—	Wallace (1951)
	Corn—entire plant	July	0.18–0.32%	0.38–0.43%	—	Melsted (1953)
	Soybeans—lower leaves	Midbloom	—	3.40%	—	C. E. Evans, et al. (1950)
	Tomato—tops	65 days old	0.79–0.96%	0.82–1.78%	—	Drake and White (1961)
Mg	Apple leaves	July	0.02–0.33%	0.21–0.53%	—	Boynton et al. (1943)
	Corn leaves	—	0.07%	0.20%	—	Jones (1929)
	Potato leaves	July	0.16–0.33%	0.40–0.86%	—	Jones and Plant (1943)
S	Alfalfa tops	Mature	0.24%	0.27%	0.75%	Neller (1925)
	Corn tops	—	0.04%	0.08%	—	Volk et al. (1945)
	Peach leaves	May–Sept.	—	0.18%	0.30–0.38%	Hayward et al. (1946)

Element	Tissue	Condition/time				Reference
	Soybean tops	—	0.14%	0.23%	—	Volk et al. (1945)
	Tobacco leaves	—	0.11%	0.15%	—	Jordan and Bardsley (1958)
	Orange leaves	Young leaves	0.08–0.10%	0.19–0.26%	—	Chapman and Brown (1941)
Fe	Sweet corn leaves	Recently matured	24–56 ppm	56–178 ppm	—	Jacobson (1945)
	Soybean tops	34 days old	28–38 ppm	44–60 ppm	—	Wallace (1956)
	Tobacco leaves	Recently matured	63–70 ppm	68–140 ppm	—	Jacobson (1945)
	Tomato leaves	Upper leaves	93–115 ppm	107–250 ppm	—	Bennett (1945)
B	Alfalfa leaves	Apr.–Sept.	—	28–654 ppm	516–996 ppm	Eaton (1944)
	Apple leaves	—	15–20 ppm	25–34 ppm	—	Askew and Chittenden (1936)
	Beet leaves	Aug.	52 ppm	104–370 ppm	637–1,263 ppm	Eaton (1944)
	Peach leaves	Spring	11–19 ppm	17–40 ppm	91–169 ppm	McLarty and Woodbridge (1950)
	Sugar beet leaves	—	6–13 ppm	10–44 ppm	—	Hale (1945)
	Tomato leaves	July–Aug.	—	34–150 ppm	253–1,416 ppm	Eaton (1944)
	Celery leaves	Oct.	15 ppm	27–48 ppm	—	Maier (1943)
	Corn leaves	Oct.	—	27–72 ppm	179 ppm	Eaton (1944)
	Tobacco leaves	Aug.	—	19–261 ppm	365–771 ppm	Eaton (1944)
	Apple leaves	—	6–7 ppm	—	25 ppm	Wilcox and Woodbridge (1943)
Mn	Apple leaves	July	2–18 ppm	25–50 ppm	—	Wiederspahn (1956)
	Orange leaves	4–7 mo old	15 ppm	25–200 ppm	1,000 ppm	Reuther et al. (1949)
	Soybean leaves	30 days old	2–3 ppm	14–102 ppm	173–999 ppm	Somers and Shive (1942)
	Tomato leaves	Oct.	5–6 ppm	70–398 ppm	—	Lyon et al. (1943)
	Tomato fruit	Oct.	0.2 ppm	2 ppm	—	Lyon et al. (1943)
	Oats tops	After flowering	6–13 ppm	14–111 ppm	—	Samuel and Piper (1928)
	Orange leaves	—	4–11 ppm	21–29 ppm	—	Levitt and Nicholson (1941)
	Tobacco leaves	—	—	160 ppm	4,000–11,000 ppm	Jacobson and Swanback (1932)
Zn	Apple leaves	—	3–22 ppm	6–40 ppm	—	Woodbridge (1951)
	Corn leaves	Tasseling	9 ppm	31–37 ppm	—	Viets et al. (1953)
	Orange leaves	—	15 ppm	20–80 ppm	—	Chapman (1949)
	Tomato leaves	Fruiting	9–15 ppm	65–198 ppm	526–1,489 ppm	Lyon et al. (1943)
	Tung leaves	Midshoot	10–26 ppm	30–229 ppm	—	Drosdoff (1950)
	Pecan leaflets	Aug.	4 ppm	4–17 ppm	—	Finch and Kinnison (1934)

Table 3.2 (continued)

Element	Species and plant part	Stage	Concentration (dry wt basis) associated with:			Investigator
			Deficiency	Normal growth	Toxicity	
Cu	Alfalfa tops	—	—	5–10 ppm	—	Pack et al. (1953)
	Apple leaves	—	2–3 ppm	5–6 ppm	—	Bould et al. (1950)
	Orange leaves	Bloom	<3.5 ppm	5–16 ppm	>23 ppm	Reuther et al. (1958)
	Pecan leaves	—	—	21–28 ppm	—	Alben and Hammar (1939)
	Tomato leaves	—	—	3–12 ppm	—	Cannel et al. (1960)
	Tung leaves	Aug.	2.6–3.1 ppm	4.8–5.7 ppm	—	Drosdoff and Dickey (1943)
	Corn leaves	—	—	1 ppm	—	Jones (1965)
Mo	Alfalfa leaves	10% bloom	0.28 ppm	0.34 ppm	—	Reisenauer (1956)
	Alfalfa leaves	Prebloom?	<0.10 ppm	0.3–0.6 ppm	—	H. J. Evans et al. (1950)
	Broccoli tops	8 wk old	0.04 ppm	16 ppm	—	Johnson et al. (1952)
	Cotton leaves	65 days old	0.5 ppm	113 ppm	—	Amin and Joham (1960)
	Tomato leaves	8 wk old	0.13 ppm	0.68 ppm	—	Johnson et al. (1952)
	Lemon leaves	—	0.01 ppm	0.03 ppm	—	Vanselow and Datta (1949)
Cl	Carrot roots	Spring cycle	—	0.44%	1.82%	Bernstein and Ayers (1953)
	Orange leaves		—	<0.3%	>0.7%	Reuther and Smith (1954)
	Corn leaves	Silking	—	0.34–0.53%	—	Younts and Musgrave (1958a, b)
	Potato leaves	130 days old	0.21%	2.58%	—	Harward et al. (1956)
	Rice shoots	Young	1.00%	1.00%	1.85–2.12%	Ehrler and Bernstein (1958)
	Sugar beet petioles	Recently matured leaves	0.04–0.1%	0.2–2.5%	—	Ulrich et al. (1959)

tion of its adequacy. Plants showing severe Cu deficiency may have somewhat greater concentrations of Cu^{2+} than those of normal plants. When plants are extremely deficient in a given element, the amount of the element present is distributed through a relatively small amount of dry matter. If a very low concentration of that element is supplied to the plants, a relatively great amount of growth occurs which may result in a lowering of the *concentration* of the previously extremely deficient element. A still greater supply of the element may finally result in an increase in concentration of that element in the plant material.

Factors Affecting Composition of Plants

Method of Culture. Regardless of whether tomato plants were grown in water culture, sand culture, or soil, Arnon and Hoagland (1943) reported that the concentrations of K^+, Ca^{2+}, Mg^{2+}, P, and N in the fruits were essentially the same.

Level of Supply to the Plant. Numerous references could of course be cited to the effect that the level of supply of a given element largely determines the concentration of that element in plants.

Plant Part. The composition of fruits is generally much less affected by inorganic nutrition than is that of leaves and other plant parts. For example, when low-K^+ and low-P conditions were provided, concentrations of K^+ and of P in leaves of tomato plants were reduced, but low levels of these elements had very little effect on their concentrations in tomato fruits (Arnon·and Hoagland, 1943).

Varietal Effects. Butler and Johnson (1957) and Johnson and Butler (1957) reported large differences in the I^- content of perennial ryegrass varieties, ranging from 18.5 to 247 and 16 to 186 $\mu g/100g$ of dry matter, respectively. They indicated that in areas where sheep are prone to develop goiter (less than 30 μg I^- per 100 g dry matter), I^- content of herbage was a significant factor in the prevention of the syndrome. The variety of ryegrass forming the pasture could greatly influence the amount of I^- received by sheep.

Later, Butler and Glenday (1962) reported that the I^- concentrations of ryegrass varieties varied from 18.5 to 247.0 $\mu g/100$ g dry weight.

Leaf Position. Position of leaves on the stem may affect the internal concentration of salts. Although older, lower leaves may draw from a more concentrated salt solution in the xylem than the upper, younger leaves, the latter tend to have a higher concentration of salt (Biddulph, 1951) (Fig. 3.2).

Genetics. In a study of the I^- concentration of varieties of perennial ryegrass, Butler and Glenday (1962) studied diallele crosses of the varieties and found that the I^- concentration in the herbage was a strongly inherited characteristic.

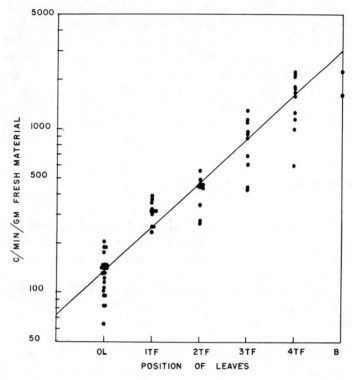

Fig. 3.2

Concentrations of ³²P in various leaves of *Phaseolus vulgaris* plants after absorption of labeled phosphate for 4 days. OL, Oldest leaf; 1TF, first trifoliolate leaf; 2TF, second trifoliolate leaf, and so on. From O. Biddulph (1951) The translocation of minerals in plants, pp. 261–275. *In* E. Truog (ed.) Mineral Nutrition of Plants, Univ. Wisconsin Press, Madison. © 1951 by the Regents of the University of Wisconsin—by permission.

ISOTOPES

Usefulness

Ion Uptake Studies. The use of isotopes for studying ion uptake and competition between ions is discussed elsewhere. Therefore, only representative generalized studies on ion uptake employing isotopes are reported here.

By obtaining daily autoradiographs of a uniformly labeled soil in which plants were growing, patterns of actual ion uptake from the soil could be established (Walker and Barber, 1961). Walker and Barber (1962) obtained autoradiographs showing the removal of ⁸⁶Rb from soil by the roots of 12-day-old corn plants (Fig. 3-3).

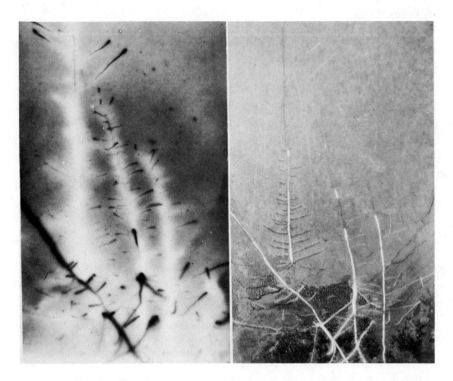

Fig. 3.3
Autoradiograph (left) and photograph (right) of a 13-day-old corn root system growing in 86Rb-labeled soil. *From* J. M. Walker and S. A. Barber (1962) Plant and Soil, 17:243–259. (Photograph courtesy of Dr. J. M. Walker, U.S. Dept. Agr., Beltsville, Md.)

Root Distribution. Radioactive tracer elements are especially useful for studying salt absorption from soils. Hall et al. (1953) placed injections of ^{32}P at various distances from and below plants to determine distribution of roots and P-absorbing capacities of roots of corn, cotton, peanuts, and tobacco. Cotton was the most shallow-rooted of the crops, while tobacco showed the greatest relative root activity of any of the crops in a 16-inch layer of soil.

Adverse Effects

Tritium oxide has been reported to be especially damaging to cell division in *Chlorella* (Porter and Knauss, 1954; Porter and Watson, 1954).

In a study of phosphate esters in duckweed *(Spirodela oligorrhiza)* adverse effects on growth and appearance of plants were noted when the nutrient solution had activities higher than 2 μc per μmole per milliliter (Bieleski, 1968).

Possible Errors

Analysis for ^{86}Rb did not properly evaluate uptake of K$^+$ by corn roots (Maas and Leggett, 1968). While Rb$^+$ was readily absorbed from solutions of 0.1 meq KCl per liter during 4-hr treatment periods, chemical analyses showed a net loss of K$^+$. As Nims (1962) has shown mathematically, the flow of a radioisotope is proportional to the flow of the ion only when the mole fraction of the species is the same on both sides of a membrane. The previously quoted results and this latter caution highlight the need for total chemical analyses to avoid error in interpretation of isotopic measurements (Maas and Leggett, 1968).

Another limitation in the use of isotopes for investigating the action of a nonlabeled isotope of the same element is indicated by the work of Weinberger and Porter (1953). They found that tritium was moved more slowly than hydrogen along anabolic pathways of growing *Chlorella pyrenoidosa* cells and, further, that the rate of tritium incorporation was between one-half and one-third that for hydrogen.

LITERATURE CITED

Alben, A. O., and H. E. Hammar. 1939. Fertilizing the pecan. Texas Pecan Growers' Ass., Proc. Annu. Meeting 19:48–54.

Amin, J. V., and H. E. Joham. 1960. Growth of cotton, as influenced by low-substrate molybdenum. Soil Sci. 89:101–107.

Arnon, D. I. 1943. Mineral nutrition of plants. Annu. Rev. Biochem. 12:493–528.

Arnon, D. I., and D. R. Hoagland. 1943. Composition of the tomato plant as influenced by nutrient supply, in relation to fruiting. Bot. Gaz. 104:576–590.

Askew, H. O., and E. T. Chittenden. 1936. The use of borax in the control of "internal cork" of apples. II. The effect of tree injection of borax solutions on the boron status of apple trees. J. Pomol. 14:239–242.

Beeson, K. C. 1941. The mineral composition of crops with particular reference to the soils in which they were grown. U. S. Dep. Agr. Misc. Pub. 369.

Bennett, J. P. 1945. Iron in leaves. Soil Sci. 60:91–105.

Bernstein, L., and A. D. Ayers. 1953. Salt tolerance of five varieties of carrots. Proc. Amer. Soc. Hort. Sci. 61:360–366.

Biddulph, O. 1951. The translocation of minerals in plants, pp. 261–275. *In* E. Truog (ed.) Mineral nutrition of plants. Univ. Wisconsin Press, Madison.

Bieleski, R. L. 1968. Levels of phosphate esters in *Spirodela*. Plant Physiol. 43:1297–1308.

Bould, C., D. J. D. Nicholas, J. M. S. Potter, J. A. H. Tolhurst, and T. Wallace. 1950. Zinc and copper deficiency of fruit trees. Annu. Rep. Agr. Hort. Res. Sta. (Bristol) 1949:45–49.

Bower, C. A., and L. M. Turk. 1946. Calcium and magnesium deficiencies in alkali soils. J. Amer. Soc. Agron. 38:723–727.

Boynton, D., and O. C. Compton. 1945. Leaf analysis in estimating the potassium, magnesium, and nitrogen needs of fruit trees. Soil Sci. 59:339–351.

Boynton, D., J. C. Cain, and J. van Geluwe. 1943. Incipient magnesium deficiency in some New York apple orchards. Proc. Amer. Soc. Hort. Sci. 42:95–100.

Briggs, G. E., A. B. Hope, and R. N. Robertson. 1961. Electrolytes and plant cells. *In* W. O. James (ed.) Bot. Monog., Vol. I. Blackwell, Oxford. 217 pp.

Brown, J. C. 1963. Interactions involving nutrient elements. Annu. Rev. Plant Physiol. 14:93−106.

Butler, G. W., and A. C. Glenday. 1962. Iodine content of pasture plants. II. Inheritance of leaf iodine content of perennial ryegrass (*Lolium perenne* L.). Australian J. Biol. Sci. 15:183−187.

Butler, G. W., and J. M. Johnson. 1957. Factors influencing the iodine content of pasture herbage. Nature 179:216−217.

Cannell, G. H., F. T. Bingham, and M. J. Garber. 1960. Effects of irrigation and phosphorus on vegetative growth and nutrient composition of tomato leaves. Soil Sci. 89:53−60.

Chapman, H. D. 1949. Citrus leaf analysis. Nutrient deficiencies, excesses, and fertilizer requirements of soil, indicated by diagnostic aid. California Agr. 3(11):10, 12, 14.

Chapman, H. D. (ed.). 1966. Diagnostic criteria for plants and soils. Div. Agr. Sci., Univ. California, Berkeley. 793 pp.

Chapman, H. D., and S. M. Brown. 1941. The effects of sulfur deficiency on citrus. Hilgardia 14:185−201.

Chapman, H. D., and S. M. Brown. 1942. Some fungal infections of citrus in relation to nutrition. Soil Sci. 54:303−312.

Chessin, M., and H. A. Scott. 1955. Mineral nutrition and the size of local lesions induced by tobacco mosaic virus. Science 121:112.

Cooil, B. J. 1948. Potassium deficiency and excess in guayule. II. Cation-anion balance in leaves. Plant Physiol. 23:403−424.

Corden, M. C. 1965. Influence of calcium nutrition on *Fusarium* wilt of tomato and polygalacturonase activity. Phytopathology 55:222−224.

Cullinan, F. P., and L. P. Batjer. 1943. Nitrogen, phosphorus, and potassium interrelationships in young peach and apple trees. Soil Sci. 55:49−60.

Cullinan, F. P., D. H. Scott, and J. G. Waugh. 1939. The effects of varying amounts of nitrogen, potassium and phosphorus on the growth of young peach trees. Proc. Amer. Soc. Hort. Sci. 36:61−68.

Drake, M., and J. M. White. 1961. Influence of nitrogen on uptake of calcium. Soil Sci. 91:66−69.

Drosdoff, M. 1950. Minor-element content of leaves from tung orchards. Soil Sci. 70:91−98.

Drosdoff, M., and R. D. Dickey. 1943. Copper deficiency of tung trees. Proc. Amer. Soc. Hort. Sci. 42:79−84.

Eaton, F. M. 1944. Deficiency, toxicity, and accumulation of boron in plants. J. Agr. Res. 69:237−277.

Edgington, L. V., and J. C. Walker. 1958. Influence of calcium and boron nutrition on development of *Fusarium* wilt of tomato. Phytopathology 48:324−326.

Ehrler, W., and L. Bernstein. 1958. Effects of root temperature, mineral nutrition, and salinity on growth and composition of rice. Bot. Gaz. 120:67−74.

Evans, C. E., D. J. Lathwell, and H. J. Mederski. 1950. Effect of deficient or toxic levels of nutrients in solution on foliar symptoms and mineral content of soybean leaves, as measured by spectrographic methods. Agron. J. 42:25−32.

Evans, H. J. 1963. Effect of potassium and other univalent cations on activity of pyruvic kinase in *Pisum sativum*. Plant Physiol. 38:397−402.

Evans, H. J., E. R. Purvis, and F. E. Bear. 1950. Molybdenum nutrition of alfalfa. Plant Physiol. 25:555−566.

Finch, A. H., and A. F. Kinnison. 1934. Zinc treatment of pecan rosette. Arizona Agr. Ext. Service Circ. 82.

Fisher, P. L. 1935. Physiological studies on the pathogenicity of *Fusarium lycopersici* Sacc. for the tomato plant. Maryland Agr. Exp. Sta. Bull. 374.

Gassner, G., and K. Hassebrauk. 1931. Untersuchungen über die Beziehungen zwischen Mineralsalzernährung und Verhalten der Getreidenpflanzen gegen Rost. Phytopathol. Z. 3:535–617.

Goodall, D. W. 1945. Studies in the diagnosis of mineral deficiency. II. A comparison of the mineral content of scorched and healthy leaves from the same apple tree. J. Pomol. Hort. Sci. 21:90–102.

Goodall, D. W., and F. G. Gregory. 1947. Chemical composition of plants as an index of their nutritional status. Imp. Bur. Hort. and Plantation Crops Tech. Commun. 17.

Hale, J. B. 1945. Deficiency diseases of the sugar beet. Unpublished report, duplicated for private circulation. Agr. Res. Council, 7828:8.

Hall, N. S., W. F. Chandler, C. H. M. van Bavel, P. H. Reid, and J. H. Anderson. 1953. A tracer technique to measure growth and activity of plant root systems. North Carolina Agr. Exp. Sta. Tech. Bull. 101.

Hallock, D. L., and O. J. Attoe, 1954. Correlation of phosphorus content of alfalfa with pH and forms of soil phosphorus. Proc. Soil Sci. Soc. Amer. 18:64–67.

Harward, M. E., W. A. Jackson, J. R. Piland, and D. D. Mason. 1956. The relationship of chloride and sulfate ions to form of nitrogen in nutrition of Irish potatoes. Proc. Soil Sci. Soc. Amer. 20:231–236.

Hayward, H. E., E. M. Long, and R. Uhvits. 1946. Effect of chloride and sulfate salts on the growth and development of the Elberta peach on Shalil and Lovell rootstocks. U. S. Dep. Agr. Tech. Bull. 922.

Howell, R. W. 1954. Phosphorus nutrition of soybeans. Plant Physiol. 29:477–483.

Hylton, L. O., A. Ulrich, and D. R. Cornelius. 1967. Potassium and sodium interrelations in growth and mineral content of Italian ryegrass. Agron. J. 59:311–314.

Jacobson, H. G. M., and T. R. Swanback. 1932. Manganese content of certain Connecticut soils and its relation to the growth of tobacco. J. Amer. Soc. Agron. 24:237–245.

Jacobson, L. 1945. Iron in the leaves and chloroplasts of some plants in relation to their chlorophyll content. Plant Physiol. 20:233–245.

Johnson, C. M., G. A. Pearson, and P. R. Stout. 1952. Molybdenum nutrition of crop plants. II. Plant and soil factors concerned with molybdenum deficiencies in plants. Plant and Soil 4:178–196.

Johnson, J. M., and G. W. Butler. 1957. Iodine content of pasture plants. I. Method of determination and preliminary investigations of species and strain differences. Physiol. Plantarum 10:100–111.

Jones, J. B., Jr. 1965. Molybdenum content of corn plants exhibiting varying degrees of potassium deficiency. Science 148:94.

Jones, J. O., and W. Plant. 1943. Note on the composition of leaves from potato manurial experiment. Rep. Agr. Hort. Res. Sta. (Bristol) 1942:44–45.

Jones, J. P. 1929. Deficiency of magnesium the cause of a chlorosis in corn. J. Agr. Res. 39:873–892.

Jones, S. S., and F. A. Long. 1952. Complex ions from iron and ethylenediaminetetra-acetate: General properties and radioactive exchange. J. Phys. Chem. 56:25–33.

Jordan, H. V., and C. E. Bardsley. 1958. Response of crops to sulfur on Southeastern soils. Proc. Soil Sci. Soc. Amer. 22:254–256.

Kamprath, E. J., and E. V. Miller. 1958. Soybean yields as a function of the soil phosphorus level. Proc. Soil Sci. Soc. Amer. 22:317–319.

Krauss, R. W. 1957. Chelates as a source of micronutrients to algae. Phycological Soc. Amer. News Bull. 10:27−28.

Krugers, J., and J. Agterdenbos. 1957. Photosensitivity of the iron(III) ethylenediamine tetraacetate complex. Nature 179:45.

Lagatu, H., and L. Maume. 1935. Variations des rapports physiologiques en correlation avec la maladie du feu sauvage chez la feuille du tabac. Compt. Rend. Acad. Sci. (Paris) 201:374−376.

Levitt, E. C., and R. I. Nicholson. 1941. Manganese deficiency in citrus in New South Wales coastal districts. Agr. Gaz., N.S.W. 52:283−286.

Lilleland, O., and J. G. Brown. 1942. The phosphate nutrition of fruit trees. IV. The phosphate content of peach leaves from 130 orchards in California, and some factors which may influence it. Proc. Amer. Soc. Hort. Sci. 41:1−10.

Lucas, R. E., and G. D. Scarseth. 1947. Potassium, calcium, and magnesium balance and reciprocal relationship in plants. J. Amer. Soc. Agron. 39:887−896.

Lyon, C. B., K. C. Beeson, and G. H. Ellis. 1943. Effects of micronutrient deficiencies on growth and vitamin content of the tomato. Bot. Gaz. 104:495−514.

Maas, E. V., and J. E. Leggett. 1968. Uptake of ^{86}Rb and K by excised maize roots. Plant Physiol. 43:2054−2056.

McLarty, H. R., and C. G. Woodbridge. 1950. Boron in relation to the culture of the peach tree. Sci. Agr. 30:392−395.

McNew, G. L., and E. L. Spencer. 1939. Effect of nitrogen supply of sweet corn on the wilt bacterium. Phytopathology 29:1051−1067.

Maier, W. 1943. Eine Bormangel Krankheit des Selleric (*Apium graveolens* var rapaceum). Gartenbauwissensehaften 18:47−48.

Melsted, S. W. 1953. Some observed calcium deficiencies in corn under field conditions. Proc. Soil Sci. Soc. Amer. 17:52−54.

Miller, D. F. 1958. Composition of cereal grains and forages. Pub. 585. Nat. Acad. Sci. − Nat. Res. Council, Washington, D.C.

Neal, D. C. 1928. Cotton wilt: A pathological and physiological investigation. Mississippi Agr. Exp. Sta. Tech. Bull. 16. (Also published in Ann. Missouri Bot. Garden (1927), 14:359−424.)

Neller, J. R. 1925. The influence of sulfur and gypsum upon the composition and yield of legumes. Washington Agr. Exp. Sta. Bull. 190.

Nims, L. F. 1962. Tracers, transfer through membranes, and coefficients of transfer. Science 137:130−132.

Pack, M. R., S. J. Toth, and F. E. Bear. 1953. Copper status of New Jersey soils. Soil Sci. 75:433−441.

Pattee, H. E., and M. R. Teel. 1967. Influence of nitrogen and potassium on variations in content of malate, citrate, and malonate in non-nodulating soybeans (*Glycine max*). Agron. J. 59:187−189.

Porter, J. W., and H. J. Knauss. 1954. Inhibition of growth of *Chlorella pyrenoidosa* by beta-emitting radioisotopes. Plant Physiol. 29:60−63.

Porter, J. W., and M. S. Watson. 1954. Gross effects of growth-inhibiting levels of tritium oxide on *Chlorella pyrenoidosa*. Amer. J. Bot. 41:550−555.

Price, C. A. 1970. Molecular approaches to plant physiology. McGraw-Hill, New York, 398 pp.

Pryor, D. E. 1940. The effect of some mineral nutrients on the development of clubroot of crucifers. J. Agr. Res. 61:149−160.

Reisenauer, H. M. 1956. Molybdenum content of alfalfa in relation to deficiency symptoms and response to molybdenum fertilization. Soil Sci. 81:237−242.

Reuther, W., and P. F. Smith. 1954. Leaf analysis of citrus, pp. 257—294. *In* N. F. Childers (ed.) Mineral nutrition of fruit crops. Somerset Press, Somerville, N.J.

Reuther, W., P. F. Smith, and A. W. Specht. 1949. A comparison of the mineral composition of Valencia orange leaves from the major producing areas of the United States. Proc. Florida State Hort. Soc. 62:38—45.

Reuther, W., T. W. Embleton, and W. W. Jones. 1958. Mineral nutrition of tree crops. Annu. Rev. Plant Physiol. 9:175—206.

Samuel, G., and C. S. Piper. 1928. Grey-speck (manganese deficiency) disease of oats. J. S. Australian Dep. Agr. 31:696—705, 789—799.

Schaffnit, E., and A. Volk. 1927. Über den Einfluss der Ernährung auf die Empfänglichkeit der Pflanzen für Parasiten. I. Forsch. Geb. Pflanzenkrank. Immunität Pflanzenr. 3:1—45.

Shear, G. M., and S. A. Wingard. 1944. Some ways by which nutrition may affect severity of disease in plants. Phytopathology 34: 603—605.

Smith, C. B., and G. A. Taylor. 1952. Tentative optimum leaf concentrations of several elements for Elberta peach and Stayman apple in Pennsylvania orchards. Proc. Amer. Soc. Hort. Sci. 60:33—41.

Somers, I. I., and J. W. Shive. 1942. The iron-manganese relation in plant metabolism. Plant Physiol. 17:582—602.

Spencer, E. L. 1935a. Effect of nitrogen supply on host susceptibility to virus infection. Phytopathology 25:178—191.

Spencer, E. L. 1935b. Influence of phosphorus and potassium supply on host susceptibility to yellow tobacco mosaic infection. Phytopathology 25:493—502.

Spencer, E. L., and G. W. McNew. 1938. The influence of mineral nutrition on the reaction of sweet-corn seedlings to *Phytomonas stewarti*. Phytopathology 28:213—223.

Stewart, I. 1963. Chelation in the absorption and translocation of mineral elements. Annu. Rev. Plant Physiol. 14:295—310.

Stoddard, D. L. 1942. *Fusarium* wilt of cantaloupe and studies on the relation of potassium and nitrogen supply to susceptibility. Trans. Peninsula Hort. Soc. 31:91—93.

Thomas, W., and W. B. Mack. 1941. Susceptibility to disease in relation to plant nutrition. Science 93:188—189.

Ulrich, A., D. Ririe, F. J. Hills, A. G. George, and M. D. Morse. 1959. Plant analysis: A guide for sugar beet fertilization. California Agr. Exp. Sta. Bull. 766.

Vanselow, A. P., and N. P. Datta. 1949. Molybdenum deficiency of the citrus plant. Soil Sci. 67:363—375.

Viets, F. G., Jr., L. C. Boawn, C. L. Crawford, and C. E. Nelson. 1953. Zinc deficiency in corn in central Washington. Agron. J. 45:559—565.

Volk, A. 1931. Beiträge zur Kenntnis der Wechselbeziehungen zwischen Kulturpflanzen, ihren Parasiten und der Umwelt. 4. Einflüsse des Bodens, der Luft und des Lichtes aus die Empfänglichkeit der Pflanzen für Krankheiten. Phytopathol. Zeit. 3:1—88.

Volk, N. J., J. W. Tidmore, and D. T. Meadows. 1945. Supplements to high-analysis fertilizers, with special reference to sulfur, calcium, magnesium, and limestone. Soil Sci. 60:427—435.

Walker, J. C. 1946. Soil management and plant nutrition in relation to disease development. Soil Sci. 61:47—54.

Walker, J. C., and W. J. Hooker. 1945a. Plant nutrition in relation to disease development. I. Cabbage yellows. Amer. J. Bot. 32:314—320.

Walker, J. C., W. J. Hooker. 1945b. Plant nutrition in relation to disease development. II. Cabbage clubroot. Amer. J. Bot. 32:487—490.

Walker, J. M., and S. A. Barber. 1961. Ion uptake by living plant roots. Science 133:

881—882.

Walker, J. M., and S. A. Barber. 1962. Absorption of potassium and rubidium from the soil by corn roots. Plant and Soil 17:243—259.

Wall, M. E. 1940a. The role of potassium in plants. Thesis, Rutgers Univ., New Brunswick, N.J.

Wall, M. E. 1940b. The role of potassium in plants. II. Effect of varying amounts of potassium on the growth status and metabolism of tomato plants. Soil Sci. 49: 315—331.

Wallace, A. 1956. Effectiveness of iron chelates in the presence of sodium bicarbonate, pp. 35—42. *In* A. Wallace (ed.) Symposium on the Use of Metal Chelates in Plant Nutrition (Seattle, Wash.), Univ. California Press, Los Angeles.

Wallace, T. 1951. The diagnosis of mineral deficiencies in plants, 2nd ed. H. M. Stationery Office, London.

Walrath, E. K., and R. C. Smith. 1952. Survey of forty apple orchards. Proc. Amer. Soc. Hort. Sci. 60:22—32.

Wander, I. W. 1946. The relation of total leaf nitrogen to the yield and color of Stayman Winesap apples at different rates of nitrogen fertilizer applications on sod. Proc. Amer. Soc. Hort. Sci. 47:1—6.

Waris, H. 1953. The significance for algae of chelating substances in the nutrient solution. Physiol. Plantarum 6:538—543.

Weinberger, D., and J. W. Porter. 1953. Incorporation of tritium oxide into growing *Chlorella pyrenoidosa* cells. Science 117:636—638.

Wiederspahn, F. E. 1956. Manganese in New Jersey apple orchards. Hort. News 37(3):3016, 3022.

Wilcox, J. C., and C. G. Woodbridge. 1943. Some effects of excess boron on the storage quality of apples. Sci. Agr. 23:332—341.

Wingard, S. A. 1941. The nature of disease resistance in plants. I. Bot. Rev. 7:59—109.

Woodbridge, C. G. 1951. A note on the incidence of zinc deficiency in the Okanagan Valley of British Columbia. Sci. Agr. 31:40.

Younts, S. E., and R. B. Musgrave. 1958a. Growth, maturity, and yield of corn, as affected by chloride in potassium fertilizer. Agron. J. 50:423—426.

Younts, S. E., and R. B. Musgrave. 1958b. Chemical composition, nutrient absorption, and stalk-rot incidence of corn, as affected by chloride in potassium fertilizer. Agron. J. 50:426—429.

CHAPTER 4

Space, Organic Acids, and Source of Nutrients

SPACE

From the time Hope and Stevens (1952) first proposed the concept of apparent free space, space has generally been considered to be resolvable into two components: the water free space (WFS) into which cations and anions freely diffuse, and the Donnan free space (DFS) (Briggs and Robertson, 1957). Inasmuch as anions must pass a large number of capillaries with varying dimensions on their way through walls, they are often restricted in their free movement by Donnan potentials (Pettersson, 1966). As a result, in certain portions of the cell wall, there may be a higher concentration of cations than of anions (Briggs and Robertson, 1957).

Free space involves that portion of the cell or tissue into which solutes and solvent move relatively freely; osmotic volume is that part into which solutes, but not the solvent, penetrate relatively slowly (Briggs et al., 1961). Neither of these spaces can be measured directly, hence such terms as apparent free space (AFS) are used. DFS involves immobile anions and, when this space is involved, there is a relatively higher concentration of mobile cations in this space than in the external medium. The concentration of cations would overestimate this space and that of the mobile anions would underestimate it; hence again, the space is often called AFS rather than free space.

Nonmetabolically absorbed ions are believed to exist in the outer space of cells—essentially the space other than that occupied by vacuoles. Free space is of such magnitude that it probably includes most of the cytoplasm (Lundegardh, 1954). Pitman (1965b) also noted that under certain conditions part of the cytoplasm may behave as free space. In one of the highest values for space, Leggett and Olsen (1964) noted that Br^-, PO_4^{3-}, and SO_4^{2-} entered yeast cells *(Saccharomyces cerevisiae)* by passive diffusion into a space which was more than 80% of the cellular volume. PO_4^{3-} and SO_4^{2-}, but not Br^-, were further accumulated from this space by a metabolically dependent uptake. Although K^+ appeared to be uniformly distributed in soybean roots, Ca^{2+} was localized in 10% of the root volume near the external solution–root interface (Leggett and Gilbert, 1967).

Epstein (1955) characterized this outer space (which Hope and Stevens (1952) called AFS) and the state of the ions in it, as:

> The apparent free space represents about 23% of the volume of barley root tissue. At equilibrium, the concentration of the ions in the "outer" space of the tissue equals their concentration in the ambient medium. There is no competition among different ions for the "space." There is no pH effect. The "space" was freely accessible to selenate, phosphate, and Ca ions, in addition to sulfate.

AFS has also been defined by Briggs and Robertson (1957): "the apparent volume of the free space equals the amount of solute in the free space divided by the amount of solute per unit volume of external solution when free space and external solution are *in equilibrium.*" In other words, the AFS is the volume by which tissue appears to dilute the solution at the end of the initial uptake period (Briggs et al., 1961). Levitt (1957) concluded that the AFS is in the cell walls and that most researchers have overestimated its magnitude. He calculated the AFS to be 8%.

WFS may be visualized according to the following statement. If the free space contains only water and diffusible ions or molecules, AFS for electrolytes and nonelectrolytes will be the same (Briggs et al., 1961).

DFS can be defined in the following terms. If the free space contains indiffusible anions, for example, uptake of cations will be greater during initial uptake than that of anions (Stiles and Skelding, 1940; Davies and Wilkins, 1951; Sutcliffe, 1954). Thus DFS refers to that portion or space of cells or tissues which contain indiffusible ions—usually anions. DFS may be located in the cytoplasm and cell walls (Briggs et al., 1961), or only in the cell walls (Dainty and Hope, 1959; Pitman, 1965). Cell walls of beet root tissue contain sufficient cation exchange sites to account for at least 95% of the DFS (Pitman, 1965).

Working with *Nitella axillaris*, Diamond and Solomon (1959) concluded that K^+, for example, may be located in three parts of the cell. About 0.1% of the K^+ exchanged rapidly (apparent half-time 23 sec) and was apparently in the cell wall; about 1.6% exchanged more slowly (apparent half-time 5 hr) and was regarded as an intracytoplasmic component; and the bulk of the K^+ exchanged very slowly (apparent half-time 40 days) and was believed to be in the vacuole. MacRobbie and Dainty (1958), working with *Nitella obtusa,* also concluded that there were three compartments: free space, protoplasmic non-free space, and the cell sap. Pitman (1963) developed methods for studying the compartmental loci of inorganic ions. There appeared to be at least two compartments of nonfree space—cytoplasm and vacuole.

In a study of transport rates of Rb^+, K^+, and Cl^- through cucumber roots, as determined by analysis of the xylem exudate, there was no indication of a carrier for Cl^- (Cooil, 1969). Further, it was noted that the time course of K^+ liberation from the tissue indicated the departure of K^+ from two separate compartments.

Brouwer (1965) reported that ions enter the WFS (cations and anions) and the exchange space (cations) of tissue by diffusion. Both spaces were regarded to be confined to the cell walls and in roots these would be the cell walls outside the endodermis. He arrived at this conclusion on the basis that *both* the plasmalemma and tonoplast are high resistance barriers against the passage of ions.

Values for AFS range from 7.8% in pea (Hylmo, 1953) to 33% in wheat (Butler, 1955) and 45.3% for *Porphyra perforata* (Eppley and Blinks, 1957). The AFS in sunflower, measured on a root volume basis, varied between 14 and 57%—depending on pretreatment of the plants and SO_4^{2-} concentration (Pettersson, 1966). Variation was attributed to differing capacities to bind SO_4^{2-} by exchange adsorption within AFS. By contrast, Ingelsten (1966), using various mannitol concentrations for wheat roots, reported that AFS was constant from full turgidity to incipient plasmolysis. Using wheat roots and labeled PO_4^{3-} and SO_4^{2-}, Butler (1959) estimated AFS at 18–20% for normal roots.

Studying SO_4^{2-} uptake by sunflower plants, Pettersson (1966) reported that exchange adsorption probably constitutes the initial stage of active ion uptake. SO_4^{2-} was assumed to migrate passively into the AFS, but SO_4^{2-} uptake within the root was shown to depend on metabolism. Ingelsten (1966) observed that a nonplasmolyzing mannitol concentration increased SO_4^{2-} retention by wheat roots, while at the same time there was an increase in SO_4^{2-} transfer to the shoot.

Uptake of salts by roots can be resolved into two main processes—a relatively rapid entry into the AFS, and a slower accumulation into vacuoles (Butler, 1959). Movement of ions into the AFS is believed to depend on diffusion and exchange (exchange adsorption); movement into vacuoles depends on endogenously produced carrier substances (Butler, 1959).

With regard to compartmentalization it has been suggested that intracellular Na^+ and K^+ are located on sites on macromolecules and capable of hopping from site to site through the icelike lattice of cell water in which they are only sparingly soluble (Bernhard, 1969). It has been suggested that cell water is ice-like and tissue water semicrystalline (Anonymous, 1969). Sodium and K^+ ions, while moving through an icelike matrix, would obey laws analogous to those governing conduction of electrons in semiconducting solids; cells would act as semiconducting crystals.

The primary endodermis of onion root tips restricted the entry of ^{46}Sc into the stele at a very early stage of development—leading to the conclusion that migration of ^{46}Sc across the root was primarily in the free space (Clarkson and Sanderson 1969).

ORGANIC ACIDS

When plants are grown in a nutrient solution, it is generally noted that pH rises with time. The direction and amount of change are determined by composition of the solution and by species of plant. If roots remove more cations than anions, they exchange H^+ (Ulrich, 1941; Van Steveninck, 1966) for the cations they absorb and the solution becomes more acid. However, if roots remove an excess of anions over cations and exchange OH^- and/or HCO_3^- for anions that were in solution, the solution becomes more basic. When NO_3^- is used as the source of N, pH usually rises; with NH_4^+ nitrogen, it usually falls. Jackson and Adams (1963) reported a release of H^+ during K^+ and Na^+ uptake in excess of anion uptake, and a release of base when anion uptake exceeded that of cation uptake.

When there is an excess of cation over anion uptake, organic acids [particularly citrate and malate (Van Steveninck, 1966)] increase in the cells (Dunne, 1932; Pucher et al., 1938; Ulrich, 1941, 1942; Overstreet et al., 1942; Pierce and Appleman, 1943; Burstrom, 1945; Jacobson and Ordin, 1954; Jacobson, 1955; Hurd, 1958; Jackson and Coleman, 1959; Wit et al., 1963; Poole and Poel, 1965; Noggle, 1966; Van Steveninck, 1966; Hiatt and Hendricks, 1967; Hiatt, 1969; Osmund and Laties, 1969). Pucher et al. (1938) first showed that the large excess of positive ions was closely correlated with the quantity of ether-soluble organic acids. When more anions than cations are absorbed, the concentration of organic acids in cells declines (Ulrich, 1941, 1942; Burstrom, 1945).

Overstreet et al. (1942) studied uptake of cations and anions by barley roots in K^+ solutions, and also in K^+–clay suspensions, and compared K^+ accumulations in the two media with CO_2 evolution and the synthesis of organic acids in the roots. Excess accumulation of cations over anions was roughly balanced by organic acid anions (other than HCO_3^-) synthesized in the roots. They deduced that the synthesized organic acids were the source of the H^+, which replaced adsorbed K^+ on the clay, and not carbonic acid.

Changes in organic acid concentration in barley roots in response to differential cation and anion uptake appeared to be associated with the low-salt component of ion uptake (Hiatt, 1967). Within cells organic acids may electrostatically bind an equivalent quantity of cations, and amino acids may bind an equivalent quantity of cations and anions (Hiatt and Lowe, 1967). Later, Hiatt (1969) demonstrated that a shift from excess anion to excess cation accumulation could be explained on the basis of relative rates of absorption of K^+, Na^+, and Cl^-. During absorption barley roots in low-K, high-Na solutions of 0.01 mM concentration showed a decrease in organic acids, whereas roots in high-K^+, low-Na^+ solutions showed an increase.

In 11 out of 12 species of plants, Pierce and Appleman (1943) found a large excess of inorganic cations over inorganic anions. When all plants were considered, this excess was highly correlated with total ether-soluble organic acids (Table 4.1). Considering the accuracy of methods at the time, it is surprising how closely the disparity between total cations and total inorganic anions was accounted for when organic acid anions were taken into consideration.

Table 4.1
Correlation between Excess Inorganic Cations and Total Organic Acids[a,b]

	Leaves				Stems and petioles			
	Total Ca, Mg, K	Total anions[c]	Excess cations[d]	Total organic acids	Total Ca, Mg, K	Total anions[c]	Excess cations[d]	Total organic acids
Lima bean	286.8	59.9	226.9	249.7	249.2	57.0	192.2	171.1
Peas	231.2	69.2	162.0	212.9	214.1	76.9	137.2	181.0
Alfalfa	243.8	78.4	165.4	185.7	205.7	66.5	139.2	140.7
Soybean	276.2	62.9	213.3	248.7	243.3	79.0	164.3	177.0
Beets	465.5	77.2	388.3	427.5	368.5	185.0	183.5	203.1
Spinach	457.7	110.5	347.2	380.4	378.7	160.4	218.3	226.2
Buckwheat	389.7	65.8	323.9	360.1	384.8	196.8	188.0	205.6
Bluegrass	212.9	124.3	88.6	108.5	—	—	—	—
Wheat	237.0	142.6	94.7	122.1	—	—	—	—
Tomato	379.2	142.1	237.1	269.8	412.7	151.3	261.4	231.6
Lettuce	306.3	105.4	200.9	220.8	—	—	—	—
Cantaloupe	557.9	143.6	414.3	99.2	320.8	174.2	146.6	119.4

[a] Pierce and Appleman (1943).
[b] Values are given in milliequivalents per 100 g of dry tissue.
[c] NO_3^-, SO_4^{2-}, $H_2PO_4^-$
[d] Total cations minus inorganic anions.

In guayule, total cations greatly exceeded total anions (inorganic plus organic) (Cooil, 1948). The remainder of the anion total was carbonate. When $CaCO_3$ is present, as in guayule leaves, an analysis for total Ca^{2+} could be misleading from a physiological point of view.

On a milliequivalent basis, cations equaled anions (inorganic, organic acids, and amino acids) in roots of sunflower, but in stems—and particularly in leaves—cations exceeded total anions about two- and threefold, respectively (Mengel, 1965). There was no explanation for the differences among roots, stems, and leaves.

Torii and Laties (1966) concluded that organic acid synthesis was increased when cation uptake exceeded anion uptake, because organic acids and cations moved into the vacuoles.

Organic acids are produced not only by respiration, but also by CO_2 fixation in roots. Fixation of CO_2 in roots is associated with an excess uptake of cations (Jacobson and Ordin, 1954). The amount of fixation was related to the extent to which cation absorption exceeded anion absorption; primarily malate was formed during fixation. Hiatt and Hendricks (1967) and Torii and Laties (1966) observed that CO_2 fixation by roots was greater with salt solutions in which cation uptake exceeded anion uptake than when the reverse occurred. Rate of CO_2 fixation was of the order required to account for the organic acid increases in the cells (Hiatt and Hendricks, 1967).

There is a compartmentation of organic acids (MacLennan et al., 1963; Steer and Beevers, 1967) in roots, since some of the malic acid appears in a cytoplasmic and some in a mitochondrial pool (Lips and Beevers, 1966a, b). Hiatt (1968) proposed that organic and amino acids play an important role in ion accumulation by providing nondiffusible charges which may bind or retain inorganic ions within cells.

During an excess of cation over anion absorption, Osmond and Laties (1969) observed an increase in organic acids in beet discs. Using techniques for studying compartmental analysis developed for inorganic ions (Pitman, 1963; Cram, 1967), they determined that organic acids left the cytoplasm in the role of counteranions during net transport of cations to the vacuole.

Wit et al. (1963) and Noggle (1966) observed a positive correlation between yield and concentration of organic acids in cells. Nutrient treatments that increased organic acids also increased yields, and vice versa. Growth is regulated in part by the organic acid concentration (Noggle, 1966). Considering a number of species, Wit et al. (1963) observed a 1:1 linear relationship between the milliequivalents of organic acid anions per kilogram and the milliequivalents of cations minus inorganic anions per kilogram when phosphate was expressed as $H_2PO_4^-$ and not as PO_4^{3-}.

Independent of the form of N nutrition for tomato, Kirkby and Mengel

(1967) reported that total cations were fairly well balanced by total anions (on a milliequivalent basis). Bound Ca^{2+} and Mg^{2+} were associated with oxalic and uronic acids. There was a close, positive relationship between the diffusible cations and diffusible anions; a similar relationship existed between the alkalinity of NO_3-free ash and the concentration of organic anions (organic acid anions and uronic acids).

In Valencia orange leaves, malate and particularly oxalate increased with increasing Ca^{2+} in the leaves (Rasmussen and Smith, 1961). Increases in K^+ increased oxalate but not malate. However, variations in internal concentrations of Ca^{2+}, Mg^{2+}, and K^+ had no effect on the concentration of citric acid in the leaves.

Wadleigh and Shive (1939) grew corn plants with the pH of the substrate varying from 3 to 8. Organic acids in the plants increased with increase in the pH of the substrate. Plants grown with a mixture of NH_4^+ and NO_3^- contained a lower concentration of organic acids and bases than did those plants receiving only NO_3. Chlorosis was almost entirely overcome when plants received equivalent concentrations of NH_4^+ and NO_3 nitrogen.

There is evidence that Fe may be carried by or associated with citrate (Tiffin, 1966a). In some instances it was noted that there was more Fe than citrate, but nevertheless Tiffin (1966b) concluded that citrate is the agent normally involved in the translocation of Fe in plants.

SOURCE OF NUTRIENTS

Solution (Nutrient Solution or Soil)

In order to study the inorganic nutrient requirements of plants, researchers quite often resort to nutrient solutions with plants grown either in water culture or aggregate culture (i.e., sand, perlite, and so on). In these experiments roots remove ions directly from the solution and yield cations (usually H^+) and anions (OH^- and/or HCO_3^-) to the solution in stochiometric proportion to the ions absorbed by roots (Jacobson, 1955; Jackson and Adams, 1963). A similar exchange occurs when roots obtain ions from the soil solution. Ions in solution can exchange with ions adsorbed on root surfaces.

Most of the early workers believed that plants obtained nutrients solely from the soil solution, and Cameron (1911) was one of the strongest proponents of this theory. When inferior yields were obtained, he believed, and seemingly demonstrated, that toxic, organic substances were present in the soil solution.

Carbonic Acid Theory

One of the earliest ideas on the mechanism of salt absorption was the car-

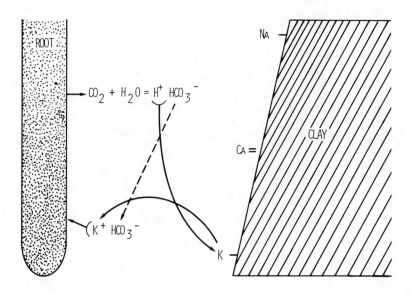

Fig. 4.1

Diagrammatic illustration of the release of ions adsorbed on colloids by CO_2 from roots (i.e., the carbonic acid theory). Redrawn from H. Jenny and R. Overstreet (1938) Proc. Nat. Acad. Sci., 24:384–392. National Academy of Sciences—by permission.

bonic acid theory. Accordingly, CO_2 produced by respiration diffuses into the soil solution where it forms carbonic acid. Next, H^+ from carbonic acid exchanges with K^+, for example, on a colloid to liberate K^+ to the soil solution (Fig. 4.1). The K^+ is then absorbed from the soil solution by the plant. Several years ago, however, convincing evidence was obtained against the carbonic acid theory. For example, barley roots absorbed more K^+ and Zn^{2+} from montmorillonitic clay than from kaolinitic clay suspensions, and this difference was in contrast with what was expected—based on the carbonic acid theory (Elgabaly et al., 1943). In the presence of carbonic acid, there was a higher concentration of bases in the kaolinite than in the montomorillonite suspension, but absorption was greater from montmorillonite than from kaolinite.

Soil Particles

It became apparent that roots can obtain some ions directly from soil particles or colloids by contact exchange or contact feeding without ions passing into solution. Devaux (1916) first called attention to the fact that root surfaces and clay colloids have adsorbed surface ions, that they are in intimate contact with each other, and that ions could be exchanged from one system to the other. If the oscillation volume of an ion adsorbed on a root surface overlaps

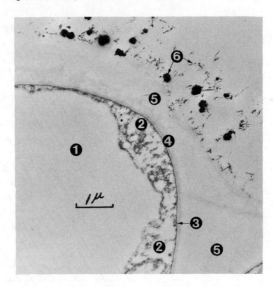

Fig. 4.2
Electron micrograph showing outer edge of a root cell in contact with soil. This is a magnified view of mucilaginous region at the root surface. 1, Large vacuole; 2, cytoplasm (dark-stained); 3, plasma membrane (thin, dark line); 4, cell wall (grayish layer); 5, mucilage; 6, black rods and larger aggregates of iron hydroxide particles in a large open macropore. (Jenny and Grossenbacher, 1962.)

with that of an ion adsorbed on a clay or organic colloid, ions may merely exchange locations (Jenny and Overstreet, 1938, 1939a). The plant may substitute one or more H^+ (depending on whether the acquired ion is monovalent or divalent) for the acquired ion, and the latter may then pass into the root. Jenny (1951) calculated that 1 mm^2 of root surface could make contact with 10^8 clay particles—each one of which might carry 6000 to 7000 exchangeable cations.

Jenny and Grossenbacher (1962) provided excellent electron micrographs of root cells in contact with soil (Fig. 4.2).

It is difficult to determine the extent to which roots obtain ions from the soil solution and the extent to which they obtain them directly from soil particles by contact feeding. There can be no doubt, however, concerning the fact that a significant portion of ion uptake from soils occurs directly from soil particles —clay and organic colloids (Jenny and Overstreet, 1939a, 1939b). Barley roots absorbed more Na^+, K^+, or Rb^+ when these ions were adsorbed on colloids than when they were present in solution (Jenny and Overstreet, 1938, 1939b; Overstreet and Jenny, 1939). Lundegardh (1947) noted that contact exchange differs quantitatively, not qualitatively, from uptake of ions from a solution, since both processes are exchange reactions.

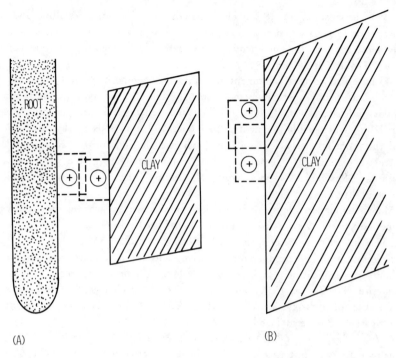

(A)   (B)

Fig. 4.3

(A) Diagrammatic illustration of exchange of adsorbed ions between root and clay colloids when the oscillation volumes (dotted lines) of the ions overlap. If the root yields an H^+ ion to the root in exchange for a nutrient ion, such as K^+, the process is called contact intake. The reverse is called contact depletion. Redrawn from H. Jenny and R. Overstreet (1939a) Soil Sci., 47:257–272. © 1939, The Williams and Wilkins Co., Baltimore, Md. (B) Surface migration of ions on a colloidal surface when oscillation volumes (dotted lines) of adsorbed ions overlap. Redrawn from H. Jenny and R. Overstreet, (1939b) J. Phys. Chem., 43:1185–1196. © 1939, The Williams and Wilkins Co., Baltimore, Md.

There is evidence that nutrients can move not only into a plant from a colloid, but also from a plant back to the colloid or into the solution (Albrecht, 1940). Intake of ions from clay colloids is called contact intake and loss of ions from the plant to the clay is known as contact depletion (Jenny and Overstreet, 1939a) (Fig. 4.3A). Ions may also move along the surface of a clay particle (Jenny and Overstreet, 1939b) (Fig. 4.3B).

Structure of Clays. Prior to the 1930s, soil clays were throught to be amorphous gels of indefinite composition (Stout and Overstreet, 1950). Russell (1926) suggested that x rays might disclose the true nature of soil colloids. Following Pauling's (1930) classic studies of the structures of micas as typical layer lattice aluminosilicates, Hendricks and Fry (1930) and Kelley et al.

(1931) demonstrated that soil clay colloids were in fact crystalline and not amorphous materials. There are several excellent reviews of the structures of clay minerals and the soil chemistry associated with them (Hendricks, 1942; Kelley, 1942, 1948; Grim, 1942).

With regard to their dominant mineralogical composition, soil clays can be classified as belonging to montmorillonitic, kaolinitic, or hydrous mica groups (Stout and Overstreet, 1950). Alkaline and neutral soils are generally dominated by montmorillonite, neutral-to-moderately acid soils by hydrous micas and the more intensely acid soils by kaolinitic aluminosilicates. Montmorillonitic clays contain large amounts of K^+ in a nonexchangeable form; apatite consists of insoluble Ca phosphates. Weathering increases the exchangeability and solubility.

Montmorillonitic clays in particular imbibe water between the cleavage planes and swell. With x-ray diffraction it has been possible to follow expansion of the layer lattice as water is absorbed (Hofmann et al., 1933).

Kaolinite is an end product of weathering and can be formed from montmorillonite by desilication (Jackson et al., 1948). Kaolinite (Gruner, 1932) and halloysite (Hendricks, 1938) are characterized by each layer lattice having an entire sheet of hydroxyl ions (Fig. 4.4). Nearly all minerals subjected to intensive weathering have surface hydroxyl ions.

Cation Exchange (Base Exchange). Base exchange capacity (BEC) is usually expressed in terms of milliequivalents per 100 g dry weight of soil. Sandy soils, with little or no clay, may have a BEC of approximately 1.0 or 2.0, whereas peat soils, with primarily organic colloids, may run as high as 80–90 meq/100 g dry soil.

Montmorillonitic clays have BEC values ranging from 70 to 120, while those of kaolinitic origin are much lower—5–20 meq/100 g of soil (Stout and Overstreet, 1950).

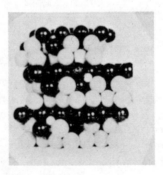

Fig. 4.4
Structure of kaolinite. Black balls, OH^- ions; larger white balls, O; smaller white balls, Al^{3+}; Si atoms surrounded by O atoms are not visible. (Stout and Overstreet, 1950.)

Clay minerals hold cations with a considerable range of bonding energies (Marshall, 1951).

Anion Exchange. Although not as pronounced as the adsorption of cations by soils, clays are also involved in anion exchange reactions. The availability of anions associated with clays is not satisfactorily understood (Overstreet and Dean, 1951).

Anions are held by positive charges originating from basic groups, hence the anion exchange capacity of soils increases with decreasing pH. In anion exchange strongly adsorbed OH^- ions are exchanged for other anions. The anion, which is substituted for an OH^- ion, must be of a size that fits into the lattice without steric hindrance (Stout and Overstreet, 1950). Anions meeting these requirements are phosphates, arsenates, molybdates, borate, fluoride, and of course hydroxyl ions. Anion exchange is particularly important with regard to phosphates and molybdates (Stout and Overstreet, 1950).

LITERATURE CITED

Albrecht, W. A. 1940. Absorbed ions on the colloidal complex and plant nutrition. Proc. Soil Sci. Soc. Amer. 5:8—16.

Anonymous. 1969. The cell — A puddle or an "ice cube?" Sci. Res. 4:35—36.

Bernhard, F. 1969. A time bomb made of simple (?) H_2O. Sci. Res. 4:36—39.

Briggs, G. E., and R. N. Robertson. 1957. Apparent free space. Annu. Rev. Plant Physiol. 8:11—30.

Briggs, G. E., A. B. Hope, and R. N. Robertson. 1961. Electrolytes and plant cells. *In* W. O. James (ed.) Bot. Mongr., Vol. I. Blackwell, Oxford, England. 217 pp.

Brouwer, R. 1965. Ion absorption and transport in plants. Annu. Rev. Plant Physiol. 16:241—266.

Burström, H. 1945. Studies on the buffer system of cells. Arkh. Bot. 32:1—18.

Butler, G. W. 1955. Minerals and living cells. J. New Zealand Inst. Chem. 19:66—75.

Butler, G. W. 1959. Uptake of phosphate and sulphate by wheat roots at low temperature. Physiol. Plantarum 12:917—925.

Cameron, F. K. 1911. The soil solution. Chem. Publ. Co., Easton, Pa.

Clarkson, D. T., and J. Sanderson. 1969. The uptake of a polyvalent cation and its distribution in the root apices of *Allium cepa*: Tracer and autoradiographic studies. Planta 89:136—154.

Cooil, B. J. 1948. Potassium deficiency and excess in guayule. II. Cation-anion balance in leaves. Plant Physiol. 23:403—424.

Cooil, B. J. 1969. Ion transport in cucumber roots. Plant Physiol. 44(Suppl.):no. 93.

Cram, W. J. 1967. Compartmentation and control of inorganic ions in higher plant cells. Ph.D. Thesis, Univ. Cambridge, Cambridge.

Dainty, J., and A. B. Hope. 1959. Ionic relations of cells of *Chara australis*. I. Ion exchange in the cell wall. Australian J. Biol. Sci. 12:395—411.

Davies, R. E., and M. J. Wilkins. 1951. The use of a double labelled salt ($K^{42}Br^{82}$) in the study of salt uptake by plant tissues. *In* Radioisotope Techniques. I. H. M. Stationery Office, London.

Devaux, H. 1916. Action rapide des solutions salines sur les plantes vivantes: Déplacement réversible d'une partie des substances basiques contenues dans la plante. Compt. Rend. Acad. Sci. (Paris) 162:561–563.

Diamond, J. M., and A. K. Solomon. 1959. Intracellular potassium compartments in *Nitella axillaris*. J. Gen. Physiol. 42:1105–1121.

Dunne, T. C. 1932. Plant buffer systems in relation to the absorption of bases by plants. Hilgardia 7:207–234.

Elgabaly, M. M., H. Jenny, and R. Overstreet. 1943. Effect of type of clay mineral on the uptake of zinc and potassium by barley roots. Soil Sci. 55:257–262.

Eppley, R. W., and L. R. Blinks. 1957. Cell space and apparent free space in the red alga *Porphyra perforata*. Plant Physiol. 32:63–64.

Epstein, E. 1955. Passive permeation and active transport of ions in plant roots. Plant Physiol. 30:529–535.

Grim, R. E. 1942. Modern concepts of clay minerals. J. Geol. 50:225–275.

Gruner, J. W. 1932. The crystal structure of kaolinite. Z. Krist. 83:75–88.

Hendricks, S. B. 1938. The crystal structure of the clay minerals: Dickite, halloysite and hydrated halloysite. Amer. Mineral. 23:295–301.

Hendricks, S. B. 1942. Lattice structure of clay minerals and some properties of clays. J. Geol. 50:276–290.

Hendricks, S. B., and W. H. Fry. 1930. The results of x-ray and microscopical examinations of soil colloids. Soil Sci. 29:457–459.

Hiatt, A. J. 1967. Relationship of cell sap pH to organic acid change during ion uptake. Plant Physiol. 42:294–298.

Hiatt, A. J. 1968. Electrostatic association and Donnan phenomena as mechanisms of ion accumulation. Plant Physiol. 43:893–901.

Hiatt, A. J. 1969. Accumulation of potassium and sodium by barley roots in a K-Na replacement series. Plant Physiol. 44:1528–1532.

Hiatt, A. J., and S. B. Hendricks. 1967. The role of CO_2 fixation in accumulation of ions by barley roots. Z. Pflanzenphysiol. 56:220–232.

Hiatt, A. J., and R. H. Lowe. 1967. Loss of organic acids, amino acids, K, and Cl from barley roots treated anaerobically and with metabolic inhibitors. Plant Physiol. 42:1731–1736.

Hofmann, U., K. Endell, and D. Wilm. 1933. The crystal structure and the swelling of montmorillonite. Z. Krist. 86:340–348.

Hope, A. B., and P. G. Stevens. 1952. Electric potential differences in bean roots and their relation to salt uptake. Australian J. Sci. Res. B5:335–343.

Hurd, R. G. 1958. The effect of pH and bicarbonate ions on the uptake of salts by discs of red beet. J. Exp. Bot. 9:159–174.

Hylmö, B. 1953. Transpiration and ion absorption. Physiol. Plantarum 6:333–405.

Ingelsten, B. 1966. Absorption and transport of sulfate by wheat at varying mannitol concentration in the medium. Physiol. Plantarum 19:563–579.

Jackson, M. L., S. A. Tyler, A. L. Willis, G. A. Bourbeau, and R. P. Pennington. 1948. Weathering sequence of clay-size minerals in soils and sediments. I. Fundamental considerations. J. Phys. Colloid Chem. 52:1237–1260.

Jackson, P. C., and H. R. Adams. 1963. Cation-anion balance during potassium and sodium absorption by barley roots. J. Gen. Physiol. 46:369–386.

Jackson, W. A., and N. T. Coleman. 1959. Ion absorption by bean roots and organic acid changes brought about through CO_2 fixation. Soil Sci. 87:311–319.

Jacobson, L. 1955. Carbon dioxide fixation and ion absorption in barley roots. Plant Physiol. 30:264–268.

Jacobson, L., and L. Ordin. 1954. Organic acid metabolism and ion absorption in roots. Plant Physiol. 29:70–75.

Jenny, H. 1951. Contact phenomena between adsorbents and their significance in plant nutrition, pp. 107–132. *In* E. Truog (ed.) Mineral nutrition of plants. Univ. Wisconsin Press, Madison.

Jenny, H., and K. Grossenbacher. 1962. Root-soil boundary zones as seen by the electron microscope. California Agr. 16:7.

Jenny, H., and R. Overstreet. 1938. Contact effects between plant roots and soil colloids. Proc. Nat. Acad. Sci. 24:384–392.

Jenny, H., and R. Overstreet. 1939a. Cation interchange between plant roots and soil colloids. Soil Sci. 47:257–272.

Jenny, H., and R. Overstreet. 1939b. Surface migration of ions and contact exchange. J. Phys. Chem. 43:1185–1196.

Kelley, W. P. 1942. Modern clay researches in relation to agriculture. J. Geol. 50:307–319.

Kelley, W. P. 1948. Cation exchange in soils. Van Nostrand Reinhold, New York.

Kelley, W. P., W. H. Dore, and S. M. Brown. 1931. The nature of the base-exchange material of bentonite, soils, and zeolites, as revealed by chemical investigation and x-ray analysis. Soil Sci. 31:25–55.

Kirkby, E. A., and K. Mengel. 1967. Ionic balance in different tissues of the tomato plant in relation to nitrate, urea, or ammonium nutrition. Plant Physiol. 42:6–14.

Leggett, J. E., and W. A. Gilbert. 1967. Localization of the Ca-mediated apparent ion selectivity in the cross-sectional volume of soybean roots. Plant Physiol. 42: 1658–1664.

Leggett, J. E., and R. A. Olsen. 1964. Anion absorption by baker's yeast. Plant Physiol. 39:387–390.

Levitt, J. 1957. The significance of "apparent free space" (AFS) in ion absorption. Physiol. Plantarum 10:882–888.

Lips, S. H., and H. Beevers. 1966a. Compartmentation of organic acids in corn roots. I. Differential labeling of two malate pools. Plant Physiol. 41:709–712.

Lips, S. H., and H. Beevers. 1966b. Compartmentation of organic acids in corn roots. II. The cytoplasmic pool of malic acid. Plant Physiol. 41:713–717.

Lundegårdh, H. 1947. Mineral nutrition of plants. Annu. Rev. Biochem. 16:503–528.

Lundegårdh, H. 1954. Anion respiration – The experimental basis of and theory of absorption, transport and exudation of electrolytes by living cells and tissues. Symp. Soc. Exp. Biol. 8:262–296.

MacLennan, D. H., H. Beevers, and J. L. Harley. 1963. Compartmentation of acids in plant tissues. Biochem. J. 89:316–327.

MacRobbie, E. A. C., and J. Dainty. 1958. Ion transport in *Nitellopsis obtusa*. J. Gen. Physiol. 42:335–353.

Marshall, C. E. 1951. The activities of cations held by soil colloids and the chemical environment of roots, pp. 57–77. *In* E. Truog (ed.) Mineral nutrition of plants, Univ. Wisconsin Press, Madison.

Mengel, K. 1965. Das Kationen-Anionen-Gleichgewicht in Wurzel, Stengel und Blatt von *Helianthuus annuus* bei K-Chlorid- und K-Sulfaternährung. Planta 65:358–368.

Noggle, J. C. 1966. Ionic balance and growth of sixteen plant species. Proc. Soil Sci. Soc. Amer. 30:763–766.

Osmond, C. B., and G. G. Laties. 1969. Compartmentation of malate in relation to ion absorption in beet. Plant Physiol. 44:7–14.

Overstreet, R., and L. A. Dean. 1951. The availability of soil anions, pp. 79–105. *In* E. Truog (ed.) Mineral nutrition of plants. Univ. Wisconsin Press, Madison.

Overstreet, R., and H. Jenny. 1939. Studies pertaining to the cation absorption mechanism of plants in soil. Proc. Soil Sci. Soc. Amer. 4:125–130.

Overstreet, R., T. C. Broyer, T. L. Isaacs, and C. C. Delwiche. 1942. Additional studies regarding the cation absorption mechanisms of plants in soil. Amer. J. Bot. 29: 227–231.

Pauling, L. 1930. The structure of the micas and related minerals. Proc. Nat. Acad. Sci. 16:123–129.

Pettersson, S. 1966. Active and passive components of sulfate uptake in sunflower plants. Physiol. Plantarum 19:459–492.

Pierce, E. C., and C. O. Appleman. 1943. Role of ether soluble organic acids in the cation-anion balance in plants. Plant Physiol. 18:224–238.

Pitman, M. G. 1963. The determination of the salt relations of the cytoplasmic phase in cells of beetroot tissue. Australian J. Biol. Sci. 16:647–668.

Pitman, M. G. 1965. The location of the Donnan free space in disks of beetroot tissue. Australian J. Biol. Sci. 18:547–553.

Poole, R. J., and L. W. Poel. 1965. Carbon dioxide and pH in relation to salt uptake by beetroot tissue. J. Exp. Bot. 16:453–461.

Pucher, G. W., H. B. Vickery, and A. J. Wakeman, 1938. Relation of the organic acids of tobacco to the inorganic basic constituents. Plant Physiol. 13:621–630.

Rasmussen, G. K., and P. F. Smith. 1961. Effects of calcium, potassium, and magnesium on oxalic, malic, and citric acid content of Valencia orange leaf tissue. Plant Physiol. 36:99–101.

Russell, E. J. 1926. Plant nutrition and crop production. Univ. California Press, Berkeley.

Steer, B. T., and H. Beevers. 1967. Compartmentation of organic acids in corn roots. III. Utilization of exogenously supplied acids. Plant Physiol. 42:1197–1201.

Stiles, W., and A. D. Skelding. 1940. The salt relations of plant tissues. II. The absorption of manganese salts by storage tissue. Ann. Bot. 4:673–700.

Stout, P. R., and R. Overstreet. 1950. Soil chemistry in relation to inorganic nutrition of plants. Annu. Rev. Plant Physiol. 1:305–342.

Sutcliffe, J. F. 1954. The exchangeability of potassium and bromide ions in cells of red beetroot tissue. J. Exp. Bot. 5:313–326.

Tiffin, L. O. 1966a. Iron translocation. I. Plant culture, exudate sampling, iron-citrate analysis. Plant Physiol. 41:510–514.

Tifflin, L. O. 1966b. Iron translocation. II. Citrate/iron ratios in plant stem exudates. Plant Physiol. 41:515–518.

Torii, K., and G. G. Laties. 1966. Dual mechanisms of ion uptake in relation to vacuolation in corn roots. Plant Physiol. 41:863–870.

Ulrich, A. 1941. Metabolism of non-volatile organic acids in excised barley roots as related to cation-anion balance during salt accumulation. Amer. J. Bot. 28:526–537.

Ulrich, A. 1942. Metabolism of organic acids in excised barley roots as influenced by temperature, oxygen tension, and salt concentration. Amer. J. Bot. 29:220–227.

Van Steveninck, R. F. M. 1966. Some metabolic implications of the tris effect in beetroot tissue. Australian J. Biol. Sci. 19:271–281.

Wadleigh, C. H., and J. W. Shive. 1939. Organic acid content of corn plants as influenced by pH of substrate and form of nitrogen supplied. Amer. J. Bot. 26:244–248.

Wit, C. T. de, W. Dijkshoorn, an d J. C. Noggle. 1963. Ionic balance and growth of plants. Verslag. Landbou. Onderzoek 69.15, 68 pp.

CHAPTER 5

Absorption

We commonly speak of salt absorption but generally salts do not enter as molecules, as once believed (Osterhout, 1930), but as cations and anions. Hypotheses concerning the entry of elements into plants now deal with attempts to explain how individual cations and anions are absorbed.

One of the more unusual concepts of absorption was suggested by Bennett (1956) when he proposed that contractile proteins of the membrane transport ions by the formation of vesicles (pinocytosis). It was assumed that the vesicle or "membrane droplet," with the occluded ions, disintegrates once it is free of the membrane and releases the transported ions.

Various articles and books (e.g., Jennings, 1963; Fried and Broeshart, 1967) cover one or more aspects of salt absorption, and Sutcliffe (1962) reviewed the literature in an excellent book on absorption.

In reading this chapter, the reader would do well to keep in mind the axiom stated by Steward and Sutcliffe (1959) concerning salt absorption: "In short, plants grow because they can absorb and accumulate salts; they also accumulate because they can grow."

THEORIES

Before discussing active and passive absorption, a distinction should be made between absorption of ions and molecules. As noted by Salisbury and Ross (1969), when ions move across membranes, the driving force is not simply chemical potential (as in the case of molecules) but the force also involves an electrochemical gradient across the membrane. The latter gradient is composed of the chemical potential difference and an electrical potential difference. The latter potential arises from the fact that one ion of a salt may diffuse at a different rate than that of the accompanying ion. Whether or not the transport of an ion is active or passive, then, depends on the contribution of the electrical potential difference and the chemical potential difference. A cation, for example, may exist in higher concentration inside a cell than outside and yet be transported inwardly purely passively if the electrical potential is sufficiently negative. Anion absorption against both a chemical potential gradient and a negative electrical potential is necessarily an active process. Salisbury and Ross (1969) describe in detail how the Nernst equation can be used to determine whether an ion is actively or passively transported across a membrane.

By use of the Nernst equation and the results from studies of the uptake of K^+ and SO_4^{2-} by roots, Higinbotham et al. (1967) concluded that only anions are actively accumulated and that cations are carried along passively. Bowling et al. (1966) came to a similar conclusion. Salisbury and Ross (1969), however, caution that more studies are required before it can be generalized that anion uptake into root cells is more typically an active process than is cation uptake.

Passive (Physical) Absorption

Introduction. Passive absorption of ions is a physical, nonmetabolic process. Metabolic energy is not required, and ions enter passively by virtue of their kinetic energy. Cellular structure is required, however, and maintenance of intact cells and organs is dependent on metabolic energy. Maintenance of structure is also important for the passive absorption of water which, incidentally, is considered to be the more important of the two types of water absorption (i.e., rather than active absorption).

Types of Passive Absorption. Diffusion is the simplest of the processes by which ions enter the cells of plants—generally through the roots. In diffusion there is a *net* movement of ions from a region of higher concentration to one of lower concentration (of the diffusing ion). Inasmuch as ions can be assimilated in the root cells or move from these cells to the top of the plant, the concentration of certain ions may be less inside the root cells than in the soil solution. By the laws of diffusion, then, these ions move from the soil into the root cells. Generally, ions move at rates commensurate with concentration differences, inside and outside, and under no circumstance can the equilibrium concentration inside cells exceed that on the outside. Usually, however, the concentrations of many ions are higher inside than outside the cells.

Inasmuch as ions in vacuoles are not generally free to diffuse out, hence to be translocated, the concentration of a given ion in the cytoplasm and cell wall is the important consideration with regard to diffusion of that ion into the cell.

In living cells protoplasm has been observed to move about or stream within certain cells, that is, protoplasmic streaming. The folding and unfolding of proteins may be involved in protoplasmic streaming (Goldacre and Lorch, 1950). It is generally agreed that such streaming or mixing of the cytoplasm facilitates the entry of ions inasmuch as it reduces the concentration of a diffusing ion just inside the cell—as compared with the external concentration. The mixing of the cytoplasm, and the concomitant reduction in the concentration of the ion immediately inside the cell, steepens the gradient and thus hastens diffusion of ions into cells. Cytoplasmic streaming is not related to the manner in which ions enter; its effect merely increases the rates of entry. Streaming may facilitate the movements of all kinds of substances in plants—ions, sugars, amino acids, organic acids, plant hormones, and so on.

In many ways roots act as a sponge in the absorption of ions. Most substances in contact with water, including roots, assume a negative charge. Whole wheat roots immersed in a dilute salt solution have a negative electrokinetic potential of approximately 60 mV (Lundegardh, 1954). The site of the charge is the root surface; the potential is caused by dissociation of acidic substances and behaves essentially as a Donnan potential.

The surfaces of roots and the cell contents have ions attached to or adsorbed onto them. These ions may be exchanged for ions in the external solution (or cells in the environment) by a physical process called ionic exchange, exchange adsorption, or adsorption exchange. Exchange adsorption has been regarded as the first step in salt uptake by algae (Schaedle and Jacobson, 1965) and by roots of higher plants (Jenny and Overstreet, 1939; Lundegardh, 1942; Elgabaly et al., 1943; Overstreet and Dean, 1951). Lundegardh (1942) stressed that there was no fundamental difference between uptake of ions from a solution and from a colloid.

According to Overstreet and Dean (1951), inasmuch as this type of absorption is an exchange process, no ion enters or leaves a cell except by exchange for another ion. These investigators noted that H^+ and OH^- or HCO_3^- are exchanged by the plant for cations and anions, respectively, from the medium. They noted that K^+, NH_4^+, Cs^+, NO_3^-, Br^-, and Cl^- enter rapidly, whereas Ca^{2+}, Mg^{2+}, Ba^{2+}, SO_4^{2-}, and $H_2PO_4^-$ enter somewhat more slowly; HCO_3^- is apparently not readily absorbed. In one of the early experiments with isotopes, barley roots immersed in $KH^{14}CO_3$ absorbed large quantities of K^+ but utilized only 4–5% of the $H^{14}CO_3^-$ (Overstreet et al., 1940).

Ca^{2+} is a slowly absorbed ion, and in most excised tissues the uptake is nonmetabolic (Maas, 1969). However, in vacuolated segments of excised corn roots, Ca^{2+} is readily and metabolically absorbed (Handley and Overstreet, 1961). Meristematic cells and vacuolated cells appear to differ, then, with accumulation by the former being nonmetabolic and by the latter metabolic. Whereas Ca^{2+} is generally absorbed nonmetabolically by roots (i.e., by ionic exchange) (Moore et al., 1961a, b), Mg^{2+} is absorbed metabolically (Moore et al., 1961b).

Sutcliffe (1954) hypothesized that easily exchanged ions are distributed throughout the intercellular spaces, cell walls, and in parts of the protoplasm, whereas those that do not readily exchange may be situated in the cell vacuoles or strongly associated with protoplasmic constitutents.

Epstein and Leggett (1954) characterized "exchange adsorption" as: (1) nonlinear with time and equilibrium approached in 30 min; (2) involving ions readily exchangeable; (3) not selective with respect to various ions; and (4) not requiring energy expenditure (i.e., uptake is attributable to Coulomb forces residing in electronegative sites). With regard to the last-mentioned point, negative binding points in cell walls have been localized and characterized by Pitman (1965).

The Donnan equuilibrium theory takes into account the situation in which there are present on one side of a membrane (e.g., inside the cell) certain nondiffusible anions. In this case cations and anions are free to diffuse into the cell, and additional cations may be accumulated in association with the nondiffusible anions inside the cell. There is therefore an unequal concentration of cations, the concentration being higher inside than outside the cell.

Although Donnan equilibrium may be readily demonstrated in a purely physical system, and there is no doubt of its operation in plants, uptake of ions by this mechanism is not considered as important as certain others for ion uptake. Donnan equilibrium cannot be regarded as being of primary importance between at least some living cells and their natural environment since in nature the ratios required at equilibrium are not usually observable (Broyer, 1947; Osterhout, 1947). With regard to K^+, Schaedle and Jacobson (1965) concluded that accumulation was dependent on oxidative metabolism and that net accumulation was a result of the creation of slowly diffusible or nondiffusible anions in *Chlorella pyrenoidosa*.

According to Münch (1927), loss of water by leaves and the relatively high OP of leaf cells may be compensated by absorption of water by roots—the cells of which have relatively lower OP values than those of the leaves. Water and materials in solution may therefore move in mass flow from the roots to the tops. In this uptake and movement, salts may also be absorbed and move up the plant (Hylmo, 1953).

If mass flow were the chief mechanism of salt uptake, there would be a rather close correspondence between the rates of transpiration and ion uptake. There is evidence supporting and contradicting (Lazaroff and Pitman, 1966) such a relationship. For barley seedlings the higher the rate of transpiration the more Ca^{2+}, and particularly Mg^{2+}, transported to the tops; there was very little effect on Na^+ and K^+. Lazaroff and Pitman (1966) concluded that mass flow, in which ions would be drawn in by the water, would be relatively nonselective. Mass flow did not appear to offer an explanation for the deifferences between the movements of monovalent and divalent ions to the tops.

Active (Metabolic) Absorption

Although confirmed by numerous investigations later, it was the early work by Hoagland and Broyer (1936, 1942) and Steward and Preston (1940) that clearly established that aerobic respiration supplies the required energy for absorption against a gradient. Active transport has been covered in detail in books by Robertson (1968) and Price (1970).

Ussing (1949, 1957) considered active transport to be the process whereby an ion is moved against an electrochemical potential gradient and is therefore dependent on a decrease in free energy in some metabolic process.

Epstein and Leggett (1954) characterized active transport as: (1) linear with time, no equilibrium being reached in the experiments; (2) involving ions essentially nonexchangeable; (3) selective with respect to various ions and groups of ions; and (4) requiring expenditure of energy (i.e., uptake is linked with metabolic processes).

Aeration (hence aerobic respiration) is important not only for the absorption of an ion, such as Br^-, but also for its retention following absorption (Fig. 5.1). When roots of tomato plants were subjected to CO_2 or N_2 rather than air, roots lost Br^- previously acquired in the presence of air.

Transport of ions across the cytoplasm and accumulation of ions in vacuoles are believed to be dependent on respiration—at a stage at which charges are separated into H^+ and electrons (Robertson, 1951). Years ago it was observed that under anaerobic conditions salts move into the cytoplasm but not into the vacuole (Hoagland and Broyer, 1942). It was concluded therefore that active transport occurred from the protoplasm into the vacuole, whereas passive transport took place from the medium into the protoplasm. By active transport ions move to the inner space (vacuole) after having come from the outer space (cytoplasm and cell walls) where they were in equilibrium with the ambient solution (Epstein, 1955).

In barley roots Ca^{2+} accumulated in xylem exudate up to 58 times its concentration in the ambient medium, and this was regarded as evidence that Ca^{2+} moved metabolically into the xylem and not by a mass flow of solution (Moore et al., 1965). The endodermis was presumed to be the barrier across which Ca^{2+} moved metabolically.

It has been assumed for years by most investigators that the endodermis is a barrier to the passive movement of ions from cortical to xylary tissue in roots. Peirson and Dumbroff (1969) noted that branch roots originate in the pericycle and that they must therefore ultimately pierce the endodermis. At the time of this disruption, they observed that ions could move passively into the xylem be-

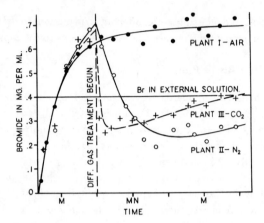

Fig. 5.1

Effects of differential gas treatments on the concentration of Br⁻ ions in exudates from decapitated plants. From T. C. Broyer (1951) The nature of the process of inorganic solute accumulation in roots, pp. 187–249. In E. Truog (ed.) Mineral Nutrition of Plants, Univ. Wisconsin Press, Madison. © 1951 by the Regents of the University of Wisconsin—by permission.

cause of the break in the endodermis. They observed that the break did not occur until the branch root was near the periphery of the cortex and that after rupture of the endodermis the endodermis of the parent and branch roots became continuous again. They concluded that the number of such breaks (i.e., branch roots breaking through) and their average duration would determine the significance of these temporary openings for the passive movement of ions into the stele.

Active and passive uptake of K^+ by pea epicotyl (*Pisum sativum* L. 'Alaska') has been studied by Macklon and Higinbotham (1970). On the basis of concentrations of K^+ in cytoplasm and in the vacuole, they concluded that movement of K^+ from cytoplasm to the vacuole is passive. Movement of K^+ into the cytoplasm from outside the cell was active and was activated by an active K^+ pump located at the plasmalemma.

In the oat coleoptile it appeared that at the plasmalemma Na^+ was pumped out and Cl^- actively transported inward (Pierce and Higinbotham, 1970). At the tonoplast there was an active inward transport of Na^+ and probably K^+, but the status of Cl^- was uncertain—depending on whether or not there is an electrical potential difference between cytoplasm and vacuole.

The mechanism of Zn^{2+} uptake by *Fontinalis antipyretica* was studied with an absorption medium containing 0.5 mM $CaSO_4$ (Pickering and Puia, 1969). About 50% of the Zn^{2+} absorbed at equilibrium was absorbed in the first half-hour. Three sections were distinguished in the uptake curve—suggesting that three successive processes (stages) were involved. The first stage was very short and not influenced by temperature, light intensity, or dinitrophenol (DNP). The second stage, lasting no more than 90 min, was very slightly affected by the same three factors. In this stage freshly killed plants absorbed more Zn^{2+} than living material. The third stage, lasting several days, was very slow and was light, temperature, and DNP dependent.

The transport mechanism in excised bean hypocotyls was practically independent of temperature and, in short-term experiments, appeared to be nonmetabolic (Waisel et al., 1970). It was reported that ion selectivity may be attributed either to specific adsorption sites or to the properties of the diffusion barrier. It was suggested that a flexible barrier (plasmalemma?) rather than fixed adsorption sites governs ion selectivity.

Rates of active Cl^- absorption by thin bean leaf slices from 0.1 mM KCl were about $\frac{1}{10}$ the rates for monovalent cation absorption and about $\frac{1}{6}$ the Cl^- absorption rates by excised bean roots (Jacoby and Plessner, 1970). In the 2–5 mM concentration range, Cl^- absorption rates for bean leaf and root tissues did not differ much and were similar to those of monovalent cation absorption. There was no ready explanation for the low rate of Cl^- absorption by system 1.

Anion or Salt Respiration (Cytochrome Pump). Although some researchers failed to observe a close relationship between salt absorption and respiration, Lundegardh and Burstrom (1933a, b) noted such a relationship. Their theory of anion respiration was expanded and further documented by Lundegardh (1940a, b, 1949b, 1950, 1954). It assumed that (1) transport of anions was mediated by a cytochrome system; (2) an oxygen gradient existed from the outer to the inner surface of the membrane which favored oxidation at the outer and reduction at the inner surface; and, (3) anion uptake was independent of cation uptake with the latter occurring by a different mechanism.

According to Lundegardh (1945), there is a sufficiency of the R•H substance in plants inasmuch as protoplasm as a whole is negatively charged and contains appreciable quantities of substances with comparatively strong acidic properties. Cations from the medium could exchange for H^+ in the plasmalemma and proceed inward through the protoplasm by exchange along paths or tracks of substances with an acid dissociation; no special energy would be required. Therefore, his theory chiefly concerns the nature of the substance, R'•OH. He reasoned that the Fe ion in the hemin group of a respiratory enzyme (such as cytochrome oxidase) would be well suited to transport anions since Fe^{3+} would attract one more anion than Fe^{2+}:

$$\begin{pmatrix} Fe^{3+} \\ \\ 3A^- \end{pmatrix} \underset{-e}{\overset{+e}{\rightleftarrows}} \begin{pmatrix} Fe^{2+} \\ \\ 2A^- \end{pmatrix} + A^-$$

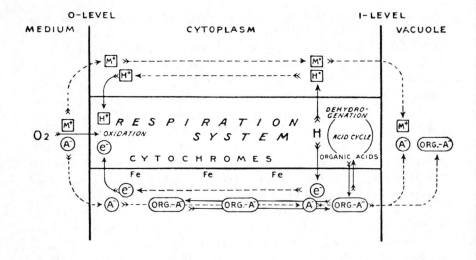

Fig. 5.2

Scheme for active absorption of ions with the salt respiration system oriented transversely across the cytoplasm. At the inner, dehydrogenation level, protons and electrons are produced, and they are moved to the outer, oxidation level in exchange for salt cations and anions, respectively. Solid arrows denote only chemical reactions or equilibria; broken arrows denote mechanical transport. From H. Burström (1951) The mechanism of ion absorption, pp. 251–260. *In* E. Truog (ed.), Mineral Nutrition of Plants, Univ. Wisconsin Press, Madison. © 1951 by the Regents of the University of Wisconsin—by permission.

The transference of an electron between two Fe atoms would move one anion along the path or chain. Respiration would be the source of electrons; the ultimate controlling factor for absorption and accumulation of ions would be the oxygen gradient in the cell (Burstrom, 1951).

Ferri-cytochrome may normally be equilibrated by malic or another organic acid (Lundegardh, 1949a). Accordingly, when an inorganic anion enters the cell, it would replace the malate ion which would then disappear in respiration. The inorganic anion could then move on to the vacuole. Incidentally, if NO_3^- were being carried by cytochrome, but the anion became removed and metabolized, its place on cytochrome could be held by a malate ion.

Wanner (1948a, b) observed different temperature coefficients for absorption of cations and anions and interpreted the differences in favor of the anion respiration theory.

Dehydrogenase reactions at the inner surface of the cell could produce protons (H^+) and electrons (e^-) (Figs. 5.2 and 5.3). Electrons would move outward along a cytochrome bridge or chain, while anions moved inward. This would be possible because, at the outer surface of the membrane, reduced Fe of the cytochrome would be oxidized, lose an electron, and pick up an anion. By the action of cytochrome oxidase, the released electron would combine with

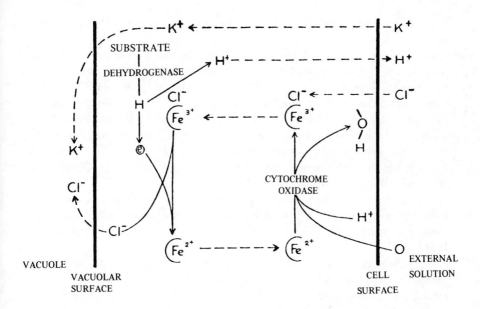

Fig. 5.3
Schematic representation of the electron and anion transport system. Solid lines represent chemical reactions and broken lines represent movements of substances. Fe^{3+}, Oxidized cytochrome; Fe^{2+}, reduced cytochrome. (Robertson and Wilkins, 1948.)

a proton and oxygen to form water. Oxidized Fe of cytochrome would become reduced again at the inner surface by the addition of an electron. During this last reaction the anion would supposedly be released inside the cell. According to this system cations would be absorbed passively.

The pronounced inhibition of ion uptake by chloramphenicol could be attributed to its action as an inhibitor in the utilization of respiratory energy (Ellis et al., 1964). Such inhibition could be interpreted as favoring the theory of anion respiration.

Tanada (1963) postulated a chain of negative sites through the membrane (rather than a moving carrier) as a possible pathway for cation entry. Lundegardh (1955) and Harris (1960) likewise believed in the existence of chains of negative sites. The latter author suggested that the concept may also apply to anions, since Boszorményi and Cseh (1961) reported that faster moving halides stimulated the uptake of slower moving halides by wheat roots.

Mitochondria, chloroplasts, nuclei, and vacuoles compete for salts in the cytoplasm. Interestingly, Lundegardh's mechanism for ion absorption could operate only for mitochrondria and for nongreen cells, since cytochrome occurs principally in the mitrochondria (Sutcliffe, 1962).

Sutcliffe (1962) listed discrepancies between the Lundegardh theory and experimental data:

(1) If there were a single carrier for anions, they should compete with each other, but competition exists between Cl^- and Br^-—not between halides, sulfate, nitrate, or phosphate. Lundegardh (1955) admitted that cytochrome was probably not the carrier of metabolically important ions; this admission in effect limits the theory to an explanation of Cl^- entry. In the experiments on which the theory was originally based, NO_3^- and Cl^- bore a similar relationship in their uptakes to that of respiration. Sutcliffe (1962) therefore raised the question whether cytochrome could be regarded as the carrier for Cl^- if it is not to be regarded as the carrier for NO_3^-.

(2) In some cases salt uptake appears to be linked to ascorbic acid oxidase instead of cytochrome–cytochrome oxidase. In anaerobic organisms some system other than that involving cytochrome and O_2 must be involved (Sutcliffe, 1962).

(3) There is evidence that certain cations, for example, K^+ (Epstein, 1954b) or Na^+ and K^+, but not Ca^{2+} or Mg^{2+} (Handley and Overstreet, 1955), may stimulate respiration. Thus the phenomenon of salt respiration is not limited to anions.

(4) Lundegardh's theory hinges on a quantitative relationship between cyanide-sensitive respiration and anion uptake. He recognized certain discrepancies (Lundegardh, 1940a). Strong evidence against this relationship was reported by Robertson et al. (1951). They noted that at certain concentrations DNP–an inhibitor of oxidative phosphorylation—increased respiration but decreased salt (KCl) uptake. However, Lundegardh (1954) suggested that only cytochrome b is involved in salt uptake, that this cytochrome is inactivated by inhibitors of oxidative phosphorylation, and that cyanide-sensitive respiration continues through other cytochromes with a bypassing of cytochrome b. At any rate, there is a possibility that phosphorylated intermediates might be important in or involved in salt uptake. Last, there is no fixed relationship between the amount of ion absorbed and the rate of respiration (Handley and Overstreet, 1955).

(5) The maximum number of anions transported for each molecule of O_2 consumed in salt respiration is four on the basis of the Lundegardh explanation. Under favorable conditions, more than four ion pairs may be transported per O_2 molecule consumed (Sutcliffe and Hackett, 1957). It was concluded, therefore, that mechanisms of ion transport, which depend on a direct linkage to electron transfer and the redox pump, must be rejected.

(6) As Russell (1954) noted, Lundegardh's concept denies the possibility that cells can store a capacity for absorbing salts as a result of prior respiration. Russell showed that previous respiration facilitated subsequent movement

of phosphate into barley roots placed under conditions of reduced respiration. P-deficient cells of *Nitzschia closterium* exhibited absorption of phosphate in the dark when previously exposed to light (Ketchum, 1939). It was concluded that this might result from a combination between the ion and phosphate-binding substances synthesized in the light. *Potomogeton* leaves absorbed Ca^{2+} in the dark following exposure to light in the absence of Ca^{2+}; cations were transported in the light but not in the dark (Lowenhaupt, 1956). When potato slices at a higher temperature and in the absence of salt were transferred to $0\,°C$ and salt, uptake of Cl^- proceeded rapidly (the absorption shoulder) and decreased only slowly (Laties, 1959b). It was suggested that a carrier precursor was formed at the higher temperature which was gradually consumed at $0\,°C$.

(7) The passive nature of absorption of this theory fails to explain the high degree of selectivity of cation uptake exhibited by many plants.

(8) Cation absorption appears to occur to a large extent independently of anion absorption in some plants—notably in yeast and in some marine algae.

Carriers. Pfeffer (1900) recognized that organisms could transport substances across a membrane in one direction even in the absence of a gradient favoring the transport. He concluded that substances might combine with certain cellular constituents during this passage. His conclusion was indeed quite similar to the present carrier concept.

Osterhout (1935) also drew a conclusion that was a forerunner of the carrier concept. He believed that electrolytes entered cells of *Valonia* by combining reversibly with a constituent, HX, of the protoplasm. He noted that there was little penetration of NH_4^+, but NH_3 or NH_4OH entered freely—apparently by combining with an hypothesized constituent HX.

The carrier concept as such was originally proposed by Honert (1936). Accordingly, the carrier was a special compound which functioned in the membrane, reacted with an ion at the surface, transported it across the membrane, and released it inside the cell (or vacuole). The carriers were regarded as ion-binding and ion-releasing compounds. This concept was later championed by Jacobson and Overstreet (1947), Hutner (1948), and Roberts et al. (1949). It was suggested that cations might be bound in plants in the form of chelated complexes; proteins, amino acids, and organic acids could form chelated compounds—particularly with polyvalent cations (Jacobson and Overstreet, 1947). Rosenberg (1948) noted that, in order for ion-binding complexes to be involved in salt absorption, there would have to be a region between the vacuole and the external surface of the cytoplasm that was relatively impermeable to free ions but permeable to the complex of ion and ion-binding compound.

Jacobson et al. (1950) also suggested that ion absorption depended on metabolically produced ion-binding compounds and visualized the phenomenon as:

$$HR + M^+ = MR + H^+ \text{ (for cations)}$$
$$R'OH + A^- = R'A + OH^- \text{ (for anions)}$$

where HR and R'OH represented the metabolically produced cationic and anionic binding substances. They suggested that the complexes were formed at the external surfaces of protoplasm and served to transport ions across the cytoplasm and into the vacuole. The results of their study on K^+ absorption were considered consistent with the above representation of ion-binding compounds. Indirect evidence was obtained indicating that K^+ was combined as salts of the hexose phosphates (Roberts et al., 1949).

According to Epstein's (1955) description of carriers, "ions combine with carrier molecules and the resulting ion–carrier complexes traverse membranes of limited permeability to free ions. At the inner surface of membranes, the ions are released from carriers" (Fig. 5.4). Internal release of the ion is the rate-limiting, essentially irreversible step (Leggett and Epstein, 1956). Active transport of ions across the plant membrane by a carrier is dependent on metabolic energy.

In castor bean leaves a perfusion study indicated that solutes in the cell walls were in dynamic equilibrium with intracellular solutes (Bernstein, 1971).

Vose (1963) listed nine factors that may affect the action of carriers: (1) chemical nature of the carrier; (2) specificity of the carrier; (3) concentration of the carrier; (4) rate of turnover; (5) certain ions may be transported by more than one carrier; (6) existence of multiple adsorption sites, that is, carriers that carry more than one type of ion; (7) relative proportion of the carriers; (8) possible adaptive synthesis; and (9) the rate of removal of ions from the sink.

Although the nature of carriers is still unknown, various suggestions have been made. Brooks (1937) championed proteins and amino acids as ampholyte carriers; Lansing and Rosenthal (1952) and Tanada (1955, 1956) favored ribonucleoproteins. The latter investigator suggested that the nucleic acid portion binds the cations, while the protein moiety binds anions. Recently, Pardee (1966) found a substance in extracts of *Salmonella typhimurium* that tightly bound SO_4^{2-}, appeared to be normally involved in SO_4^{2-} absorption, and was almost certainly a protein. Price (1970) noted that, if carriers are protein, one would expect to find mutants that lack or have reduced ability to absorb certain ions, and that Pardee (1968) has indeed observed such mutants. Proteins are constantly degraded and synthesized and, if carriers are protein in nature, this would explain the requirement for ATP in ion absorption, since ATP is essential for protein synthesis. Steward and Street (1947) postulated that carriers are phosphorylated, energy-rich compounds formed by aerobic respiration. They attempted to rationalize their postulate with respiration and protein synthesis. Accordingly, the carrier is broken down in the process of its incorporation into protein (at the seat of protein synthesis); simultaneously, the asso-

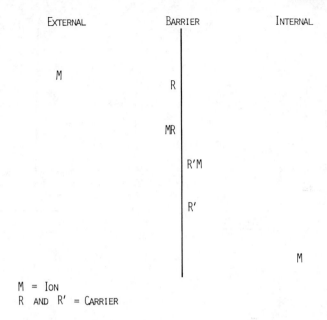

Fig. 5.4
Schematic representation of the process of ion absorption. (Epstein and Leggett, 1954.)

ciated ion or ions (cation, anion, or both) would be released in free solution into the vacuole. Bennet-Clark (1956) suggested that lecithin might be an ion carrier (Fig. 5.5). Lecithin has been reported to be present in all living cells.

In any event carriers appear to possess or to need to possess the following characteristics (Overstreet and Jacobson, 1952): (1) intermediate metabolic products or closely related substances; (2) lack of stability *in vitro*; (3) ability to undergo chemical alteration in the course of their carrier function; and, (4) probably function as chelating agents.

Foulkes (1956) reported evidence he interpreted to indicate the presence of a carrier for K^+ uptake in yeast (but not one for Na^+ extrusion). Concentration of the K^+ carrier was calculated to be 0.1 μmole/g of yeast.

The existence and functioning of carriers would explain many of the experimental findings with regard to absorption of various cations and anions, and interrelationships among cations and among anions. For example, K^+, Rb^+, and Cs^+ have been shown to compete for the same site, whereas Na^+ and Li^+ do not (Epstein and Hagen, 1952). With Na^+ and K^+ traversing membranes over different sites (Epstein and Hagen, 1952; Scott and Hayward, 1954), selectivity with regard to these ions rests on a mechanistic basis. Similarly, there is an explanation for the interaction (antagonism) between ions; ions competing for the same site could, among other effects, interfere with each other during absorption.

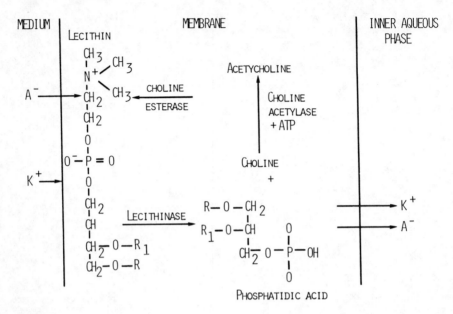

Fig. 5.5

A cyclic mechanism of active transport involving lecithin as an amphoteric carrier. (Bennet-Clark, 1956.)

Epstein and Leggett (1954) found that Ca^{2+}, Ba^{2+}, and Sr^{2+} competed for identical sites; Mg^{2+} passed over a different site.

Briggs et al. (1961) believed that in competition studies it is necessary to determine: (1) rate of uptake of salt A alone; (2) rate of uptake of salt B alone; (3) rate of uptake of salt A with constant concentration of A and varying B; and (4) rate of uptake of salt B with constant concentration of B and varying A. They argued that in competition studies special care should be taken in interpreting the effect of one ion on uptake of another ion since generally only net uptake or net flux is measured. For example, they noted that if a tissue had a concentration-dependent Na^+ pump that extruded Na^+ at the same time K^+ was being accumulated the interpretation of an apparent competitive effect could be misleading.

MacRobbie and Dainty (1958) published the first efflux and influx data for cells of *Nitellopsis obtusa*. Their data were interpreted as indicating active transport of Cl^- by the tonoplast into the vacuole and of Na^+ out of the cell by the plasmalemma.

Hagen and Hopkins (1955) and Hagen et al. (1957) showed that $H_2PO_4^-$ and HPO_4^{2-} enter over different sites but that OH^- interfers competitively with both sites, hence both forms of phosphate. According to Leggett and Epstein (1956), $SO_4{}^{2-}$ and $SeO_4{}^{2-}$ compete for the same site in barley roots. Phosphate, $NO_3{}^-$, and Cl^- had no measurable affinity for the SO_4–SeO_4 binding site. $SO_4{}^{2-}$–$SeO_4{}^{2-}$ competition has also been reported for wheat (Hurd-Karrer, 1938), corn (Trelease and Beath, 1949), yeast (Fels and Cheldelin, 1949), *Chlorella vulgaris* (Shrift, 1954), and *Aspergillus niger* (Weissman and Trelease, 1955). There is an active uptake of $SeO_4{}^{2-}$ into the roots of *Astragalus crotalariae* (Shrift and Ulrich, 1969).

Epstein (1953) reported that Cl^- competed with Br^-, but that $NO_3{}^-$ did not compete for the Br^- site. However, competition between $NO_3{}^-$ and Br^- has been reported (MacDonald and Laties, 1963). Honert and Hooymans (1955) postulated that there is a separate carrier for $NO_3{}^-$, that its absorption is much greater at pH 6 than at 8, and that $HCO_3{}^-$ has no effect on $NO_3{}^-$ uptake. They believe that the $NO_3{}^-$ carrier acts at the surface of the epidermal root protoplasm.

In discussing an interrelationship between Na^+ and Rb^+, Sutcliffe (1959) reported that, in storage tissues and barley plants, Rb^+ uptake was increased, decreased, or unaffected by the addition of Na^+—depending on the concentrations of the salts.

Two carriers have been reported for yeast (Conway, 1954). One appears to be specific for Na^+ and removes Na^+ from the cell, whereas the other is specific for K^+ and introduces K^+ into the cell.

A carrier is believed to be involved in the outward movement of K^+ from beet tissue (Poole, 1969). It has been suggested that the active efflux of H^+ (Poole, 1966) and the active uptake of K^+ (Osmund and Laties, 1968) may result from the action of a single carrier. The relative importance of the influx and efflux would depend on the ratio of ATP to ADP and P_i (Poole, 1969). Active transport of K^+ is believed to occur at the plasmalemma (Pitman and Saddler, 1967; Osmund and Laties, 1968).

In a study designed to exclude diffusion as an explanation, active uptake of K^+ was demonstrated (Etherton, 1963). Potassium was actively transported inward at low external concentrations and outward at high external concentrations (Etherton, 1963, 1967); Na^+ was actively transported from pea roots (Etherton, 1967).

The pretreatment of bean stem tissue in $CaSO_4$ solution (i.e., aging) generally increases ion absorption and respiration (Rains, 1969a). Na^+ was freely absorbed by freshly sliced bean stem tissue over the range 0.02–50 mM; in aged tissue Na^+ absorption was nil below 0.2 mM. However, K^+ absorption by fresh tissue was nil below 0.5 mM; in aged tissue K^+ absorption was substantial from 0.01 to 50 mM. Absorption of Na^+ by fresh tissue and of K^+ by aged tissue was sensitive to antimetabolites. Rains (1969a) noted that in freshly sliced tissue type-1 and -2 mechanisms operated in Na^+ transport, whereas only the type-2 mechanism was involved in K^+ transport. In aged tissue the situation was reversed, K^+ being absorbed by two mechanisms and Na^+ by the type-2 mechanism which is operative only at high Na^+ concentrations. It was postulated that these differences might explain the removal of Na^+ from the transpiration stream in intact bean stem tissue (Jacoby, 1965; Rains, 1969b).

The mechanisms of uptake of micronutrients have only recently received attention (Page and Dainty, 1964; Broda, 1965; Schmid et al., 1965; Bowen, 1968, 1969; Maas et al., 1968). Absorption of Cu^{2+}, Zn^{2+}, and Mn^{2+} was shown to be coupled to oxidative phosphorylation and specifically to energy conservation site I (Bowen, 1969). Cu^{2+} and Zn^{2+} were absorbed by the same carrier site(s), as shown by competition; Mn^{2+} was absorbed by a second, independent mechanism. These findings are consistent with those of Schmid et al. (1965) to the effect that Cu^{2+} competitively interfered with the transport of Zn^{2+} into the inner space; Mn^{2+} was without effect. Cu^{2+} and Zn^{2+} uptake were significantly inhibited by Fe^{3+}, but Mn^{2+} absorption was unaffected. The most of the experiments Ca^{2+} was included, since it has been shown to be essential for selective ion transport systems in cellular membranes (Rains et al., 1964).

There is evidence that Mn^{2+} is absorbed by a fast phase (exchangeable Mn^{2+}) with equilibrium attained in about 30 min, and that it is also absorbed by a slow phase (adsorbed Mn^{2+}) which may also involve exchange sites. The latter sites are different from those of the former phase or are less accessible, and equilibrium is not reached in 3 hr (Page and Dainty, 1964). Maas et al. (1968) essentially agree, and add that Mn^{2+} is absorbed in very much the same general manner as other cations. They found that, after initial, nonmetabolic equilibration within barley roots, Mn^{2+} was absorbed at a slower, steady-state rate for at least several hours.

Borate is reported to be absorbed into sugarcane leaf mesophyll cells by a rapid and reversible influx within 20 min (phase I), followed by a slower and irreversible accumulatory phase (phase II) (Bowen, 1968). It was noted that highly specific mechanisms 1 and 2 transport borate across the initial barrier into the cells, and that reaction 3 carries borate across the vacuolar membrane. Mechanisms 1 and 2 were linked to respiratory electron transport, mechanism 3 to oxidative phosphorylation.

Ca has often been reported to increase the absorption of various cations and associated anions, and this stimulating effect of Ca^{2+} on absorption has been termed the "Viets effect"—first discovered by Viets (1944). In view of the effect of Ca^{2+} on absorption of ions, reports dealing with Ca^{2+} have been assembled in this section.

Ca is especially promotive of Cl^- uptake by beetroot tissue according to Pitman (1964). As a mechanism for Ca^{2+} stimulation of Cl^- uptake, he suggested that uptake of Cl^- is limited by diffusion through a negatively charged surface or membrane which is more permeable to Cl^- when divalent cations, rather than univalent cations, are the counterions.

Ca had been reported to increase the rate of SO_4^{2-} absorption and, later, Leggett et al. (1965) reported that the increase was caused by the effect of Ca^{2+} on the turnover rate of the carrier. Ca competitively inhibits absorption of Li^+ (Epstein, 1960). He noted the similarities of Li^+ and Ca^{2+} that might explain why these quite dissimilar ions compete with each other for a given carrier site.

Ca, Sr, La, Mn, Al, but not Mg^{2+}, decreased Li^+ (or Na^+) absorption but increased K^+ absorption (Jacobson et al., 1961). Ca may be associated with the cell surface region, and it would be that portion of the Ca^{2+}, localized on the surface, that affects the entry of other ions (Moore et al., 1961a).

Rb was absorbed by excised barley roots containing Ca^{2+} strictly as a linear function of time (for 1 hr), and the absorption was temperature dependent (Epstein et al., 1962). There was no evidence of an initial nonmetabolic exchange uptake. It was concluded that when Ca^{2+} was present, the general, nonselective, cation exchange capacity of the roots was largely satisfied by Ca^{2+} ions, and that the time course of absorption for monovalent cations was the same as that for anions. Inasmuch as they found no evidence for nonselective cation exchange, Epstein et al. (1962) disagreed with the portrayal by Briggs et al. (1961) of the time course of absorption of monovalent cations. Rb absorption by barley roots was regarded as an adsorption exchange process (Tanada, 1964). Calcium promoted the uptake of Rb^+ (Tanada, 1955) by mung bean roots—presumably by increasing the affinity of the binding sites for Rb^+ (Tanada, 1962). It was noted that time studies indicated Ca^{2+} had to penetrate the membrane before its effect became manifested. The "Viets effect" was also observed for phosphate (Tanada, 1955).

Ca decreases the uptake of some ions such as Zn^{2+} (Schmid et al., 1965) and Na^+ (Handley et al., 1965a), but increases that of Cl^- (Pitman, 1964) and other ions (Handley, et al., 1965b). Handley et al. explained that Ca^{2+} may decrease uptake by stabilizing the membrane—with a consequent decline in permeability. With regard to increases Ca^{2+} may be involved in metabolic accumulation mechanisms of certain alkali cations (e.g., Rb^+), increasing their uptake while increasing K^+ depletion.

The inhibitory effect of Ca^{2+} on K^+ absorption, from dilute K^+ concentrations, has been attributed to a decrease in permeability of the membrane (Elzam and Hodges, 1967). Stimulatory effects of Ca^{2+} on absorption have been attributed to its postulated role in maintaining the integrity of cellular membranes (Marinos, 1962; Steveninck, 1965). Ca deficiency has been shown to result in a disorganization of membrane structure (Marinos, 1962). Hooymans (1964) concluded that the effect of Ca^{2+} on the uptake of cations and anions, was based on two completely different mechanisms, namely, screening of electronegative charges in the close neighborhood of the carrier sites in the latter and, in the former, of a lowering of the permeability of the plasmalemma. In this connection the counterion has been shown to alter Cl^- uptake drastically (Laties et al., 1964). With K^+ the Cl^- absorption isotherm was hyperbolic under certain conditions, but with Ca^{2+} hyperbolic under all conditions. An electrochemical model (or theory) was therefore proposed for the transport of ions across plant membranes.

Selectivity in nutrient uptake can occur on the outer cell membrane with the action of Ca^{2+} being primarily on selectivity according to Waisel (1962). He reported that Ca^{2+} inhibited the uptake of highly hydrated cations (Li^+, Na^+, and probably H^+) but stimulated uptake of cations with smaller hydration shells (K^+, Rb^+, and Ca^{2+}). He reasoned that Ca^{2+} caused a change in the actual concentrations of the ions available to the carrier sites.

In moving across membranes ions are subject not only to concentration gradients but also to electrical fields (Higinbotham et al., 1964). Several studies have shown that an electropotential difference (PD) of approximately 100 mV (interior negative) is present across plant cell membranes (Hill and Osterhout, 1938; Blinks, 1949; Walker, 1955; Etherton and Higinbotham, 1960; Blount and Levedahl, 1960; Hope and Walker, 1961; Higinbotham et al., 1961; Etherton, 1963). Electrical potential gradients are just as important as concentration gradients in determining passive ion movements (Dainty, 1962). Inasmuch as Ca^{2+} increases the transmembrane electropotential gradient, Ca^{2+} constitutes a nonspecific force tending to drive cations into the cell (Higinbotham et al., 1964). This action of Ca^{2+} would explain the frequently observed increase in absorption rates of various ions in the presence of Ca^{2+}, but it would not of course explain the role of Ca^{2+} in maintaining the selectivity characteristics of membranes.

Single-cell electropotentials of barley *(Hordeum vulgare* L. 'Compana') root cortex were measured at different external concentrations of KCl in the presence of Ca^{2+} (Pitman et al., 1971). Roots were low in salt and thus conditions were similar to those used to demonstrate dual mechanisms. In 0.5 mM $CaSO_4$ there was an increase with time of cell negativity from about -65 mV 15 min after cutting the segments to approximately -185 mV in 6–8 hr. It was reasoned that changes in selective ion transport would accompany these changes. It was concluded that whether or not there is a clear relationship between cell potential and mechanisms 1 and 2 of cation transport depends upon whether cell potentials of freshly cut or of aged tissue represent values relevant to intact roots.

One of the fundamental problems of ion transport is that of differentiating between those ion species transported by metabolically driven ion pumps and those that enter or leave the cell by diffusion (Macklon, 1970). Electrochemical theory proposes models that allow relatively precise analysis of the ionic relationships of the cell and aids in distinguishing between actively and passively transported ions. It was noted that an ion may move against its chemical concentration gradient if the electrochemical gradient is favorable.

Shone (1970) has provided an excellent review of the uptake of ions by roots, including the role of gradients in electrochemical potential, and the translocation of ions within a plant.

Solutions with increasing IAA concentrations (from 10^{-9} to 10^{-7} M) cause membrane potentials of coleoptile cells to become increasing more negative (Etherton, 1970).

An attempt has been made to combine electrochemistry and enzyme kinetics into a unified approach to the problems of salt absorption (Thellier, 1970). The parameters K_m and V_m are replaced by log $(G/[P])$ and A/r, which have simple biological meanings. Applied to the problem of absorption, this leads to an explanation of experimental results without the necessity of invoking the presence of two carrier systems.

Apparently, Ca^{2+} can increase, or have no effect on, the uptake of monovalent ions—depending on the pH of the solution (Jacobson et al., 1960), Ca^{2+} completely changed the interrelationship in absorption of such monovalent cations as K^+ and Li^+; with Ca^{2+} there was no Li^+ interference during K^+ absorption. Ca^{2+} was considered to create a barrier—probably at the cell surface. The stimulating effect of Ca^{2+} was believed to be essentially a blocking of the interfering ions; the main effect of Ca^{2+} was deemed to be on the membrane per se.

Ca^{2+} has been reported to affect two absorption processes—one of which is stimulated and one inhibited (Kahn and Hanson, 1957). It was reported that Ca^{2+} increased the affinity between the K^+ ion and the postulated carrier; in a second reaction, independent of the first, Ca^{2+} decreased the velocity of the metabolic phase of K^+ uptake. The first reaction is relatively greater in corn, whereas the second is relatively greater in soybean. In the presence of Ca^{2+}, K^+ uptake was more reduced in soybean than in corn roots; the first effect of Ca^{2+} was greater with corn, while the second was greater with soybeans. It was concluded that these effects were consistent with the known differences in K^+/Ca^{2+} ratios in corn and soybean plants. Absorption of K^+ is especially promoted by Ca^{2+}; Na^+ and Li^+ absorption, however, is reduced (Epstein, 1961).

Even at a given concentration of Ca^{2+}, the element may increase or decrease absorption of K^+—depending on the external concentration of K^+ (Overstreet et al., 1952). The depressing action of Ca^{2+} on K^+ uptake is related to the competition between these cations for the carrier. The stimulating effect of Ca^{2+} or K^+ uptake indicates that Ca^{2+} functions as a cofactor in the utilization of the K^+ complex produced during absorption.

Drew and Biddulph (1969) were interested in determining whether or not Ca^{2+} and K^+ uptake into roots and shoots of bean (*Phaseolus vulgaris* 'Red Kidney') were metabolic processes, that is, sensitive to respiratory inhibitors and temperature. They noted that movements of K^+ and Ca^{2+} into the shoot were affected by inhibitors and low temperature. In the roots K^+ accumulation was sensitive to inhibitors and low temperature, but that of Ca^{2+} was not. These investigators concluded that Ca^{2+} accumulation by bean roots is not directly controlled by metabolic processes.

The effect of Ca salts on accumulation of K^+ (by excised barley roots) from 5 mN KCl or K_2SO_4 depended on whether Ca^{2+} was added as the Cl^- or SO_4^{2-} salt (Tadano et al., 1969). Pretreatment of roots with $CaCl_2$ stimulated subsequent K^+ uptake from K_2SO_4, but pretreatment with $CaSO_4$ did not. In contrast, accumulation of Na^+ and K^+ from solutions containing these ions as SO_4^{2-} or Cl^- salts was changed from preferential uptake of Na^+ to preferential uptake of K^+ by addition of either $CaCl_2$ or $CaSO_4$. Tadano and co-workers concluded that Ca salts may influence K^+ accumulation partly as a result of the effects on anion uptake, but that selectivity for K^+ depends on the presence of Ca^{2+} and is influenced little by the accompanying anion.

Ultraviolet radiation and ribonuclease treatment produce effects on absorption similar to Ca^{2+} removal; treatments of roots with EDTA for 30–40 min removes two-thirds of the Ca^{2+} according to Foote and Hanson (1964). They also reported that anion uptake appeared to be critically linked to the presence of divalent cations in roots.

In vetch, stable Ca^{2+} transport was reduced to one-fifth when Sr^{2+} concentration was increased from 0.5 mM to 2.5 mM; yet stable Sr^{2+} transport did not change (Hutchin and Vaughan, 1968). This was interpreted to indicate a lack of competitive inhibition.

Ca is essential for maintenance of selective ion transport in cellular membranes (Epstein, 1961; Rains et al., 1964). Mn is also effective but, strangely, Mg^{2+} is largely ineffective (Epstein, 1961). Ca^{2+} ions minimize root injury resulting from Na^+ and H^+ (Rains et al., 1964).

Calcium increased and Mg^{2+} decreased Mn^{2+} uptake by barley roots (Maas et al., 1969). It was concluded that selectivity in ion absorption results from cation-induced conformational changes in the structure of the carrier molecule. As a result of such changes, access to potential transport sites would be permitted for some ions and denied to others. In this connection, cation activation of enzymes is believed to arise from cation-induced structural and configurational changes in the enzyme protein (Evans and Sorger, 1966).

Ion-binding compounds apparently must be capable of readily undergoing oxidation and reduction, or of undergoing some change in energy level such as would be involved with phosphorylated compounds as carriers.

There is general agreement with Epstein's (1953, 1954a) concept that active transport of all ions involves their combination with protoplasmic constituents, and that this forms the chemical basis for selectivity.

Ion uptake measurements are amenable to kinetic analysis, and this constitutes in part a basis for credence of the carrier theory. Useful as the carrier theory has been, it has not been completely free of criticism. In fact, the general applicability of competition phenomena in ion uptake is not a guarantee for the involvement of carriers, according to Wit et al. (1963). In addition, these authors had difficulty interpreting concentration experiments in which cation uptake in some way depended on anion uptake. Briggs et al. (1961) questioned the evidence for competition (by ions on carriers) when there were insufficient data with regard to various combinations of competing ions during uptake. The situation is further complicated by the effect of Ca^{2+} on the uptake of various ions and by its effect on cell permeability. Even when the uptake of Ca^{2+} is negligible, it may influence the nonmetabolic uptake of K^+, Na^+, or Rb^+ (El Nashar et al., 1966; Epstein, 1966; Handley et al., 1965a, b; Hooymans, 1964; Marschner, 1964; Rains and Epstein, 1967a, b; Waisel, 1962). This interaction is not considered a chemical interaction among K^+, Na^+, and Ca^{2+} on the hypothetical carrier but is extracted as a change in the constant in the kinetic equation (Leggett, 1968). Noggle et al. (1964) were unable to treat competition because of the enhancement effect of Ca^{2+} on Rb^+ uptake.

According to the carrier theory, accumulation of ions assumes a negligible back reaction, but efflux of an ion does become evident as the cell approaches saturation. The suggestion that the entry process is identical for net accumulation, isotopic exchange, and ion interchange indicates that the usual mathematical formulation of the carrier theory should take into consideration the back reaction. When this is considered, the equation becomes analogous to a mass action exchange reaction with diffusion as the rate-limiting step (Leggett, 1968). Even though ion transport can be reduced by metabolic inhibitors, it does not necessarily follow that the cation or anion is being actively transported (Leggett, 1968).

Unquestionably, there is a rate-limiting step in absorption, but the position of this step, or steps, in the overall process of ion uptake by plants is unknown (Fried and Shapiro, 1961). It was reported that it could occur during the movement of an ion to the vicinity of the root, at the turnover of the carrier-ion combination, or at the transport step to the plant top.

There are therefore some logical reasons to question the existence and functioning of carriers despite the fact that uptake measurements are amenable to kinetic analysis. The case for ion carriers has so far been circumstantial, that is, their presence has been deduced solely from kinetic considerations (Laties, 1959a). Briggs et al. (1961) concluded that the evidence for ions combining with carriers and for competition between or among ions for carriers was inadequate.

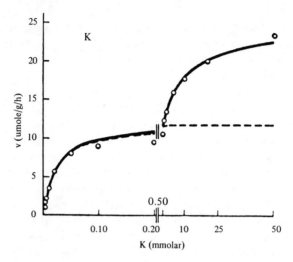

Fig. 5.6

Bimodal response curve for the absorption of K$^+$ by excised barley roots indicating the operation of two mechanisms for the absorption of K$^+$. Rate (v) of absorption of K labeled with ^{42}K as a function of the concentration of KCl. The solid line at the low concentrations, continued by the dashed line, is a plot of the Michaelis–Menten equation. (Epstein, 1966.)

One of the most fascinating developments in salt absorption research involves the discovery of dual mechanisms of absorption of ions. When a bimodal curve is obtained for the absorption of an ion, as shown for K$^+$ in Fig. 5.6, the implication is that there are two mechanisms operating for the absorption of that ion.

For example, with regard to the absorption of Na$^+$, at a low concentration of Na$^+$ (from 0.005 to 0.2 mM), mechanism 1 operates in barley roots; the mechanism is severely inhibited by Ca^{2+} and rendered virtually inoperative for Na$^+$ absorption in the presence of Ca^{2+} and K$^+$ (Rains and Epstein, 1967a). However, from 0.5 to 50 mM, a second, low-affinity mechanism (2) comes into play. In the presence of Ca^{2+} and K$^+$, mechanism 2 is the only one to absorb Na$^+$ effectively, since Na$^+$ absorption by mechanism 1 is virtually abolished by Ca^{2+} and K$^+$.

Thus there appear to be two mechanisms for the absorption of alkali cations such as Na$^+$ and K$^+$ (Epstein, 1961, 1966; Epstein et al., 1963; Rains and Epstein, 1965, 1967b; Rains, 1969a). The high-affinity mechanism (1), operative even at very low K$^+$ concentrations, is highly selective for K$^+$ as compared with Na$^+$, whereas the low-affinity mechanism (2), which only comes into play at concentrations of 0.5 to 1.0 mM and above, transports Na$^+$ as well as K$^+$. Mechanism-2 sites have a higher affinity for Na$^+$ than for K$^+$; the reverse is true for mechanism-1 sites (Rains and Epstein, 1967b). For plants growing in a relatively high concentration of Na$^+$ and a low concentration of K$^+$, Epstein (1966) stressed that mechanism 1 would be extremely important. The high-affinity mechanism 1, of K–Rb absorption, operates at half-maximal velocity at an external K^+ concentration of approximately 0.018 mM (0.7 ppm K^+—a concentration of the order of that in soil solutions); mechanism 2 operates at one-half maximal at 16 mM (Epstein et al., 1963).

Mechanism 2 is reported to be heterogeneous with several transport sites with different affinities for various cations (Rains and Epstein, 1965).

Dual mechanisms of absorption have been reported for: K^+ (Fried and Noggle, 1958; Epstein et al., 1963; Smith and Epstein, 1964; Rains, 1969a; Jeschke, 1970); Rb^+ (Fried and Noggle, 1958; Smith and Epstein, 1964); Cs^+ (Bange and Overstreet, 1960); NH^{4+} (Fried et al., 1965); Na^+ (Fried and Noggle, 1958; Epstein, 1961, 1966; Rains and Epstein, 1967b); Mg^{2+} (Yoshida, 1964); Ca^{2+} (Osmund, 1966); Sr^{2+} (Fried and Noggle, 1958); Cl^- (Elzam et al., 1964; Boszorményi and Cseh, 1964; Elzam and Epstein, 1965); Br^- (Boszorményi and Cseh, 1964); I^- (Boszorményi and Cseh, 1964); $H_2PO_4^-$ (Carter and Lathwell, 1967); SO_4^{2-} (Leggett and Epstein, 1956); choline sufate (Nissen and Benson, 1964)

For choline sulfate, Nissen and Benson (1964) proposed a two-site carrier with an anionic site binding the quaternary N and a cationic site binding the sulfate group.

For Cl^- absorption the high-affinity mechanism (1) operates at external Cl^- concentrations between 0.005 and 0.20 mM (Elzam et al., 1964), and Br^- competitively inhibits Cl^- uptake (Elzam and Epstein, 1965), while F^- and I^- have no effect. The low-affinity mechanism (2) operates when the external Cl^- concentration is in the range of 0.5–50.0 mM (Elzam et al., 1964). In this connection it has been noted that it is often possible to calculate two K_m values and V_{max} values for ion absorption, and it is generally assumed then that two carriers or carrier sites are involved in absorption (Hodges and Vaadia, 1964b). One site is assumed to be functional at low and the other at high external concentrations. In the case of Cl^- uptake by onion roots, it was concluded that the greater uptake of ions, when high external concentrations were employed, could simply be attributable to an additional diffusion component of uptake, and not necessarily to the action of a separate carrier site for high external Cl^- concentrations.

There appear to be two sets of carrier sites for Rb^+—one of them specific for Rb^+ (and K^+) and the other with appreciable affinity for Na^+ as well (Epstein, 1966). The evidence indicates that the first, high-affinity mechanism may be located in the plasmalemma and the second, low-affinity mechanism in the tonoplast. This view of the spatial separation of mechanisms 1 and 2 was shared by Torii and Laties (1966a, b) and Lüttge and Laties (1966).

For beet discs, the type-2 mechanisms—those that become apparent at concentrations above 1.0 mM—were considered to be located in the tonoplast and reached by the ions after diffusion across the plasmalemma (Osmund and Laties, 1968). The latter membrane was considered to be permeable in this range of concentrations. System 1 controlled movement of ions into the cytoplasm; system 2, movement into the vacuole. However, Welch and Epstein (1969) presented arguments against the tonoplast (as the barrier for system 2), and indicated that the plasmalemma was the seat of system-1 and -2 mechanisms of alkali cation absorption, and extended their conclusion to the absorption of Cl^-.

It has recently been reported that the low- and high- concentration salt transport mechanisms operate in parallel across the plasmalemma (Welch and Epstein, 1968). More recently, a computing technique was used to simulate the course of uptake of K^+, Na^+, and Cl^- by low-salt roots (Pitman, 1969). On the basis of the permeability data, it was suggested that anion transport at the plasmalemma must be greater than that provided by mechanism 1 and that there is a need for transport of both cations and anions at the tonoplast.

For the dual mechanisms for absorption of Cs^+, mechanism 1 is affected competitively by K^+, Rb^+, and, to a lesser extent, NH_4^+ (Bange and Overstreet, 1960). H^+ ion competes in mechanism 1 with Rb^+, K^+, Na^+, and Sr^{2+} (Fried and Noggle, 1958).

In wheat roots Cl^-, Br^-, and I^- are absorbed by two mechanisms; Cl^- and Br^- are transported by the same carrier, and I^- by a different one (Boszorményi and Cseh, 1964).

In *Atriplex spongiosa*, increasing K^+ concentrations had a depressive effect on leaf Ca^{2+} only when there was a high concentration (50,000 μM) of Ca^{2+} in the nutrient solution; K^+ did not affect the internal Ca^{2+} level when the external Ca^{2+} concentration was low (500–5000 μM) (Osmund, 1966). It was proposed that there are two independent mechanisms for Ca^{2+} absorption—a highly specific, K-insensitive mechanism at low Ca^{2+} concentrations, and a nonspecific, K-sensitive mechanism at higher external Ca^{2+} concentrations.

In excised corn roots there appear to be two mechanisms (sites) for the absorption of orthophosphate, with one dominating at high-P concentrations and one at low-P concentrations (Carter and Lathwell, 1967).

Absorption of phosphate by roots and simultaneous transport to the tops of intact wheat seedlings were studied over a concentration range of 0.001–50 mM (Edwards, 1970). In the low-concentration range (<1.0 mM), phosphate absorption showed two hyperbolic isotherms representing sites of different affinity. It was suggested that both sites may be at the plasmalemma and that both are involved in absorption of the predominant $H_2PO_4^-$ ion. Simultaneous absorption and transport studies established that the low-concentration mechanisms of phosphate absorption do not correspond to systems 1 and 2 of the dual absorption mechanism.

Absorption isotherms indicated dual mechanisms of Mg^{2+} and Cl^- absorption for excised corn roots from solutions above 1 meq/liter (Maas and Ogata, 1971). In the physiological pH range, Ca^{2+} greatly suppressed the rate of Mg^{2+} absorption but had little effect on Cl^-. The influence of Ca^{2+} on

Mg^{2+} appeared to be noncompetitive and independent of the effect of Ca^{2+} on membrane permeability.

In citrus leaf slices there was no evidence for two distinct Cl^- transport systems (Robinson and Smith, 1970) as reported by other workers. It was proposed that Cl^- influx is normally dependent on oxidative phosphorylation, but that cyclic photophosphorylation may provide an alternative energy source.

Ion absorption by excised rice roots is reported to follow the dual mechanism pattern (Kannan, 1971).

The foregoing examples of dual mechanisms concern roots of plants, but dual mechanisms for absorption of K^+ and Rb^+ have also been shown for corn leaf tissue (Smith and Epstein, 1964).

Significantly, Epstein (1966) suggested that the development of two mechanisms for the absorption of a given ion might very well have survivial or evolutionary significance. This development would, for example, enable a plant to obtain sufficient K^+ in the presence of a large excess of Na^+ ions. He listed the many species of plants in which dual mechanisms of absorption have been observed and studied.

In many studies of salt absorption, translocation to the tops may affect the results of an absorption study. Conversely, features of the salt absorption process may affect the results of a study of translocation. When excised roots are used in absorption studies, there is always a question of relevancy of the results to those that would be obtained with intact roots (i.e., plants). In a recent, excellent review, Laties (1969) considered dual mechanisms of salt uptake in relation to compartmentation and long-distance transport. He noted that concepts of ion absorption have been altered and expanded with the discovery of dual or multiple mechanisms. Nonvacuolate root tips absorb ions by system 1, while vacuolate root sections absorb by systems 1 and 2 (Torii and Laties, 1966a).

Salt levels in fertile soils are generally in the range of system 1 according to Laties (1969). He reported that when plant tissue is subjected to high external concentrations of salt, as in certain experiments, cytoplasm rapidly becomes filled with salt. Subsequently, steady-state uptake reflects delivery of ions to the vacuole, and kinetically it appears as though only system 2 is operating. In other words, systems 1 and 2 are discernible only in short-term absorption periods with cytoplasm initially depleted of salt.

In this connection shoots tend to have a high K/Na ratio, and it seems likely that the high K/Na ratio in the shoots must have its origin in the root. Some researchers have postulated that the high K/Na ratio, which generally prevails in roots, is caused by an inwardly directed K^+ pump (Pitman et al., 1968; Jennings, 1967) and/or an outwardly directed Na^+ pump (Scott et al., 1968; Pitman, et al., 1968).

As Laties (1969) noted, however, the disparity between these observations and those of Epstein and his colleagues (Epstein et al., 1963; Rains and Epstein, 1967a, b) emphasize the importance of time and concentration in studies of K^+ selectivity in the range of system 2. The latter authors used short-term experiments with low-salt roots rather than long-term ones, hence essentially used high-salt roots; therefore, operation of system 2 (at the tonoplast) was discernible in their experiments.

Further, with regard to K/Na ratios in shoots and roots, Bange (1959, 1965) and Bange and Vliet (1961) postulated that two carriers are involved in K^+ and Na^+ absorption with both carriers favoring K^+ uptake. Later, Bange and Hooymans (1967) suggested that separate carriers mediate K^+ and Na^+ transport. For the latter suggestion, Laties (1969) noted that Na^+ and K^+ entry are affected by system 1 and diffusion. He postulated that Ca^{2+} "serves to knit the plasma membrane," minimizes diffusion penetration (of Na^+), and emphasizes the role of system 1. Inasmuch as system 1 markedly favors K^+ over Na^+, Na^+ uptake would be reduced to practically nil in the presence of Ca^{2+} and K^+ (Rains and Epstein, 1967a). Laties (1969) noted that in the absence of Ca^{2+} diffusive penetration of Na^+ into the cytoplasm is considerable.

Endodermal cells are generally regarded to constitute a barrier to the passive diffusion of ions from the cortex to the stele, and the controlling feature of these cells is the Casparian strip. Owing to the presence of this continuous suberized structure, ions cannot pass either between or through the walls of the endodermal cells. Materials must pass through the protoplasm of the endodermal cells. Inasmuch as there may be a decreasing O_2 tension from the epidermis to the stele (Crafts and Broyer, 1938), the innermost layer of cortical cells might lose salts to the endodermis (Crafts, 1961); back-diffusion of salts from the xylem vessels would be prevented by the impervious Casparian strip.

Several years ago, it was proposed that ions are accumulated in the cytoplasm of cortical cells which then deliver the ions to the stele by means of the cytoplasmic continuum—the symplast (Crafts and Broyer, 1938). This is the essence of the symplasm theory for radial transport of salts across a root. Movements across the plasma membranes were regarded as active and ion movements to the xylem as passive. The alternative view states that ions move freely through cortical space and that they are actively transported to the xylem by means of the endodermis (Steward and Sutcliffe, 1959) or by cells within the stele (Hylmo, 1953; Anderson and House, 1967; Anderson and Reilly, 1968). Oertli (1967) postulated that the absorption process consisted of a pump (inwardly directed active transport) and a leak system (carrier-mediated bidirectional passive transport). A pump–leak formulation yields biphasic absorption curves with concentration; unequivocally, both systems would operate at the plasmalemma.

The Crafts and Broyer (1938) symplasm concept has recently been championed by Laties (1969). He regards the symplasm as the major pathway of radial ion movement in the cortex. He visualized a gathering function of ions by the cortex and a passive function of xylary conduits in the stele. He regarded a "leaky stelar parenchyma" as the "linchpin" of the symplasm theory of radial transport (across a root). Further, at relatively low concentrations of salt, radial transport would occur predominantly through the symplasm, and selective plasma membrane absorption along with Na^+ extrusion would account for the ultimate K^+ enrichment in the shoot. He considered the endodermis a barrier to free diffusion between the external solution and the stele when ions are moving through free space. He stressed that system 2 has no connection with ion transfer to the shoot; this system operates at the tonoplast and controls only the movement of ions into the vacuole.

With regard to the interplay between dual absorption mechanisms, rate of transpiration, and K/Na ratios, Laties (1969) concluded that a high rate of transpiration would be accompanied by hydrodynamic flow through the symplasm and the withdrawal of Na^+ into vacuoles, or that its extrusion from roots would be lessened. Hence more Na^+ would reach the tops of plants under high rates of transpiration.

In addition, dual mechanisms are involved in the question of whether long-distance salt transport in plants is active or passive, since the question rests on the external concentration of salts being considered (Laties, 1969). If a low external salt concentration exists, system 1 is in kinetic control, and translocation appears to be an active process; at high external salt concentrations, a diffusive, nonselective component is superimposed on the active component, and translocation appears to be more-or-less passive.

Uptake of K^+ by exuding root systems of *Ricinus communis* was measured and shown to be dependent on the solute potential of the medium (Minchin and Baker, 1969). Also, the optimum was shown to exhibit a dual absorption isotherm—the kinetics of which indicated a low K_m system (system 1) and a high K_m system (system 2).

Phosphorylated Intermediates. Although no one has yet succeeded in isolating or demonstrating directly the postulated high-energy intermediate (Chance et al., 1967), there is strong evidence that a high-energy intermediate or ATP may function in ion transport (Budd and Laties, 1964). In experiments with corn roots under anaerobic conditions, electrons were bled off from the cytochrome chain, and yet this abbreviated chain mediated Cl^- transport strongly. This evidence indicates that oxidative phosphorylation is primarily responsible for ion transport. Dehydrogenase, and not cytochrome c, was believed to be involved in Cl^- uptake. Additional support for the role of oxidative phosphorylation in ion uptake comes from research by Robertson et al. (1951). Oxygen consumption by carrot root tissue was increased by DNP, whereas accumulation of KCl was strongly inhibited. With barley roots, Ordin and Jacobson (1955) found that DNP at 1.1×10^{-6} M reduced K^+ and Br^- absorption without an effect on O_2 uptake. Using several kinds of plant materials and studying the effects of DNP and arsenate on Rb^+ uptake, Higinbotham (1959) concluded that Rb^+ absorption was dependent on the availability of ATP. A similar conclusion was made with regard to the effect of arsenate on the absorption of $SO_4{}^{2-}$ and Cl^- by corn roots (Weigl, 1963, 1964).

ATP is the energy source for P uptake at the plasmalemma (Bledsoe et al., 1969).

High-energy intermediates of oxidative phosphorylation were believed to participate directly in Ca^{2+} binding and uptake by corn mitochondria (Elzam and Hodges, 1968). The substrate-driven system was more efficient in Ca^{2+} accumulation than the ATP-driven system. For large accumulations of Ca^{2+}, P_i was essential, P_i was accumulated along with Ca^{2+}, and the ratio of accumulated Ca/P_i (1.6:1) indicated that hydroxyapatite was precipitated in the mitochondria.

In other work with corn mitochondria, the ATP-driven Ca^{2+} uptake required exogenous Mg^{2+} and was stimulated by P_i (Hodges and Hanson, 1965). They concluded that a high-energy intermediate of oxidative phosphorylation participated in the accumulation of divalent ions and P_i. It was suggested (Hanson et al., 1965) that Ca^{2+} accomplishes its effect on P by diverting high-energy phosphate from ATP formation into P_i uptake, and later work (Truelove and Hanson, 1966) implicated this role of Ca.

The potential energy of the contracted state of corn mitochondria is available to drive Ca^{2+} and P_i accumulation according to Kenefick and Hanson (1966). They noted that it was immaterial whether Ca^{2+} or P_i reacted first with the contracted mitochondria.

Sutcliffe and Hackett (1957) stated that a phosphorylation mechanism provides a unifying concept which applies to ion transport in anaerobic as well as aerobic systems. Further, it is a mechanism that provides a basically uniform scheme for the transport of neutral molecules as well as of positively or negatively charged ions. This is interesting inasmuch as there is increasing evidence that the kinetic and metabolic patterns of transport of neutral organic molecules resemble those of inorganic ion transport (Bieleski, 1960, 1962; Nissen and Benson, 1964).

In sugarcane leaf tissue, accumulation of exogenously supplied sucrose, fructose, or glucose was a hyperbolic function of sugar concentration—suggesting intermediate compound formation between the sugar and some receptor or carrier within the leaf cells (Bieleski, 1962). In addition, it was noted that sucrose and glucose interacted competitively.

Adenosinetriphosphatases (ATPases) have been implicated in transport of Na^+ and K^+ across animal membranes (Skou, 1957). Animal enzymes are activated by Na^+ and K^+ and are sensitive to a cardiac glycoside, ouabain. In the freshwater alga *Nitella translucens*, influx of K^+ and efflux of Na^+ were sensitive to ouabain, but absorption of Cl^- was not (MacRobbie, 1962). Thus the systems for Cl^- and for Na^+-K^+ were quite different. It was concluded that influx of K^+ and efflux of Na^+ were regulated by the plasmalemma and influx of Cl^- into the vacuole by the tonoplast. It appears that ouabain-sensitive ATPases may be important in reciprocal movements of Na^+ and K^+. Reportedly, ATPase is involved in the active extrusion of Na^+ from nerve fibers (Skou, 1957). Brown and Altschul (1964) demonstrated a glycoside (ouabain)-sensitive transport ATPase in peanut seedlings, but its presence was not studied in relation to ion transport.

With regard to the enzymic Na–K pump, it has been suggested that there may be a linkage between conformational changes in a center for phosphorylation and one for translocation; this linkage would be a basic feature of the mechanism of the pump (Post et al., 1969).

Either electron transfer phenomena or ATP could represent the form of energy utilized in salt uptake (Robertson, 1960). In this connection, Fisher and Hodges (1969) found there was sufficient Mg^{2+}- (or Mn^{2+}-) requiring ATPase additionally stimulated by K^+ and other monovalent ions to account for the observed rates of K^+ influx into oat roots.

Ion accumulation can be driven by the hydrolysis of ATP, and ion flow in reverse might conceivably be harnessed to synthesize ATP. The latter is the basis for the chemiosmotic hypothesis. According to this hypothesis (Mitchell, 1966), electron flow in mitochondria and chloroplasts is coupled to the formation of a proton gradient (from H^+ accumulation by chloroplasts during illumination) which would provide potential energy for the formation of ATP from ADP by a dehydration mechanism involving an anisotropic membrane-bound ATPase. The intermediate between electron transport and phosphorylation is a pH gradient and/or membrane potential (Jagendorf, 1967). A large membrane potential would be sufficient to drive ATP formation.

Energy-linked ion movements in chloroplasts, particularly uptake of H^+ on illumination, clearly relates ion uptake to photosynthesis, electron transfer, and photophosphorylation (Packer and Crofts, 1967). It was noted that NH_4^+ uptake in the dark was coupled to light-activated ATPase, and that there was a close correlation between light-induced structural changes of chloroplasts and ion transport. The active, light-dependent ion movement was H^+ uptake. The "chemical" hypothesis for ion uptake invokes the high-energy phosphate bond (or at least some kind of high-energy intermediate), whereas the chemiosmotic hypothesis involved H^+, that is, H^+ uptake by illuminated chloroplasts produces a membrane potential which could cause the formation of ATP from ADP and thus mediate ion uptake. Packer and Crofts (1967) concluded, however, that none of the evidence clarifies the *mechanism* by which H^+ transport is coupled to electron flow.

It has been reported that a cell membrane potential based on the premise that intracellular Na^+ and K^+ ions migrate freely in liquid cell water is thermodynamically impossible (Bernhard, 1969). In addition, existence of a Na pump has been questioned, since calculations indicate that the energy required to operate the pump at rates fast enough to maintain intracellular ion concentrations would be 350 times the energy cellular metabolism can produce (Bernhard, 1969).

Protein Synthesis. There is evidence that may be interpreted to show a direct **relationship between protein synthesis and ion uptake (Steward and Millar,** 1954; Brenner and Maynard, 1966). It is hypothesized that new binding sites (e.g., at a nucleic acid template surface for protein synthesis) are produced or multiplied in leaves in the light.

Protein synthesis and phosphorylated, energy-rich carriers have been linked by Steward and Street (1947). These investigators postulated that carriers are phosphorylated, energy-rich N compounds. Being amphoteric, they could conceivably carry both cations and anions. It was suggested that the carrier is degraded in the process of its incorporation into protein (at the seat of protein synthesis), and that simultaneously the associated ion or ions are released in free solution into the vacuole.

In dividing cells it is postulated that ions (particularly cations) are bound to sites being synthesized. When division ceases, ions are released from the binding sites into the vacuole during the period of cell enlargement (Steward and Sutcliffe, 1959). However, even fully-mature, vacuolated cells may have "loaded" cytoplasmic sites prior to discharge of the ions into the vacuoles; in such cells protein synthesis is supposedly minimal (MacDonald and Laties, 1963; Hodges and Vaadia, 1964a).

Chloramphenicol reportedly inhibits protein synthesis and reduces ion uptake very markedly (Sutcliffe, 1960; Jacoby and Sutcliffe, 1962), or has only slightly inhibitory effects on ion absorption (Balogh et al., 1961). Pronounced effects of chloramphenicol on ion uptake could be ascribed to its action as an inhibitor in the utilization of respiratory energy (Stoner et al., 1964; Ellis et al., 1964).

In this connection, Jacoby and Sutcliffe (1962) made the very interesting suggestion that accumulation of K^+ may be related to the synthesis of a particular kind of protein. This protein might be located in the surface membrane and act as an ion carrier through a cycle of breakdown and resynthesis. The rest of the protein of the cell would not be involved in ion absorption.

Therefore, although there often appears to be a relationship between ion uptake and protein synthesis (and growth), it is not established whether the relationship is a direct or an indirect one. Further, such a relationship would shed no light on specificity or selectivity in ion uptake.

Amino Acids. It was postulated that amino acids might be involved in ion absorption, since they possess properties of the hypothesized R•H and R'•OH

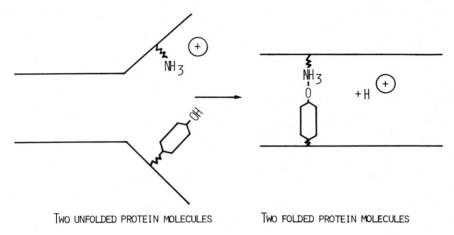

TWO UNFOLDED PROTEIN MOLECULES   TWO FOLDED PROTEIN MOLECULES

Fig. 5.7
Hypothetical scheme for the driving of a chemical reaction in which protein molecules fold up. Redrawn from R. J. Goldacre (1952) Int. Rev. Cytol., 1:135–164. Academic Press, Inc., New York, N.Y.—by permission.

substances with the H^+ of the $-COOH$ and the OH^- of the $-NH_3OH$ groups being exchangeable for cations and anions, respectively (Brooks, 1937). By an orientation of amino acids in the protoplasm or plasmalemma Brooks (1937) envisaged the entrance of cations and anions along different paths by a series of exchanges involving H^+ and OH^-.

Folding and Unfolding of Proteins. It has been hypothesized that the folding and unfolding of proteins could provide a basis for osmotic work, absorption of ions, and driving reactions (Goldacre, 1952). In ameba, protein molecules fold at one end of the cell and unfold at the other—a phenomenon resulting in osmotic work. Also, when protein molecules are unfolded, they have more surface area for adsorption; the side chains and other groups, which were tied up in the peptide chain, become free and can then absorb other substances. Last, the unfolding of protein molecules could drive a chemical reaction. That is, the products of a reaction could be squeezed off the surface of an enzyme when it folded (Fig. 5.7).

LITERATURE CITED

Anderson, W. P., and C. R. House. 1967. A correlation between structure and function in the root of *Zea mays*. J. Exp. Bot. 18:544–555.

Anderson, W. P., and E. J. Reilly. 1968. A study of the exudation of excised maize roots after removal of the epidermis and outer cortex. J. Exp. Bot. 19:19–30.

Balogh, E., Z. Böszörményi, and E. Cseh. 1961. The effect of chloramphenicol on the amino acid metabolism and ion uptake of isolated wheat roots. Biochim. Biophys.

Acta 52:381–383.

Bange, G. G. J. 1959. Interactions in the potassium and sodium absorption by intact maize seedlings. Plant and Soil 11:17–29.

Bange, G. G. J. 1965. Upward transport of potassium in maize seedlings. Plant and Soil 22:280–306.

Bange, G. G. J., and R. Overstreet. 1960. Some observations on absorption of cesium by excised barley roots. Plant Physiol. 35:605–608.

Bange, G. G. J., and J. J. M. Hooymans. 1967. *In* Isotopes in plant nutrition, pp. 249–263. Int. At. Energy Agency, Vienna.

Bange, G. G. J., and E. van Vliet. 1961. Translocation of potassium and sodium in intact corn seedlings. Plant and Soil 15:312–328.

Bennet-Clark, T. A. 1956. Salt accumulation and mode of action of auxin. A preliminary hypothesis, pp. 284–291. *In* R. L. Wain, and F. Wightman (eds.) The chemistry and mode of action of plant growth substances. Butterworths, London.

Bennett, H. S. 1956. The concept of membrane flow and membrane vesiculation as mechanisms for active transport and ion pumping. J. Biophys. Biochem. Cytol. 2:99–103.

Bernhard, R. 1969. A time bomb made of simple (?) H_2O. Sci. Res. 4:36–39.

Bernstein, L. 1971. Method for determining solutes in the cell walls of leaves. Plant Physiol. 47:361–365.

Bieleski, R. L. 1960. The physiology of sugar-cane. IV. Effects of inhibitors on sugar accumulation in storage tissue cells. Australian J. Biol. Sci. 13:221–231.

Bieleski, R. L. 1962. The physiology of sugar-cane. V. Kinetics of sugar accumulation. Australian J. Biol. Sci. 15:429–444.

Bledsoe, C., C. V. Cole, and C. Ross. 1969. Oligomycin inhibition of phosphate uptake and ATP labeling in excised maize roots. Plant Physiol. 44:1040–1044.

Blinks, L. R. 1949. The source of the bioelectric potentials in large plant cells. Proc. Nat. Acad. Sci. 35:566–575.

Blount, R. W., and B. H. Levedahl. 1960. Active sodium and chloride transport in the single celled marine alga *Halicystis ovalis*. Acta Physiol. Scand. 49:1–9.

Böszörményi, Z., and E. Cseh. 1961. The uptake of halide ions and their relationships in absorption. Physiol. Plantarum 14:242–252.

Böszörményi, Z., and E. Cseh. 1964. Studies of ion-uptake by using halide ions changes in the relationships between ions depending on concentration. Physiol. Plantarum 17:81–90.

Bowen, J. E. 1968. Borate absorption in excised sugarcane leaves. Plant Cell Physiol. 9:467–478.

Bowen, J. E., 1969. Absorption of copper, zinc, and manganese by sugarcane leaf tissue. Plant Physiol. 44:255–261.

Bowling, D. J. F., A. E. S. Macklon, and R. M. Spanswick. 1966. Active and passive transport of the major nutrient ions across the root of *Ricinus communis*. J. Exp. Bot. 17:410–425.

Brenner, M. L., and D. N. Maynard. 1966. A study of rubidium accumulation in *Euglena gracillis*. Plant Physiol. 41:1285–1288.

Briggs. G. E., A. B. Hope, and R. N. Robertson. 1961. Electrolytes and plant cells. *In* W. O. James (ed.) Bot. Mongr., Vol. I. Blackwell, Oxford. 217 pp.

Broda, E. 1965. Mechanism of uptake of trace elements by plants (experiments with radiozinc). Isotopes Radiation Soil-Plant Nutr. Studies, Proc. Symp., Ankara 1965:207–215.

Brooks, S. C. 1937. Selective accumulation with reference to ion exchange by the protoplasm. Trans. Faraday Soc. 33:1002–1006.

Brown, H. D., and A. M. Altschul. 1964. Glycoside-sensitive ATPase from *Arachis hypogaea*. Biochem. Biophys. Res. Commun. 15:479—483.

Broyer, T. C. 1947. The movement of materials into plants. II. The nature of solute movement into plants. Bot. Rev. 13:125—167.

Broyer, T. C. 1951. The nature of the process of inorganic solute accumulation in roots, pp. 187—249. *In* E. Truog (ed.) Mineral nutrition of plants. Univ. Wisconsin Press, Madison.

Budd, K., and G. G. Laties. 1964. Ferricyanide-mediated transport of chloride by anaerobic corn roots. Plant Physiol. 39:648—654.

Burström, H. 1951. The mechanism of ion absorption, pp. 251—260. *In* E. Truog (ed.) Mineral nutrition of plants. Univ. Wisconsin Press, Madison.

Carter, O. G., and D. J. Lathwell. 1967. Effects of temperature on orthophosphate absorption by excised corn roots. Plant Physiol. 42:1407—1412.

Chance, B., C. P. Lee, and L. Mela. 1967. Control and conservation of energy in the cytochrome chain. Fed. Proc. 26:1341.

Conway, E. J. 1954. Some aspects of ion transport through membranes. Symp. Soc. Exp. Biol., 8:297—324.

Crafts, A. S. 1961. Translocation in plants. Holt, New York. 182 pp.

Crafts, A. S., and T. C. Broyer. 1938. Migration of salts and water into xylem of the roots of higher plants. Amer. J. Bot. 25:529—535.

Dainty, J. 1962. Ion transport and electrical potentials in plant cells. Annu. Rev. Plant Physiol. 13:379—402.

Drew, M. C., and O. Biddulph. 1969. Non-metabolic uptake of calcium by intact bean roots. Plant Physiol. 44(Suppl.):no. 94.

Edwards, D. G. 1970. Phosphate absorption and long-distance transport in wheat seedlings. Australian J. Biol. Sci. 23:255—264.

Elgabaly, M. M., H. Jenny, and R. Overstreet. 1943. Effect of type of clay mineral on the uptake of zinc and potassium by barley roots. Soil Sci. 55:257—262.

Ellis, R. J., K. W. Joy, and J. F. Sutcliffe. 1964. The inhibition of salt uptake by D-serine. Phytochemistry 3:213—219.

el Nashar, M., A. K. Helmy, M. N. Hassan, and M. M. Elgabaly. 1966. Effect of divalent and other ions on potassium uptake by barley roots. Plant and Soil 25:290—296.

Elzam, O. E., and E. Epstein. 1965. Absorption of chloride by barley roots: Kinetics and selectivity. Plant Physiol. 40:620—624.

Elzam, O. E., and T. K. Hodges. 1967. Calcium inhibition of potassium absorption in corn roots. Plant Physiol. 42:1483—1488.

Elzam, O. E., and T. K. Hodges. 1968. Characterization of energy-dependent Ca^{2+} transport in maize mitochondria. Plant Physiol. 43:1108—1114.

Elzam. O. E., D. W. Rains, and E. Epstein. 1964. Ion transport kinetics in plant tissue: Complexity of the chloride absorption isotherm. Biochem. Biophys. Res. Commun. 15:273—276.

Epstein, E. 1953. Mechanism of ion absorption by roots. Nature 171:83—84.

Epstein, E. 1954a. Cells run a ferry system — How plants get nutrients. Agr. Res. 3(6):6—7.

Epstein, E. 1954b. Cation-induced respiration in barley roots. Science 120:987—988.

Epstein, E. 1955. Passive permeation and active transport of ions in plant roots. Plant Physiol. 30:529—535.

Epstein, E. 1960. Calcium-lithium competition in absorption by plant roots. Nature 185:705—706.

Epstein, E. 1961. The essential role of calcium in selective cation transport by plant cells. Plant Physiol. 36:437—444.

Epstein, E. 1966. Dual pattern of ion absorption by plant cells and by plants. Nature 212:1324–1327.

Epstein, E., and C. E. Hagen. 1952. A kinetic study of the absorption of alkali cations by barley roots. Plant Physiol. 27:457–474.

Epstein, E., and J. E. Leggett. 1954. The absorption of alkaline earth cations by barley roots: Kinetics and mechanism. Amer. J. Bot. 41:785–791.

Epstein, E., D. W. Rains, and W. E. Schmid. 1962. Course of cation absorption by plant tissue. Science 136:1051–1052.

Epstein, E., D. W. Rains, and O. E. Elzam. 1963. Resolution of dual mechanisms of potassium absorption by barley roots. Proc. Nat. Acad. Sci. 49:684–692.

Etherton, B. 1963. Relationship of cell transmembrane electropotential to potassium and sodium accumulation ratios in oat and pea seedlings. Plant Physiol. 38:581–585.

Etherton, B. 1967. Steady state sodium and rubidium effluxes in *Pisum sativum* roots. Plant Physiol. 42:685–690.

Etherton, B. 1970. Effect of indole-3-acetic acid on membrane potentials of oat coleoptile cells. Plant Physiol. 45:527–528.

Etherton, B., and N. Higinbotham. 1960. Transmembrane potential measurements of cells of higher plants as related to salt uptake. Science 131:409–410.

Evans, H. J., and G. J. Sorger. 1966. Role of mineral elements with emphasis on the univalent cations. Annu. Rev. Plant Physiol. 17:47–76.

Fels, I. G., and V. H. Cheldelin. 1949. Selenate inhibition studies. III. The role of sulfate in selenate toxicity in yeast. Arch. Biochem. 22:402–405.

Fisher, J., and T. K. Hodges. 1969. Monovalent ion stimulated adenosine triphosphatase from oat roots. Plant Physiol. 44:385–395.

Foote, B., and J. B. Hanson. 1964. Ion uptake by soybean tissue depleted of calcium by ethylenediaminetetraacetic acid. Plant Physiol. 39:450–460.

Foulkes, E. C. 1956. Cation transport in yeast. J. Gen. Physiol. 39:687–704.

Fried, M., and H. Broeshart. 1967. The soil-plant system in relation to inorganic nutrition. Academic, New York.

Fried, M., and J. C. Noggle. 1958. Multiple site uptake of individual cations by roots as affected by hydrogen ion. Plant Physiol. 33:139–144.

Fried, M., and R. E. Shapiro. 1961. Soil-plant relationships in ion uptake. Annu. Rev. Plant Physiol. 12:91–112.

Fried, M., F. Zsoldos, P. B. Vose, and I. L. Shatokhin. 1965. Characterizing the NO_3 and NH_4 uptake process of rice roots by use of ^{15}N labelled NH_4NO_3. Physiol. Plantarum 18:313–320.

Goldacre, R. J. 1952. The folding and unfolding of protein molecules as a basis of osmotic work. Int. Rev. Cytol. 1:135–164.

Goldacre, R. J., and T. J. Lorch. 1950. Folding and unfolding of protein molecules in relation to protoplasmic streaming, amoeboid movement and osmotic work. Nature 166:497–500.

Hagen, C. E., and H. T. Hopkins. 1955. Ionic species in orthophosphate absorption by barley roots. Plant Physiol. 30:193–199.

Hagen, C. E., J. E. Leggett, and P. C. Jackson. 1957. The sites of orthophosphate uptake by barley roots. Proc. Nat. Acad. Sci. 43:496–503.

Handley, R., and R. Overstreet. 1955. Respiration and salt absorption by excised barley roots. Plant Physiol. 30:418–426.

Handley, R., and R. Overstreet. 1961. Uptake of calcium and chlorine in roots of *Zea mays*. Plant Physiol. 36:766–769.

Handley, R., A. Metwally, and R. Overstreet. 1965a. Divalent cations and the permeability to Na of the root meristem of *Zea mays*. Plant and Soil 22:200–206.

Handley, R., A. Metwally, and R. Overstreet. 1965b. Effects of Ca upon metabolic and nonmetabolic uptake of Na and Rb by root segments of *Zea mays*. Plant Physiol. 40:513—520.

Hanson, J. B., S. S. Malhotra, and C. D. Stoner. 1965. Action of calcium on corn mitochondria. Plant Physiol. 40:1033—1040.

Harris, E. J. 1960. Transport and accumulation in biological systems. Academic, New York.

Higinbotham, N. 1959. The possible role of adenosine triphosphate in rubidium absorption as revealed by the influence of external phosphate, dinitrophenol and arsenate. Plant Physiol. 34:645—650.

Higinbotham, N., B. Etherton, and R. J. Foster. 1961. The source and significance of the electropotential of higher plant cells. Plant Physiol. 36:xxxv.

Higinbotham, N., B. Etherton, and R. J. Foster. 1964. Effect of external K, NH_4, Na, Ca, Mg, and H ions on the cell transmembrane electropotential of *Avena* coleoptile. Plant Physiol. 39:196—203.

Higinbotham, N., B. Etherton, and R. J. Foster. 1967. Mineral ion contents and cell transmembrane electropotentials of pea and oat seedling tissue. Plant Physiol. 42:37—46.

Hill, J. E., and W. J. V. Osterhout. 1938. Calculations of bioelectric potentials. II. The concentration potential of KCl in *Nitella*. J. Gen. Physiol. 21:541—556.

Hoagland, D. R., and T. C. Broyer. 1936. General nature of the process of salt accumulation by roots with description of experimental methods. Plant Physiol. 11:471—507.

Hoagland, D. R., and T. C. Broyer. 1942. Accumulation of salt and permeability in plant cells. J. Gen. Physiol. 25:865—880.

Hodges, T. K., and J. B. Hanson. 1965. Calcium accumulation by maize mitochondria. Plant Physiol. 40:101—109.

Hodges, T. K., and Y. Vaadia. 1964a. Chloride uptake and transport in roots of different salt status. Plant Physiol. 39:109—114.

Hodges, T. K., and Y. Vaadia. 1964b. The kinetics of chloride accumulation and transport in exuding roots. Plant Physiol. 39:490—493.

Honert, T. H. van den. 1936. Vergadering Vereeninging Proefstations Personnel (Djember, Java) 16:85—93. (Original not seen; title unavailable.)

Honert, T. H. van den, and J. J. M. Hooymans. 1955. On the absorption of nitrate by maize roots in water culture. Acta Bot. Neerl. 4:376—384.

Hooymans, J. J. M. 1964. The role of calcium in the absorption of anions and cations by excised barley roots. Acta Bot. Neerl. 13:507—540.

Hope, A. B., and N. A. Walker. 1961. Ionic relations of cells of *Chara australis* R. Br. IV. Membrane potential differences and resistances. Australian J. Biol. Sci. 14:26—44.

Hurd-Karrer, A. M. 1938. Relation of sulphate to selenium absorption by plants. Amer. J. Bot. 25:666—675.

Hutchin, M. E., and B. E. Vaughan. 1968. Relation between simultaneous Ca and Sr transport rates in isolated segments of vetch, barley, and pine roots. Plant Physiol. 43:1913—1918.

Hutner, S. H. 1948. Essentiality of constituents of sea water for growth of a marine diatom. Trans. N. Y. Acad. Sci. 10:136—141.

Hylmö, B. 1953. Transpiration and ion absorption. Physiol. Plantarum 6:333—405.

Jacobson, L., and R. Overstreet. 1947. A study of the mechanism of ion absorption by plant roots using radioactive elements. Amer. J. Bot. 34:415—420.

Jacobson, L., R. Overstreet, H. M. King, and R. Handley. 1950. A study of potassium absorption by barley roots. Plant Physiol. 25:639—647.

Jacobson, L., D. P. Moore, and R. J. Hannapel. 1960. Role of calcium in absorption of monovalent cations. Plant Physiol. 35:352—358.

Jacobson, L., R. J. Hannapel, D. P. Moore, and M. Schaedle. 1961. Influence of calcium on selectivity of ion absorption process. Plant Physiol. 36:58—65.

Jacoby, B. 1965. Sodium retention in excised bean stems. Physiol. Plantarum 18: 730—739.

Jacoby, B., and O. E. Plessner. 1970. Some aspects of chloride absorption by bean leaf tissue. Ann. Bot. 34:177—183.

Jacoby, B., and J. F. Sutcliffe. 1962. Connection between protein synthesis and salt absorption in plant cells. Nature 195:1014.

Jagendorf, A. T. 1967. The chemiosmotic hypothesis of photophosphorylation, pp. 69—78. In A. S. Pietro, F. A. Greer, and T. J. Army (eds.) Harvesting the sun — Photosynthesis in plant life. Academic, New York.

Jennings, D. H. 1963. The absorption of solutes by plant cells. Iowa State Univ. Press, Ames.

Jennings, D. H. 1967. Electrical potential measurements, ion pumps and root exudation — A comment and a model explaining cation selectivity by the root. New Phytol. 66:357—369.

Jenny, H., and R. Overstreet. 1939. Cation interchange between plant roots and soil colloids. Soil Sci. 47:257—272.

Jeschke, W. D. 1970. The influx of potassium ions in leaves of Elodea densa, dependence on light, potassium concentration, and temperature. Planta 91:111—128.

Kahn, J. S., and J. B. Hanson. 1957. The effect of calcium on potassium accumulation in corn and soybean roots. Plant Physiol. 32:312—316.

Kannan, S. 1971. Kinetics of iron absorption by excised rice roots. Planta 96:262—270.

Kenefick, D. G., and J. B. Hanson. 1966. Contracted state as an energy source for Ca binding and Ca + inorganic phosphate accumulation by corn mitochondria. Plant Physiol. 41:1601—1609.

Ketchum, B. H. 1939. The absorption of phosphate and nitrate by illuminated cultures of Nitzschia closterium. Amer. J. Bot. 26:399—407.

Lansing, A. I., and T. B. Rosenthal. 1952. The relationship between ribonucleic acid and ion transport across the cell surface. J. Cellular Comp. Physiol. 40:337—345.

Laties, G. G. 1959a. Active transport of salt into plant tissue. Annu. Rev. Plant Physiol. 10:87—112.

Laties, G. G. 1959b. The generation of latent-ion-transport capacity. Proc. Nat. Acad. Sci. 45:163—172.

Laties, G. G. 1969. Dual mechanisms of salt uptake in relation to compartmentation and long-distance transport. Annu. Rev. Plant Physiol. 20:89—116.

Laties, G. G., I. R. MacDonald, and J. Dainty. 1964. Influence of the counter-ion on the absorption isotherm for chloride at low temperature. Plant Physiol. 39:254—262.

Lazaroff, N., and M. G. Pitman. 1966. Calcium and magnesium uptake by barley seedlings. Australian J. Biol. Sci. 19:991—1005.

Leggett, J. E. 1968. Salt absorption by plants. Annu. Rev. Plant Physiol. 19:333—346.

Leggett, J. E., and E. Epstein. 1956. Kinetics of sulfate absorption by barley roots. Plant Physiol. 31:222—226.

Leggett, J. E., R. A. Galloway, and H. G. Gauch. 1965. Calcium activation of ortho-phosphate absorption by barley roots. Plant Physiol. 40:897—902.

Lowenhaupt, B. 1956. The transport of calcium and other cations in submerged aquatics. Biol. Rev. 31:371—395.

Lundegårdh, H. 1940a. Salt absorption of plants. Nature 145:114—115.

Lundegårdh, H. 1940b. Investigations as to the absorption and accumulation of inorganic ions. Ann. Roy. Agr. Coll. Sweden 8:233.

Lundegårdh, H. 1942. Electrochemical relations between the root system and the soil. Soil Sci. 54:177–189.

Lundegådh, H. 1945. Absorption, transport and exudation of inorganic ions by the roots. Arkh. Bot. 32A(12):1–139.

Lundegårdh, H. 1949a. Growth, bleeding, and salt absorption of wheat roots as influenced by substances which interfere with glycolysis and aerobic respiration. Ann. Roy. Agr. Coll. Sweden 16:339–371.

Lundegårdh, H. 1949b. Quantitative relations between respiration and salt absorption. Ann. Roy. Agr. Coll. Sweden 16:372–403.

Lundegårdh, H. 1950. The translocation of salts and water through wheat roots. Physiol. Plantarum 3:103–151.

Lundegårdh, H. 1954. Anion respiration – The experimental basis of and theory of absorption, transport, and exudation of electrolytes by living cells and tissues. Symp. Soc. Exp. Biol. 8:262–296.

Lundegårdh, H. 1955. Mechanisms of absorption, transport, accumulation, and secretion of ions. Annu. Rev. Plant Physiol. 6:1–24.

Lundegårdh, H., and H. Burström. 1933a. Untersuchungen über die Salzaufnahme der Pflanzen. III. Quantitative Beziehungen zwischen Atmung und Anionenaufnahme. Biochem. Z. 261:235–251.

Lundegårdh, H., H. Burström. 1933b. Atmung und Ionenaufnahme. Planta 18:683–699.

Luttge, U., and G. G. Laties. 1966. Dual mechanisms of ion absorption in relation to long distance transport in plants. Plant Physiol. 41:1531–1539.

Maas, E. V. 1969. Calcium uptake by excised maize roots and interactions with alkali cations. Plant Physiol. 44:985–989.

Maas, E. V., and G. Ogata. 1971. Absorption of magnesium and chloride by excised corn roots. Plant Physiol. 47:357–360.

Maas, E. V., D. P. Moore, and B. J. Mason. 1968. Manganese absorption by excised barley roots. Plant Physiol. 43:527–530.

Maas, E. V., D. P. Moore, and B. J. Mason. 1969. Influence of calcium and magnesium on manganese absorption. Plant Physiol. 44:796–800.

MacDonald, I. R., and G. G. Laties. 1963. Kinetic studies of anion absorption by potato slices at $0°$ C. Plant Physiol. 38:38–44.

Macklon, A. E. S. 1970. Electrochemical aspects of ion transport in plants. J. Sci. Food Agr. 21(4):178–181.

Macklon, A. E. S., and N. Higinbotham. 1970. Active and passive transport of potassium in cells of excised pea epicotyls. Plant Physiol. 45:133–138.

MacRobbie, E. A. C. 1962. Ionic relations of *Nitella translucens*. J. Gen. Physiol. 45:861–878.

MacRobbie, E. A. C., and J. Dainty. 1958. Ion transport in *Nitellopsis obtusa*. J. Gen. Physiol. 42:335–353.

Marinos, N. G. 1962. Studies on submicroscopic aspects of mineral deficiencies. I. Calcium deficiency in the shoot apex of barley. Amer. J. Bot. 49:834–841.

Marschner, H. 1964. Einfluss von Calcium auf die Natriumaufnahme und die Kaliumabgabe isolierter Gerstenwurzeln. Z. Pflanzenernähr. Düng. Bodenk. 107:19–32.

Minchin, F. R., and D. A. Baker. 1969. Water dependent and water independent fluxes of potassium in exuding root systems of *Ricinus communis*. Planta 89:212–223.

Mitchell, P. 1966. Chemiosmotic coupling in oxidative and photosynthetic phosphorylation. Biol. Rev. 41:445–502.

Moore, D. P., L. Jacobson, and R. Overstreet. 1961a. Uptake of calcium by excised barley roots. Plant Physiol. 36:53–57.

Moore, D. P., R. Overstreet, and L. Jacobson. 1961b. Uptake of magnesium and its interaction with calcium in excised barley roots. Plant Physiol. 36:290–295.

Moore, D. P., B. J. Mason, and E. V. Maas. 1965. Accumulation of calcium in exudate of individual barley roots. Plant Physiol. 40:641–644.

Münch, E. 1927. Versuche über den Saftkreislauf. Ber. Deut. Bot. Ges. 45:340–356.

Nissen, P., and A. A. Benson. 1964. Active transport of choline sulfate by barley roots. Plant Physiol. 39:586–589.

Noggle, J. C., C. T. de Wit, and A. L. Fleming. 1964. Interrelationships of calcium and rubidium absorption by excised roots of barley and plantain. Proc. Soil Sci. Soc. Amer. 28:97–100.

Oertli, J. J. 1967. The salt absorption isotherm. Physiol. Plantarum 20:1014–1026.

Ordin, L., and L. Jacobson. 1955. Inhibition of ion absorption and respiration in barley roots. Plant Physiol. 30:21–27.

Osmond, C. B. 1966. Divalent cation absorption and interaction in *Atriplex*. Australian J. Biol. Sci. 19:37–48.

Osmond, C. B., and G. G. Laties. 1968. Interpretation of the dual isotherm for ion absorption in beet tissue. Plant Physiol. 43:747–755.

Osterhout, W. J. V. 1930. The accumulation of electrolytes. II. Suggestions as to the nature of accumulation in *Valonia*. J. Gen. Physiol. 14:285–300.

Osterhout, W. J. V. 1935. How do electrolytes enter the cell? Proc. Nat. Acad. Sci. 21:125–132.

Osterhout, W. J. V. 1947. The absorption of electolytes in large plant cells. II. Bot. Rev. 13:194–215.

Overstreet, R., and L. A. Dean. 1951. The availability of soil anions, pp. 79–105. *In* E. Troug (ed.), Mineral nutrition of plants. Univ. Wisconsin Press, Madison.

Overstreet, R., and L. Jacobson. 1952. Mechanisms of ion absorption by roots. Annu. Rev. Plant Physiol. 3:189–206.

Overstreet, R., S. Ruben, and T. C. Broyer. 1940. The absorption of bicarbonate ion by barley plants as indicated by studies with radio-active carbon. Proc. Nat. Acad. Sci. 26:688–695.

Overstreet, R., L. Jacobson, and R. Handley. 1952. The effect of calcium on the absorption of potassium by barley roots. Plant Physiol. 27:583–590.

Packer, L., and A. R. Crofts. 1967. The energized movement of ions and water by chloroplasts, pp. 23–64. *In* D. R. Sanadi (ed.) Current topics in bioenergetics. Academic, New York.

Page, E. R., and J. Dainty. 1964. Manganese uptake by excised oat roots. J. Exp. Bot. 15:428–443.

Pardee, A. B. 1966. Purification and properties of a sulfate-binding protein from *Salmonella typhimurium*. J. Biol. Chem. 241:5886–5892.

Pardee, A. B. 1968. Membrane transport proteins. Science 162:632–637.

Peirson, D. R., and E. B. Dumbroff. 1969. Probable sites for passive movement of ions across the endodermis. Plant Physiol. 44(Suppl.):no. 104.

Pfeffer, W. 1900. The physiology of plants, Vol. I. Transl. and edited by A. J. Ewart, Oxford Univ. Press, Oxford.

Pickering, D. C., and I. L. Puia. 1969. Mechanism for the uptake of zinc by *Fontinalis antipyretica*. Physiol. Plantarum 22:653–661.

Pierce, W. S., and N. Higinbotham. 1970. Compartments and fluxes of K^+, Na^+ and Cl^- in *Avena* coleoptile cells. Plant Physiol. 46:666–673.

Pitman, M. G. 1964. The effect of divalent cations on the uptake of salt by beetroot. J. Exp. Bot. 15:444–456.

Pitman, M. G. 1965. The location of the Donnan free space in disks of beetroot tissue.

Australian J. Biol. Sci. 18:547–553.

Pitman, M. G. 1969. Stimulation of Cl⁻ uptake by low-salt barley roots as a test of models of salt uptake. Plant Physiol. 44:1417–1427.

Pitman, M. G., and H. D. W. Saddler. 1967. Active sodium and potassium transport in cells of barley roots. Proc. Nat. Acad. Sci. 57:44–49.

Pitman, M. G., A. C. Courtice, and B. Lee. 1968. Comparison of potassium and sodium uptake by barley roots at high and low salt status. Australian J. Biol. Res. 21: 871–881.

Pitman, M. G., S. M. Mertz, Jr., J. S. Graves, W. S. Pierce, and N. Higinbotham. 1971. Electrical potential differences in cells of barley roots and their relation to ion uptake. Plant Physiol. 47:76–80.

Poole, R. J. 1966. The influence of the intracellular potential on potassium uptake by beetroot tissue. J. Gen. Physiol. 49:551–563.

Poole, R. J. 1969. Carrier-mediated potassium efflux across the cell membrane of red beet. Plant Physiol. 44:485–490.

Post, R. L., S. Kume, T. Tobin, B. Orcutt, and A. K. Sen. 1969. Flexibility of an active center in sodium-plus-potassium adenosine triphosphatase, pp. 306–326. *In* Membrane Porteins (Proc. Symp. sponsored by the New York Heart Ass.). Little, Brown, Boston.

Price, C. A. 1970. Molecular approaches to plant physiology. McGraw-Hill, New York, 398 pp.

Rains, D. W. 1969a. Sodium and potassium absorption by bean stem tissue. Plant Physiol. 44:547–554.

Rains, D. W. 1969b. Cation absorption by slices of stem tissue of bean and cotton. Experientia 25:215–216.

Rains, D. W., and E. Epstein. 1965. Transport of sodium in plant tissue. Science 148: 1611.

Rains, D. W., and E. Epstein. 1967a. Sodium absorption by barley roots: Role of the dual mechanisms of alkali cation transport. Plant Physiol. 42:314–318.

Rains, D. W., and E. Epstein. 1967b. Sodium absorption by barley roots: Its mediation by mechanism 2 of alkali cation transport. Plant Physiol. 42:319–323.

Rains, D. W., W. E. Schmid, and E. Epstein. 1964. Absorption of cations by roots. Effects of hydrogen ions and essential role of calcium. Plant Physiol. 39:274–278.

Roberts, R. B., I. Roberts, and D. B. Cowie. 1949. Potassium metabolism in *Escherichia coli*. II. Metabolism in the presence of carbohydrates and their metabolic derivatives. J. Cellular Comp. Physiol. 34:259–292.

Robertson, R. N. 1951. Mechanism of absorption and transport of inorganic nutrients in plants. Annu. Rev. Plant Physiol. 2:1–24.

Robertson, R. N. 1960. Ion transport and respiration. Biol. Rev. 35:231–264.

Robertson, R. N. 1968. Protons, electrons, phosphorylation, and active transport. Cambridge Monogr. Exp. Biol., no. 15. Cambridge Univ. Press, Cambridge. 96 pp.

Robertson, R. N., and M. J. Wilkins. 1948. Studies in the metabolism of plant cells. VII. The quantitative relation between salt accumulation and salt respiration. Australian J. Sci. Res. B1:17–37.

Robertson, R. N., M. J. Wilkins, and D. C. Weeks. 1951. Studies in the metabolism of plant cells. IX. The effect of 2,4-dinitrophenol on salt accumulation and salt respiration. Australian J. Sci. Res. B4:248–264.

Robinson, J. B., and F. A. Smith. 1970. Chloride influx into citrus leaf slices. Australian J. Biol. Sci. 23:953–960.

Rosenberg, T. 1948. On accumulation and active transport in biological systems. Acta Chem. Scand. 2:14–33.

Russell, R. S. 1954. The relationship between metabolism and the accumulation of ions by plants. Symp. Soc. Exp. Biol. 8:343–366.

Salisbury, F. B., and C. Ross. 1969. Plant physiology. Wadsworth, Belmont, Calif. 747 pp.

Schaedle, M., and L. Jacobson. 1965. Ion absorption and retention by *Chlorella pyrenoidosa*. I. Absorption of potassium, Plant Physiol. 40:214–220.

Schmid, W. E., H. P. Haag, and E. Epstein. 1965. Absorption of zinc by excised barley roots. Physiol. Plantarum 18:860–869.

Scott, B. I. H., H. Gulline, and C. K. Pallaghy. 1968. The electrochemical state of cells of broad bean roots. I. Investigations of elongating roots of young seedlings. Australian J. Biol. Sci. 21:185–200.

Scott, G. T., and H. R. Hayward. 1954. Evidence for the presence of separate mechanisms regulating potassium and sodium distribution in *Ulva lactuca*. J. Gen. Physiol. 37:601–620.

Shone, M. G. T. 1970. Physico-chemical aspects and ion uptake by plants. Sci. Progress 58(230):183–201.

Shrift, A. 1954. Sulfur-selenium antagonism. I. Antimetabolite action of selenate on the growth of *Chlorella vulgaris*. Amer. J. Bot. 41:223–230.

Shrift, A., and J. M. Ulrich. 1969. Transport of selenate and selenite into *Astragalus* roots. Plant Physiol. 44:893–896.

Skou, J. C. 1957. The influence of some cations on an adenosine triphosphatase from peripheral nerves. Biochim. Biophys. Acta 23:394–401.

Smith, R. C., and E. Epstein. 1964. Ion absorption by shoot tissue: Kinetics of potassium and rubidium absorption by corn leaf tissue. Plant Physiol. 39:992–996.

Steveninck, R. F. M. van. 1965. The significance of calcium on the apparent permeability of cell membranes and the effects of substitution with other divalent ions. Physiol. Plantarum 18:54–69.

Steward, F. C., and F. K. Millar. 1954. Salt accumulation in plants: A reconsideration of the role of growth and metabolism. A. Salt accumulation as a cellular phenomenon. Symp. Soc. Exp. Biol. 8:367–393.

Steward, F. C., and G. Preston. 1940. Metabolic processes of potato discs under conditions conducive to salt accumulation. Plant Physiol. 15:23–61.

Steward, F. C., and H. E. Street. 1947. The nitrogenous constituents of plants. Annu. Rev. Biochem. 16:471–502.

Steward, F. C., and J. F. Sutcliffe. 1959. Plants in relation to inorganic salts, pp. 253–478. *In* F. C. Steward (ed.) Plant physiology – A treatise, Vol. II. Academic, New York.

Stoner, C. D., T. K. Hodges, and J. B. Hanson. 1964. Chloramphenicol as an inhibitor of energy-linked processes in maize mitochondria. Nature 203:258–261.

Sutcliffe, J. F. 1954. The exchangeability of potassium and bromide ions in cells of red beetroot tissue. J. Exp. Bot. 5:313–326.

Sutcliffe, J. F. 1959. Salt uptake in plants. Biol. Rev. 34:159–218.

Sutcliffe, J. F. 1960. New evidence for a relationship between ion absorption and protein turnover in plant cells. Nature 188:294–297.

Sutcliffe, J. F. 1962. Mineral salts absorption in plants. Pergamon, Oxford. 194 pp.

Sutcliffe, J. F., and D. P. Hackett. 1957. Efficiency of ion transfer in biological systems. Nature 180:95–96.

Tadano, T., J. H. Baker, and M. Drake. 1969. Role of the accompanying anion in the effect of calcium salts on potassium uptake by excised barley roots. Plant Physiol. 44:1639–1644.

Tanada, T. 1955. Effects of ultraviolet radiation and calcium and their interaction on salt absorption by excised mung bean roots. Plant Physiol. 30:221–225.

Tanada, T. 1956. Effect of ribonuclease on salt absorption by excised mung bean roots. Plant Physiol. 31:251–253.

Tanada, T. 1962. Localization and mechanism of calcium stimulation of rubidium absorption in the mung bean root. Amer. J. Bot. 49:1068—1072.

Tanada, T. 1963. Kinetics of Rb absorption by excised barley roots under changing Rb concentrations. I. Effects of other cations on Rb uptake. Plant Physiol. 38:422—425.

Tanada, T. 1964. Kinetics of Rb absorption by excised barley roots under changing Rb concentrations. II. An interpretation of kinetic data. Plant Physiol. 39:593—597.

Thellier, M. 1970. Electrokinetic interpretation of the functioning of biological systems and its application to the study of mineral salts absorption. Ann. Bot. 34:983—1009.

Torii, K., and G. G. Laties. 1966a. Dual mechanisms of ion uptake in relation to vacuolation in corn roots. Plant Physiol. 41:863—870.

Torii, K., and G. G. Laties. 1966b. Organic acid synthesis in response to excess cation absorption in vacuolate and non-vacuolate sections of corn and barley roots. Plant Cell Physiol. 7:395—403.

Trelease, S. F., and O. A. Beath. 1949. Selenium. Publ. by the authors, New York, 292 pp.

Truelove, B., and J. B. Hanson. 1966. Calcium-induced phosphate uptake in contracting corn mitochondria. Plant Physiol. 41:1004—1013.

Ussing, H. H. 1949. The distinction by means of tracers between active transport and diffusion. Acta Physiol. Scand. 19:43.

Ussing, H. H. 1957. General principles and theories of membrane transport, pp. 39—56. *In* Q. R. Murphy (ed.) Metabolic aspects of transport across cell membranes. Univ. Wisconsin Press. Madison.

Viets, F. G. 1944. Calcium and other polyvalent cations as accelerators of ion accumulation by excised barley roots. Plant Physiol. 19:466—480.

Vose, P. B. 1963. Varietal differences in plant nutrition. Herbage Abstr. 33:1—13.

Waisel, Y. 1962. The effect of Ca on the uptake of monovalent ions by excised barley roots. Physiol. Plantarum 15:709—724.

Waisel, Y., R. Neumann, and Z. Kuller. 1970. Selectivity and ion transport in excised bean hypocotyls. Physiol. Plantarum 23:955—963.

Walker, N. A. 1955. Microelectrode experiments on *Nitella*. Australian J. Biol. Sci. 8:476—489.

Wanner, H. von, 1948a. Untersuchungen über die Temperaturabhängigkeit der Salzaufnahme durch Pflanzenwurzeln. I. Die relative Grösse der Temperaturkoeffizienten (Q_{10}) von Kationen- und Anionenaufnahme. Ber. Schweiz. Bot. Ges. 58:123—130.

Wanner, H. von. 1948b. Untersuchungen über die Temperaturabhängigkeit der Salzaufnahme durch Pflanzenwurzeln. II. Die Temperaturkoeffizienten von Kationen- und Anionenaufnahme in Abhängigkeit von der Salzkonzentration. Ber. Schweiz. Bot. Ges. 58:383-390.

Weigl, J. 1963. Die Bedeutung der energiereichen Phosphate bei der Ionenaufnahme durch Wurzeln. Planta 60:307—321.

Weigl, J. 1964. Zur Hemmung der aktiven Ionenaufnahme durch Arsenat. Z. Naturforsch. 19b:646—648.

Weissman, G. S., and F. S. Trelease. 1955. Influence of sulfur on the toxicity of selenium to *Aspergillus*. Amer. J. Bot. 42:489—495.

Welch, R. M., and E. Epstein. 1968. The dual mechanisms of alkali cation absorption by plant cells: Their parallel operation across the plasmalemma. Proc. Nat. Acad. Sci. 61:447—453.

Welch, R. M., and E. Epstein. 1969. The plasmalemma: Seat of the type 2 mechanisms of ion absorption. Plant Physiol. 44:301—304.

Wit, C. T. de, W. Dijkshoorn, and J. C. Noggle. 1963. Ionic balance and growth of plants. Verslad. Landbou. Onderzoek 69.15. 68 pp.

Yoshida, F. 1964. Thesis. State Agr. Univ., Wageningen, Netherlands. (Original not seen; title not available.)

CHAPTER 6

Factors Affecting Absorption

FACTORS

Mass Action

As shown by the distribution of Rb^+ and Na^+ in baker's yeast (*Saccharomyces cerevisiae*), Leggett et al. (1965b) concluded that the principles of mass action are applicable in salt uptake regardless of the kinetics and selectivities.

Portion of Root Tip

In evaluating salt absorption it is imperative that one consider the detailed anatomy of the root tip (Fig. 6.1). It is interesting to note that phloem elements are present very near the tip of the root, whereas xylem elements are differentiated some distance (approximately 500 μ) from the tip (Esau, 1941, 1943). Thus there is provision for bringing an energy supply from the tops to the very tip of the root, but not for carrying absorbed salts away from this region. Owing to the high metabolic activity and the lack of differentiated xylem elements, approximately the first millimeter of the root behind the tip may absorb a large quantity of salt, but usually there is little or no transport away from this area (Brouwer, 1954; Wiebe and Kramer, 1954). For barley roots there was greater translocation to the tops of ions absorbed approximately 27–30 mm from the tip of the root than from the first few millimeters of the root (Wiebe and Kramer, 1954). In contrast with most reports, it was noted that ^{86}Rb absorbed by the terminal 5 mm of intact bean roots was translocated to other parts of the plant even though this apical portion of the root was de-

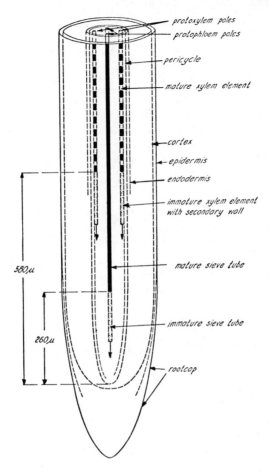

protoxylem poles
protophloem poles
pericycle
mature xylem element
cortex
epidermis
endodermis
immature xylem element
with secondary wall
mature sieve tube
immature sieve tube
rootcap

580μ

260μ

Fig. 6.1
Diagram of a tobacco root tip showing spatial relations of different regions of the root and of the first vascular elements. The endodermis, as a cell layer, becomes defined before the sieve tubes differentiate but develops Casparian strips slightly below the level where the first xylem elements mature. (In this diagram the endodermis is indicated only at the level where Casparian strips were present.) (Esau, 1941.)

void of functional xylem tissue (Steward and Koontz, 1968). Wiebe and Kramer (1954) suggested that apically accumulated salts would be translocated once the erstwhile apical portion was no longer apical but instead had become part of the differentiated and matured portion of the root.

As with Na^+, Ca^{2+}, and Cl^-, uptake of Sr^{2+} is a nonmetabolic process in the meristematic portion of the root tip, but nevertheless depends on metabolism farther back in the vacuolated cells (Handley and Overstreet, 1963).

A decreasing rate of net accumulation of $^{36}Cl^-$ occurs with time and also a reduced uptake and transport in high-Cl onion roots. Hodges and Vaadia (1964) noted that the terminal 3 cm was the most active in $^{36}Cl^-$ accumulation and transport to the exudate.

Salt accumulation occurs most actively near the root tip (0–15 mm), but maximal translocation across the root takes place some distance from the apex (25–60 mm). Sutcliffe (1959) regarded the plasmalemma as a cation-permeable, anion-impermeable membrane and concluded that cations could move passively into the cytoplasm.

Xylem exudates were collected at hourly intervals from segments excised from two portions of the primary root of corn *(Zea mays* L.) partially immersed in salt solution containing ^{86}Rb (Smith, 1970). Basal segments exudated earlier than did apical segments, and the former reached a steady state sooner than the apical ones. However, during their respective steady states, apical segments produced three times the volume of exudate per hour and translocated eight times as much Rb^+ per hour as did basal segments. These differences in the time course of exudation and the large differences in output were interpreted as indicating two independent systems of ion transport operating simultaneously in intact roots.

Metabolic State or Past History of Roots

Past history of roots, prior to their use in salt absorption studies, has a profound bearing on the experimental results, hence the conclusions that may be drawn (Pitman, 1969). There is general agreement that roots with the greatest potential for salt absorption are those possessing a high-carbohydrate and low-salt status (Broyer, 1951; Russell and Barber, 1960).

Barley roots showed a strong selectivity for K^+ over Na^+ uptake whether grown in aerated or unaerated nutrient solution (Pitman, 1969). However, roots from nonaerated $CaSO_4$ had a stronger selectivity for K^+ over Na^+ than did those from an aerated $CaSO_4$ solution.

As for metabolic status, including the internal concentration of salts, barley roots grown on dilute $CaSO_4$ are low-salt roots (Pitman, 1969). High-salt barley roots show a stronger preference for K^+ over Na^+ than do low-salt roots (Pitman, 1965; Pitman et al., 1968).

Previously absorbed K^+ in plants caused a marked reduction in the short-term influx of ^{86}Rb-labeled K^+ into roots of barley seedlings (Johansen et al., 1970). Influx values agreed with net K^+ absorption rates into intact plants—suggesting that K^+ efflux was negligible in comparison with influx. Earlier interpretations of a large K^+ efflux component from excised roots, approaching equilibrium K^+ concentrations, are considered to result from an underestimation of net K^+ absorption rates—resulting from xylem exudation as the K^+ status of the roots increased.

Using low-salt barley roots, the net loss of H^+ was studied by Pitman, (1970) during salt absorption. Comparison of rates of loss from roots in different concentrations of KCl showed that H^+ loss increased in the same way as the mechanism-2 component of salt uptake. This H^+ loss appeared to be coupled to salt absorption, and was not a result of increased respiration or metabolic breakdown of sugars. In view of the large negative potential of the cells (-60 mV), this investigator suggested that the H^+ loss was attributable to an outward proton transport process. Further, he suggested that H^+ release may be underestimated owing to diffusion of HCO_3^- out of the cells with HCO_3^- dissociating into OH^- and CO_2 in the solution.

Respiration

Numerous workers have reported a positive relationship between rates of aerobic respiration and salt absorption (Hoagland and Broyer, 1936; Arrington and Shive, 1936; Broyer and Overstreet, 1940; Pepkowitz and Shive, 1944). There is also a direct relationship between utilization (i.e., loss) of sugars and absorption of cations and anions (Hoagland and Broyer, 1936).

In a study of phosphorylation by mitochondria and absorption of phosphate by barley roots, Jackson et al. (1962) observed that absorption of ions in general, and of phosphate in particular, occurs against concentration gradients so great as to require free energy changes of the order of 8000 cal/mole. They noted that the rate-limiting reactions for esterification of phosphate by mitochondria and for phosphate uptake by barley roots were identical; the mitochondrion was the site of phosphate uptake. Their kinetic analyses indicated that the sites of phosphorylation and phosphate absorption, respectively, are those of oxidative phosphorylation coupled to cytochrome b and diphosphopyridine nucleotide oxidation or reduction in the respiratory chain.

Phosphate uptake by Baker's yeast, *S. cerevisiae*, had three components (Leggett, 1961). The minor one was associated with oxidative phosphorylation coupled to cytochrome b. The other two components were associated with oxidative phosphorylation coupled to the action of glyceraldehyde-3-phosphate dehydrogenase.

The energy-dependent influx of Rb^+ into excised roots of corn, wheat, and barley was determined and compared with the Rb^+-stimulated ATPase activity of membrane fractions obtained from root homogenates of these species (Fisher et al., 1970). External Rb^+ concentrations were in the range of 1–50 mM. The ratio of Rb^+ influx/Rb^+-stimulated ATPase was approximately 0.85 and was nearly constant for all species and Rb^+ concentrations. The correlation coefficient for Rb^+ influx versus Rb^+-activated ATPase was 0.94. The results support the concept that ATP is the energy source for ion transport in roots, and that an ATPase participates in the energy transduction process involved in energy-dependent ion transport.

By heating, freezing, or prolonged exposure to DNP, it was shown that there is an energy-requiring step in the transport of Ca^{2+} from the external solution to xylem vessels of barley roots (Barber and Koontz, 1963).

Yeast cells appear to have an efficient energy-releasing glycolytic system, since uptake of Zn^{2+}, Co^{2+}, and Ni^{2+} (Fuhrmann and Rothstein, 1968), P (Leggett, 1961), and Rb^+ (Leggett, et al., 1962) was the same under aerobic and anaerobic conditions. In many ways yeast cells act as cation exchange resins during cation binding by the cells (Leggett et al., 1965b). The rate of cation entry into yeast cells was in agreement with a diffusion-limited, passive exchange process (Leggett et al., 1962, 1965b).

Temperature

Absorption of NO_3^- was shown to be affected much more by temperature than certain other ions; almost no NO_3^- was absorbed at 13°C (Williams and Vlamis, 1962). The decrease in NO_3^- absorption at the lower temperature was so much greater than that observed for any other ion that a secondary mechanism was suggested as the limiting factor in NO_3^- absorption, namely, that the reductase enzyme system is temperature dependent and reacts very slowly at 13°C. There would thus be a failure of NO_3^- utilization, and the accumulation of NO_3^- in the tissues would further reduce its absorption.

Increases in root temperature (12, 18, 24, and 30°C) progressively increased uptake (and translocation) of Fe by tomato plants (Riekels and Lingle, 1966). As Mn^{2+} was increased in the nutrient solution, ^{59}Fe uptake and translocation were increased to a point and then decreased.

A soil-temperature-dependent plant disorder was observed in maize shoots with symptoms resembling Ca^{2+} deficiency (Walker, 1969). Plants grown at 21°C appeared normal; however, plants grown at higher soil temperatures up to 35°C had a progressively greater tendency for emerging leaves to stick together and remain rolled. Blades of older leaves were torn, notched, and twisted. Chang et al. (1968) reported a similar finding for tobacco.

For 16 out of 17 elements studied, Walker (1969) reported that calculated total uptakes increased with increased soil temperature to maximal values between 26 and 34°C—depending on the element. Uptake of B, however, was uniquely different. Total B in shoots did not vary appreciably in plants grown at soil temperatures from 12 to 20°C, even though shoot dry weight increased about sevenfold. B uptake then increased from approximately 190 to 1900 μg per pot in plants grown at soil temperatures from 20 to 31°C. Concurrently, B concentration in shoots decreased from 91 to 16 ppm between 12 and 20°C, and then increased up to 272 ppm at 35°C.

In some soil temperature studies in the past, the desired soil temperatures were not attained—at least not uniformly throughout the mass of the soil. Recognizing the difficulty of achieving a given temperature throughout the soil mass, Walker (1967) devised a special apparatus to achieve this goal and proved the reliability of the apparatus by thermocouples at various locations in the soil.

Aeration

Under low O_2 conditions barley roots become adapted by development of an extensive air path which facilitates gaseous exchange with the air via intercellular spaces in the shoot (Pitman, 1969). Barley seedlings are thus able to grow whether the solution is aerated or not (Bryant, 1934; Heide et al., 1963; Vlamis and Davis, 1944).

Hopkins et al. (1950), working with tomato, soybean, and tobacco, reported that ion accumulation occurred with as low as 0.5% O_2 in the gas mixture around roots; in general, accumulation of ions increased as O_2 concentration was increased up to 6.4%. Phosphate absorption was independent of O_2 tension over the range 3–100%; at an O_2 tension of 0.3%, the phosphate absorption rate was half-maximal (Hopkins, 1956).

Leonard and Pinckard (1946) concluded that perhaps translocation of O_2 from the tops to the roots enabled cotton plants to withstand anaerobic soil conditions. There may be a significant transfer of O_2 from the shoots to the roots under low soil O_2 conditions, since Pepkowitz and Shive (1944) reported appreciable absorption of Ca, P, and K by tomatoes and soybeans in a zero O_2 treatment.

Under anaerobic conditions soybean roots absorbed greater amounts of NO_3^- than under aerobic conditions (Gilbert and Shive, 1945). When roots are in an anaerobic environment, O_2 may be made available to the roots when NO_3^- is reduced (Arnon, 1937).

Two percent CO_2-in-air increased cation uptake by beet root tissue at pH 7.4, as compared with lower pH values (Poole and Poel, 1965).

One reason for an interest in movement of O_2 to the stele is the theory of root pressure exudation which assumes the existence of an O_2 deficiency in the stele. However, radial diffusion of O_2 into the stele of corn or jack bean roots was observed to be too slow to provide enough O_2 for a rate of respiration equal to that of well-aerated tissue (Fiscus and Kramer, 1970).

Moisture Stress

[32]P uptake by intact tomato plants was measured as a function of varying external water potentials created by mannitol (Greenway et al., 1969). Down to −5.4 atm the amount of labeled P in the roots remained constant, but the amount transported to shoots was reduced. Potentials of −10.4 atm reduced the amount of labeled P in both root and shoot. Results were the same after an immediately applied stress or if measured after water potentials were gradually lowered.

For tomato plants, lowering the water potential of culture solutions from -0.4 to -5.4 atm reduced both P and Br^- transport to the shoot, but the concentration in roots was unaffected (Greenway and Klepper, 1969). The main effect of water flow on anion transport to shoots occurred after ions had been actively absorbed by roots and was not attributable to mass flow increasing ion delivery to sites of active uptake. The amount of P transported was much less at low external water potential in the light than at high water potential in the dark—even when water flow was the same.

Chemicals

Efflux of Rb^+ from preloaded corn roots across the plasmalemma to root cortical cells was stimulated by 5×10^{-4} M UO_2^{2+}, whereas efflux of Rb^+ into xylem was inhibited (Weigl, 1970). Efflux of Cl^- across the plasmalemma was stimulated by 10^{-5} M CCCP (carbonylcyanid-m-chlorophenylhydrazone); efflux of Cl^- via the xylem was inhibited.

Diffusion of Ions and Mass Flow of Soil Solution

Movement of ions to the root surface is one important aspect of salt absorption according to Barber (1962). He pointed out that, depending on the nature of the ions and their relative concentrations, ions may move to the root surface primarily by diffusion (e.g., K^+) or primarily by mass flow of soil solution (e.g., Ca^{2+}) through the soil.

Of the three ways (Barber et al., 1963) that ions may reach roots—mass flow of soil solution, diffusion through the soil solution, and growth of roots into new regions of soil—the first two are effective with ions such as NO_3^- which exchange between colloids and the soil solution (Milthorpe and Moorby, 1969). It was noted that perhaps Ca^{2+} and Mg^{2+} could be supplied by mass flow of soil solution. Ions such as $H_2PO_4^-$, K^+, Mn^{2+}, Fe^{3+}, and Zn^{2+} were regarded as moving only very short distances, so that sustained root growth is required for their absorption. It has been reported (Lewis and Quirk, 1967) that P is absorbed only by the root hair zone. Soil type can alter the extent to which plant roots obtain ions by diffusion or by mass flow of soil solution (Barber et al., 1962).

Element Concentration

For a given element the steepness of the gradient of concentration, from outside to inside of the root cells, has an influence on the rate at which an element ordinarily enters the root. The steeper the gradient, the greater the *amount* that enters. There may, however, be a greater *percentage* uptake when the external concentration (of that element) is relatively low.

Uptake of monovalent cations by barley roots is largely independent of external concentration, whereas uptake of divalent cations is proportional to external concentration (Lazaroff and Pitman, 1966).

Using bean (*Phaseolus vulgaris*) plants, the uptake of Zn^{2+} and P were studied with intact leaves, enzymically isolated leaf cells, leaf discs, excised roots, and stem callus tissue (Rathmore et al., 1970). Rate of Zn^{2+} absorption was a function of external Zn^{2+} concentration and pH. Absorption of Zn^{2+} was not inhibited by respiratory inhibitors (DNP, azide, cyanide, and amytal), and was not light or temperature dependent. Q_{10} values for Zn^{2+} uptake ranged between 1.0 and 1.2. However, the uptake of P was temperature and light dependent, and drastically reduced by metabolic inhibitors.

With storage tissues and barley plants, Rb^+ uptake may, depending on the concentration of salts, be increased, decreased, or unaffected by the addition of Na^+ (Sutcliffe, 1959).

Presence or Absence of Tops (Transpiration)

Although research with excised root systems has elucidated many facets of salt absorption, *continued*, long-time absorption of salts may be quite different for intact plants than for excised roots. With intact plants ions may move out of the roots, through the stems, and on into the tops with varying degrees of rapidity. Removal of ions from roots may alter many considerations with regard to salt absorption by roots.

Solutes may accumulate in free space as water is selectively absorbed by cortical and endodermal cells during transpiration (Bernstein and Nieman, 1960). Calculations indicated that as much as a doubling in solute concentration could occur, and this may account for increased salt uptake by plants under conditions of high transpiration.

In a study of the kinetics of Rb^+ absorption and translocation by barley plants, the data were consistent with the existence of two active processes—one leading to accumulation of Rb^+ by the roots and the other to accumulation by the tops (Fried et al., 1961).

Stem-root grafts of seedling plants showed that genotypic differences in P, Mg, Mn, and B accumulations in soybean seeds are controlled by the scion of the plant (Kleese and Smith, 1970).

Hylmo (1953), Epstein (1960b), and Kramer (1957) agreed that there is an active accumulation of ions in roots, but a passive movement of ions through free space to the xylem and thence to shoots in the transpiration stream. However, with intact pea plants translocation of Fe to shoots was dependent on metabolic activity of root cells (Branton and Jacobson, 1962).

Inasmuch as salt absorption may be affected by water absorption, any factor affecting the latter could possibly alter salt absorption. In this connection,

Falk's (1966c) results are of interest. He studied the effect of rapidly increasing or decreasing the osmotic potential of the solution on water uptake and transpiration of wheat plants. Transpiration was measured by a microwave hygrometer (Falk, 1966a, b) and water uptake by a specially designed recording potometer. Rapidly increasing the osmotic potential caused a temporary increase and then a decrease in transpiration; water transport was immediately reduced but was followed by an increase. The course of events, after the rapid withdrawal of mannitol, were just the reverse of those following the addition of mannitol. The results indicated that the primary effect of rapidly changing the osmotic potential of the solution was localized at the root surface.

Plant physiologists have considered the relationship between rate of transpiration and uptake of salts; salts might enter passively with water absorbed by the plant. Some researchers found a relationship between rate of transpiration (or absorption of water) and salt absorption (Schmidt, 1936; Hylmo, 1953, 1955, 1958; Kramer, 1957; Russell and Barber, 1960; Lopushinsky, 1964), while others conceded that transpiration was effective in moving ions upward in the plant only after they had been metabolically delivered to the vascular tissue (Broyer and Hoagland, 1943; Brouwer, 1954, 1956; Honert, 1933; Honert, et al., 1955; Arisz, 1956; Russell and Shorrocks, 1957, 1959). There was no relationship between the transpiration rates of two barley genotypes and their accumulation of Sr^{2+} (Pinkas and Smith, 1966). Russell and Barber (1960) reviewed the evidence for and against a relationship between transpiration and salt absorption. They concluded that there is a site external to the vascular tissue which constitutes a barrier to the free movement of ions and makes it possible for ions to be much more concentrated within vascular tissue than in tissues adjacent to the stele. Further, transfer of ions across this barrier depends on an expenditure of metabolic energy.

When little water is being transpired, the concentration of ions in the transpiration stream may exceed that in the external medium by a factor of 100 or more according to Russell and Shorrocks (1959). They decided that ions are not transferred passively across the roots of intact plants. Pettersson (1960) noted that the retention of ions in roots was independent of transpirational intensity.

Hylmo (1955) offered a possible reconciliation of the evidence for and against a relationship between transpiration and salt absorption. He reported that for those ions not preferentially taken into the vaculoes of root cells there is a relationship between transpiration and salt translocation. For those ions not moved into vacuoles, an increase in transpiration might result in a greater uptake because of removal of ions from the roots to the shoots. For those ions accumulated preferentially in vacuoles of root cells, there is no relationship between transpiration and either absorption or translocation.

Lazaroff and Pitman (1966) reported that transpiration influences Ca^{2+}

absorption from high, but not from low external concentrations of Ca^{2+}. This might explain why high K^+ inhibited absorption of Ca^{2+} by *Atriplex spongiosa* at high external Ca^{2+} concentrations but not at low external Ca^{2+} concentrations (Osmond, 1966).

Uptake of Ca^{2+} into barley roots was unaffected by transpiration rate, but translocation of Ca^{2+}—and particularly of Mg^{2+}—to the tops was greater under a high than under a low rate of transpiration according to Lazaroff and Pitman (1966). They also noted that transpiration rate had very little effect on the translocation of Na^+ or K^+ to tops.

In the absorption of Ca^{2+} by bean plants, the xylem cylinder has been reported to operate essentially as an exchange column for Ca (Biddulph et al., 1961). Later, Bell and Biddulph (1963) suggested that this exchange mechanism for Ca^{2+} might operate to regulate the amount of Ca^{2+} reaching various plant parts. They concluded that "this permits the various tissues to acquire nutrient ions in proportion to their metabolic utilization rather than in proportion to their transpirational rates, as in the case with mass flow." Translocation of Ca^{2+} into the primary leaves of *P. vulgaris* was essentially independent of the transpiration rate at low (0.5 mM) Ca^{2+} levels (Koontz and Foote, 1966).

Light

Light versus darkness can have a profound effect on salt absorption by green tissues or cells (Lookeren Campagne, 1957; MacRobbie, 1962). Inasmuch as salt absorption is affected by metabolism, loss of an energy source in darkness could profoundly affect ion uptake. However, when P-deficient cells of *Nitzschia closterium* were previously exposed to light, they absorbed appreciable phosphate in the dark (Ketchum, 1939). In a study of the influence of light on phosphate metabolism of lettuce seed, Surrey and Gordon (1962) noted that under continuous irradiation red light accelerated and far red light depressed the uptake and esterification of phosphate.

The concept that high-energy phosphorylated compounds may be involved in ion uptake encompasses those systems in which the energy for transport is furnished not only through respiration but also through photophosphorylation (Lookeren Campagne, 1957; MacRobbie, 1962). Assuming that photophosphorylation may be involved in ion uptake, it is obvious that darkness would reduce ion absorption. There is some reason to believe that photophosphorylation may be important, rather than respiration, inasmuch as the sources for respiration would be essentially unchanged in concentration when darkness was first imposed.

Using radioisotopes, Nobel and Packer (1965) found that Ca^{2+}, Na^+, and PO_4^{3-} absorption by spinach chloroplasts was triggered by light and mediated by substrate and ATP; other nucleoside triphosphates were relatively ineffective. Light did not result in an appreciable uptake of ^{42}K, ^{86}Rb, ^{54}Mn, ^{65}Zn, ^{59}Fe, $^{35}SO_4$, ^{131}I, ^{82}Br, or ^{36}Cl, and these investigators therefore concluded that chloroplasts contain an energy-dependent light-stimulated mechanism for regulation of their ionic composition. Although Nobel and Packer (1965) observed no effect of light on Cl^- uptake, Lookeren Campagne (1957) reported that Cl^- absorption was light-dependent in *Vallisneria* leaves. Action spectra of Cl^- absorption and photosynthesis were identical; thus chlorophyll was implicated in the light-dependent Cl^- absorption. However, less light was required to saturate the light-dependent Cl^- uptake than was required for saturation of photosynthesis.

Light has been reported to increase the permeability of membranes, hence uptake of Na^+ (MacRobbie and Dainty, 1958), NO_3^- (Nagai and Tazawa, 1962), and other ions (Hope and Walker, 1960), and the loss of K^+ eightfold (MacRobbie and Dainty, 1958). Nagai and Tazawa (1962) reported that light increased the resting electropotential of cell membranes and that this change accelerated ion uptake. They divided the effect of light on ion uptake into two processes: (1) movement across the plasmalemma which is controlled by the potential gradient, and (2) movement across the cytoplasm and tonoplast which is controlled by photosynthesis.

Light of relatively low intensity increased absorption of K^+ by slices of corn leaf tissue (Rains, 1968). ATP was involved in K^+ absorption in light and in dark. In light, ATP came from cyclic photophosphorylation, and in the dark from oxidative phosphorylation. Thus in the light the ultimate source of energy for active ion uptake involved a process other than one linked directly with transfer of electrons in the mitochondrial chain. The light–response curve was very similar to the one obtained by Lookeren Campagne (1957) for Cl^- absorption by *Vallisneria*. Rains (1968) concluded, however, that uptake of K^+ was independent of absorption of Cl^-, since Smith and Epstein (1964) had shown that absorption of K^+ by corn leaf slices was independent of the accompanying anion—Cl^- or SO_4^{2-}.

In *Nitella clavata*, Na^+ influx was strongly correlated with light intensity even though light had no effect on membrane potential or ionic composition of the vacuolar sap as observed by Barr and Broyer (1964). They were unable, however, to distinguish between passive movement of Na^+ into the cell as accelerated by a light-induced increase in permeability and an active transport mechanism operative in the same direction as the electrochemical gradient.

In *Chlorella* and *Ankistrodesmus*, light stimulated the assimilation of NO_3^- and NO_2^- in the absence of CO_2 under both aerobic and anaerobic conditions (Morris and Ahmed, 1969). This finding is of interest in connection with studies with isolated chloroplasts, which have shown that NO_3^- and NO_2^- can replace CO_2 as the ultimate acceptor for the photochemically produced reducing potential.

There was a light-dependent K^+ uptake into chopped pea leaves of 7.5 μmoles/g fresh weight per hour when the external bathing solution contained 5 mM KCl—the rate increasing to 56 μmoles/g fresh weight per hour when HCO_3^- replaced Cl^- (Nobel, 1969). Stimulation of K^+ uptake by HCO_3^- appears to represent a nonspecific anion effect that apparently depends on the pK_a of the acid (Nobel, 1970).

Light causes the uptake of H^+ in the chloroplast, and some ion moves to balance the charge as a consequence of this event—resulting in either an efflux of charged cations or an uptake of charged anions, or some combination between these two charges occurs to balance charge neutrality (Packer et al., 1970). An increase in concentration of associated acids occurs inside the chloroplast.

In leaves of *Elodea densa,* light increases K^+ influx by a factor of 30 to 50 as compared to dark (Jeschke, 1970). Influx was inhibited by uncoupling agents and inhibitors of energy transfer—suggesting a dependence on ATP production. On the basis of the carrier concept, and using the equations of enzyme kinetics, a change of the apparent K'_m and V'_{max} values caused by light can be predicted in the direction found experimentally. However, the necessary rise of ATP concentration in light is higher than can be anticipated *in vivo.*

Cl^- transport (i.e., active uptake) of the bladder vacuole of *A. spongiosa* in the light was apparently dependent on noncyclic photosynthetic electron transport (Lüttge and Osmond, 1970).

Membrane Permeability

There is a factor which must be considered, however, in any attempt to differentiate between the significance of respiration and photophosphorylation. That factor involves the effect of light on permeability of membranes. In general, light is associated with an increase in permeability (Hope and Walker, 1960), and darkness with a decrease.

In *Chlorella pyrenoidosa,* permeability was regarded as a limiting factor for Rb^+ but not for Na^+ accumulation (Schaedle and Jacobson, 1966).

Organic acids were studied with regard to their effects on uptake and retention of ions and on respiration in barley roots having low and high KCl concentrations (Jackson and Taylor, 1970). Absorptions of K^+, Na^+, Ca^{2+}, Cl^-, and O_2 were measured. Organic acids with high pK_a increased the permeability of roots to ions and decreased the respiration when present in sufficient concentration at pH 5, but had no inhibitory effect at pH 7. Effects of formate, acetate, propionate, and glutarate were attributed to entry of undissociated acid molecules into the effective membranes. Lack of a permeability increase with succinate, which has lower distribution coefficients to lipid solvents than do the aliphatic acids, was explained by failure of sufficient amounts of the hydrophilic succinic acid molecules to penetrate the membranes. Jackson and Taylor (1970) concluded that undissociated organic acids in root membranes can increase the permeability of roots to ions.

Entry of formate, acetate, succinate, and salts of several other organic acids into barley roots occurred mainly as ions from solutions at pH 5–7 (Jackson et al., 1970). Thus entry of organic acids at pH 5 occurs as ions through hydrophilic regions of the limiting membranes rather than by distribution of the undissociated acid to lipid-rich regions, as had been previously believed.

Hendricks and Jackson (1970) suggested that lipoprotein membranes have scattered "watery pinholes or pores." Contrary to the idea that organic acids pass membranes as undissociated molecules, they found that the acids ionize and that the negatively charged organic acid anion passes into the cell through the water pores in the membrane. They estimated the diameter of the pores to be about 8 A, which compares quite closely with pore sizes estimated by other scientists for animal membranes.

Cycloheximide drastically reduced the rate of root pressure exudation in detopped tobacco, and cation uptake was reduced by relatively low levels of cycloheximide. With regard to the effect on ion transport, Wallace et al. (1970) suggested that a damaging effect of cycloheximide on structure of membranes might explain many of their results.

The $(Na^+$ plus $K^+)$-stimulated activity of membrane-bound ATPase can be considered a biochemical expression of active transport of Na^+ and K^+ across a membrane (Kylin and Gee, 1970). With homogenates from salt-excreting leaves of the mangrove, *Avicennia nitida*, the effect of Na/K ratios varied with ionic strength of the test medium.

Inorganic sulfate enters the mycellia of *Aspergillus nidulans*, *Penicillium chrysogenum*, and *P. notatum* by a temperature-, energy-, pH-, ionic strength-, and concentration-dependent transport system (permease)—the transport being unidirectional (Bradfield et al., 1970). With excess external SO_4^{2-}, ATP sulfurylase-negative mutants accumulated inorganic SO_4^{2-} intracellularly to about 0.04 M, and the intracellular SO_4^{2-} was retained against a concentration gradient. Sulfate permease is under metabolic control. L-Methionine is a metabolic repressor of sulfate permease, while intracellular SO_4^{2-} and possibly L-cysteine (or a derivative of L-cysteine) are feedback inhibitors.

Electropotential (or Electrochemical Gradient)

In pea and oat seedlings, the electrochemical gradient for Na^+, Mg^{2+}, and Ca^{2+} was from the external solution to the interior of the cell, and thus passive diffusion should occur in an inward direction (Higinbotham et al., 1967). For anions (Cl^-, NO_3^-, $H_2PO_4^-$, and SO_4^{2-}), however, the gradient was observed to be in the opposite direction, and it was concluded that an active influx pump for anions must exist.

During uptake of K^+ and NO_3^- by excised pea epicotyls, the PD increased prior to rapid ion accumulation and thus appeared to be a prerequisite for accumulation (Macklon and Higinbotham, 1968).

In coleoptiles and roots of *Avena sativa* L. var. Victory and roots of *Pisum sativum* L. var. Alaska, K^+ was accumulated actively at low external K^+ levels and extruded at high external K^+ levels, while Na^+ was extruded at most external Na^+ concentrations (Etherton, 1963). Active transport was assumed when distribution of a given ion (inside and outside the cells) did not conform to the Nernst equation. The electropotential across membranes was approximately 100 mV (Walker, 1955; Higinbotham et al., 1961).

The membrane PD is generally considered to arise from the diffusion of ions (Scott, 1967), and the diffusion potential can be described by the Nernst equation. Davis and Higinbotham (1969) observed that the PD between the exudate and the external solution was in partial accord with predictions from the Nernst equation based on K^+ concentration changes. Anions, and cations other than K^+, did not appear to be as directly responsible (as judged by fitting to the Nernst equation). These investigators concluded that the potential is complex and does not appear to be attributable simply to ionic diffusion. It is a function of the living root and is dependent on concentration gradients as well as rates of active and passive transport. They caution that this puts limitations on the use of the Nernst equation in establishing the identity of ions actively transported to the xylem vessels in a complex structure such as the root.

Surfactants

Surfactants may cause certain biochemical effects in addition to lowering the surface tension. Parr and Norman (1964) noted that Tween 20 and 80, nonionic surfactants, were not only inhibitory to growth but also reduced K^+ uptake by cucumber roots.

Lipids

Lipids of roots of five grape rootstocks were studied in relation to the extent the rootstocks permitted Cl^- accumulation in the leaves (Kuiper, 1968a). Monogalactose diglyceride concentration was directly related to Cl^- accumulation, and a striking negative correlation existed between the concentration of lignoceric acid and Cl^- accumulation. The variety with the highest Cl^- accumulation contained an unusually low concentration of sterols. Chloride accumulation, then, was related to the concentration of charged rather than neutral lipids. It was also noted that Cl^- absorption capacity of lipids in membranes of grape root cells accounted for the differences in Cl^- transport to the leaves (Kuiper, 1968b). Monogalactose diglyceride had a high Cl^- transport capacity compared with phosphatidylcholine, but the latter compound had a greater capacity for exchange of Na^+ against K^+. It was concluded that phosphatidyl-choline had the properties of a cation exchanger and monogalactose diglyceride those of an anion transporter.

Auxin

Stimulation by auxin of K^+ (or Rb^+) absorption appeared to be dependent on Ca^{2+} in the external solution or, perhaps, the Ca^{2+} status of the tissue (Higinbotham et al., 1962).

Gibberellic acid application to the trifoliolate leaf of bean (*P. vulgaris* L.) plants enhanced absorption of Fe applied to the primary leaf (Kannan and Mathew, 1970). 2-Chloroethyltrimethylammonium chloride increased absorption by the primary leaf, while 6-furfurylaminopurine (kinetin) increased transport of Fe from the primary leaf to other parts. When the roots were pretreated with gibberellic acid, absorption of Fe by the primary leaf and subsequent transport to the trifoliolate leaves were increased. Triiodobenzoic acid reduced absorption and transport of Fe.

pH

In solutions pH is one of the factors that can alter the relative rates of uptake of NH_4^+ nitrogen versus NO_3 nitrogen. In general, a high pH favors uptake of NH_4^+ and a low pH the uptake of NO_3^-. Considering NO_3^- alone, Honert and Hooymans (1955) reported that NO_3^- absorption was much greater at pH 6 than at 8. With corn roots there was about 50% as much NO_3^- absorbed per hour at pH 7.4 as at 6.0; HCO_3^- had no influence on NO_3^- absorption (Fig. 6.2). Similarly, Fried et al. (1965) found that rice roots absorbed the NH_4^+ ion at a rate 5 to 20 times as fast as for NO_3^-—depending on pH. NH_4^+ was absorbed most rapidly at pH 7.0–8.5 and NO_3^- at pH 4.0–5.5. They were unable to distinguish more than one site for the absorption of NO_3^-.

By using excised barley roots in KBr, and by the appropriate selection of pH and temperature, it was possible to achieve excess K^+ absorption, excess Br^- absorption, or equal absorption of K^+ and Br^- (Jacobson et al., 1957).

With excised barley roots there was no profound effect of a pH range from 4 to 8 on the absorption of K^+, NO_3^-, and halide as observed by Hoagland and Broyer (1940). Their findings are at variance with those of Hurd (1958), who reported that as the pH was raised from 6 to 9 there was an increase in the rate of K^+ uptake and an increase in the amount of K^+ held by beet discs at saturation. Increased K^+ uptake was correlated with uptake of HCO_3 ions in the alkaline range—the additional K^+ accompanying the HCO_3 ions. Bicarbonate ions were converted primarily into malic acid; pH had no effect on Cl^-

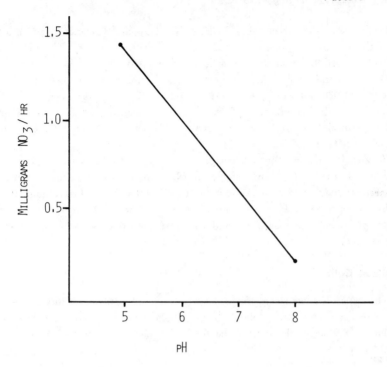

Fig. 6.2
Relation between pH and NO_3^- absorption rate. NO_3^- concentration approximately 10 ppm; 20° C. (Honert and Hooymans, 1955.)

uptake. Supporting the work by Hurd (1958), Poole and Poel (1965) reported increased uptake of cations with an increase in pH. However, the latter investigators concluded that cation uptake was not primarily determined by external HCO_3^- concentration and that the pH effect was genuine.

With intact plants of tomato, lettuce, and Bermuda grass, Arnon et al. (1942) observed that roots were injured at pH 3.0, that there was no ion absorption, and that Ca^{2+} and P were lost to the solution. P absorption showed a marked decrease at pH 9, but no other untoward effects. Ca^{2+} uptake was diminished at pH 4 and 5—as compared with higher pH values. Between pH 4 and 9, there were no profound effects of pH on absorption of Mg^{2+}, K^+, and NO_3. Experience has shown that plants may grow well in soils having pH values from as low as pH 4.5 to as high as 8.5 (Stout and Overstreet, 1950). The earlier view, that most plants have distinct pH optima for good growth, has not been supported by solution culture studies (Arnon and Johnson, 1942; Arnon et al., 1942).

Owing to H^+ in the surfaces of roots, they may have a considerably lower pH value than the surrounding medium. Williams and Coleman (1950) measured the pH of root surfaces placed in CO_2-saturated water. The pH of the water was 4.10, but root surfaces of lettuce measured 2.91.

Low pH and/or low Ca^{2+} increase the permeability of cell membranes (Epstein, 1961; Pitman, 1964, 1969; Siegel and Daly, 1966; Leggett and Gilbert, 1967). The interaction of Ca^{2+} and H^+, causing an increase in membrane permeability, has been discussed by Marschner (1964) and Marschner and Mengel (1966).

Benzimidazole was reported to enhance the uptake of cations by plant roots (Klingensmith, 1961; Klingensmith and Norman, 1960), but the apparent enhancement may have been attributable to the activity of this compound as a buffer as reported by Parr and Norman (1962). These investigators found that when pH was controlled, benzimidazole actually inhibited K^+ absorption by barley roots.

Type of Colloid

The type of colloid can affect the base exchange capacity (BEC) of a soil with, in general, organic colloids having higher exchange capacities than clay colloids. Obviously, the greater the BEC, the more nutrient cations a soil can hold.

Amino Acids

Uptake of SO_4^{2-} by cultured cells of tobacco appears to be regulated by the end products—the S amino acids (Hart and Filner, 1969). In fact, both the uptake of NO_3^- and of SO_4^{2-} appear to be regulated by amino acids.

Base Saturation of Colloids

Graham and Albrecht (1943) and Jenny (1946) reported that NO_3^- adsorbed onto Amberlite was available for the growth of plants.

Arnon and Grossenbacher (1947) reported on the suitability of synthetic ion-exchange materials, known as Amberlites, as sources for adsorbed cations and anions. However, they noted that when Ca–Mg–K and NO_3–PO_4–SO_4 were adsorbed on Amberlites, Ca^{2+} and Mg^{2+} were unavailable for plant growth. When only Ca–Mg and PO_4–SO_4 were on Amberlites, and a solution of KNO_3 was supplied daily, they obtained good growth and fruiting of plants.

In studying the unavailability of Ca^{2+} and Mg^{2+} when K^+ was also on Amberlite, Arnon and Meagher (1947) found that Ca^{2+} and Mg^{2+} were readily leachable from Amberlites when the concentration of divalent bases was about eight times that of K^+. They also noted that PO_4^{3-} and SO_4^{2-}

were unavailable to plants when NO_3^- was also present on the Amberlite. The depressing action of NO_3^- on availability of PO_4^{3-} and SO_4^{2-} could be overcome when the concentration of PO_4^{3-} (computed as $H_2PO_4^-$) plus SO_4^{2-} was roughly twice that of the nitrate.

Cation availability (as determined by leaching) was similar whether studied with Amberlites or with Otaylite hydrogen–bentonite no. 2 clay. Thus, with Amberlites and with natural clay (Allaway, 1945), there is evidence that the availability of adsorbed Ca^{2+} to plants is related to the degree of saturation with Ca^{2+}. Soils with different types of colloids may show marked differences in availability of their replaceable Ca^{2+} (Allaway, 1945). Exchangeable Ca^{2+} is available for plant growth only when its degree of saturation (on the exchange complex) is relatively high. Calcium availability to plants is greatly reduced if about 50% of the BEC is satisfied with Na^+ or K^+ (Ratner, 1944; Bower and Turk, 1946; Thorne, 1946; Itallie, 1948). High levels of adsorbed K^+ are just as effective as Na^+ in preventing the intake of Ca^{2+} and Mg^{2+} by plants (Bower and Turk, 1946). The highest level of Na^+ tolerated by tomato plants was between 60 and 70% of the BEC of bentonite clay mixed with sand (Thorne, 1944). Jenny and Cowan (1933) reported that growth of soybeans was sharply reduced when Ca^{2+} saturation was below 30% of the total exchange capacity. According to Ratner (1935), oats and wheat failed to grow in a Na–Ca system unless Ca^{2+} saturation was greater than 30–40%. Working with tomatoes, Thorne (1944) reported that growth was markedly reduced with less than 50% Ca^{2+} saturation. The amount of Ca^{2+} absorbed by soybean plants was a function of the percentage saturation of clay with Ca^{2+} (40–100%)—even when the same *total* amounts of Ca^{2+} were supplied (Albrecht, 1940) Alkali soils, high in exchangeable Na^+, cannot supply sufficient Ca^{2+} to plants even in the presence of $CaCO_3$ in the soil (Bower and Turk, 1946). Soils saturated with NH_4^+, Na^+, or K^+ do not support plant growth even when $CaCO_3$ is added to the soil (Gedroiz, 1931).

Per unit weight, organic colloids have a higher BEC than inorganic (clay) colloids. In this connection, Albrecht (1941) reported that, as organic matter in the soil increased, total BEC and percentage saturation with Ca^{2+} and Mg^{2+} increased.

At comparable levels of adsorption, Ca^{2+} on kaolinite clay may be 10 to 20 times as readily available to the plant as Ca^{2+} held by montmorillonite clay (Marshall, 1948).

Inasmuch as K^+ is held less strongly by colloids than Mg^{2+}, Mg^{2+} less than Ca^{2+}, and Ca^{2+} less than H^+, Baker (1968) indicated that the percent saturation of the BEC, satisfied by K^+ need not be very high—if maintained. For field crops he recommended a K^+ saturation of 1% and a Mg^{2+} saturation of 10% of the total BEC.

Stage of Plant Growth

There have been numerous studies involving uptake of various essential elements as a function of the stage of growth. It is generally reported that the bulk of P is absorbed during early stages of growth. However, B must be absorbed continuously throughout the life cycle of the plant.

Loehwing (1951) studied the relationship between the ontogeny of plants and absorption (or release) of nutrients. In general, he observed that many plants absorb a large portion of their nutrients in the early stages of growth. At the time of flowering, nutrients (especially N and K) were often lost from the plant via the roots; and, following flowering, anabolic activities and nutrient absorption were renewed.

Cation Exchange Properties of Roots

Devaux (1916) first mentioned the fact that cations might be held by roots (cation exchange property) and by colloids, and that because of intimate contact with each other ions might move from one system to the other.

According to Williams and Coleman (1950), plant root surfaces have cation exchange properties that can be measured by the adsorption and release of various cations. Slightly different capacities are obtained with various cations, but agreement is reasonably close in most cases. The measurement involves the replacement of H^+ on root surfaces with other cations. It was observed that exchange capacity was not dependent on metabolism in roots, since the same values were obtained at 0 and 25°C and with living or ether-killed roots.

The following cation exchange capacities (CEC) of plant roots (in milliequivalents per 100 g fresh weight), as determined with Ba^{2+}, were obtained: 0.257, tobacco; 0.633, soybean; 1.308, sugar beet. Tobacco had the lowest exchange capacity, sugar beet the highest, and soybean was selected here to represent one of the intermediate values Williams and Coleman (1950) reported.

Leguminosae have approximately twice the CEC of the Gramineae. Within either group the higher the CEC the greater the uptake of nutrients by plants (Drake et al., 1951; McLean et al., 1956; Mouat, 1960; Heintze, 1961).

The higher the CEC, the greater the preferential uptake of Ca^{2+} over K^+ (or Na^+), and it was therefore concluded (Elgabaly and Wiklander, 1949) that this might explain why legumes, with a relatively high CEC, have relatively high requirements for and internal concentrations of Ca^{2+}.

The CEC of roots has been shown to be closely correlated with percent N in roots (McLean et al., 1956). Negatively charged sites, which have been attributed to carboxyl groups in pectic material of the cell wall, are primary responsible for root CEC (Keller and Deuel, 1957).

Constancy of Total Cations

It has been generally observed that the sum of cations per 100 g of dry plant material tends to be a constant for a given species of plant (McLean, 1956; Hylton et al., 1967), so that, for example, if there is a decreased uptake of Mg^{2+}, absorption of one or more other cations tends to be increased. For alfalfa the sum of equivalents of Ca^{2+}, Mg^{2+}, and K^+, per unit of material, was reasonably constant—170–187 meq/100 g dry matter (Bear and Prince, 1945).

Interrelationships Among Ions

Working with pineapple, Nightingale (1942) found that when plants required large amounts of NO_3^- extra K^+ was required for NO_3^- absorption —amounts over and above those required for other functions of K^+. NO_3^- level affected P absorption; P level affected NO_3^- absorption.

Low levels of Al^{3+} (1.85×10^{-4} M) increased the total amount and concentration of P in roots and shoots of 8-wk-old perennial ryegrass according to Randall and Vose (1963). They postulated that the increase was related to the known stimulation by Al^{3+} of the reduction of cytochrome. They further suggested that subsequent to the uptake of P the element was bound by Al^{3+} within the plant, since Al^{3+} toxicity had been shown to cause P deficiency symptoms. In a study of the effect of Al^{3+} on the uptake and metabolism of P by barley seedings, it was noted that Al^{3+} inhibited hexokinase, so that there was a decreased rate of phosphorylation of sugar (Clarkson, 1966). Further, at the cell surface or in the free space, Al^{3+} was believed to cause a fixation of phosphate by an adsorption–precipitation reaction. It was concluded that Al^{3+} was not interfering with translocation of P per se.

Two navy bean (*P. vulgaris* L.) varieties, Sanilac and Saginaw, were grown in soil, solution culture, and split medium (Ambler and Brown, 1969). In general, Fe^{3+} and P aggravated Zn^{2+} deficiency. The Sanilac variety took up more Fe^{3+} and P than did Saginaw, and Zn^{2+} deficiency symptoms were more pronounced in the former variety.

Faster-moving halides stimulated the uptake of slower-moving halides by wheat roots (Boszorményi and Cseh, 1961).

In the absence of interfering ions, the quantity of Rb^+ disappearing from the solution agreed with that absorbed by barley roots as observed by Tanada (1963). Believing in a chain of negative sites through the membrane (rather than a moving carrier), he reasoned that, should an ion such as Na^+ or Ca^{2+}, whose rate of diffusion is less than that of Rb^+, get ahead of it on the chain, movement of Rb^+ from the external sites along the chain would be slowed down.

Absorption of NH_4^+ by corn plants reduced the rate of absorption of other cations—particularly Ca^{2+} (Wadleigh and Shive, 1939). There is considerable evidence that increasing external concentrations of K^+ progressively decrease the concentrations of Ca^{2+} in plants (Beeson et al., 1944; York et al., 1953; McEvoy, 1955; Lazaroff and Pitman, 1966; Osmond, 1966; Freeman, 1967; Hylton et al., 1967; Johansen et al., 1968a). Johansen et al. suggested that this phenomenon may be important in Ca^{2+} nutrition of plants under field conditions. Average concentrations of K^+ and Ca^{2+} in soil extracts are quite low—80–1500 μM and 200–2000 μM, respectively (Barber, 1962). For alfalfa increasing the Ca^{2+} supply did not depress K^+ uptake (York et al., 1953).

Working with *A. spongiosa*, Osmond (1966) noted that increasing K^+ concentrations had a depressive effect on leaf Ca^{2+} only at high levels of Ca^{2+} in solution (50,000 μM), with no effect of K^+ at low Ca^{2+} levels (500 and 5000 μM). He therefore proposed that there were two independent mechanisms of Ca^{2+} absorption—a highly specific, K-insensitive mechanism at low Ca^{2+} concentrations and a nonspecific, K-sensitive mechanism at higher Ca^{2+} concentrations. Dual absorption mechanisms have been proposed for other ions (Epstein, 1966). However, Johansen et al. (1968a) did not support Osmond's (1966) hypothesis that there exists a Ca-specific, K-insensitive mechanism of Ca^{2+} absorption at low external Ca^{2+} concentrations.

Lazaroff and Pitman (1966) reported that transpiration influences Ca^{2+} absorption from high Ca^{2+} concentrations, but not from low concentrations, and suggested that this difference might explain the divergent results of different investigators. Transpirational effects could not explain the effect of K^+ on Ca^{2+} noted by Johansen et al. (1968a), however, since concentrations of both K^+ and Ca^{2+} were low. They noted that two levels of external Ca^{2+} (250 and 2500 μM) had no efect on rate of absorption, concentration, or total amount of K^+ absorbed when three levels of external K^+ were used (20, 200, 2000 μM) (Johansen et al., 1968b). For all concentrations of K^+ in solution, the high-affinity mechanism dominated the absorption of K^+ by intact barley plants.

In soybeans Mg^{2+} uptake was inhibited when both Ca^{2+} and K^+ were present in the solution, but not by K^+ or Ca^{2+} alone (Leggett and Gilbert, 1969).

The effect of Ca^{2+} on the absorption of other ions by plant roots is complicated by a range of effects from promotion to inhibition. Viets (1944) first reported on the promotive effect of Ca^{2+} and other polyvalent cations on the absorption of other ions, and this effect has come to be known as the "Viets effect." This effect is pronounced with K^+ (Viets, 1944; Overstreet et al., 1952; Kahn and Hanson, 1957; Epstein, 1961), Rb^+ (Tanada, 1955, 1956, 1962; Epstein, 1960a), Br^- (Viets, 1944), Cl^- (Pitman, 1964; Elzam and Epstein,

1965), SO_4^{2-} (Leggett and Epstein, 1956), and PO_4^{3-} (Tanada, 1955, 1956; Leggett et al., 1965a). In contrast, Ca^{2+} markedly depresses uptake of Na^+ and Li^+ (Epstein, 1960a, 1961; Jacobson et al., 1960; Rains and Epstein, 1967a).

Research by Elzam and Hodges (1967) shed additional light on the complexity of the effect of Ca^{2+} on the uptake of other ions. With corn roots Ca (or Mg) SO_4 or Cl^- inhibited K^+ transport during initial phases of transport, but as the absorption period was lengthened the Ca^{2+} effect changed to a promotive one (i.e., the "Viets effect"). With barley roots $CaSO_4$ had no effect on K^+ absorption, whereas $CaCl_2$ stimulated K^+ absorption. In view of the different responses of corn and barley roots, these investigators concluded that it is hazardous to apply results from one species to all species.

Ca^{2+} has been reported to divert energy away from the phosphorylation of ADP to that of phosphate uptake (Hanson et al., 1965).

Inasmuch as lichens generally have much higher concentrations of fallout radionuclides than higher plants, and Lapps and Alaskan Eskimos have 30 to 100 times the average body burden of $^{137}Cs^+$ (Salo and Miettinen, 1964), Handley and Overstreet (1968) studied uptake of carrier-free $^{137}Cs^+$ by a lichen, *Ramalina reticulata* Kremp. Uptake did not appear to be directly linked to metabolism; the presence of a barrier to uptake, stabilized by Ca^{2+}, was indicated.

While studying pretreated excised roots of *Hordeum vulgare, Z. mays,* and *Glycine max,* Franklin (1969) noted that the rate of P absorption was greatest for roots pretreated with trivalent cations, intermediate with divalent cations, and lowest with monovalent cations. He hypothesized that acceleration of P uptake by roots in the case of polyvalent cations was attributable partly or entirely to a greater reduction in the electrical potential at the root surface or within pores of the negatively charged cell wall with polyvalent than with monovalent cations.

Epstein (1961) noted that, in the absence of Ca^{2+}, Na^+ interfered with K^+ absorption; in its presence there was no effect of Na^+ on K^+ absorption.

High concentrations of monovalent anions and low concentrations of monovalent cations inhibited accumulation of Mg^{2+} and phosphate by beet mitochondria according to Millard et al. (1965). Therefore they stressed the importance of considering total ionic composition of the medium when studying uptake of particular ions.

Extrusion of Na^+ from *Scenedesmus* cells was stimulated by low concentrations of phosphate that kept cells practically free of Na^+ (Kylin, 1966). Increasing the external phosphate level stimulated Na^+ uptake more than Na^+ extrusion, and a net uptake occurred.

In a study of the effects of various cations (Mn^{2+}, Zn^{2+}, Cu^{2+}, Ca^{2+}, Mg^{2+}, and K^+ plus Rb^+) on uptake and transport of Fe^{2+} in soybeans,

Zn^{2+} interfered most with Fe—reducing not only uptake but, to a greater extent, translocation of Fe to the tops (Lingle et al., 1963). Ambler (1969) and Ambler et al. (1970) noted a similar effect of Zn^{2+} on Fe transport in soybean plants.

Radioiron did not move into the tops of sweet orange seedlings when HCO_3^- was present in the nutrient solution (Wallihan, 1961). However, the low concentration of radioiron in the roots when HCO_3^- was present suggested that absorption rather than translocation was affected by bicarbonate.

Up to 0.1 meq/liter Na^+, Li^+, Ca^{2+}, Ba^{2+}, and Mg^{2-} had essentially no depressive effect on Cs^+ absorption, whereas K^+, Rb^+, and NH_4^+ markedly depressed Cs^+ uptake (Handley and Overstreet, 1961a).

The influence of Ca^{2+} on the aging of bean stem *(P. vulgaris)* slices and on absorption of K^+ and Na^+ by fresh and aged slices was investigated (Rains and Floyd, 1970). In the presence of Ca^{2+}, fresh tissue showed a preferential Na^+ uptake. Preference for Na^+ over K^+ resulted from a differential depressive effect of Ca^{2+} on absorption of these two ions. In aged tissue Na^+ uptake was also depressed, but K^+ absorption was accelerated—with a net result of a much greater absorption of K^+ than Na^+. Presence of Ca^{2+} in the aging medium promoted development of K^+-absorbing capacity as well as an increase in rate of respiration but did not influence the loss of capacity to absorb Na^+ as tissue aged. This, along with the demonstration that protein synthesis was involved in the development of K^+-absorbing capacity by aging tissue, suggests that Ca^{2+} may have an effect on basic physiological processes concerned with development of ion absorption by aging tissue.

Excised barley roots accumulated 40–50% more K^+ from 0.04 mM KCl than from 0.06 mM KCl when incubated for 24 hr in KCl solutions containing 0.2 mM $CaSO_4$ or in solutions containing KCl and NaCl at a total concentration of 0.1 mM with the mole fractions of K^+ and Na^+ varied in replacement series (Hiatt, 1970a). This phenomenon was observed with Cl^-, NO_3^-, or SO_4^{2-} as the counterion. Inasmuch as Cl^- and NO_3^-, but not SO_4^{2-}, are readily absorbed by barley roots, concurrent anion absorption did not appear to be an important factor. Changes in Na^+ concentration altered the magnitude of K^+ accumulation but did not significantly change the K^+ concentration at which the accumulation peak occurred. Since the phenomenon occurred even in the absence of Na^+, it is apparent that the anomaly was not the result of an interaction between K^+ and Na^+.

When excised barley roots were incubated in 0.01–0.1 mM KCl solutions containing 0.2 mM $CaSO_4$, there was a peak in the K^+ accumulation–concentration curve at 0.02–0.04 mM KCl (Hiatt, 1970b). The peak in the K^+ accumulation curve was shifted to lower K^+ concentrations when Ca^{2+} concentration was decreased, and to higher K^+ concentrations when Ca^{2+} concentration was increased. Increasing Ca^{2+} concentration in the treatment solution was observed, then, to be stimulatory, inhibitory, or neutral depending on K^+ concentration. When ^{86}Rb was used as a tracer for K^+, accumulation of K^+ was grossly overestimated, and the apparent K^+ accumulation curve, as estimated with ^{86}Rb-labeled K^+, was hyperbolic over the concentration range 0.01–0.1 mM. It was concluded that ^{86}Rb is a poor tracer for K^+ over the concentration range 0.01–0.1 mM.

In a study of the foliar uptake of Na^+, K^+, Rb^+ and Cs^+ by bean plants, no competition was found in the speed of initial penetration, from the outer leaf surface to the protoplasts of epidermal cells, between any two of the cations considered separately (Levi, 1970). The magnitude of penetration of Na^+, Rb^+, and Cs^+ was increased byK^+, while that of K^+ was increased by Rb^+ or Cs^+. A given cation could influence the retention of another cation. In general, total quantities of cation transported were influenced differently by the various cations and, in turn, determined the total quantities absorbed.

Culture solution and field experiments indicated that P-deficient sugar beets, particularly young plants, do not absorb NO_3^- as well as plants adequately supplied with P (Hills et al., 1970). Decreased absorption of NO_3^- by P-deficient plants in aerated culture solution with all roots exposed continuously to high concentrations of NO_3^- indicates that the phenomenon reflects not merely increased root extension, but a physiological aberration brought about by P deficiency. Plants low in NO_3^- and in phosphate may be low in NO_3^- because of the effect of P on NO_3^- absorption.

P absorption by excised roots of 10 plant species was stimulated by a short pretreatment with polyvalent cations, or by raising the ionic strength of the absorbing solution (Franklin, 1970). Pretreatment consisted of a 1-min rinse with 10^{-3} N Cl^- salt solutions prior to exposure to 2×10^{-5} M KH_2PO_4. Results supported the hypothesis that the cell wall can act as a negatively charged membrane regulating P absorption.

El-Sheikh et al. (1971) studied selective absorption of K^+ and Rb^+ or of K^+ and Na^+ by intact sugar beet plants from modified conventional nutrient solutions over an extended period of growth. K^+ and Rb^+ were mutually competitive in absorption; high selectivity of K^+ relative to Na^+ absorption was observed. Na was excluded during the early growth of sugar beet plants.

Ca^{2+} increased phosphate uptake relatively more at low than at high phosphate concentrations (Robson et al., 1970). Pretreatment at various Ca^{2+} concentrations had no effect on phosphate absorption, while transfer to solutions of different Ca^{2+} concentrations caused an immediate response in phosphate absorption. They concluded that Ca^{2+} increased phosphate absorption by screening electronegative charges on the roots, and thus increased the accessibility of absorption sites to phosphate.

Rate of uptake of Ca^{2+} from various salts decreased in the order: NO_3^-

$> Cl^- = Br^- > SO_4^{2+}$. K^+ and H^+ greatly interfered with Ca^{2+} absorption, while Li^+ and Na^+ had only slight effects (Maas, 1969).

Fried and Broeshart (1967) compiled an extensive list of reports that deal with effects of one ion on the uptake of another ion.

Recycling of Ions from Tops Back to Roots

Collander (1941) suggested that, varying with the species and the ion, there may be a recycling of certain ions from tops to roots. Should this happen with a given ion, it is conceivable that absorption of that ion might be decreased over that which would occur if the ion were not returning from the tops to the roots. Commenting on the generally observed low concentrations of Na^+ in tops of most plants, Cooil et al. (1965) concluded that downward movement of Na^+ in living squash cells might partially account for the appearance that upward transport of Na^+ is restricted. Na has also been reported to move downward in *P. vulgaris* (Wallace and Hemaidan, 1963).

The ecological aspects of mineral nutrition have been given emphasis in a recent book on the subject (Rorison, 1969).

Microorganisms

Barber (1968) has fully documented various ways in which microorganisms may affect the mineral nutrition of plants and, importantly, the interpretation placed on certain studies of nutrition. Deficiencies may be induced by microorganisms—the immobilization of Zn^{2+} and the oxidation of Mn under certain conditions (Bowen and Rovira, 1966). In other situations oxidized Mn can be made available to plants by microbial action (Mulder and Gerretsen, 1952; Heilman, 1967).

Uptake of P by roots and the transfer of P to shoots increased in nonsterile conditions for tomato and clover plants—as compared with sterile conditions (Bowen and Rovira, 1966).

The effect of bacteria and fungi on absorption and translocation may vary with the concentration of the element under consideration as observed by Barber and Loughman (1967). They found that, at a very low concentration of P in the solution and under nonsterile conditions, the microorganisms absorbed enough P, so that roots absorbed less P under nonsterile than under sterile conditions. At a higher level of P, there was no difference in the amount of P absorbed and translocated whether the plants were grown under sterile or nonsterile conditions. Microorganisms have also been shown to affect the absorption of Na^+ and Cl^- by washed beet root discs—the discs having higher concentrations of these ions under sterile than under nonsterile conditions (MacDonald, 1967).

Barber (1968) cautions that microorganisms must be considered when examining uptake of ions over a wide range of concentrations. At low concentrations uptake by microorganisms is dominant, while at high concentrations the entry of ions into roots is dominant. These effects must be considered before concluding, on the basis of kinetic analysis of data, that there are two mechanisms of ion uptake. The effects of microorganisms at low and high concentrations of a given ion are such as to suggest that two mechanisms of ion uptake are involved (Barber, 1968).

Mycorrhiza

Mycorrhiza means "fungus-root" (Kelley, 1950), and certain trees normally have a fungus in intimate, symbiotic association with the roots. Pine trees, in particular, are noted for the ubiquitous association of mycorrhiza with their root systems. When pines were first introduced into Puerto Rico and Australia, they grew poorly until soil containing mycorrhizal fungi was introduced (Briscoe, 1960). Roots of some pine trees lack root hairs (Went and Stark, 1968). Other trees, including citrus, may have mycorrhiza associated with the roots. Nutrient status of the soil may affect invasion of roots by mycorrhiza, and Reed and Fremont (1935) noted that roots of citrus trees receiving fertilizer developed a resistance to invasion by endophytic mycorrhizal fungi. Roots of trees in the temperate zone have ectotrophic mycorrhiza, whereas tropical trees generally have endotrophic mycorrhiza in which the fungi penetrate living cells (Went and Stark, 1968). Endotrophic mycorrhiza, characteristic of orchids, had been thought to be rare in trees.

In the Mesopotamian desert near Baghdad, mycorrhiza were associated with roots of cultivated and native trees and shrubs-such as *Phoenix dactylifera* L. and *Zizyphus spina-christi* Willd. (Khudairi, 1969).

Went and Stark (1968) advanced the theory of "direct mineral cycling." According to this theory the bulk of minerals available in the tropical rain forest ecosystem is bound in dead and living organic systems. Little available mineral ever occurs free in the soil at any one time. Mycorrhiza, which are extremely abundant in the surface litter and thin humus of the forest floor, are believed to be capable of digesting dead organic litter and passing minerals and food substances through their hyphae to living root cells. In this manner little soluble mineral leaks into the soil where it can be leached. This is contrary to conditions that exist when bacteria are the main agents of decay, since minerals are released directly into the soil and can be leached.

Mycorrhiza are associated with many saprophytes (Went and Stark, 1968), and they are especially important for orchids since noninfected orchid seedlings cannot grow (Meyer, 1966).

Psilotum nudum (Bierhorst, 1953) and beech trees (Harley and McCready, 1950) have mycorrhiza. Beech roots absorbed $^{32}PO_4^{3-}$ more rapidly with mycorrhiza than without the fungus.

It is generally believed that saprophytic, higher plants, which lack chlorophyll, must rely on mycorrhiza to bring in their food supply. Routien and Dawson (1943) have suggested that mycorrhiza aid in mineral uptake by creating an excess of H^+ near the root. These investigators observed, using prepared clays with adsorbed Ca^{2+}, Mg^{2+}, K^+, and Fe^{3+}, that mycorrhizal plants absorbed greater quantities of these nutrients at low levels of base saturation than did nonmycorrhizal plants. At higher levels of base saturation, differences attributable to the fungus were slight.

Species

Cl^- content was remarkably higher in leaves of *Metasequoia glyptostroboides* than in other conifers despite the fact that there was practically no difference in total osmotic pressure or percent moisture. In *Metasequoia*, about 25% of the total OP of the sap could be accounted for by the Cl^- fraction, whereas in other conifers it was approximately 5% of the total (Takada and Nagai, 1953).

Collander (1941) grew 21 species of plants with a solution containing equivalent concentrations of Na^+, K^+, and Rb^+, and found that there was a striking diversity in selectivity. The most striking effect was that the species with the highest concentration of Na^+ exceeded that of the one with the lowest concentration by some 60-fold (Fig. 6.3). Such differences with regard to Na^+ were interpreted by Briggs et al. (1961) to indicate that: (1) some plants have more binding sites for Na^+ than other plants, or (2) some plants have more resistance to accumulation of Na^+ in their vacuoles than do other plants, or (3) that Na^+ is actively extruded by some plants.

For soybeans a genetic factor appeared to be involved in the extent to which varieties accumulated Cl^- (McCollum, 1960).

Some species may show a preference for certain ions over other ions. For example, *C. pyrenoidosa* showed a selective preference of alkali metal cations of the order: $Rb^+ > K^+ >>> Na^+$ (Schaedle and Jacobson, 1967). They also demonstrated that a cation of higher preference could replace a cation of lower preference in the cell.

Similar striking differences in plant composition, as a function of the medium, have been reported for several species of algae by Hober (1946) (Table 6.1) and for *Chara ceratophylla* (Collander, 1942) (Table 6.2).

Root Specificity

By grafting between P-tolerant and P-sensitive soybean varieties, the critical genotypic difference in P nutrition was shown to reside in the roots (Foote and Howell, 1964). Tolerance of the tolerant variety was attributable primarily to a reduction in P accumulation.

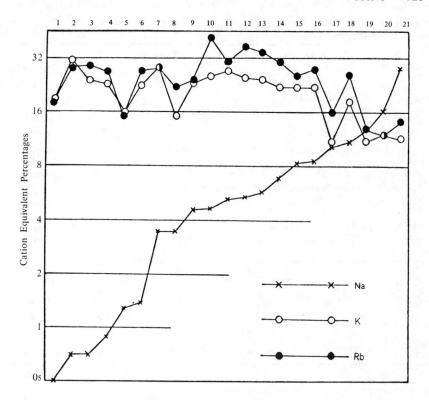

Fig. 6.3

Equivalent percentages of Na^+, K^+, and Rb^+ in plants cultivated in a solution containing these cations in equivalent amounts. Plants are arranged according to increasing Na^+ content. 1, *Fagopyrum*; 2, *Zea*; 3, *Helianthus*; 4, *Chenopodium*; 5, *Salsola*; 6, *Pisum*; 7, *Nicotiana*; 8, *Solanum*; 9, *Spinacia*; 10, *Avena*; 11, *Aster*; 12, *Papaver*; 13, *Lactuca*; 14, *Plantago lanceolata*; 15, *Melilotus*; 16, *Vicia*; 17, *Atriplex litorale*; 18, *Sinapis*; 19, *Salicornia*; 20, *Plantago maritima*; 21, *Atriplex hortense*. (Collander, 1941.)

In derooted wax bean plants in low concentrations of NaCl, Na^+ was retained in the basal portion of the stem, whereas Cl^- was distributed throughout the entire plant (Jacoby, 1964). It was suggested that Na^+ was retained in the basal parts of the stems by active accumulation of Na^+ by cells lining the translocation path (Jacoby, 1965).

Hawkeye (HA) and A62–9 (E–9) are Fe-efficient genotypes, while PI-54619–5–1 (PI) and A62–10 (1–10) are Fe-inefficient (Elmstrom and Howard, 1970). Absorption of Fe by 17-day-old plants indicated that solutions in which either HA or PI plants had been previously cultured contained a heat-labile factor which increased Fe accumulation by Fe-stressed HA plants. When PI plants were grown with HA plants, the PI plants apparently excreted a sub-

Table 6.1
Concentrations of Ions in the Sap and Milieu of Cells[a,b]

Ion	Valonia macrophysa	Halicystis sp.	Seawater
Cl	597	603	580
Na	90	557	498
K	500	6.4	12
Ca	1.7	8	12
Mg	Trace?	16.7	57
SO_4	Trace?	Trace	36

[a]from Höber, R., 1946: Physical Chemistry of Cells and Tissues, P. Blakiston C., Philadelphia, Pa. McGraw-Hill Book Co.—by permission.
[b]Concentrations are millimolar.

Table 6.2
Composition of the Sap of *Chara ceratophylla,* and of the Media in which the Plants were Grown[a,b]

Portion analyzed	Na	K	Mg	Ca	Cl
Sap	152	66	26	13	233
Medium	68	1.4	14	3.8	80
Sap	126	61	20	11	208
Medium	31	0.6	6.5	2.0	36
Sap	84	77	—	13	176
Medium	0.21	0.04	—	3.3	0.13

[a] Collander (1942).
[b] Values are given in milliequivalents per liter.

stance that inhibited Fe accumulation. The inhibiting factor was not evident when HA plants grown separately were placed in solution used only for growing PI plants. Instead, a heat-labile factor was found that stimulated Fe accumulation. Accumulation of Fe by Fe-inefficient plants (PI or 1–10) was unaffected or slightly reduced in the presence of Fe-efficient plants (HA or E–9).

Mass of Roots to Volume of Solution

The ratio of mass of roots to the volume of solution can have a profound effect on the results of absorption studies (Jacobson et al., 1961). Liberation of Ca^{2+} and other absorption-modifying substances from the roots was regarded as the most important factor with a large mass of roots and a small volume of solution.

Age of Cells

The rate of Fe^{3+} absorption by cells enzymically isolated from tobacco leaves was correlated with age of leaves from which the cells were derived

(Kannan, 1969). Cells from young leaves absorbed Fe^{3+} more rapidly than did those from old leaves.

Jacoby and Dagan (1969) studied the effects of aging on Na^+ fluxes in primary leaf sections from control and decapitated bean plants. Net- and influxes to vacuoles increased during leaf expansion and net chlorophyll synthesis. Both fluxes and amount of chlorophyll decreased rapidly in senescing leaves.

Ion uptake per unit of O_2 absorption indicated that metabolic energy is utilized almost exclusively for Na^+ transport in fresh tissue, but that this energy is diverted to K^+ transport as the slices age (Floyd and Rains, 1971).

Interpretation of Isotopic Measurements

Based on analysis for $^{86}Rb^+$, Maas and Leggett (1968) found that $^{86}Rb^+$ uptake did not evaluate uptake of K^+ by corn roots. $^{86}Rb^+$ was readily absorbed, whereas chemical analyses actually indicated a net loss of K^+. Nims (1962) showed mathematically that the flow of a radioisotope is proportional to flow of the ion only when the mole fraction of the species is the same on both sides of the membrane. His caution and the results by Maas and Leggett (1968) highlight the need for total chemical analyses to avoid error in the interpretation of isotopic measurements. Helder (1958) also cautioned that results with Rb^+ may not be indicative of what would be obtained with K.

Plants can discriminate against Cs^+ uptake at low K^+ nutrient concentrations (Cline and Hungate, 1960). Discrimination of $^{86}Rb^+$ from K^+ approximated that noted for $^{137}Cs^+$. It was concluded that there may be potential errors in the use of ratios for predicting uptake of $^{137}Cs^+$. Depending on the concentrations of K^+ and Cs^+ in the nutrient solution, absorption may favor K^+ in certain cases and Cs^+ in others (Middleton et al., 1960). Thus K^+ does not behave exactly as Cs^+, nor Cs^+ as K^+.

In a study of translocation in Z. *mays* roots, the ratio of Sr^{2+} to Ca^{2+} translocated was exactly equal to the ratio of their concentrations in the nutrient solution, hence there was no evidence of discrimination (Hutchin and Vaughan, 1967). Similarly, in simultaneous double-tracer determinations, using isolated corn roots, radiostrontium showed polar transport characteristics extraordinarily similar to those of radiocalcium (Vaughan et al., 1967).

Reduction

Apparently, Fe^{3+} must be reduced to Fe^{2+} before being absorbed by roots. Working with an Fe-efficient soybean variety, Hawkeye, Ambler et al. (1971) noted that Fe reduction was most pronounced between the regions of root elongation and root maturation at both the epidermis and endodermis. Reducing capacity was greatest in the young diarch roots, indicating that these particular roots contribute significantly in the absorption of Fe. An Fe reductant was exuded from the roots into the solution. Reduction at or in the root and the amount of reductant exuded by the roots were greater with Fe-deficient than with Fe-sufficient plants.

LITERATURE CITED

Albrecht, W. A. 1940. Adsorbed ions on the colloidal complex and plant nutrition. Proc. Soil Sci. Soc. Amer. 5:8–16.

Albrecht, W. A. 1941. Soil organic matter and ion availability for plants. Soil Sci. 51:487–494.

Allaway, W. H. 1945. Availability of replaceable calcium from different types of colloids as affected by degree of calcium saturation. Soil Sci. 59:207–217.

Ambler, J. E. 1969. Effect of zinc on the translocation of iron in soybean plants. Ph.D. Thesis, Univ. Maryland, College Park.

Ambler, J. E., and J. C. Brown. 1969. Cause of differential susceptibility to zinc deficiency in two varieties of navy beans (*Phaseolus vulgaris* L.). Agron. J. 61:41–43.

Ambler, J. E., J. C. Brown, and H. G. Gauch. 1970. Effect of zinc on translocation of iron in soybean plants. Plant Physiol. 46:320–323.

Ambler, J. E., J. C. Brown, and H. G. Gauch. 1971. Sites of iron reduction in soybean plants. Agron. J. 63:95–97.

Arisz, W. H. 1956. Significance of the symplasm theory for transport across the root. Protoplasma 46:5–62.

Arnon, D. I. 1937. Ammonium and nitrate nitrogen nutrition of barley at different seasons in relation to hydrogen-ion concentration, manganese, copper, and oxygen supply. Soil Sci. 44:91–113.

Arnon, D. I., and K. A. Grossenbacher. 1947. Nutrient culture of crops with the use of synthetic ion-exchange materials. Soil Sci. 63:159–182.

Arnon, D. I., and C. M. Johnson. 1942. Influence of hydrogen ion concentration on the growth of higher plants under controlled conditions. Plant Physiol. 17:525–539.

Arnon, D. I., and W. R. Meagher. 1947. Factors influencing availability of plant nutrients from synthetic ion-exchange materials. Soil Sci. 64:213–221.

Arnon, D. I., W. E. Fratzke, and C. M. Johnson. 1942. Hydrogen ion concentration in relation to absorption of inorganic nutrients by higher plants. Plant Physiol. 17:515–524.

Arrington, L. B., and J. W. Shive. 1936. Oxygen and carbon dioxide content of culture solutions in relation to cation and anion nitrogen absorption by tomato plants. Soil Sci. 42:341–356.

Baker, D. E. 1968. Crop production related to the capacity of soil to hold plant nutrients. Vegetable Growers Messenger 20:8.

Barber, D. A. 1968. Microorganisms and the inorganic nutrition of higher plants. Annu. Rev. Plant Physiol. 19:71–88.

Barber, D. A., and H. V. Koontz. 1963. Uptake of nitrophenol and its effect on transpiration and calcium accumulation in barley seedlings. Plant Physiol. 38:60–65.

Barber, D. A., and B. C. Loughman. 1967. The effect of micro-organisms on the absorption of inorganic nutrients by intact plants. II. Uptake and utilization of phosphate by barley plants grown under sterile and non-sterile conditions. J. Exp. Bot. 18: 170–176.

Barber, S. A. 1962. A diffusion and mass-flow concept of soil nutrient availability. Soil Sci. 93:39–49.

Barber, S. A., J. M. Walker, and E. H. Vasey. 1962. Principles of ion movement through the soil to the plant root. Int. Soil Conf. (New Zealand), A-3:3—6.

Barber, S. A., J. M. Walker, and E. H. Vasey. 1963. Mechanism for the movement of plant nutrients from the soil and fertilizer to the plant root. Agr. Food Chem. 11:204—207.

Barr, C. E., and T. C. Broyer. 1964. Effect of light on sodium influx, membrane potential, and protoplasmic streaming in *Nitella*. Plant Physiol. 39:48—52.

Bear, F. E., and A. L. Prince. 1945. Cation-equivalent constancy in alfalfa. J. Amer. Soc. Agron. 37:217—222.

Beeson, K. C., C. B. Lyon, and M. W. Barrentine. 1944. Ionic absorption by tomato plants as correlated with variations in the composition of the nutrient medium. Plant Physiol. 19:258—277.

Bell, C. W., and O. Biddulph. 1963. Translocation of calcium. Exchange versus mass flow. Plant Physiol. 38:610—614.

Bernstein, L., and R. H. Nieman. 1960. Apparent free space of plant roots. Plant Physiol. 35:589—598.

Biddulph, O., F. S. Nakayama, and R. Cory. 1961. Transpiration stream and ascension of calcium. Plant Physiol. 36:429—436.

Bierhorst, D. W. 1953. Structure and development of the gametophyte of *Psilotum nudum*. Amer. J. Bot. 40:649—658.

Böszörményi, Z., and E. Cseh. 1961. The uptake of halide ions and their relationships in absorption. Physiol. Plantarum 14:242—252.

Bowen, G. D., and A. D. Rovira. 1966. Microbial factor in short-term phosphate uptake studies with plant roots. Nature 211:665—666.

Bower, C. A., and L. M. Turk. 1946. Calcium and magnesium deficiencies in alkali soils. J. Amer. Soc. Agron. 38:723—727.

Bradfield, G., P. Somerfield, T. Meyn, M. Holby, D. Babcock, D. Bradley, and I. H. Segel. 1970. Regulation of sulfate transport in filamentous fungi. Plant Physiol. 46:720—727.

Branton, D., and L. Jacobson. 1962. Iron transport in pea plants. Plant Physiol. 37: 539—545.

Briggs, G. E., A. B. Hope, and R. N. Robertson. 1961. Electrolytes and plant cells. *In* W. O. James (ed.) Bot. Monogr., Vol. I. Blackwell, Oxford. 217 pp.

Briscoe, C. B. 1960. Early results of mycorrhizal inoculation of pine in Puerto Rico. Caribbean Forester 20:73—77.

Brouwer, R. 1954. The regulating influence of transpiration and suction tension on the water and salt uptake by roots of intact *Vicia faba* plants. Acta Bot. Neerl. 3: 264—312.

Brouwer, R. 1956. Investigations into the occurrence of active and passive components in the ion uptake by *Vicia faba*. Acta Bot. Neerl. 5:287—314.

Broyer, T. C. 1951. The nature of the process of inorganic solute accumulation in roots, pp. 187—249. *In* E. Truog (ed.), Mineral nutrition of plants. Univ. Wisconsin Press, Madison.

Broyer, T. C., and D. R. Hoagland. 1943. Metabolic activities of roots and their bearing on the relation of upward movement of salts and water in plants. Amer. J. Bot. 30:261—273.

Broyer, T. C., and R. Overstreet. 1940. Cation exchange in plant roots in relation to metabolic factors. Amer. J. Bot. 27:425—430.

Bryant, A. E. 1934. Comparison of anatomical and histological differences between roots of barley seedlings grown in aerated and in non-aerated solutions. Plant Physiol. 9:389—391.

Chang, S. Y., R. H. Lowe, and A. J. Hiatt. 1968. Relationship of temperature to the development of calcium deficiency symptoms in *Nicotiana tabacum.* Agron. J. 60:435—436.

Clarkson, D. T. 1966. Effect of aluminum on the uptake and metabolism of phosphorus by barley seedlings. Plant Physiol. 41:165—172.

Cline, J. R., and F. P. Hungate. 1960. Accumulation of potassium, cesium[137], and rubidium[86] in bean plants grown in nutrient solution. Plant Physiol. 35:826—829.

Collander, R. 1941. Selective absorption of cations by higher plants. Plant Physiol. 16:691—720.

Collander, R. 1942. Die Elektrolyt-Permeabilität und Salz-Akkumulation pflanzlicher Zellen. Tabul. Biol. Hague 19:313—333.

Cooil, B. J., R. K. de la Fuente, and R. S. de la Pena. 1965. Absorption and transport of sodium and potassium in squash. Plant Physiol. 40:625—632.

Davis, R. F., and N. Higinbotham. 1969. Effects of external cations and respiratory inhibitors on electrical potential of the xylem exudate of excised corn roots. Plant Physiol. 44:1383—1392.

Devaux, H. 1916. Action rapide des solutions salines sur les plantes vivantes: Déplacement réversible d'une partie des substances basiques contenues dans la plante. Compt. Rend. Acad. Sci. (Paris) 162:561—563.

Drake, M., J. Vengris, and W. G. Colby. 1951. Cation-exchange capacity of plant roots. Soil Sci. 72:139—147.

Elgabaly, M. M., and L. Wiklander. 1949. Effect of exchange capacity of clay mineral and acidoid content of plant on uptake of sodium and calcium by excised barley and pea roots. Soil Sci. 67:419—424.

Elmstrom, G. W., and F. D. Howard. 1970. Promotion and inhibition of iron accumulation in soybean plants. Plant Physiol. 45:327—329.

El-Sheikh, A. M., T. C. Broyer, and A. Ulrich. 1971. Interaction of rubidium or sodium with potassium in absorption by intact sugar beet plants. Plant Physiol. 47:709—712.

Elzam, O. E., and E. Epstein. 1965. Absorption of chloride by barley roots: Kinetics and selectivity. Plant Physiol. 40:620—624.

Elzam, O. E., and T. K. Hodges. 1967. Calcium inhibition of potassium absorption in corn roots. Plant Physiol. 42:1483—1488.

Epstein, E. 1960a. Calcium-lithium competition in absorption by plant roots. Nature 185:705—706.

Epstein, E. 1960b. Spaces, barriers, and ion carriers: ion absorption by plants. Amer. J. Bot. 47:393—399.

Epstein, E. 1961. The essential role of calcium in selecting cation transport by plant cells. Plant Physiol. 36:437—444.

Epstein, E. 1966. Dual pattern of ion absorption by plant cells and by plants. Nature 212:1324—1327.

Esau, K. 1941. Phloem anatomy of tobacco affected with curly top and mosaic. Hilgardia 13:437—490.

Esau, K. 1943. Origin and development of primary vascular tissues in seed plants. Bot. Rev. 9:125—206.

Etherton, B. 1963. Relationship of cell transmembrane electropotential to potassium and sodium accumulation ratios in oat and pea seedlings. Plant Physiol. 38:581—585.

Falk, S. O. 1966a. Quantitative determinations of the effect of excision on transpiration. Physiol. Plantarum 19:493—522.

Falk, S. O. 1966b. A microwave hydrometer for measuring plant transpiration. Z. Pflanzenphysiol. 55:31—37.

Falk, S. O. 1966c. Effect on transpiration and water uptake by rapid changes in the osmotic potential of the nutrient solution. Physiol. Plantarum 19:602–617.

Fiscus, E. L., and P. J. Kramer. 1970. Radial movement of oxygen in plant roots. Plant Physiol. 45:667–669.

Fisher, J. D., D. Hansen, and T. K. Hodges. 1970. Correlation between ion fluxes and ion-stimulated adenosine triphosphatase activity of plant roots. Plant Physiol. 46:812–814.

Floyd, R. A., and D. W. Rains. 1971. Investigations of respiratory and ion transport properties of aging bean stem slices. Plant Physiol. 47:663–667.

Foote, B. D., and R. W. Howell. 1964. Phosphorus tolerance and sensitivity of soybeans as related to uptake and translocation. Plant Physiol. 39:610–613.

Franklin, R. E. 1969. Effect of adsorbed cations on phosphorus uptake by excised roots. Plant Physiol. 44:697–700.

Franklin, R. E. 1970. Effect of adsorbed cations of phosphorus absorption by various plant species. Agron. J. 62:214–216.

Freeman, G. G. 1967. Studies on potassium nutrition of plants. I. Effects of potassium concentration on growth and mineral composition of vegetable seedlings in sand culture. J. Sci. Food Agr. 18:171–176.

Fried, M., and H. Broeshart. 1967. The soil-plant system in relation to inorganic nutrition. Academic, New York.

Fried, M., H. E. Oberländer, and J. C. Noggle. 1961. Kinetics of rubidium absorption and translocation by barley. Plant Physiol. 36:183–191.

Fried, M., F. Zsoldos, P. B. Vose, and I. L. Shatokhin. 1965. Characterizing the NO_3 and NH_4 uptake process of rice roots by use of ^{15}N labelled NH_4NO_3. Physiol. Plantarum 18:313–320.

Fuhrmann, G., and A. Rothstein. 1968. The transport of Zn^{2+}, Co^{2+} and Ni^{2+} into yeast cells. Biochim. Biophys. Acta 163:325–330.

Gedroiz, K. K. 1931. Exchangeable cations of the soil and the plant: I. Relation of plant to certain cations fully saturating the soil exchange capacity. Soil Sci. 32:51–63.

Gilbert, S. G., and J. W. Shive. 1945. The importance of oxygen in the nutrient substrate for plants -- Relation of the nitrate ion to respiration. Soil Sci. 59:453–460.

Graham, E. R., and W. A. Albrecht. 1943. Nitrate absorption by plants as an anion exchange phenomenon. Amer. J. Bot. 30:195–198.

Greenway, H., and B. Klepper. 1969. Relation between anion transport and water flow in tomato plants. Physiol. Plantarum 22:208–219.

Greenway, H., P. G. Hughes, and B. Klepper. 1969. Effects of water deficit on phosphorus nutrition of tomato plants. Physiol. Plantarum 22:199–207.

Handley, R., and R. Overstreet. 1961. Effect of various cations upon absorption of carrier-free cesium. Plant Physiol. 36:66–69.

Handley, R., and R. Overstreet. 1963. Uptake of strontium by roots of *Zea mays*. Plant Physiol. 38:

Handley, R., and R. Overstreet. 1968. Uptake of carrier-free ^{137}Cs by *Ramalina reticulata*. Plant Physiol. 43:1401–1405.

Hanson, J. B., S. S. Malhotra, and C. D. Stoner. 1965. Action of calcium on corn mitochondria. Plant Physiol. 40:1033–1040.

Harley, J. L., and C. C. McCready. 1950. The uptake of phosphate by excised mycorrhizal roots of the beech. New Phytol. 49:388–397.

Hart, J. W., and P. Filner. 1969. Regulation of sulfate uptake by amino acids in cultured tobacco cells. Plant Physiol. 44:1253–1259.

Heide, H. van der, B. M. de Boer-Bolt, and M. H. van Raalte. 1963. The effect of a low oxygen content of the medium on the roots of barley seedlings. Acta Bot. Neerl. 12:231–247.

Heilman, P. E. 1967. Manganese deficiency in cauliflower and broccoli induced by soil fumigation with dichloropropenes. Soil Sci. 103:401–403.

Heintze, S. G. 1961. Studies on the cation-exchange capacities of roots. Plant and Soil 13:365–383.

Helder, R. J. 1958. Studies on the absorption, distribution and release of labelled rubidium ions in young intact barley plants. Acta Bot. Neerl. 7:235–249.

Hendricks, S. B., and P. C. Jackson. 1970. Probing the gatekeeper's secrets. Agr. Res. 18(12):8–9.

Hiatt, A. J. 1970a. An anomaly in potassium accumulation by barley roots. I. Effect of anions, sodium concentration, and length of absorption period. Plant Physiol. 45:408–410.

Hiatt, A. J. 1970b. An anomaly in potassium accumulation by barley roots. II. Effect of calcium concentration and rubidium-86 labeling. Plant Physiol. 45:411–414.

Higinbotham, N., B. Etherton, and R. J. Foster. 1961. The source and significance of the electropotential of higher plant cells. Plant Physiol. 36:xxxv.

Higinbotham, N., M. J. Pratt, and R. J. Foster. 1962. Effects of calcium, indoleacetic acid, and distance from stem apex on potassium and rubidium absorption by excised segments of etiolated pea epicotyl. Plant Physiol. 37:203–214.

Higinbotham, N., B. Etherton, and R. J. Foster. 1967. Mineral ion contents and cell transmembrane electropotentials of pea and oat seedling tissue. Plant Physiol. 42:37–46.

Hills, F. J., R. L. Salisbery, A. Ulrich, and K. M. Sipitanos. 1970. Effect of phosphorus on nitrate in sugar beet (Beta vulgaris L.). Agron. J. 62:91–92.

Hoagland, D. R., and T. C. Broyer. 1936. General nature of the processes of salt accumulation by roots with description of experimental methods. Plant Physiol. 11:471–507.

Hoagland, D. R., and T. C. Broyer. 1940. Hydrogen-ion effects and the accumulation of salt by barley roots as influenced by metabolism. Amer. J. Bot. 27:173–185.

Hodges, T. K., and Y. Vaadia. 1964. Uptake and transport of radiochloride and tritiated water by various zones of onion roots of different chloride status. Plant Physiol. 39:104–108.

Höber, R. 1946. Physical chemistry of cells and tissues. Blakiston, Philadelphia.

Honert, T. H. van den. 1933. The phosphate absorption by sugar cane. Verslag 13e Bijeenkomst van de Vereeniging van Proefstations-Personeel, Buitenzorg, Java, 7–20.

Honert, T. H. van den, and J. J. M. Hooymans. 1955. On the absorption of nitrate by maize roots in water culture. Acta Bot. Neerl. 4:376–384.

Honert, T. H. van den, J. J. M. Hooymans, and W. S. Volkers. 1955. Experiments on the relation between water absorption and mineral uptake by plant roots. Acta Bot. Neerl. 4:139–155.

Hope, A. B., and N. A. Walker. 1960. Ionic relations of cells of Chara australis. III. Vacuolar fluxes of sodium. Australian J. Biol. Sci. 13:277–291.

Hopkins, H. T. 1956. Absorption of ionic species of orthophosphate by barley roots: Effects of 2,4-dinitrophenol and oxygen tension. Plant Physiol. 31:155–161.

Hopkins, H. T., A. W. Specht, and S. B. Hendricks. 1950. Growth and nutrient accumulation as controlled by oxygen supply to plant roots. Plant Physiol. 25:193–209.

Hurd, R. B. 1958. The effect of pH and bicarbonate ions on the uptake of salts by discs of red beet. J. Exp. Bot. 9:159–174.

Hutchin, M. E., and B. E. Vaughan. 1967. Relation between calcium and strontium transport rates as determined simultaneously in isolated segments of the primary root of *Zea mays*. Plant Physiol. 42:644—650.

Hylmö, B. 1953. Transpiration and ion absorption. Physiol. Plantarum 6:333—405.

Hylmö, B. 1955. Passive components in the ion absorption of the plant. I. The zonal ion and water absorption in Brouwer's experiments. Physiol. Plantarum 8:433—449.

Hylmö, B. 1958. Passive components in the ion absorption of the plant. II. The zonal water flow, ion passage and pore size in roots of *Vicia faba*. Physiol. Plantarum 11:382—400.

Hylton, L. O., A. Ulrich, and D. R. Cornelius. 1967. Potassium and sodium interrelations in growth and mineral content of Italian ryegrass. Agron. J. 59:311—314.

Itallie, T. B. van. 1948. Cation equilibria in plants in relation to the soil: II. Soil Sci. 65:393—415.

Jackson, P., and J. M. Taylor. 1970. Effects of organic acids on ion uptake and retention in barley roots. Plant Physiol. 46:538—542.

Jackson, P., S. B. Hendricks, and B. M. Vasta. 1962. Phosphorylation by barley root mitochondria and phosphate absorption by barley roots. Plant Physiol. 37:8—17.

Jackson, P., J. M. Taylor, and S. B. Hendricks. 1970. Entry of organic acid anions into roots. Proc. Nat. Acad. Sci. 65:176—184.

Jacobson, L., R. Overstreet, R. M. Carlson, and J. Chastain. 1957. The effect of pH and temperature on the absorption of K and Br by barley roots. Plant Physiol. 32:658—662.

Jacobson, L., D. P. Moore, and R. J. Hannapel. 1960. Role of calcium in absorption of monovalent cations. Plant Physiol. 35:352—358.

Jacobson, L., R. J. Hannapel, M. Schaedle, and D. P. Moore. 1961. Effect of root to solution ratio in ion absorption experiments. Plant Physiol. 36:62—65. 5.

Jacoby, B. 1964. Function of bean roots and stems in sodium retention. Plant Physiol. 39:445—449.

Jacoby, B. 1965. Sodium retention in excised bean stems. Physiol. Plantarum 18:730—739.

Jacoby, B., and J. Dagan. 1969. Effects of age on sodium fluxes in primary bean leaves. Physiol. Plantarum 22:29—36.

Jenny, H. 1946. Adsorbed nitrate ions in relation to plant growth. J. Colloid Sci. 1:33—47.

Jenny, H., and E. W. Cowan. 1933. Über die Bedeutung der im Boden adsorbierten Kationen für das Pflanzenwachstum. Z. Pflanzenernähr. Düng. Bodenk. A31:57—67.

Jeschke, W. D. 1970. The influx of potassium ions in leaves of *Elodea densa*, dependence on light, potassium concentration, and temperature. Planta 91:111—128.

Johansen, C., D. G. Edwards, and J. F. Loneragan. 1968a. Interactions between potassium and calcium in their absorption by intact barley plants. I. Effects of potassium on calcium absorption. Plant Physiol. 43:1717—1721.

Johansen, C., D. G. Edwards, and J. F. Loneragan . 1968b. Interactions between potassium and calcium in their absorption by intact barley plants. II. Effects of calcium and potassium concentrations on potassium absorption. Plant Physiol. 43:1722—1726.

Johansen, C., D. G. Edwards, and J. F. Loneragan. 1970. Potassium fluxes during potassium absorption by intact barley plants of increasing potassium content. Plant Physiol. 45:601—603.

Kahn, J. S., and J. B. Hanson. 1957. The effect of calcium on potassium accumulation in corn and soybean roots. Plant Physiol. 32:312—316.

Kannan, S. 1969. Factors related to iron absorption by enzymatically isolated leaf cells. Plant Physiol. 44:1457—1460.

Kannan, S., and T. Mathew. 1970. Effects of growth substances on the absorption and transport of iron in plants. Plant Physiol. 45:206–209.

Keller, P. von, and H. Deuel. 1957. Kationenaustauschkapazität und Pektingehalt von Pflanzenwurzeln. Z. Pflanzenernähr. Düng. Bodenk. 79:119–131.

Kelley, A. P. 1950. Mycotrophy in plants. Chronica Botanica, Waltham, Mass. 223 pp.

Ketchum, B. H. 1939. The absorption of phosphate and nitrate by illuminated cultures of *Nitzschia closterium*. Amer. J. Bot. 26:399–407.

Khudairi, A. K. 1969. Mycorrhiza in desert soils. Bioscience 19:598–599.

Kleese, R. A., and L. J. Smith. 1970. Scion control of genotypic differences in mineral salts accumulation in soyabean (*Glycine max* L. Merr.) seeds. Ann. Bot. 34:183–189.

Klingensmith, M. J. 1961. The effect of benzimidazole on cation uptake by plant roots. Amer. J. Bot. 48:711–716.

Klingensmith, M. J., and A. G. Norman. 1960. Benzimidazole enhancement of ion uptake by plant roots. Science 131:354–355.

Koontz, H. V., and R. E. Foote. 1966. Transpiration and calcium deposition by unifoliate leaves of *Phaseolus vulgaris* differing in maturity. Physiol. Plantarum 19:313–321.

Kramer, P. J. 1957. Outer space in plants. Science 125:633–635.

Kuiper, P. J. C. 1968a. Lipids in grape roots in relation to chloride transport. Plant Physiol. 43:1367–1371.

Kuiper, P. J. C. 1968b. Ion transport characteristics of grape root lipids in relation to chloride transport. Plant Physiol. 43:1372–1374.

Kylin, A. 1966. Uptake and loss of Na^+, Rb^+, and Cs^+ in relation to an active mechanism for extrusion of Na^+ in *Scenedesmus*. Plant Physiol. 41:579–584.

Kylin, A., and R. Gee. 1970. Adenosine triphosphatase activities in leaves of the mangrove *Avicennia nitida* Jacq. Plant Physiol. 45:169–172.

Lazaroff, N., and M. G. Pitman. 1966. Calcium and magnesium uptake by barley seedlings. Australian J. Biol. Sci. 19:991–1005.

Leggett, J. E. 1961. Entry of phosphate into yeast cell. Plant Physiol. 36:277–284.

Leggett, J. E., and E. Epstein. 1956. Kinetics of sulfate absorption by barley roots. Plant Physiol. 31:222–226.

Leggett, J. E., and W. A. Gilbert. 1967. Localization of the Ca-mediated apparent ion selectivity in the cross-sectional volume of soybean roots. Plant Physiol. 42:1658–1664.

Leggett, J. E., and W. A. Gilbert. 1969. Magnesium uptake by soybeans. Plant Physiol. 44:1182–1186.

Leggett, J. E., R. A. Olsen, and B. D. Spangler. 1962. Cation absorption by baker's yeast as a passive process. Proc. Nat. Acad. Sci. 48:1949–1956.

Leggett, J. E., R. A. Galloway, and H. G. Gauch. 1965a. Calcium activation of orthophosphate absorption by barley roots. Plant Physiol. 40:897–902.

Leggett, J. E., W. R. Heald, and S. B. Hendricks. 1965b. Cation binding by baker's yeast and resins. Plant Physiol. 40:665–671.

Leonard, O. A., and J. A. Pinckard. 1946. Effect of various oxygen and carbon dioxide concentrations on cotton root development. Plant Physiol. 21:18–36.

Levi, E. 1970. The influence of accompanying cations on the foliar uptake of Na, K, Rb, and Cs. Physiol. Plantarum 23:871–877.

Lewis, D. G., and J. P. Quirk. 1967. Phosphate diffusion in soil and uptake by plants. III. P^{31} movement and uptake by plants as indicated by P^{32} autoradiography. Plant and Soil 26:454–468.

Lingle, J. C., L. O. Tiffin, and J. C. Brown. 1963. Iron uptake-transport of soybeans as influenced by other cations. Plant Physiol. 38:71–76.

Loehwing, W. F. 1951. Mineral nutrition in relation to the ontogeny of plants, pp.

343—358. *In* E. Truog (ed.) Mineral nutrition of plants, Univ. Wisconsin Press, Madison.

Lookeren Campagne, R. N. van. 1957. Light-dependent chloride absorption in *Vallisneria* leaves. Acta Bot. Neerl. 6:543—582.

Lopushinsky, W. 1964. Effect of water movement on ion movement into the xylem of tomato roots. Plant Physiol. 39:494—501.

Lüttge, U., and C. B. Osmond. 1970. Ion absorption in *Atriplex* leaf tissue. III. Site of metabolic control of light-dependent chloride secretion to epidermal bladders. Australian J. Biol. Sci. 23:17—25.

Maas, E. V., and J. E. Leggett. 1968. Uptake of ^{86}Rb and K by excised maize roots. Plant Physiol. 43:2054—2056.

McCollum, R. E. 1960. Evidence for implication of a genetic factor for chloride accumulation by soybeans. Agron. Abstr., p. 21.

MacDonald, I. E. 1967. Bacterial infection and ion absorption capacity in beet disks. Ann. Bot. 31:163—172.

McEvoy, E. T. 1955. Interaction of sodium and potassium on growth and mineral content of flue-cured tobacco. Can. J. Agr. Sci. 35:294—299.

Macklon, A. E. S., and N. Higinbotham. 1968. Potassium and nitrate uptake and cell transmembrane electropotential in excised pea epicotyls. Plant Physiol. 43:888—892.

McLean, E. O. 1956. Uptake of sodium and other cations by five crop species. Soil. Sci. 82:21—28.

McLean, E. O., D. Adams, and R. E. Franklin. 1956. Cation exchange properties of plant roots as related to their nitrogen contents. Proc. Soil Sci. Soc. Amer. 20: 345—347.

MacRobbie, E. A. C. 1962. Ionic relations of *Nitella translucens*. J. Gen. Physiol. 45: 861—878.

MacRobbie, E. A. C., and J. Dainty. 1958. Sodium and potassium distribution and transport in the sea weed *Rhodymenia palmata* (L.) Brev. Physiol. Plantarum 11: 782—801.

Marschner, H. 1964. Einfluss von Calcium auf die Natriumaufnahme und die Kaliumabgabe isolierter Gerstenwirzeln. Z. Pflanzenernähr. Düng. Bodenk. 107:19—32.

Marschner, H., and K. Mengel. 1966. Der Einfluss von Ca- und H-Ionen bei unterschiedlichen Stoffwechselbedingungen auf die Membranepermeabilitat junger Gerstenwurzeln. Z. Pflanzenernähr. Düng. Bodenk. 112:39—49.

Marshall, C. E. 1948. Ionization of calcium from soil colloids and its bearing on soil-plant relationships. Soil Sci. 65:57—68.

Meyer, F. H. 1966. Mycorrhiza and other plant symbioses, pp. 171—255. *In* S. M. Henry (ed.) Symbiosis. Vol. I, Associations of microorganisms, plants, and marine organisms. Academic, New York. 478 p.

Middleton, L. J., R. Handley, and R. Overstreet. 1960. Relative uptake and translocation of potassium and cesium in barley. Plant Physiol. 35:913—918.

Millard, D. L., J. T. Wiskich, and R. N. Robertson. 1965. Ion uptake and phosphorylation in mitochondria: Effect of monovalent ions. Plant Physiol. 40:1129—1135.

Milthorpe, F. L., and J. Moorby. 1969. Vascular transport and its significance in plant growth. Annu. Rev. Plant Physiol. 20:117—138.

Morris, I., and J. Ahmed. 1969. The effect of light on nitrate and nitrite assimilation by *Chlorella* and *Ankistrodesmus*. Physiol. Plantarum 22:1166—1174.

Mouat, M. C. H. 1960. Interspecific differences in strontium uptake by pasture plants as a function of root cation-exchange capacity. Nature 188:513—514.

Mulder, E. G., and F. C. Gerretsen. 1952. Soil manganese in relation to plant growth. Advance. Agron. 4:221—277.

Nagai, R., and M. Tazawa. 1962. Changes in the resting potential and ion absorption induced by light in a single cell. Plant and Cell 3:323—339.

Nightingale, G. T. 1942. Potassium and phosphate nutrition of pineapple in relation to nitrate and carbohydrate reserves. Bot. Gaz. 104:191—223.

Nims, L. F. 1962. Tracers, transfer through membranes, and coefficients of transfer. Science 137:130—132.

Nobel, P. S. 1969. Light-dependent potassium uptake by *Pisum sativum* leaf fragments. Plant Cell Physiol. 10:597—605.

Nobel, P. S. 1970. Relation of light-dependent potassium uptake by pea leaf fragments to the pK of the accompanying organic acid. Plant Physiol. 46:491—493.

Nobel, P. S., and L. Packer. 1965. Light-dependent ion translocation in spinach chloroplasts. Plant Physiol. 40:633—640.

Osmond, C. B. 1966. Divalent cation absorption and interaction in *Atriplex*. Australian J. Biol. Sci. 19:37—48.

Overstreet, R., L. Jacobson, and R. Handley. 1952. The effect of calcium on the absorption of potassium by barley roots. Plant Physiol. 27:583—590.

Packer, L., S. Murakami, and C. W. Mehard. 1970. Ion transport in chloroplasts and plant mitochondria. Annu. Rev. Plant Physiol. 21:271—304.

Parr, J. F., and A. G. Norman. 1962. Buffering effects of benzimidazole in absorption of potassium by excised barley roots. Plant Physiol. 37:821—825.

Parr, J. F., and A. G. Norman. 1964. Effects of nonionic surfactants on root growth and cation uptake. Plant Physiol. 39:502—507.

Pepkowitz, L. P., and J. W. Shive. 1944. The importance of oxygen in the nutrient substrate for plants — Ion absorption. Soil Sci. 57:143—154.

Pettersson, S. 1960. Ion absorption in young sunflower plants. I. Uptake and transport mechanisms for sulphate. Physiol. Plantarum 13:133—147.

Pinkas, L. L. H., and L. H. Smith. 1966. Physiological basis of differential strontium accumulation in two barley genotypes. Plant Physiol. 41:1471—1475.

Pitman, M. G. 1964. The effect of divalent cations on the uptake of salt by beetroot. J. Exp. Bot. 15:444—456.

Pitman, M. G. 1965. Sodium and potassium uptake by seedlings of *Hordeum vulgare*. Australian J. Biol. Sci. 18:10—24.

Pitman, M. G. 1969. Adaptation of barley roots to low oxygen supply and its relation to potassium and sodium uptake. Plant Physiol. 44:1233—1240.

Pitman, M. G. 1970. Active H^+ efflux from cells of low-salt barley roots during salt accumulation. Plant Physiol. 45:787—790.

Pitman, M. G., A. C. Courtice, and B. Lee. 1968. Comparisons of potassium and sodium uptake by barley roots at high and low salt status. Australian J. Biol. Res. 21:871—881.

Poole, R. J., and L. W. Poel. 1965. Carbon dioxide and pH in relation to salt uptake by beetroot tissue. J. Exp. Bot. 16:453—461.

Rains, D. W. 1968. Kinetics and energetics of light-enhanced potassium absorption by corn leaf tissue. Plant Physiol. 43:394—400.

Rains, D. W., and E. Epstein. 1967. Sodium absorption by barley roots: Role of the dual mechanisms of alkali cation transport. Plant Physiol. 42:314—318.

Rains, D. W., and R. A. Floyd. 1970. Influence of calcium on sodium and potassium absorption by fresh and aged bean stem slices. Plant Physiol. 46:93—98.

Randall, P. J., and P. B. Vose. 1963. Effect of aluminum on uptake and translocation of phosphorus by perennial ryegrass. Plant Physiol. 38:403—409.

Rathmore, V. S., S. H. Wittwer, W. H. Jyung, Y. P. S. Bajaj, and M. W. Adams. 1970. Mechanisms of zinc uptake in bean (*Phaseolus vulgaris*) tissues. Physiol. Plantarum 23:908—919.

Ratner, E. I. 1935. The influence of exchangeable sodium in the soil on its properties as a medium for plant growth. Soil Sci. 40:459–468.

Ratner, E. I. 1944. Interaction between roots and soil colloids as a problem on the physiology of mineral nutrition of plants. I. Unstable equilibria in the cation exchange between the roots of plants and the soil colloids. Comp. Rend. (Doklady) Acad. Sci. 42:313–317.

Reed, H. S., and T. Fremont. 1935. Factors that influence the formation and development of mycorrhizal associations in citrus roots. Phytopathology 25:645–647.

Riekels, J. W., and J. C. Lingle. 1966. Iron uptake and translocation by tomato plants as influenced by root temperature and manganese nutrition. Plant Physiol. 41: 1095–1101.

Robson, A. D., D. G. Edwards, and J. F. Loneragan. 1970. Calcium stimulation of phosphate absorption by annual legumes. Australian J. Agr. Res. 21:601–612.

Rorison, I. H. 1969. Ecological aspects of mineral nutrition in plants. V. A. Davis, Philadelphia.

Routien, J. B. and R. F. Dawson. 1943. Some interrelationships of growth, salt absorption, respiration, and mycorrhizal development in *Pinus echinata* Mill. Amer. J. Bot. 30:440–451.

Russell, R. S., and D. A. Barber. 1960. The relationship between salt uptake and the absorption of water by intact plants. Annu. Rev. Plant Physiol. 11:127–140.

Russell, R. S., and V. M. Shorrocks. 1957. The effect of transpiration on the absorption of inorganic ions by intact plants. Radioisotopes Sci. Res. 4:286.

Russell, R. S., and V. M. Shorrocks. 1959. The relationship between transpiration and the absorption of inorganic ions by intact plants. J. Exp. Bot. 10:301–316.

Salo, A., and J. K. Miettinen. 1964. Strontium 90 and cesium 137 in Arctic vegetation during 1961. Nature 201:1177–1178.

Schaedle, M., and L. Jacobson. 1966. Ion absorption and retention by *Chlorella pyrenoidosa*. II. Permeability of the cell to sodium and rubidium. Plant Physiol. 41:248–254.

Schaedle, M., and L. Jacobson. 1967. Ion absorption and retention by *Chlorella pyrenoidosa*. III. Selective accumulation of rubidium, potassium, and sodium. Plant Physiol. 42:953–958.

Schmidt, O. 1936. Die Mineralstoffaufnahme der höheren Pflanze als Funktion einer Wechselbeziehung zwischen inneren und äusseren Faktoren. Z. Bot. 30:289–334.

Scott, B. I. H. 1967. Electric fields in plants. Annu. Rev. Plant Physiol. 18:409–418.

Siegel, S. M., and O. Daly. 1966. Regulation of betacyanin efflux from beet root by poly-L-lysine, Ca-ion, and other substances. Plant Physiol. 41:1429–1434.

Smith, R. C. 1970. Time course of exudation from excised corn root segments of different stages of development. Plant Physiol. 45:571–575.

Smith, R. C., and E. Epstein. 1964. Ion absorption by shoot tissue: Kinetics of potassium and rubidium absorption by corn leaf tissue. Plant Physiol. 39:992–996.

Steward, K. K., and H. V. Koontz. 1968. Transport of rubidium absorbed by the apex of the bean root. Plant Physiol. 43:583–588.

Stout, P. R., and R. Overstreet. 1950. Soil chemistry in relation to inorganic nutrition of plants. Annu. Rev. Plant Physiol. 1:305–342.

Surrey, K., and S. A. Gordon. 1962. Influence of light on phosphate metabolism in lettuce seed: spectral response red, far-red interaction. Plant Physiol. 37:327–332.

Sutcliffe, J. F. 1959. Salt uptake in plants. Biol. Rev. 34:159–218.

Takada, H., and S. Nagai. 1953. Notes on the rich chloride content and the osmotic pressure of *Metasequoia glyptostroboides*. Proc. Japan Acad. 29:274–278.

Tanada, T. 1955. Effects of ultraviolet radiation and calcium and their interaction on salt absorption by excised mung bean roots. Plant Physiol. 30:221—225.

Tanada, T. 1956. Effect of ribonuclease on salt absorption by excised mung bean roots. Plant Physiol. 31:251—253.

Tanada, T. 1962. Localization and mechanism of calcium stimulation of rubidium absorption in the mung bean root. Amer. J. Bot. 49:1068—1072.

Tanada, T. 1963. Kinetics of Rb absorption by excised barley roots under changing Rb concentrations. I. Effects of other cations on Rb uptake. Plant Physiol. 38: 422—425.

Thorne, D. W. 1944. Growth and nutrition of tomato plants as influenced by exchangeable sodium, calcium, and potassium. Proc. Soil Sci. Soc. Amer. 9:185—189.

Thorne, D. W. 1946. Calcium carbonate and exchangeable sodium in relation to the growth and composition of plants. Proc. Soil Sci. Soc. Amer. 11:397—401.

Vaughan, B. E., E. C. Evans, III, and M. E. Hutchin. 1967. Polar transport characteristics of radiostrontium and radiocalcium in isolated corn root segments. Plant Physiol. 42:747—750.

Viets, F. G. 1944. Calcium and other polyvalent cations as accelerators of ion accumulation by excised barley roots. Plant Physiol. 19:466—480.

Vlamis, J., and A. R. Davis. 1944. Effects of oxygen tension on certain physiological responses of rice, barley, and tomato. Plant Physiol. 19:33—51.

Wadleigh, C. H., and J. W. Shive. 1939. Base content of corn plants as influenced by pH of substrate and form of nitrogen supply. Soil Sci. 47:273—283.

Walker, J. M. 1967. Soil temperature patterns in surface-insulated containers in water baths related to maize behavior. Proc. Soil Sci. Soc. Amer. 31:400—403.

Walker, J. M. 1969. One-degree increments in soil temperatures affect maize seedling behavior. Proc. Soil Sci. Soc. Amer. 33:729—736.

Walker, N. A. 1955. Microelectode experiments on *Nitella*. Australian J. Biol. Sci. 8:476—489.

Wallace, A., and N. Hemaidan. 1963. Sodium transport from roots to shoots in bush bean and radish. Plant Physiol. 38:viii.

Wallace, A., R. T. Ashcroft, M. W. M. Leo, and G. A. Wallace. 1970. Effect of cyclohexamide, gamma irradiation, and phosphorus deficiency on root pressure exudation in tobacco. Plant Physiol. 45:300—303.

Wallihan, E. F. 1961. Effect of sodium bicarbonate on iron absorption by orange seedlings. Plant Physiol. 36:52—53.

Weigl, J. 1970. Effect of CCCP and UO_2^{++} on ion fluxes in roots. Planta 91:270—273.

Went, F. W., and N. Stark. 1968. Mycorrhiza. Bioscience 18:1035—1039.

Wiebe, H. H., and P. J. Kramer. 1954. Translocation of radioactive isotopes from various regions of roots of barley seedlings. Plant Physiol. 29:342—348.

Williams, D. E., and N. T. Coleman. 1950. Cation exchange properties of plant root surfaces. Plant and Soil 2:243—256.

Williams, D. E., and J. Vlamis. 1962. Differential cation and anion absorption as affected by climate. Plant Physiol. 37:198—202.

York, E. T., Jr., R. Bradfield, and M. Peech. 1953. Calcium-potassium interactions in soils and plants: II. Reciprocal relationship between calcium and potassium in plants. Soil Sci. 76:481—491.

CHAPTER 7

Permeability and Membranes

INTRODUCTION

The ease, or lack of it, with which ions or molecules move into or out of cells is called permeability, and cell membranes control this phenomenon. Walls of parenchyma cells are freely permeable to most substances, and they offer negligible resistance to water movement (Levitt et al., 1936; Russell and Woolley, 1961).

Permeability is a property of the membrane and not of substances traversing it (Kramer, 1969). Cells of certain plants may be freely permeable to a given ion, for example, Na^+, whereas cells of other plants may be relatively impermeable. Rates at which substances traverse membranes are sometimes but not always related to size, solubility in water or lipid, or any other given characteristic of the substance. For that reason we conclude that membranes are differentially permeable since, for cells of any given plant, they permit rapid passage of some substances, but only slow passage of others. Semipermeable is sometimes used, but it is an ambiguous term which is less descriptive than the former term.

Prior to discussing permeability and membranes (Davson and Danielli, 1943, 1952), it should be noted that if it were possible to explain permeability data in terms of membranes, a discussion of membranes should logically precede a discussion of permeability. Unfortunately, it is not possible to achieve this goal at this time and, therefore, general information on permeability is presented first in order that information on membranes may be viewed in the light of reported observations and data on permeability.

PERMEABILITY

Permeability pertains to the "the *capacity* of a membrane to permit movement of a given substance through it." Terms in this definition are important. "Capacity" includes all rates of entry or exit from nearly zero to very high rates. Permeability is not an all-or-none proposition since, even for some substance whose movement through a membrane is highly restricted, there is some movement. "Membrane" denotes and emphasizes the fact that permeability is a property of membranes. "Movement" is deemed more appropriate than "diffusion," for example, since ions may possibly move through membranes by some process other than diffusion. "A given substance" must be included in any proper definition of permeability, since permeability to various substances may differ widely, and one must, therefore, indicate what substance is being considered.

Permeability should be expressed in terms of "the *amount* of a *given substance* moving through a *given area* of membrane in a *given time* at a *given temperature* in response to a *given concentration gradient* across the membrane." The number of particles moving per unit time, through a given area of a membrane, is called flux. Flux (J) is equal to the permeability (L_a) of the membrane multiplied by the driving force (ΔG) causing diffusion:

$$J = L_a \Delta G$$

Criticism of Certain Studies of "Permeability"

In numerous reports an investigator refers to the effect of a given factor or treatment on the permeability of roots, for example, to some substance or substances. All too frequently, plants from two or more treatments are compared with regard to concentration of K^+, for example, which plants contain at the end of the experiment. Considerable time may have elapsed between initiation of treatments and the time the plants were sampled for analysis. Treatments may have had very diverse effects on growth and extent of root systems and particularly on amounts of absorbing surface on the root systems under study. Plants with a more extensive root system and root surface are usually larger plants than those in the other treatments. Let us suppose that the smaller plants showed a higher concentration of K^+ in the roots, stems, and leaves than did those same parts of larger plants from another treatment. Does it necessarily follow, then, that membranes of root cells of the smaller plants were more permeable to K^+? Many investigators have drawn such conclusions, but in the two treatments they obviously have not truly studied permeability of the roots to K^+. If the total dry weight of roots, stems, and leaves, of the two kinds of plants is also known, it would be possible to multiply each of these values by the percentage of K^+ in each part and to determine the total amount of K^+ in the two kinds of plants. Although the smaller plants, stunted in comparison with the larger plants because of the treatment they received, had higher *concentrations* of K^+ in their plant parts, it is wholly possible that these parts might contain lesser *amounts* of K^+ than the corresponding parts of the larger plants from another treatment. Percentage composition of plants is often the basis of erroneous conclusions in plant nutrition studies and, certainly in the case just stated, it would be no true measure of permeability of roots to potassium.

Farhoomand and Peterson (1968) recently called attention to the difference between "concentration" and "content"—terms so often used incorrectly.

Rates of Entry or Absorption

Permeability pertains to the potential rate at which a substance can enter a cell, and absorption to the actual rate of entry of the substance into a cell. There may even be a negative relationship between permeability of a cell to a given substance and rate of absorption. If the cell already contains a high concentration of a substance to which it is highly permeable, the *rate* of absorption may be quite low. However, if a substance of low permeability is being used within the cell, the substance may enter at a more rapid rate than one of high permeability which is not being utilized within the cell.

Factors Affecting Permeability

Membrane permeability may change (1) immediately after application of a suitable external factor (e.g., ionizing radiation, chemicals, water shortage, temperature) or (2) after a latent period—probably by changing metabolism which in turn modifies membrane structure. Stadelmann (1969a) suggested that permeability changes could be explained by a shift in the ratio between leaflet and micelle membrane surface area. Changes from lipid layer to the more stable leaflet form would reduce the number of membrane pores and thereby lower the permeability.

Stadelmann (1969b) noted that a substance or factor that affects permeability does not necessarily cause a progressive increase in permeability when the substance or factor is increased. At a certain intensity of the causative factor, permeability may be decreased. Such a reversal would explain contradictory reports in the literature with regard to whether a given substance or factor increases or decreases permeability.

Several factors have been studied with respect to their effects on the permeability of membranes, and Slatyer (1967) discusses the measurement of cell and tissue permeability. Kramer (1969) presents the errors inherent in measurements of permeability. The factors discussed below have received the most attention:

Temperature. Within limits permeability generally increases with increase in temperature until 50–60°C is reached. At this point death of most cells occurs and there is an irreversible increase in permeability. Temperatures low enough to produce ice crystals generally result in disruption of protoplasm, damage to the cells and to the membranes, and an increase in permeability. This increase, however, is no longer characteristic of a living cell.

Honert and Hooymans (1955) stated that a "solidification" of lipoid membranes presumably inhibits NO_3^- absorption at temperatures below 10° C.

Judged by betacyanin efflux, beet root tissue differs in stability toward O_2 at low and high temperatures (45–60°C and 60–100°C, respectively) (Siegel, 1969). The effect of temperature can be divided into a high activation energy (93 kcal/mole) process in the lower temperature range and a low activation energy process (19 kcal/mole) in the higher range (greater than 60°C). It was suggested that elevating the temperature initially brings about reversible conformational changes in the membrane. With continuing increase in temperature in the presence of O_2, membrane chemical groups susceptible to oxidation are exposed, and upon oxidation render conformational changes irreversible.

Effects of Various Ions. The effects of various ions on permeability have been studied extensively since the time of Osterhout's (1936) earliest studies. In general, it has been reported that monovalent cations increase permeability of membranes to substances, whereas divalent and trivalent cations decrease it. According to Osterhout (1922), who used conductivity measurements of *Laminaria* to determine changes in overall permeability, Li^+, Na^+, K^+, NH_4^+, Cs^+, and Rb^+ increase permeability. The cations are listed here in their probable order of effectiveness in increasing permeability, and this is approximately the order of the lyotropic series.

A low concentration of $NaNO_3$ increases permeability much more than the same concentration of KNO_3 (Kuiper, 1963).

Although a later effect may be to increase permeability, initially permeability may be reduced by bivalent and trivalent cations. The ultimate effect on permeability appears to depend on whether the bivalent and trivalent cations are associated with monovalent or polyvalent anions.

Antagonism of Ions. When one ion reverses an effect that another ion ordinarily produces, the interaction between these two ions is called antagonism, that is, one ion antagonizes the action of the other ion in some phenomenon such as the permeability of beet root cell membranes to the red, vacuolar pigment, betacyanin. For example, Bayliss (1924) reported that, when freshly cut red beet tissue was immersed in 0.31 M NaCl, betacyanin diffused out of the tissue. It did not do so in water, and thus NaCl had increased the permeability of membranes to the pigment. However, when $CaCl_2$ was added to the 0.31 M NaCl to bring $CaCl_2$ to a concentration of 0.17 M, the pigment did not diffuse out of the cells. Evidently, then, Ca^{2+} ions in some way reversed the usual effect of Na^+ ions on permeability to betacyanin.

In view of the extremely low concentration at which a given ion may antagonize the action or effect of another ion, Salisbury and Ross (1969) concluded that antagonism remains a mystery.

Nonelectrolytes. Some nonelectrolytes, such as ether and chloroform, generally decrease overall permeability when used at extremely low concentrations. In slightly higher concentrations their use results in an increase in permeability. In still higher concentrations there is a further increase which is irreversible when either of these agents is removed.

Hydration. The degree of water availability in cells determines the degree of hydration of cell colloids—other factors such as the quality and quantity of electrolytes remaining constant. The availability of water also affects the degree of hydration of membranes. Usually, the more highly hydrated, the more permeable the membranes become to substances in general, and vice versa. However, it has also been reported that at water potentials even as low as -20 atm the mechanisms of ion uptake and integrity of the tonoplast appear to be maintained, but permeability of the plasmalemma and leakage from cells are greatly increased (Greenway et al., 1968).

Light. Lepeschkin (1930), among others, showed that light generally increases permeability. In his review of the effect of light on the mineral nutrition of plants, Withrow (1951) reported that most of the early work indicated that light favored uptake of various ions and/or salts. He noted, however, that many of the conclusions were drawn before the work of Hoagland et al. (1926) which showed the dependence of the salt uptake process on a source of respiratory energy such as sugars.

More recent research on the effect of light has corrected for the effect of light in providing respiratory substrate, hence the results appear to be specific effects of light on permeability per se. Light was reported to increase the permeability of cells to Na^+ (Hope and Walker, 1960), Na^+ and K^+ (MacRobbie and Dainty, 1958), and NO_3^- (Beevers et al., 1965). For a seaweed, *Rhodymenia palmata* (L.) Brev., light increased the permeability of the cell membranes to K^+ efflux about 8-fold (MacRobbie and Dainty, 1958). Light increased the resting electropotential of cell membranes of *Nitella flexilis* and resulted in an accelerated ion uptake according to Nagai and Tazawa (1962). These investigators divided the effect of light on ion uptake into two processes: (1) movement across the plasmalemma which was controlled by the potential (chemical) gradient; and (2) movement across the cytoplasm and tonoplast which was controlled by photosynthesis—presumably by respiratory substrates.

Stimulation, Fertilization, and Injury. Stimulation is a term that has been applied to various nonrelated events that result in increases in permeability. Permeability of egg cells of certain species of invertebrate animals increases when fertilization occurs (Meyer and Anderson, 1939). Injury or wounding of cells and tissues, whether attributable to mechanical or other causes, also usually results in an increase in permeability (Meyer and Anderson, 1939).

In animals excitable membranes have the special ability to change, rapidly and reversibly, their permeability to ions; these changes control ion movements which carry electric currents propagating the nerve impulses (Nachmansohn, 1969). Further, acetylcholine is the specific signal released upon excitation, and it is recognized by a specific protein—the acetylcholine receptor. The latter induces a conformational change in membranes which results in increased permeability to ions.

Auxins. IAA has been reported to decrease markedly the Na^+ efflux from *Vallisneria* and *Ruppia* when the leaves are placed in a hypotonic solution of SO_4^{2-} or Cl^- salts of K^+ or Mg^{2+} (Kawahara and Takada, 1958). However, under hypertonic conditions (by adding mannitol), in which cells were plasmolyzed, Na^+ efflux was greater in the presence of auxin than in its absence.

Calcium. Several workers reported that Ca^{2+} is required for maintaining the normal permeability characteristics of membranes—particularly the plasmalemma (Viets, 1944; Helder, 1958; Laties, 1959). Electron microscopy showed that Ca^{2+} deficiency resulted in a disorganization of membrane structure (Marinos, 1962). With regard to certain reports of K^+ loss and Na^+ uptake, the promotive effect of Ca^{2+} has been attributed to its effect on the permeability of membranes (Marschner, 1964). Calcium may affect permeability by bonding negative charges of the plasma surface and cell wall; other divalent cations are predominantly associated with the protoplasm (Steveninck, 1965), hence do not so generally alter the permeability characteristics of membranes.

It has been reported that Ca^{2+} decreases the permeability of the membrane to Na^+ but increases it for a very similar ion, Rb^+ (Waisel, 1962). Mg^{2+} was about as effective as Ca^{2+} in preventing the harmful effects of low pH on the accumulation mechanism in that Mg^{2+} provided the same barrier to H^+ as provided by Ca^{2+}.

It is generally believed that divalent cations usually decrease the permeability of membranes (Collander, 1959). Jacobson et al. (1960) reported that Ca^{2+} creates a barrier—probably at the cell surface. Ca^{2+} can, however, increase or decrease the permeability of membranes. The stimulating effect of Ca^{2+} (on ion uptake) was considered to result from a blocking of interfering ions; the main effect of Ca^{2+} was on the membrane.

Diseases. Increases in permeability of host cell membranes are an early symptom of many plant diseases (Gottlieb, 1944; Williams and Keen, 1967).

These changes sometimes occur in advance of observable lesions (Thatcher, 1939, 1942; Wheeler and Black, 1963). Victorin, the toxin from *Helminthosporium victoriae,* induces an increase in the permeability of cells (host) within 5 min (Wheeler and Black, 1963). The toxin causes a loss of electrolytes from host cells, inability of root cells to plasmolyze, stoppage of protoplasmic streaming, and the breaking of all plasma membranes (Samaddar and Scheffer, 1968). During infection by *Rhizoctonia solani* Kuhn, permeability of membranes of *Phaseolus aureus* increases prior to symptom expression; the permeability increase was regarded as the initial host response to infection (Lai et al., 1968). Host cells of squash hypocotyls, contiguous to fungal hyphae of *Hypomyces solani* f. sp. Cucurbitae, are completely permeable to solutes and do not accumulate neutral red or exhibit cyclosis (Hancock, 1968).

Enzymes. Carbonic anhydrase was found in 10 species of marine algae belonging to the *Chlorophyta, Rhodophyta,* and *Phaeophyta* (Bowes, 1969). Although the role of carbonic anhydrase in plants is not established, it could possibly increase the transport of CO_2 across membranes (Enns, 1967). In solution, HCO_3^- diffuses almost as fast as CO_2, but in a membrane CO_2 diffuses about 100 times faster than ions. With carbonic anhydrase present, it could convert HCO_3^- to CO_2 and greatly facilitate the diffusion of carbon. If CO_2 were coming through the membrane, the enzyme could convert it to HCO_3^- (on the inside), hence steepen the gradient for diffusion (Bowes, 1969).

For many animal tissues a Na-activated ATPase associated with membranes is involved in active Na^+ transport out of the cells against an electrochemical gradient (DuPraw, 1968; Dowben, 1969). In fact, plasma membranes are major obstacles to the entry of molecules, and thus permeation must occur metabolically with the assistance of chemical energy (ATP) (Frey-Wyssling and Mühlethaler, 1965).

In a study of the entry of sugars into tomato roots, it was observed that sucrose, but not glucose or levulose, entered (Dormer and Street, 1949; Street and Lowe, 1950). The hypothesis was advanced that the mechanism of sucrose utilization involved the operation of a specific sucrose phosphorylase at the surface of the absorbing cells.

Carbon Dioxide or Oxygen. Accumulation of CO_2 appears to be more detrimental to water absorption than a decrease in oxygen in soil (Devlin, 1969). An increase in CO_2 increases the viscosity of protoplasm and decreases permeability (Fox, 1933; Seifriz, 1942). In hypocotyls of *Helianthus annuus* L. CO_2 rapidly decreased, and O_2 increased the permeability of cells to water (Glinka and Reinhold, 1962). Although a pronounced accumulation of CO_2 could affect permeability and in turn water absorption, Kramer (1959) reported that toxic accumulations of CO_2 in soils are of rare occurrence.

MEMBRANES

Evidence for Existence of Membranes

Theoretical. On purely theoretical grounds, substances that lower surface or interfacial tension should accumulate at interfaces (e.g., between the cytoplasm and the cell wall, between the cytoplasm and the vacuole), and the presence of certain of these substances might conceivably account for the actual "membrane" and the "differentially permeable" properties of membranes.

Experimental. Although cytoplasm contains approximately 80% or more of water, there is a sharp line of demarcation between the cytoplasm and vacuole. That is, the portion of cytoplasm next to the vacuole appears to be immiscible with water of the vacuole. Isolated protoplasts suspended in water do not mix with the water; they retain their identity as though bounded by membranes.

Microdissection studies have visually demonstrated what appear to be membranes that can be stretched by micromanipulation or even removed from the protoplast.

In recent years techniques have been developed for isolating a variety of membranes from various types of cells (Dowben, 1969). From a given type of cell, these membranes have proven remarkably constant in chemical composition.

Electron micrographs have provided strong proof of the existence of membranes and revealed minute details about them. There is, however, some disagreement as to interpretation of electron micrographs, and sometimes there is a question as to whether or not the "structure" of the membrane is an artifact. This question arises from the technique of preparing membranes and from the conditions under which electron micrographs are made.

When a series of steps in isolation and preparation of membranes is followed rigidly, an experimenter usually obtains the same appearance of "membranes" in his electron micrographs. However, in a given preparation the appearances of membranes of various cells may be dissimilar. There is the possibility that what is often observed may represent only one of several states a given membrane may assume. Although many investigators consider this last point quite seriously, it is interesting to note that Robertson (1962, 1968) reported that the same kind of 75-A unit membrane, found at the surface of the Schwann cell, could be demonstrated at the surfaces of many plant, animal, and bacterial cells.

Names and Locations of Membranes

The membranes that have received the most attention are the plasmalemma (Mast, 1924) (exoplast, ectoplast, or plasma membrane) and the tonoplast (endoplast or vacuolar membrane).

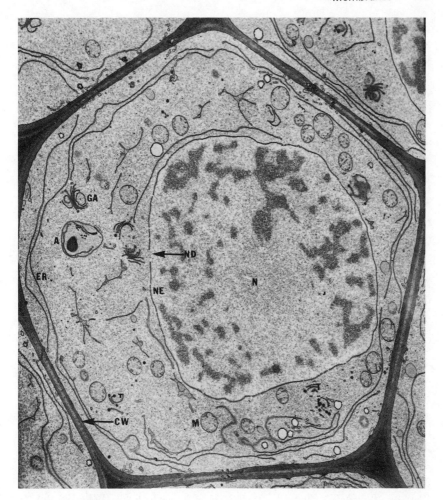

Fig. 7.1
Transverse section through a meristematic rootcap cell showing the nucleus (N), nuclear envelope (NE), endoplasmic reticulum (ER), mitochondrion (M), Golgi apparatus (GA), an amyloplast (A), and unidentified inclusions. Note discontinuity (ND) in the nuclear envelope. The dark bounding region represents the cell wall (CW). Approximately ✕ 8000. From W. G. Whaley, et al. (1960) Am. J. Botan., 47:401–419. Botanical Society of America—by permission.

The plasmalemma is the membrane that lies at the junction of the cytoplasm and cell wall. For various organelles and membranes in the cell, see Fig. 7.1. It is deeply invaginated and may be continuous with the endoplasmic reticulum (ER) (Sutcliffe, 1962)—a system of intercommunicating canals, vesicles, or cisterns that occur throughout the cytoplasm. Ribosomes—centers of protein synthesis—are associated with the plasma side of the ER. If the plasmalemma and ER are continuous, cytoplasm would present a greater surface area to the external medium than has previously been thought to be the case. The plasma-lemma appears to be cation permeable but anion impermeable; therefore cations can move passively into the cytoplasm (Sutcliffe, 1959).

The tonoplast is the membrane that lies at the junction of cytoplasm and the vacuole (assuming one central vacuole). It has been regarded as a considerable barrier to the free diffusion or exchange of ions (Sutcliffe, 1954). If a cell is multivacuolate, there is a membrane in the portion of the cytoplasm that immediately borders each vacuole. The nucleus, plastids, and mitochrondria also have membranes. Each mitochondrion is bounded by two parallel membranes, each 40–60 A, separated by a distance of 60–90 A. The inner membrane is invaginated to give rise to a system of internal membranes—the cristae mitochondriales (Sutcliffe, 1962). Single-unit membranes are associated with plasmalemma, tonoplast, and spherosomes; double membranes are characteristic of nucleus, mitochondria, plastids, ER, and the Golgi apparatus (Salisbury and Ross, 1969).

In addition, cytoplasm is interlaced by the ER or "ergastoplasm." The extensiveness of the ER in cells is indicated in Fig. 7.2. It may be visualized as a deflated sac extensively folded and extending throughout the cytoplasm. The cavity or lumen of the ER is bounded on each side by a single membrane. The ER has received considerable attention in recent years. In certain cell types in which the ER is well developed, the membrane systems of the cell, plus attached components, may account for perhaps as much as 80–90% of the total cell mass (Hechter, 1968).

In some meristematic cells, the volume of the ER is small or possibly nil; mature cells generally contain an ER (Mercer, 1960). In contrast with the report by Mercer (1960), it has been reported that the ER is extensive and clear in young cells, but that in older, less active cells it appears to have dissolved into the cytoplasm (Salisbury and Ross, 1969). The nucleus may have direct contact with the cytoplasm of the cell. The ER is believed to be lipoprotein and to divide the cytoplasm into two phases; one is continuous with the nucleus, and the other, perhaps, with the external medium. In fact, the ER appears to be continuous with the nuclear membrane or nuclear envelope, extends to the cell surface (Watson, 1959; Whaley et al., 1960), and may even permeate the cell wall (Whaley, et al., 1959) or neighboring cells (Whaley et al., 1960). Based on their study of corn root cells, Whaley et al. (1959) concluded that the ER of a cell appears to be part of an intercellular system. It may also be considered to divide the cytoplasm into cavities, which might be the mechanistic basis for the compartmentalization of enzymes and other cellular constituents. Ribosomes are attached to the outside of each of the membranes of the ER.

In view of the attachment of ribosomes to the outside of each of the ER membranes, the involvement of the ER in protein synthesis is obvious. In addition, small vesicles appear to be pinched off from the ER during mitosis; the

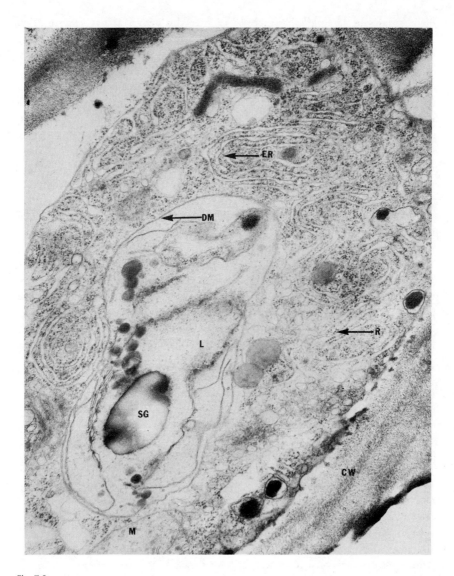

Fig. 7.2
Electron micrograph of an ultrathin section of a cell of dodder *(Cuscuta campestris)* stem showing a well-developed endoplasmic reticulum (ER), starch grain (SG), leucoplast (L), mitochondrion (M), double membrane (DM), ribosome (R), and cell wall (CW). $\times$ 15,300. Courtesy of Dr. M. Kenneth Corbett, Department of Botany, University of Maryland.

vesicles move to the cell plate where they are believed to be involved in synthesis of the middle lamella, and they may possibly be contributory to synthesis of the primary wall. In yeast cells and nectar hairs of *Abutylon* flowers, anaerobiosis results in a remarkable increase in ER; this suggests that fermentation and glycolysis occur in the ER (Salisbury and Ross, 1969). Last, the ER may be involved in fat synthesis, since tiny vesicles pinched off from ends of the ER have been observed to develop into spherosomes.

The Golgi apparatus, or dictyosomes, consists of several stacks of adjacent, flattened vesicles; each stack is a dictyosome. The Golgi apparatus may secrete materials other than cellulose to the middle lamella, primary wall and, perhaps, even to the secondary wall. The Golgi apparatus has been observed to accumulate in the region of cell plate formation (Whaley et al., 1960). Also, there is the possibility that the typical, large, central vacuole of most parenchymatous cells may be formed by fusion of smaller vacuoles which arose from dictyosomes (Salisbury and Ross, 1969).

The nucleus is enclosed by an enveloping system of two roughly parallel membranes spaced apart by 200–300 A (Watson, 1959). There are discontinuities, that is, pores, in the nuclear envelope in corn root cells (Whaley et al., 1959) and animal cells (Watson, 1959). In the latter, the circular pores, about 1000 A in diameter, are present in large numbers, and the pores are formed by the junction of the inner and outer membranes around the periphery of the pore. Watson (1959) concluded that the nuclear envelope is actually a specialized cytoplasmic structure and not properly a part of the nucleus and, further, that the classic distinction between nucleus and cytoplasm is unsatisfactory in view of recent findings.

In addition to the plasmalemma, tonoplast, ER, and nuclear membranes, there are mitochondrial, plastid, and other membranes. Mitochondria are normally found in all cells except those of bacteria and blue-green algae (Salisbury and Ross, 1969). Apparently, all organelles have membranes.

Structure, Composition, and Function

The composition and structure of membranes must be related to the varying rates at which substances enter or leave cells and to the various physiological capabilities membranes exhibit—including active transport (of ions and molecules), differential permeability, quantum conversion in photosynthesis, electron transport, and oxidative phosphorylation. In general, it is agreed that no structure so far proposed adequately accounts for all these properties.

Numerous techniques have been used to elucidate the presence and nature of membranes. Among the physical approaches, the following techniques have been used: microdissection (stretchability indicating the presence of at least some protein), x-ray diffraction, infrared spectroscopy, optical rotatory dispersion (ORD), nuclear magnetic resonance (NMR), circular dichroism (CD), and electron microscopy involving both shadowing and freeze-etch preparations. Chemically, membranes have been analyzed for lipids, proteins, carbohydrates, and other constituents. Many of these techniques are "averaging" approaches and therefore do not necessarily pinpoint the existence or nature of specific functional groups within the membrane. In addition, certain of these techniques may modify the true nature of the membrane as a result of the technique itself. Last, results from most of these techniques are subject to different interpretations by various investigators.

Sieve Theory. Although it is of historical significance only, it was once thought that membranes acted as sieves through the pores of which small molecules could pass, whereas larger ones could not (Traube, 1867). In addition, it was assumed that relative rates of entry of various substances were a function of their sizes (Ruhland, 1912). The correlation between permeability coefficients and molecular size is particularly striking when a series of homologous compounds is examined (Ruhland and Hoffmann, 1925). Evidence was soon accumulated, however, that could not be explained by this simplest of all theories. In contrast with Ruhland (1912), Collander and Barlund (1933) noted very little correlation between rate of solute entry and size of molecule. They regarded lipoid solubility as important, but recognized its limitation as an explanation for the entry of all substances. In a fatty acid series, rate of entry *increased* with increase in size of molecules. Molecules of glycerol and monoacetin are very nearly the same size, and yet the latter enters cells 15 times more rapidly than the former (Miller, 1938).

Lipoid Theory. While studying permeability of plant membranes, Overton (1899) noted that nonpolar substances (such as diethyl ether) entered rapidly, and that rate of entry appeared to be correlated with solubility in lipid. As a result, he proposed his "lipoid solubility" theory (Overton, 1895, 1897, 1899, 1900, 1902, 1904) in which lipids, possibly lecithin and cholesterol, were assumed to be constitutents of membranes. For certain homologous series rates of entry were related to lipid solubility, but later exceptions were noted. The theory did not explain the entry of polar substances—such as water and inorganic salts. Polar compounds, which are relatively lipid insoluble, must and do enter and leave cells. It is possible according to this theory, that polar compounds could traverse membranes, but at a very slow rate. These compounds would presumably pass as uncharged molecules, since little or no dissociation of these compounds in a lipid would be expected (Osterhout, 1936).

The immiscibility of cytoplasm with the vacuole, for instance, indicates the presence of lipid on the outermost portion of the cytoplasm bordering the vacuole. The low surface tension is also regarded as indicative of lipids. Naturally occurring lipids, for example, phospholipids, galactolipids, and sulfolipids, would not only be expected to concentrate at interfaces but also to lower the surface tension. Benson (1964) reviewed plant membrane lipids in considerable detail.

Treatment of living protoplasts from the *Avena* coleoptile with selected detergents and polyene antibiotics indicates that little sterol is present in the external surface of the plasma membrane (Ruesink, 1971). Lysis of protoplasts in carboxymethyl–RNAse, which is enzymatically almost inactive, provided strong evidence that the lysis previously observed in RNAse is not an indication of RNA in the membrane.

Protein–Lipid "Mosaic" Theory. According to this theory (Nathansohn, 1904), membranes are composed of lipid and hydrated protein in a mosaic pattern with some portions being lipid and other portions a hydrated protein. The nonmiscibility of cytoplasm with the vacuole has already been cited as evidence for the presence of lipid in membranes; the stretchability of membranes observed in microdissection studies strongly implicated protein chains which might slide over each other, that is, be capable of "stretching." The presence of both lipid and hydrated protein would account for the entry of both polar (lipid-insoluble) and nonpolar (lipid-soluble) compounds. Generally, polar compounds are lipid insoluble, and these would pass through the hydrated protein portion of the membrane, and the nonpolar compounds through the lipid portion.

A study of oil drops in eggs or egg extracts revealed a surface tension of only 0.8 dyne/cm rather than the 15 dynes/cm that was expected for an oil–water interface (Harvey and Shapiro, 1934; Danielli and Harvey, 1935), and it was obvious that some surface-active agent was lowering the surface or interfacial tension. Danielli and Davson (1935) concluded that the interface has oriented lipids whose polar ends face the water and are covered with a layer of protein. This concept led to models of membranes composed of lipid and protein.

Adsorption Theory. This theory rested on the assumption that the more strongly molecules were adsorbed by the membrane, the more readily they would pass through it. Adsorption is generally regarded as the first step in ion uptake, but the theory, as originally proposed, no longer receives serious consideration.

Chemical Reaction Theory. According to this theory, a penetrating ion or molecule combines with some substance at the outer boundary of the membrane to form an intermediate compound, the latter moves across the membrane, and the ion or molecule is released by a second reaction at the inner boundary. Such an idea was first proposed by Pfeffer (1900). This is the essence of the "carrier" theory.

Protein–Lipid–Protein Membrane (PLP). Danielli and Davson (1935) proposed that the interface has oriented lipid molecules whose polar ends face the exterior and interior (aqueous) environments; the two surfaces are covered with layers of protein (Fig. 7.3). This concept became the classic Danielli–Davson model for membrane structure, and Robertson (1959, 1968) termed it a "unit membrane." With an electron microscope he observed a three-layered membrane about 75 A thick; there were two dense lines about 20

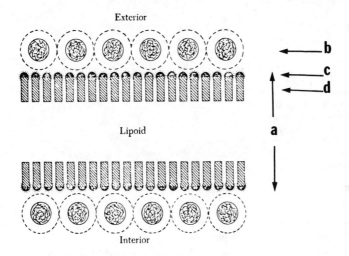

Exterior

Lipoid

Interior

Fig. 7.3
Danielli–Davson model of a cell membrane composed of a bilayer of lipid (a) and coated with layers of globular proteins (b) on the surfaces of the bilayer. The polar ends (c) of the lipid molecules are in contact with the hydrated globular proteins; the nonpolar ends (d) face the center of the lipid bilayer. From J. F. Danielli and H. Davson (1935) J. Cellular Comp. Physiol., 5:495–508. The Wistar Inst. Press—by permission.

A wide separated by a lighter space 35 A wide. He concluded that the dense lines were the proteins and polar groups of the membrane and that the lighter interzone spaces were the nonpolar groups. Later, he suggested that the outer layer is composed of protein and carbohydrate, the core of a bimolecular layer of phospholipid, and the inner layer a coating of protein. Perfect symmetry of the membrane appears unlikely since polarity of the plasmalemma, for example, is indicated in physiological processes (e.g., resorption and secretion) and by the differential staining of the outer and inner surfaces of the membrane by osmic acid (Frey-Wyssling and Mühlethaler, 1965).

Additional evidence of the protein–lipid composition of membranes has come from analyses of plant and animal membranes (Table 7.1). Although lipid and protein are found, the percentages of the two may vary from a preponderance of one to a preponderance of the other (Korn, 1966; Branton and Park, 1968).

Later, the Danielli–Davson model was modified by Danielli (1954) to a membrane with defined thickness, pore structure (actually a channel of hydrated protein rather than just a pore), and a bimolecular lipid leaflet stabilized by adsorbed protein monolayers (Fig. 7.4).

We now turn to a discussion of the evidence for and against a PLP membrane. Much of the evidence for the PLP model was provided by Branton (1969) in his recent, excellent review of membrane structure.

Table 7.1

Percent Lipid and Protein of Plant and Animal Membranes[a]

Membrane	Lipid (%)	Protein (%)
Myelin (O'Brien, 1965)	80	20
Chloroplast lamellae (Park and Pon, 1961)	50	50
Erythrocyte (Dodge et al., 1963)	40	60
Mitochrondrial inner membranes		
(Fleischer, et al., 1961)	25	75

[a] From Branton, D. and R. B. Park, (eds.), 1968: Papers on Biological Membrane Structure, Little, Brown and Co., Boston, Mass. 311 p.—by permission.

(1) Positive birefringence of intact nerve myelin indicates that the lipid molecules are oriented with their long axes radial to the axis of the nerve fiber and interspersed with sheets of protein.

(2) X-ray diffraction patterns are consistent with the concept that membranes are composed of a continuous bimolecular lipid layer with nonlipid components on both surfaces.

(3) Regardless of the protein/lipid ratios of various membranes (as presented in Table 7.1), there is enough lipid to form a central bilayer.

(4) Differential scanning calorimetry indicates that extensive hydrophobic lipid–protein associations are unlikely.

(5) Electron microscopy shows two positively stained lines 20 A wide separated by an approximately 35-A unstained space (i.e., a unit membrane). Presumably, the dark lines could be considered to be the lipid portions and the lighter line the protein, that is, a lipid–protein–lipid (LPL) membrane. However, the addition of protein to solutions in which lipids have been dispersed increases the density and width of the outer dark bands, indicating that the arrangement is PLP and not LPL.

(6) Model systems made with lipid and protein have been observed to form vesicles permeable to water and anions but much less so to cations; the thickness, electrical resistance, and interfacial tension, among other properties, are quite similar to those of natural membranes.

Electrical conductivity and permeability to small molecules are comparable in lipid bilayers and membranes, but not equal (Henn and Thompson, 1969).

However, according to Bernhard (1969), reporting on a conference on membrane structure, Korn has criticized the Danielli–Davson model, as follows. (1) There is insufficient protein to cover the two surfaces of lipid bilayer membranes, since a major portion of the protein of membranes appears to be in α-helical form and little or none in the β form; and (2) the red shift, obtained in optical rotation data, could be interpreted to indicate that the α-helical membrane proteins are buried in a relatively hydrophobic region. Interactions between mitochondrial structural protein and phospholipids have proved

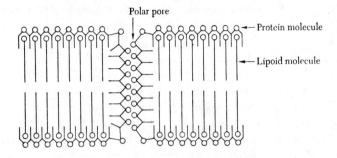

Fig. 7.4
Modified Danielli–Davson model of a membrane showing defined thickness, pore structure, and a bimolecular lipid leaflet stabilized by adsorbed protein monolayers. From J. F. Danielli (1954) The present position in the field of facilitated diffusion and selective active transport, pp. 1–32. *In* J. A. Kitching (ed.) Recent Developments in Cell Physiology (Proc. 7th Symp. Colston Res. Soc. held at the Univ. of Bristol, March 29–April 1, 1953), Academic, New York. Colston Research Society and Academic Press, Inc.—by permission.

to be nonionic. He also reported that, according to Rothstein, bilayer membranes have been observed in osmium-fixed mitochondria from which over 90% of the lipid has been removed.

Green and Perdue (1966b) raised the following objections to the PLP model proposed by Danielli and Davson. (1) Lipid and protein are bound by hydrophobic bonding and not electrostatic interaction, as predicted by the unit membrane hypothesis; (2) phospholipid can be removed and the membrane appears unaltered, as indicated by electron microscopy; the framework, then, appears to be protein and not phospholipid, and the spacings as observed are determined by protein; (3) infrared absorption analysis has not confirmed the presence of extended proteins; and (4) there is evidence that membranes may be composed of closely nestled, repeating units.

It is also pertinent to note that, on the basis of theoretical considerations of surface free energy, Danielli (1966, 1968a, b) concluded that a bimolecular membrane had a surface free energy lower than one that was thinner or thicker than a bilayer. Danielli (1968b) also concluded that free energy changes were unfavorable to the formation of micelles and that these changes were sufficiently large for the bilayer structure to be predominant. In fact, he suggested (Danielli, 1967) that the micellar structure of membranes sometimes seen in electron microscopy is an artifact arising from removal of the membrane from its natural aqueous environment. He maintained, however, that the membrane contains a variety of macromolecules which are required for specialized functions. That is, that there are specialized regions concerned with special permeation mechanisms, such as glucose and Na^+ transport, and that these regions are enzymelike. Frey-Wyssling and Mühlethaler (1965) concluded that, for the enzymic activity of the plasmalemma (i.e., digestion and synthesis), the presence of a mosaic structure of different desmoenzymes must be assumed.

In order to account for sudden changes in permeability, it has been suggested that phase reversal of membranes might occur. This would involve a change such as an oil-in-water system converting to a water-in-oil system. Phase reversal of membranes, as suggested by Clowes (1918), could not possibly operate reversibly in the thin lipid layer of membranes (Danielli and Davson, 1968). However, Bangham and Horne (1964) reported that the bimolecular leaflet of lipids is a liquid-crystalline structure capable of reversible structural modifications.

Lipid-Protein-Lipid Membrane (LPL). As indicated earlier, electron micrographs might conceivably be interpreted in favor of either a PLP or an LPL arrangement. These two concepts differ not only with regard to the arrangement of the lipid and protein, but also with regard to the type of bonding. The PLP arrangement is generally assumed to rest on extensive polar bonding, while the LPL model suggests apolar bonding between lipid and protein.

Branton (1969) evaluated the LPL model as follows. (1) Surface tension measurements are equally compatible with the LPL model since naturally occurring membrane lipids (e.g., phospholipids, galactolipids, and sulfolipids) are amphiphilic and would form a low surface tension with water; and (2) ORD and CD of numerous membranes show spectra not usually found in simple systems. Owing to the red shift, the optical rotation data can best be explained in terms of a substantial amount of protein in α-helix and random conformations with little or none in the β-configuration. Helical proteins would be consistent with the LPL, but not with some PLP models based on unfolded polypeptide chains; (3) the observed presence of special, hydrophobic protein in membranes argues against the PLP membrane and in favor of the LPL model; (4) NMR spectroscopy indicates an immobilization of fatty acid chains; however, the immobilization could result from association with protein or from tight packing with other lipids; and (5) polar groups of lipids appear to be susceptible to lipase which indicates their ready accessibility.

Subunit Membrane (Composed of Repeating Units or Elementary Particles): According to Green and Perdue (1966a), Fernández-Morán discovered the repeating unit in the inner membrane of the mitochondrion in 1961 and called it an elementary particle. He published his observations a year later (Fernández-Morán, 1962). However, he had seen only the headpiece and, later, Fernandez-Morán et al. (1964) showed that the elementary particle (EP) was tripartite and consisted of a headpiece, a stalk, and a basepiece (see diagram at left in Fig. 7.5 and Fig. 7.6B). Green and Perdue (1966a) list numerous investigators who have corroborated the tripartite nature of most elementary particles. The quantasomes from chloroplasts were reported by (Fer-

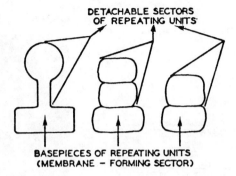

Fig. 7.5
Diagrammatic representation of possible forms of repeating units or elementary particles. From D. E. Green and J. F. Perdue (1966a) Ann. N.Y. Acad. Sci., 137:667–684. New York Acad. of Sciences—by permission.

nández-Morán et al., 1968) to be similar in naure and in dimensions to the EP of mitochondria. Weier and Benson (1967) reported that four subunits comprise a quantasome.

According to this theory, a membrane is composed entirely of repeating units the basepieces of which form the continuum of the membrane (Fig. 7.5); the knobs (headpiece and stalk) can be removed without affecting the integrity of the membrane. In fact, the headpiece and stalk of the EP have been separated by detergents or ultrasonic vibration and then reassembled (Kopaczyk et al., 1968b; Green and MacLennan, 1969). The tripartite EP of the inner mitochondrial membrane has a mass, on the basis of measured size, of about 800,000 daltons according to Green and MacLennan (1969), or 1.4×10^6 molecular weight as reported by Green and Hechter (1965). The latter investigators noted that the EP may exist in the extended or in the compressed form; the former is tripartite and the latter is a spherical body about 150 A in diameter. They hypothesized that templates rather than polysomes are involved in the assembly of the membrane subunits.

According to Green and Perdue (1966a), each membrane has its own distinctive repeating unit and, although the tripartite EP is widespread, not all repeating units are tripartite. Further, they noted that tripartite units from different membranes may be of different sizes. They added that the basepieces in a membrane are apparently all alike but that the knobs can vary; apparently there are some five kinds of elementary particles in the inner mitochondrial membrane. The outer mitochondrial membrane has repeating units without stalks (Fig. 7.6A).

Membranes are made of repeating units, but all membranes do not necessarily have the *same* repeating units, since four types of units have been observed (Green, 1969).

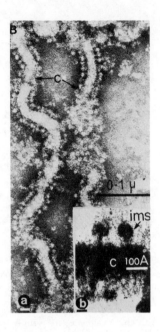

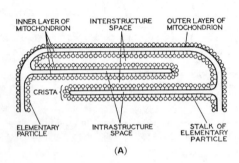

(A)

Fig. 7.6

(A) Diagrammatic representation of the outer and inner membranes of mitochondria. Note that the elementary particles of the inner membrane have stalks, whereas those of the outer membrane do not. From D. E. Green and J. F. Perdue (1966a) Ann. N.Y. Acad. Sci., 137:667–684. New York Acad. of Sciences—by permission. (B and **b**) Projecting subunits (elementary particles or oxysomes) associated with the inner membrane of rat liver mitochondria ✕192,000. (c) Two of the subunits. The subunits consist of a stem 30–35 A long and a round head 75–80 A in diameter. Center-to-center spacing; C, 100 A. From D. F. Parsons (1963) Science. 140:985–987. Copyright 1963 by the American Association for the Advancement of Science—by permission.

The electron transfer chain is believed to reside in the basepieces, and phospholipid is mandatory for this activity (Green and Tzagoloff, 1966; Harris et al., 1968).

ATPase is localized in the headpiece-stalk sectors (Racker et al., 1965; Kagawa and Racker, 1966; Harris et al., 1968; Kopaczyk et al., 1968a), which have a molecular weight of approximately 280,000 (Penefsky and Warner, 1965).

There is striking evidence that mitochondrial membranes may exist in three forms—nonenergized, energized, and energized-twisted configurations (Fig. 7.7) (Green et al., 1968; Penniston et al., 1968; Green and MacLennan, 1969). By appropriate treatment, Harris et al. (1968) caused the membranes of most of a mitochondrial preparation to exist, at any one time, in any one of the three configurations.

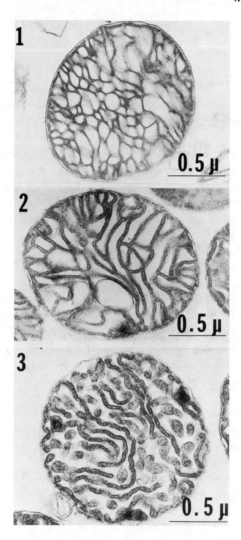

Fig. 7.7

Nonenergized (1), energized (2), and energized-twisted (3) configurations of membranes of beef heart mitochondria. ×31,000. From R. A. Harris et al. (1968) Proc. Nat. Acad. Sci., 59:830–837. National Academy of Sciences—by permission.

Conformational energy may provide the mechanism by which work (e.g., ion transport) is performed (Green et al., 1968; Harris et al., 1968; Green and MacLennan, 1969). Either electron transfer or hydrolysis of ATP can lead to the energized state (of the membrane) required for the performance of work (Harris et al., 1968), as follows.

Electron transfer = Energized state = ATP
                   (of membrane)
                          ↓
                   Work performances

The energized state may be required for the transport of ions across the membrane against a concentration gradient (Penniston and Green, 1968). In fact, it has been proposed that the energized state of the membrane is the functional equivalent of the elusive high-energy intermediates (Green et al., 1968) which have never been isolated (Harris et al., 1968); the energized state of membranes, however, has been amply demonstrated directly by electron microscopy.

Since the original discovery of subunits, repeating units, or elementary particles, many membranes have been shown to possess such units—particularly chloroplasts (Park and Biggins, 1964; Weier and Benson, 1967) (Fig. 7.8) and mitochondria. Chloroplast membranes do not show substantial birefringence—as would be expected if they contained a highly-oriented lipid bilayer (Frey-Wyssling and Steinmann, 1948, 1953).

In contrast with the PLP or the LPL models of membrane structure, then, it has been argued that there is no distinct separation of protein into one layer and lipid into another (Green and Perdue, 1966b; Korn, 1966). It has been suggested that protein and lipids are bound partly by ionic bonds between oppositely charged portions and partly by van der Waal forces between hydrophobic regions of the two substances (Salisbury and Ross, 1969).

Branton (1969) criticized the subunit model on the basis that: (1) Subunit structure is not seen in all membranes prepared by the same technique or even in all membranes in the same electron micrograph. He contended that globular subunits may appear because of a surface mosaic rather than because of lipoprotein or globular subunits within the membrane; and (2) freeze-etching, prior to electron microscopy, exposes the inner, hydrophobic membrane faces. This feasibility is considered by him to be consistent with a PLP or an LPL model but difficult to visualize for a subunit membrane in which cleavage of hydrophobically bonded subunits should occur at right angles with the plane of the membrane rather than within the plane. However, Weier and Benson (1967) observed subunit structure in chloroplast membranes following shadowing and freeze-etching.

Instead of a double membrane Crane et al. (1969) proposed a binary membrane; in fact, they suggested that the single layer seen in unit membranes may be an artifact resulting from failure of one side of the membrane to stain or from an intercalation of two layers into one layer. They contended that the binary structure could explain fractionation of membranes through a central layer by the freeze-etching procedure. Two layers in freeze-etch cross-section would simply be two layers of a binary membrane.

Fig. 7.8

Cr shadowed preparation of a spinach chloroplast lamella showing the micellar nature of this membrane. The crystalline array of quantasome subunits shown here is the least common arrangement, linear and random arrays being the more common. $\times$71,000. From R. B. Park and J. Biggins (1964) *Science*, 144:1009–1011. Copyright 1964 by the American Association for the Advancement of Science—by permission.

Subunit Membrane (Disc and Pillar Arrangement). A radically different type of membrane was visualized by Kavanau (1965), which consists of disc and pillar forms of subunit lipid micelles. Chemically induced cycles, of collapse of the pillars into discs and of opening out of the discs into pillars, would provide broad potentialities for doing work—such as facilitated diffusion and active transport (Kavanau, 1968).

Lattice-Type Membrane. A novel concept of membrane structure and function was advanced by Hechter (1968). This model retains the basic features of the unit membrane concept, but the protein layers form a system of interlocked

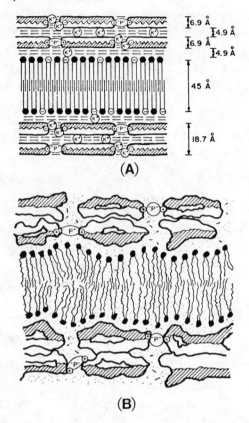

(A)

(B)

Fig. 7.9
(A) Schematic representation of a resting, polarized membrane in which the basic features of the unit membrane concept are retained; the protein layers are represented as a system of interlocked hexagonal discs cemented together by water layers in an icelike arrangement to form a precisely ordered lattice system. Icelike water is represented by dashed lines. The 45-Å lipid bilayer forms the central portion of the membrane with protein and icelike water on both surfaces. Although other ions may be involved, K^+ is shown as the principal counterion for fixed negative sites and phosphate for positive sites. Selectivity for K^+ over Na^+ depends on this organization of the membrane units to form a precise lattice. (B) Depolarized membrane devoid of lattice structure and no longer exhibiting high selectivity for K^+ over Na^+. The water, now as mobile water molecules, is shown as small dots. Note that membrane structure is not entirely lost. From O. Hechter (1968) Role of water structure in the molecular organization of cell membranes, pp. 300–311. In D. Branton and R. B. Park (eds.) Papers on Biological Membrane Structure, Little, Brown, Boston. Little, Brown and Co.—by permission.

hexagonal discs cemented together by water layers in an icelike arrangement to form a precisely ordered lattice system (Fig. 7.9). When the membrane is highly organized in a lattice structure (Fig. 7.9A), it is polarized, and selectivity of K^+ over Na^+ is apparent. When the membrane becomes depolarized (Fig. 7.9B), it no longer exhibits high selectivity for K^+ over that for Na. If true, there would be a possible mechanistic explanation for the high-affinity and low-affinity mechanisms, 1 and 2, discussed earlier.

NMR spectroscopy indicates that tissue water may be semicrystalline or ice-like (Anonymous, 1969). It was further reported that intracellular water exists as multiple, polarized layers adsorbed onto cellular proteins.

King (1969) indicated that some scientists regard membranes as having a "lamellar or liquid crystalline phase," and living systems as being liquid crystals or, more correctly, as being in the paracrystalline state.

Amino Groups and Fixed Charges in Membranes. Rothstein (1962) used chemical "probes" to interact with ligands of the membrane, hence to reveal features of its structure and function. For example, the uranyl ion, which cannot penetrate a cell, reacts with phosphoryl and carboxyl groups of the membrane and prevents, respectively, (1) the transport of sugar and (2) the action of invertase.

According to Rothstein and Steveninck (1966), there are two modes of transport of hexoses through membranes—one an equilibrating system (facilitated diffusion) which depends on a concentration gradient of sugar across the membrane), and the other active transport. They reported that active transport of sugar depends on phosphoryl groups in the membrane and, when uranyl combines with these groups, there is no transport of sugar. They also noted that during transport of glucose 80% of the phosphoryl groups disappeared from the outer face of the membrane but reappeared when all the glucose had been transported.

Rothstein (1968) reported that permeability to ions was related to a pore structure with fixed charges, and he suggested that amino groups (of membranes) may be responsible for the charges. When the charges are positive, anion uptake would be favored over cation uptake and, when negative, cation over that of anion uptake. A change in pH, on either side of the isoelectric point, could switch the charge from positive to negative, and vice versa.

From an evolutionary point of view, Rothstein (1964) concluded that the advent of cell walls made it possible for cells to move to a diluted (as compared with seawater) environment without bursting, but that drastic changes in membrane properties had to occur. The membrane had to be tighter with permeability to ions minimized. He noted that the K–Na discriminating system appeared to be an old one, in an evolutionary sense, but that walled cells retained the system and adopted it for producing the osmotic gradients necessary for growth and for maintaining electrolyte levels in a diluted environment. Additional information on membranes is available in books by Kleinzeller and Kotyk (1961), Kavanau (1965), Branton and Park (1968), Rogers and Perkins (1968), Dalton and Haguenau (1968), and Dowben (1969); and, recent major articles by Preston (1970), Anonymous (1970), Hendler (1971), and Green and Young (1971). The fine structure of plant cells and membranes is presented in a well-illustrated recent book (Ledbetter and Porter, 1970).

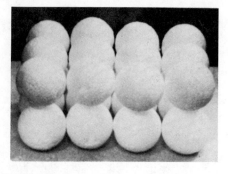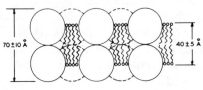

Fig. 7.10

The protein crystal model for membranes. Left: Photograph of a model showing the double-layer arrangement of proteins in membranes. Right: A diagrammatic cross section showing the relationship of the phospholipid molecules to the proteins. The large circles are proteins; the dashed circles represent proteins behind the plane of the section. The small circles are the polar lipid heads, and the wavy lines are the nonpolar lipid tails. From G. Vanderkooi and D. E. Green (1971) Bioscience, 21:409–415. Copyright 1971 by the American Institute of Biological Science—by permission.

Protein–Liquid Crystal. In a radical departure from other theories, it has been proposed that a double layer of protein molecules makes up the membrane continuum, and that a bilayer of lipids fills the spaces between the proteins (Fig. 7.10) (Vanderkooi and Sundaralingam, 1970; Vanderkooi and Green, 1970, 1971). Thickness of the membrane is determined by proteins and not lipids. In fact, this new model explains why 90% or more of the lipid can be removed from a membrane although the thickness of the membrane remains unchanged. Vanderkooi et al. believe that their model satisfactorily accounts for all experimental observations on properties of membranes and the interrelationships of lipid and protein.

General Conclusions About Membranes

Although it is dangerous to make predictions in a field developing as rapidly as that of membrane structure, it is possible to state some of the concepts rather generally agreed upon at this time.

Membranes do exist; the surfaces are believed to be hydrophilic and the interiors to be hydrophobic. The PLP arrangement is favored over the LPL arrangement. Subunit structure (repeating units or elementary particles) may be a feature of at least certain membranes—particularly chloroplasts and mitochondria —but not necessarily of all membranes. In fact, the structures of various membranes may vary one from the other. Finally, no structure has yet been proposed which would explain the various capabilities of membranes—such as active transport, differential permeability, quantum conversion in photosynthesis, and oxidative phosphorylation, among others.

The concepts of subunits or of the particulate nature of membranes do not currently mesh with the three-layered appearance generally observed for membranes in electron micrographs, but Salisbury and Ross (1969) note that this apparent layering may be better understood when the chemistry of fixing and staining for electron microscopy is more fully elucidated.

A concept of Frey-Wyssling and Mühlethaler (1965) reconciles several theories of permeability and membrane structure. They hypothesized that wedge-shaped pores may occur between the globular macromolecules of membranes. If these pores were lined with lipophilic compounds, there would still be room for water molecules to pass through, but larger molecules would pass in relation to their lipid solubility. Their concept explains the ready permeation of membranes by water and suggests a structural basis for the operation of the sieve and lipoid solubility theories.

The newest model of membrane structure, that is, the protein–liquid crystal model, may be regarded as the one that best accounts for the vast number of experimental observations and become generally accepted.

LITERATURE CITED

Anonymous. 1969. Icelike cell water shakes biotheories. Sci. Res. 4(13):9–10.

Anonymous. 1970. New model for lipid and protein membranes sparks strong debate. Chem. Eng. News 48:40–42.

Bangham, A. D., and R. W. Horne. 1964. Negative staining of phospholipids and their structural modification by surface-active agents as observed in the electron microscope. J. Mol. Biol. 8:660–668.

Bayliss, W. M. 1924. Principles of general physiology, 4th ed. Longmans, Green, London.

Beevers, L., L. E. Schrader, D. Flesher, and R. H. Hageman. 1965. The role of light and nitrate in the induction of nitrate reductase in radish cotyledons and maize seedlings. Plant Physiol. 40:691–698.

Benson, A. A. 1964. Plant membrane lipids. Annu. Rev. Plant Physiol. 15:1–16.

Bernhard, R. 1969. Conference on cell membranes: Seeds of rebellion. Sci. Res. 4(5):25–26.

Bowes, G. W. 1969. Carbonic anhydrase in marine algae. Plant Physiol. 44:726–732.

Branton, D. 1969. Membrane structure. Annu. Rev. Plant Physiol. 20:209–238.

Branton, D., and R. B. Park (eds.). 1968. Papers on biological membrane structure. Little, Brown, Boston. 311 pp.

Clowes, G. H. A. 1918. On the action exerted by antagonistic electrolytes on the electrical resistance of emulsion membranes. Proc. Soc. Exp. Biol. Med. 15:107.

Collander, R. 1959. Cell membranes: Their resistance to penetration and their capacity for transport, pp. 3–102. *In* F. C. Steward C. (ed.) Plant physiology – A treatise, Vol. II. Academic, New York.

Collander, R., and H. Bärlund. 1933. Permeabilitätsstudien an *Chara ceratophylla*. Acta Bot. Fennica 11:1–114.

Crane, F. L., J. D. Hall, and F. J. Ruzicka. 1969. Binary membranes: A new concept in membrane structure. Plant Physiol. 44(Suppl.):no. 21.

Dalton, A. J., and F. Haguenau (eds.). 1968. The membranes, ultrastructure in biological systems, Vol. IV. Academic, New York. 223 pp.

Danielli, J. F. 1954. The present position in the field of facilitated diffusion and selective active transport, pp. 1–32. *In* J. A. Kitching (ed.), Recent developments in cell physiology (Proc. 7th Symp. Colston Res. Soc., Univ. of Bristol, March 29–April 1, 1953). Academic, New York.

Danielli, J. F. 1966. On the thickness of lipid membranes. J. Theoret. Biol. 12:439–441.

Danielli, J. F. 1967. The problem of receptors in relation to regulation of the "milieu interieur." *In* Les Concepts de Claude Bernard sur le Milieu Interieur (Proc. Colloqium, The Foundation Singer-Polignac). Masson, Paris.

Danielli, J. F. 1968a. Phospholipid membranes are necessarily bimolecular, pp. 529–537. *In* B. Pullman (ed.) Molecular associations in biology. Academic, New York.

Danielli, J. F. 1968b. The formation, physical stability, and physiological control of paucimolecular membranes, pp. 239–253. *In* Symp. Int. Soc. Cell Biol., Vol. VI. Academic, New York.

Danielle, J. F. and H. A. Davson. 1935. A contribution to the theory of permeability of thin films. J. Cellular Comp. Physiol. 5:495–508.

Danielli, J. F. and H. A. Davson. 1968. A contribution to the theory of permeability of thin films, pp. 69–82. *In* D. Branton and R. B. Parks (eds.) Papers on biological membrane structure. Little, Brown, Boston.

Danielli, J. F., and E. N. Harvey. 1935. The tension at the surface of mackerel egg oil, with remarks on the nature of the cell surface. J. Cellular Comp. Physiol. 5: 483–494.

Davson, H., and J. F. Danielli. 1943. The permeability of natural membranes. Macmillan, New York. 361 pp.

Davson, H., and J. F. Danielli. 1952. The permeability of natural membranes. 3rd ed. Cambridge Univ. Press, Cambridge.

Devlin, R. M. 1969. Plant physiology. 2nd ed. Van Nostrand Reinhold, New York. 446 pp.

Dodge, J. T., C. Mitchell, and D. J. Hanahan. 1963. The preparation and chemical characteristics of hemoglobin-free ghosts of human erythrocytes. Arch. Biochem. Biophys. 100:119–130.

Dormer, K. J., and H. E. Street. 1949. The carbohydrate nutrition of tomato roots. Ann. Bot. 13:199–217.

Dowben, R. M. 1969. General physiology: A molecular approach. Harper, New York, 619 pp.

DuPraw, E. J. 1968. Cell and molecular biology. Academic, New York, 739 pp.

Enns, T. 1967. Facilitation by carbonic anhydrase of carbon dioxide transport. Science 155:44–47.

Farhoomand, M. B., and L. A. Peterson. 1968. Concentration and content. Agron. J. 60:708–709.

Fernández-Morán, H. 1962. Cell membrane ultrastructure. Low temperature electron microscopy and x-ray diffraction studies of lipoprotein components in lamellar systems. Circulation 26:1039.

Fernández-Morán, H., T. Oda, P. V. Blair, and D. E. Green. 1964. A macromolecular repeating unit of mitochondrial structure and function. Correlated electron micro-scopic and biochemical studies of isolated mitochondria and submitochondrial particles of beef heart muscle. J. Cell Biol. 22:63.

Fernández-Morán, H., T. Oda, P. V. Blair, and D. E. Green. 1968. A macromolecular repeating unit of mitochondrial structure and function – Correlated electron microscopic and biochemical studies of isolated mitochondria and submitochondrial particles of beef heart muscle, pp. 219–256. *In* D. Branton and R. B. Park (eds.) Papers on biological membrane structure. Little, Brown, Boston.

Fleischer, S., H. Klouwen, and G. Brierley. 1961. Studies of the electron transfer system. XXXVIII. Lipid composition of purified enzyme preparations derived from beef heart mitochondria. J. Biol. Chem. 236:2936–2941.

Fox, D. G. 1933. Carbon dioxide narcosis. J. Cellular Comp. Physiol. 3:75.

Frey-Wyssling, A., and K. Mühlethaler. 1965. Ultrastructural plant cytology (with an introduction to molecular biology). Elsevier, Amsterdam. 377 pp.

Frey-Wyssling, A., and E. Steinmann. 1948. Die Schichtendoppelbrechung grosser Chloroplasten. Biochim. Biophys. Acta 2:254–259.

Frey-Wyssling, A., and E. Steinmann. 1953. Ergebnisse der Feinbauanalyse der Chloroplasten. Viertelj. Naturforsch. Ges. (Zürich) 98:20–29.

Glinka, Z., and L. Reinhold. 1962. Rapid changes in permeability of cell membranes to water brought about by carbon dioxide and oxygen. Plant Physiol. 37:481–486.

Gottlieb, D. 1944. The mechanism of wilting caused by *Fusarium bubigenum* var. *lycopersici*. Phytopathology 34:41–59.

Green, D. E. 1969. Membranes. Sci. Res. 4(4):5–6.

Green, D. E., and O. Hechter. 1965. Assembly of membrane subunits. Proc. Nat. Acad. Sci. 53:318–325.

Green, D. E., and D. H. MacLennan. 1969. Structure and function of the mitochondrial cristae membrane. Bioscience 19:213–222.

Green, D. E., and J. F. Perdue. 1966a. Correlation of mitochondrial structure and function. Ann. New York Acad. Sci. 137:667–684.

Green, D. E., and J. F. Perdue. 1966b. Membranes as expressions of repeating units. Proc. Nat. Acad. Sci. 55:1295–1302.

Green, D. E., and A. Tzagoloff. 1966. The mitochondrial electron transfer chain. Arch. Biochem. Biophys. 116:293–304.

Green, D. E., and J. H. Young. 1971. Energy transduction in membrane systems. Amer. Sci. 59:92–100.

Green, D. E., J. Asai, R. A. Harris, and J. T. Penniston. 1968. Conformational basis of energy transformations in membrane systems. III. Configurational changes in the mitochondrial inner membrane induced by changes in functional states. Arch. Biochem. Biophys. 125:684–705.

Greenway, H., B. Klepper, and P. G. Hughes. 1968. Effects of low water potential on ion uptake and loss for excised roots. Planta 80:129–141.

Hancock, J. G. 1968. Effect of infection by *Hypomyces solani* f. sp. Cucurbitae on apparent free space, cell membrane permeability, and respiration of squash hypocotyls. Plant Physiol. 43:1666–1672.

Harris, R. A., J. T. Penniston, J. Asai, and D. E. Green. 1968. The conformational basis of energy conservation in membrane systems. II. Correlation between conformational change and functional states. Proc. Nat. Acad. Sci. 59:830–837.

Harvey, E. N., and H. Shapiro. 1934. The interfacial tension between oil and protoplasm within the living cells. J. Cellular Comp. Physiol. 5:255–267.

Hechter, O. 1968. Role of water structure in the molecular organization of cell membranes, pp. 300–311. *In* D. Branton and R. B. Park (eds.) Papers on biological membrane structure. Little, Brown, Boston.

Helder, R. J. 1958. Studies on the absorption, distribution and release of labelled rubidium ions in young intact barley plants. Acta Bot. Neerl. 7:235–249.

Hendler, R. W. 1971. Biological membrane ultrastructure. Physiol. Rev. 51:66.

Henn, F. A., and T. E. Thompson. 1969. Synthetic lipid bilayer membranes. Annu. Rev. Biochem. 38:241.

Hoagland, D. R., P. L. Hibbard, and A. R. Davis. 1926. The influence of light, temperature, and other conditions on the ability of Nitella cells to concentrate halogens in the cell sap. J. Gen. Physiol. 10:121—146.

Honert, T. H. van den, and J. J. M. Hooymans. 1955. On the absorption of nitrate by maize roots in water culture. Acta Bot. Neerl. 4:376—384.

Hope, A. B., and N. A. Walker. 1960. Ionic relations of cells of Chara australis. III. Vacuolar fluxes of sodium. Aust. J. Biol. Sci. 13:277—291.

Jacobson, L., D. P. Moore, and R. J. Hannapel. 1960. Role of calcium in absorption of monovalent cations. Plant Physiol. 35:352—358.

Kagawa, Y., and E. Racker. 1966. Partial resolution of the enzymes catalyzing oxidative phosphorylation. X. Correlation of morphology and function in submitochondrial particles. J. Biol. Chem. 241:2475.

Kavanau, J. L. 1965. Structure and function in biological membranes. Holden-Day, San Francisco.

Kavanau, J. L. 1968. Membrane structure and function, pp. 288—299. In D. Branton and R. B. Park (eds.) Papers on biological membrane structure. Little, Brown, Boston.

Kawahara, A., and H. Takada. 1958. Further experiments on auxin-induced Na efflux by Vallisneria and Ruppia leaves. J. Inst. Polytech. D9:19—26.

King, L. J. 1969. Biocrystallography — An interdisciplinary challenge. Bioscience 19:505—518.

Kleinzeller, S., and A. Kotyk (eds.). 1961. Membrane transport and metabolism. Academic, New York.

Kopaczyk, K., J. Asai, D. W. Allmann, T. Oda, and D. E. Green. 1968a. Resolution of the repeating unit of the inner mitochondrial membrane. Arch. Biochem. Biophys. 123:602.

Kopaczyk, K., J. Asai, and D. E. Green. 1968b. Reconstitution of the repeating unit of the mitochondrial inner membrane. Arch. Biochem. Biophys. 126:358—379.

Korn, E. D. 1966. Structure of biological membranes. Science 153:1491—1498.

Kramer, P. J. 1959. Transpiration and the water economy of plants, p. 607—726. In F. C. Steward (ed.) Plant physiology, Vol. II. Academic, New York.

Kramer, P. 1969. Plant and soil water relationships — A modern synthesis. McGraw-Hill, New York. 482 pp.

Kuiper, P. J. C. 1963. Some considerations of water transport across living cell membranes, pp. 59—68. In I. Zelitch (ed.) Stomata and water relations in plants. Connecticut Agr. Exp. Sta. Bull. 664.

Lai, M. T., A. R. Weinhold, and J. G. Hancock. 1968. Permeability changes in Phaseolus aureus associated with infection by Rhizoctonia solani. Phytopathology 58:240—245.

Laties, G. G. 1959. Active transport of salt into plant tissue. Annu. Rev. Plant Physiol. 10:87—112.

Ledbetter, M. C., and K. R. Porter. 1970. Introduction to the fine structure of plant cells. Springer, New York. 188 pp.

Lepeschkin, W. W. 1930. Light and the permeability of protoplasm. Amer. J. Bot. 17:953—970.

Levitt, J., G. W. Scarth, and R. D. Gibbs. 1936. Water permeability of isolated protoplasts in relation to volume change. Protoplasma 26:237—248.

MacRobbie, E. A. C., and J. Dainty. 1958. Sodium and potassium distribution and transport in the sea weed Rhodymenia palmata (L.) Brev. Physiol. Plantarum 11:782—801.

Marinos, N. G. 1962. Studies on submicroscopic aspects of mineral deficiencies. I.

Calcium deficiency in the shoot apex of barley. Amer. J. Bot. 49:834−841.

Marschner, H. 1964. Einfluss von Calcium auf die Natriumaufnahme und die Kaliumabgabe isolierter Gerstenwirzeln. Z. Pflanzenernähr. Düng. Bodenk. 107:19−32.

Mast, S. O. 1924. Structure and locomotion in *Amoeba proteus*. Anat. Rec. 29:88.

Mercer, F. 1960. The submicroscopic structure of the cell. Annu. Rev. Plant Physiol. 11:1−24.

Meyer, B. S., and D. B. Anderson. 1939. Plant physiology. Van Nostrand, Reinhold, New York. 696 pp.

Miller, E. C. 1938. Plant physiology (with reference to the green plant). 2nd ed. McGraw-Hill, New York. 1201 pp.

Nachmansohn, D. 1969. Proteins of excitable membranes, pp. 187−224. *In* Membrane proteins (Proc. Symp. sponsored by the New York Heart Ass.). Little, Brown, Boston.

Nagai, R., and M. Tazawa. 1962. Changes in resting potential and ion absorption induced by light in a single cell. Plant Cell Physiol. 3:323−339.

Nathansohn, A. 1904. Weitere Mitteilungen über die Regulation der Stoffaufnahme. Jahrb. Wiss. Bot. 40:403−442.

O'Brien, J. S. 1965. Stability of the myelin membrane. Science 147:1099−1107.

Osterhout, W. J. V. 1922. Injury, recovery, and death in relation to conductivity and permeability. Lippincott, Philadelphia.

Osterhout, W. J. V. 1936. The absorption of electrolytes in large plant cells. Bot. Rev. 2:283−315.

Overton, E. 1895. Über die osmotischen Eigenschaften der lebenden Pflanzen- und Tierzelle. Viertelj. Naturforsch. Ges. (Zürich) 40:159−201.

Overton, E. 1897. Über die osmotischen Eigenschaften der Zelle in ihrer Bedeutung für die Toxikologie und Pharmakologie. Viertelj. Naturforsch. Ges. (Zürich) 41:383−406.

Overton, E. 1899. Über die allgemeinen osmotischen Eigenschaften der Zelle, ihre vermutlichen Ursachen und ihre Bedeutung für die Physiologie. Viertelj. Naturforsch. Ges. (Zürich) 44:88−135.

Overton, E. 1900. Studien über die Aufnahme der Anilinfarben durch die lebende Zelle. Jahrb. Wiss. Bot. 34:669−701.

Overton, E. 1902. Bieträge zur allgemeinen Muskel- und Nervenphysiologie. Arch. Gesammte Physiol. 92:115−280.

Park, R. B., and J. Biggins. 1964. Quantasomes: Size and composition. Science 144:1009−1011.

Park, R. B., and N. G. Pon. 1961. Correlation of structure with function in *Spinacea oleracea* chloroplasts. J. Mol. Biol. 3:1−10.

Parsons, D. F. 1963. Mitochondrial structure: Two types of subunits on negatively stained mitochondrial membranes. Science 140:985−987.

Penefsky, H. S., and R. C. Warner. 1965. Partial resolution of the enzymes catalyzing oxidative phosphorylation. VI. Studies on the mechanism of cold inactivation of mitochondrial adenosine triphosphatase. J. Biol. Chem. 240:4694.

Penniston, J. T., and D. E. Green. 1968. The conformational basis of energy transformations in membrane systems. IV. Energized states and pinocytosis in erythrocyte ghosts. Arch. Biochem. Biophys. 128:339−350.

Penniston, J. T., R. A. Harris, J. Asai, and D. E. Green. 1968. The conformational basis of energy transformations in membrane systems. I. Conformational changes in mitochondria. Proc. Nat. Acad. Sci. 59:624−631.

Pfeffer, W. 1900. The physiology of plants, Vol. I. Transl. and edited by A. J. Ewart. Oxford Univ. Press, Oxford.

Preston, R. D. (ed.). 1970. Advances in botanical research, Vol. III. Academic, New York, 322 pp.

Racker, E., D. D. Tyler, R. W. Estabrook, T. E. Conover, D. F. Parsons, and B. Chance. 1965. Correlations between electron transport activity, ATPase, and morphology of submitochondrial particles, p. 1077. *In* T. E. King, H. S. Mason, and M. Morrison (eds.) Oxidases and related redox systems. Wiley, New York.

Robertson, J. D. 1959. The ultrastructure of cell membranes and their derivatives. Biochem. Soc. Symp. 16:3—43.

Robertson, J. D. 1962. The membrane of the living cell. Sci. Amer. 206(4):64—72.

Robertson, J. D. 1968. The ultrastructure of cell membranes and their derivatives, pp. 162—209. *In* D. Branton, and R. B. Park (eds.) Papers on biological membrane structure. Little, Brown, Boston.

Rogers, H. J., and H. R. Perkins. 1968. Cell walls and membranes. Spon, London. 436 pp.

Rothstein, A. 1962. Functional implications of interactions of extracellular ions with ligands of the cell membrane. Circulation 26:1189—1200.

Rothstein, A. 1964. Membrane function and physiological activity of microorganisms, pp. 23—39. *In* J. F. Hoffman (ed.) The cellular functions of membrane transport. Prentice-Hall, Englewood Cliffs, N.J.

Rothstein, A. 1968. Stoffwechsel und Membranpermeabilität von Erythrocyten und Thrombocyten, pp. 407—415. I. Int. Symp. *Wien), 17—20 Juni, 1968.

Rothstein, A., and J. van Steveninck. 1966. Phosphate and carboxyl ligands of the cell membrane in relation to uphill and downhill transport of sugars in the yeast cell. Ann. New York Acad. Sci. 137:606—623.

Ruesink, A. W. 1971. The plasma membrane of *Avena* coleoptile protoplasts. Plant Physiol. 47:192—195.

Ruhland, W. 1912. Die Plasmahaut als Ultrafilter bei der Kolloidaufnahme. Ber. Deut. Bot. Ges. 30:139—141.

Ruhland, W., and C. Hoffman. 1925. Permeability of *Beggiatoa mirabilis* to organic nonelectrolytes. Planta 1:1.

Russell, M. B., and J. T. Woolley. 1961. Transport processes in the soil-plant system, pp. 695—722. *In* M. X. Zarrow (ed.), Growth in living systems. Basic Books, New York.

Salisbury, F. B., and C. Ross. 1969. Plant physiology. Wadsworth, Belmont, Calif. 747 pp.

Samaddar, K. R., and R. P. Scheffer. 1968. Effect of the specific toxin in *Helminthosporium victoriae* on host cell membranes. Plant Physiol. 43:21—28.

Seifriz, W. 1942. Some physical properties of protoplasm and their bearing on structure. The structure of protoplasm. Iowa State Coll. Press, Ames. 245 pp.

Siegel, S. M. 1969. Further studies on factors affecting the efflux of betacyanin from beet root: A note on thermal effects. Physiol. Plantarum 22:327—331.

Slatyer, R. O. 1967. Plant-water relationships. Academic, New York.

Stadelmann, E. J. 1969a. Cell membrane permeability changes and their measurement. Plant Physiol. 44(Suppl.):no. 20.

Stadelmann, E. J. 1969b. Permeability of the plant cell. Annu. Rev. Plant Physiol. 20:585—606.

Steveninck, R. F. M. van. 1965. The significance of calcium on the apparent permeability of cell membranes and the effects of substitution with other divalent ions. Physiol. Plantarum 18:54—69.

Street, H. E., and J. S. Lowe. 1950. The carbohydrate nutrition of tomato roots. II. The mechanism of sucrose absorption by excised roots. Ann. Bot. 14:307—329.

Sutcliffe, J. F. 1954. The exchangeability of potassium and bromide ions in cells of red beetroot tissue. J. Exp. Bot. 5:313—326.

Sutcliffe, J. F. 1959. Salt uptake in plants. Biol. Rev. 34:159—218.

Sutcliffe, J. F. 1962. Mineral salts absorption in plants. Pergamon, Oxford. 194 pp.

Thatcher, F. S. 1939. Osmotic and permeability relations in the nutrition of fungus parasites. Amer. J. Bot. 26:449—458.

Thatcher, F. S. 1942. Further studies of osmotic and permeability relations in parasitism. Can. J. Res. C20:283—311.

Traube, M. 1867. Experimente zur Theorie der Zellbildung und Endosmose. Arch. Anat. Physiol. Wiss. Med. 87:128—165.

Vanderkooi, G., and D. E. Green. 1970. Biological membrane structure. I. The protein crystal model for membranes. Proc. Nat. Acad. Sci. 66:615.

Vanderkooi, G., and D. E. Green. 1971. New insights into biological membrane structure. Bioscience 21:409—415.

Vanderkooi, G., and M. Sundaralingam. 1970. Biological membrane structure. II. A detailed model for the retinal rod outer segment membrane. Proc. Nat. Acad. Sci. 67:233.

Viets, F. G., Jr. 1944. Calcium and other polyvalent cations as accelerators of ion accumulation by excised barley roots. Plant Physiol. 19:466—480.

Waisel, Y. 1962. The effect of Ca on the uptake of monovalent ions by excised barley roots. Physiol. Plantarum 15:709—724.

Watson, M. 1959. Further observations on the nuclear envelope of the animal cell. J. Biophys. Biochem. Cytol. 6:147—156.

Weier, T. E., and A. A. Benson. 1967. The molecular organization of chloroplast membranes. Amer. J. Bot. 54:389—402.

Whaley, W. G., H. H. Mollenhauer, and J. Leech. 1960. The ultrastructure of the meristematic cell. Amer. J. Bot. 47:401—419.

Whaley, W. G., H. H. Mollenhauer, and J. Kephart. 1959. The endoplasmic reticulum and the Golgi structures in maize root cells. J. Biophys. Biochem. Cytol. 5:501—506.

Wheeler, H., and H. S. Black. 1963. Effects of *Helminthosporium victoriae* and victorin upon permeability. Amer. J. Bot. 50:686—693.

Williams, P. H., and N. T. Keen. 1967. Relation of cell permeability alterations to water congestion in cucumber angular leaf spot. Phytopathology 57:1378—1385.

Withrow, R. B. 1951. Light as a modifying influence on the mineral nutrition of plants, pp. 389—410. *In* E. Truog (ed.) Mineral nutrition of plants. Univ. Wisconsin Press, Madison.

CHAPTER 8

Translocation

In the following discussion, emphasis is placed on translocation of inorganic ions and molecules, but in some cases it is necessary to discuss the translocation of organic molecules. In the translocation of sugar, sucrose is generally the major component, as reported by Zimmermann (1957). He noted in some species, for example, white ash, elm, and linden, that raffinose and stachyose may also be present in phloem exudate. The total sugar concentrated of sieve tube exudate is usually between 10 and 25% (w/v) (Zimmermann, 1957, 1960), and that of amino acids and amides between 0.03 and 0.4% (w/v) (Zimmermann, 1960).

It is interesting to note some of the properties of phloem and xylem exudates of sugar beet roots (Table 8.1). With regard to inorganic ions in xylem sap of plants in general, Milthorpe and Moorby (1969) indicated values generally in the range of 400–2500 ppm.

Translocation, transport, or conduction may be defined as "a movement of solutes or water from one part of a plant to another part—generally through specialized conducting tissues." However, substances can move from cell to cell, such as from one parenchymatous cell to another, without moving in specialized conducting tissues (i.e., xylem and phloem). The greater rates and quantities of translocation occur in the specialized conducting tissues.

In general, water and inorganic ions and molecules move in the xylem; organic molecules move in the phloem. There is one striking exception; all substances moving *out of leaves*—even water—move outward through the phloem. There are cases of unusual events or apparent irregularities in translocation; at least there are exceptions to some of the general scheme of events. For exam-

Table 8.1
Properties of Phloem and Xylem Exudate of Sugar Beet Roots[a]

Property	Phloem	Xylem
Viscosity, at 20° C	1.58	0.98
Surface tension, at 20°C (dynes/cm)	50.3	57.6
Density, at 20°C	1.059	1.004
Total solids (%)	16.0	0.8
Conductivity ($\times 10^3$)	10.79	2.30
Osmotic pressure		
Freezing point (atm)	18.04	1.63
Conductance (atm)	3.86	0.82
Electrolytes (% of total solutes)	21.5	50.3
pH	8.02	6.12
Sucrose (%)	8.8	0.10
Reducing sugars (%)	0.33	0.03
Soluble N, excluding NO_3^- (g/liter)	5.65	0.10
Total amino acids, calculated as		
glutamic acids (%)	1.10	—
Heat-coagulable protein N (%)	0.48	—

[a] Fife et al. (1962).

ple, some organic compounds may move in the xylem, and at least some inorganic ions apparently move in the phloem (Bollard, 1960).

For additional details on translocation, the reader may consult Crafts (1938, 1961), Curtis (1935), and Richardson (1968). Additional coverage on translocation is also available in three recent textbooks of plant physiology (Devlin, 1969; Salisbury and Ross, 1969; Price, 1970).

While reading the rest of this chapter it should be remembered that in translocation the two main problems (Richardson, 1968) are: (1) What pathway (i.e., tissue) is involved in the translocation of substances? (2) What is the mechanism or driving force responsible for movement?

THEORIES

Simple Diffusion

Ordinary diffusion has long been recognized as inadequate for explaining the rates at which substances diffuse and the amounts that diffuse. For example, within 40 min newly absorbed ^{32}P was detected in the leaves and stem tips of tomato plants over 6 ft tall (Arnon et al., 1940). Radioactive Br^- moved from the culture solution to the leaves at the tips of 15-ft-long cucurbits in 5 min (Stout et al., 1947). The observed diffusion constant of sugar in sieve tubes is approximately 40,000 times as great as the diffusion constant for sugar in a 2% solution of sucrose in water (Mason and Maskell, 1928).

Protoplasmic Streaming

The significance of protoplasmic streaming in circulating and distributing substances within a cell was first suggested by Vries (1885) and was later supported by Curtis (1925, 1935) and Mitchell and Worley (1964). If, for example, a given substance were rapidly distributed uniformly throughout a sieve tube, its movement to the next sieve tube would presumably be expedited.

Mitchell and Worley (1964) observed rotational streaming of protoplasm in phloem and xylem fibers of bean stems, and in cells that appeared to be fibers in stems of young cucumber, tomato, sunflower, and flax. In immature bean phloem fibers, plastids were carried a distance of approximately 2 mm—the average length of a fiber—in about 3 min (3.6 cm/hr).

Working with epidermal, cortical, and fiber cell types of broadbean, bean, and pea, Worley (1968) reported that protoplasmic streaming was reversibly stopped by DNP; streaming renewed when DNP was washed out. He concluded that rotational streaming accelerated the translocation of soluble substances in fiber cells.

It has been suggested by Sutcliffe (1960) that protoplasmic streaming probably depends on a rhythmic folding and unfolding of protein molecules. He noted that streaming was stopped by chloramphenicol, whereas salt absorption was not. In view of this observation, this investigator concluded that streaming was not important in salt absorption per se. Although streaming does not appear to be a mechanism of absorption, it could aid in steepening the gradient from outside to the inside of the cell.

The addition of ATP increases streaming in a transient way, with the rate soon returning to normal (Kamiya, 1960), but the reason for this effect is unknown. It was also noted that the force of streaming increased when respiration was inhibited and decreased when glycolysis was inhibited. The energy source for streaming, then, appeared to lie in fermentation and not in respiration. ATP produced by fermentation may possibly be available to drive the streaming, since it is produced in the ground plasm where streaming occurs. Presumably, ATP produced by respiration is not readily available to the mechanochemical system of contractile proteins.

In barley root hairs rotational streaming was stimulated approximately 1.3–1.8 times by the addition of *myo*-inositol (Soran and Lazar, 1969). The stimulation was not attributable to a direct effect of *myo*-inositol, but to ATP formed in the reaction: inositol hexaphosphate + ADP $\rightleftarrows$ ATP + inositol. In *Nitella clavata*, protoplasmic streaming was dependent on light only at intensities below 700 ergs/cm^2 sec (Barr and Broyer, 1964). Even at the lowest light intensities, rates were at least 60% of the maximal rate.

There are two main objections to this theory as regards explaining the observed rates and amounts of translocation of materials throughout the plant.

First, protoplasmic streaming has not been universally observed in sieve tubes. Second, this phenomenon would not explain the rapid rates of translocation that have been recorded.

Interfacial Diffusion

This theory (Honert, 1932) is based on the principle that substances that lower the interfacial tension between two immiscible liquids tend to spread *rapidly* by diffusional gradients over the interface between them. Substances move in this interface from the region of their introduction or formation, where surface tension is lowered, toward regions not containing the substance or where surface tension is relatively higher. This type of movement is approximately 68,000 times as rapid as that occurring by simple diffusion.

There are numerous interfaces in cells where this type of rapid movement could occur. Inasmuch as lowering of surface tension is dependent on the concentration of the interface-active substance, the rate of transport would be a function of concentration gradient. This is in accord with the observed relationship between rate of translocation and concentration gradient. Incidentally, Honert (1932) suggested that protoplasmic streaming is the result of surface tension phenomena and not the vehicle of transport.

Translocation of protosynthate has been reported to occur by interfacial flow (Vernon and Aronoff, 1952). Nelson (1963) has also championed this theory for explaining the rapid movement of materials in phloem.

Activated Diffusion

Mason and Phillis (1937) proposed that transport occurred through stationary cytoplasm by a process they called "activated diffusion." They proposed that living cytoplasm was capable of hastening diffusion either by activating the diffusing molecule or by decreasing the resistance to diffusion through the cytoplasm.

Mass Flow or Pressure Hypothesis

According to this theory, as originally proposed by Münch (1927) and most recently championed by Zimmermann (1961) and Crafts (1961), there is a turgor pressure gradient between supplying and receiving cells (or tissues). Sugars are formed in leaves, and the osmotic pressure of the cells is increased; sugars are used by the roots with a concomitant lowering of osmotic pressure. These phenomena, then, tend to establish the gradient on which this theory rests. As water is moved from the roots to the tops, salts would be carried along by the mass flow of solution. Downward movement of this mass flow of solution in the phloem would carry sugars from the leaves to the roots.

There is experimental evidence consistent with the mass flow hypothesis. In many cases, although not in all, the required gradient exists and is in the right direction. Phloem cells have been observed to be under pressure; phloem exudes when it is pierced; and a volume of exudate from a given sieve tube may exceed manyfold the volume of the sieve tube cell.

Working with aphids attached to three stems, Zimmermann (1961) cut off the stylets and collected phloem exudate. He observed that a single sieve element 20–30 μ in diameter and 0.4 mm long could be continuously refilled 3 to 10 times per second with a concentrated sugar solution for hours or days without any visible injury. He concluded that mass flow was the only reasonable explanation for this rapid rate of refilling. He also noted that the exudate concentration decreased approximately 0.01 M per meter in a downward direction of a normal tree during the summer. According to Poiseulle's equation, he determined that the pressure gradient, to which this would correspond, was fully sufficient to force solution at the observed rates through capillaries of the dimensions of the sieve tube lumen and the combined sieve pores.

Zimmermann (1961) reported that exudates of over 250 species, from 55 plant families, are known. They have been shown to contain 10–25% dry matter, of which 90% or more is sugar—occasionally accompanied by a sugar alcohol. Sucrose may be as high as 0.5 M; hexoses are not present in most exudates and, in the few species that do contain them, they occur in very low concentrations.

With a plasmolyzing mannitol concentration around wheat roots, Ingelsten (1966) reported that transport of SO_4^{2-} appeared to be strictly a mass flow phenomenon with a diffusion barrier no longer restricting transport.

Trip and Gorham (1968b) introduced [^{14}C] sugar and tritium oxide (THO) through a cut side vein or flap of a squash leaf and followed the translocation patterns along with that of T photosynthate. They concluded that if THO moved as liquid water in the phloem along with the [^{14}C] sugars, as blockage by steam-girdling suggested, mass flow could be a possible mechanism of translocation.

In *Vicia faba*, bidirectional translocation of ^{14}C assimilate occurred, but it was shown with a fluorescent dye (K fluorescein) that each bundle and sieve tube translocated the dye in one direction only (Peterson and Currier, 1969). Research with the dye was interpreted as support for the mass flow theory. Similar results were obtained with *Phaseolus vulgaris*, *Vinca rosea*, and *Pelargonium hortum*.

There is some evidence that would appear to discredit mass flow as a satisfactory explanation of translocation. Using an approach similar to that just discussed for Trip and Gorham (1968b), Gage and Aronoff (1960) and Choi and Aronoff (1966) concluded that mass flow was not a dominant process in the translocation of photosynthate in their experiments.

Validity of the mass flow concept was tested by bark strips of willow sealed to polyethylene tubes having two compartments (Peel et al., 1969). It was then possible to study transport along the sieve tubes of tritiated water, ^{14}C-labeled sugars, and ^{32}P phosphates from one compartment toward a stylet situated in bark above the other compartment. Lack of rapid longitudinal movement of tritiated water along the sieve tubes indicated that transport is unlikely to involve a mass flow of solution.

A study of the translocation of acid fuchsin and of ^{45}Ca^{2+} injected into dogwood trees indicated that Ca^{2+} did not move primarily by mass flow as the dye apparently did (Thomas, 1967).

Kaufmann and Kramer (1967) studied the phloem and leaf water potential of yellow poplar *(Liriodendron tulipifera* L.) and red maple *(Acer rubrum* L.) using a thermocouple psychrometer (Richards and Ogata, 1958) and the vapor equilibration technique (Slatyer, 1958). Values of water potential were such that a gradient in turgor pressure apparently did not exist at a time when the rate of translocation was expected to be high. They concluded that their results did not support the mass flow theory.

Using tritiated and ^{14}C-labeled sugars, a bidirectional translocation of sugars was observed within the same sieve tube by Trip and Gorham (1968a). They concluded that this observation did not support the mass flow theory which would require a unidirectional flow of sugar in the lumen of a given sieve tube. Others have also demonstrated bidirectional translocation—the simultaneous movement of one or more substances upward and downward in the same section of a stem (Biddulph and Cory, 1960, 1965; Webb and Gorham, 1964). For example, Eschrich (1967) demonstrated simultaneous bidirectional translocation of fluorescein and [^{14}C] urea in the phloem of a stem of *V. faba.* Dye was placed on a lower leaf, [^{14}C] urea on an upper leaf, and aphids on the stem between these leaves obtained both compounds from the phloem.

Labeled Ca^{2+} did not ascend the stem of bean plants with the labeled water (THO) of the transpiration stream—as would be expected by the mass flow concept (Biddulph et al., 1961). In a later study Ca^{2+} moved up the stem of bean *(P. vulgaris* L. var. Red Kidney) plants by a process of exchange, and these results were regarded as inconsistent with mass flow (Bell and Biddulph, 1963).

There would be considerable resistance to the flow of a solution in a tissue such as the phloem, and many have doubted that the magnitude of turgor pressure in leaves is great enough to overcome this resistance. In addition, one of the weakest points in the mass flow theory is the anatomy of sieve plates and the degree of permeability of sieve plates under undisturbed conditions (Zimmermann, 1960). In fact, Crafts (1961), a champion of the mass flow concept, reported that only one aspect lacked a satisfactory explanation, namely, the observed extraordinarily high permeability of the sieve plates.

In order for mass flow to occur, supplying cells in leaves would have to have higher turgor pressure values than receiving cells of roots. This situation does not exist in certain cases—such as potato tubers and sugar beets. In sugar beets sucrose moves against a concentration gradient from sieve tubes to storage cells (without transformation despite the fact that all necessary enzymes for transformation are present) (Kursanov, 1963).

Mass flow would account for translocation in the phloem, for example, in only one direction at a given time. Phillis and Mason (1936) showed that organic N compounds may be traveling upward in cotton plants at the same time carbohydrates are moving downward in the phloem. Also, observed rates of movement of various substances have often been much too high to be explained by the mass flow concept. The retarding effects of low temperature and oxygen on translocation would substantiate the proposition that an active role of the cytoplasm of sieve tubes is involved in translocation.

In his early work Crafts (1931) suggested that an osmotic system prevailed in plants, that the total phloem was the channel for conduction with the majority of the transport occurring in cell walls. Later, Crafts (1932) stated that the exudation from cut peduncles of cucurbits could not be explained on the basis of mass flow through perforations in the sieve plates. He concluded that movement occurred through the sieve tube lumina and partly through walls of the phloem. In 1933, he concluded (Crafts, 1933) that theories of protoplasmic streaming and pressure flow through phloem walls were inadequate to explain the observed rate of translocation in potato. He reported that the sieve tube lumina apparently afforded the most available channels for translocation and thus accepted the Münch mass flow concept. In a very interesting and thorough book, Crafts (1961) reviewed the pertinent literature on translocation and attempted to explain the reported findings in terms of the mass flow concept even though certain authors had concluded that their results did not corroborate the concept.

Pitman (1965) concluded that measured free space was inadequate for the transport of inorganic ions at rates that have been observed, and Milthorpe and Moorby (1969) claimed that most transport of ions occurs through the symplast.

Polarized Transport

Polarized transport pertains to those situations in which a given substance moves only in one direction—often against a gradient. Loomis (1945) studied this type of transport in some detail in the corn plant. Sucrose moved from leaves to stalks against a 2- to 50-fold gradient. The polarizing action of developing grain (corn) is illustrated by removing all leaves from the main stalk with well-developed tillers or sucker branches at the base. If pollination has oc-curred and leaves near the ear are not removed until the polarizing action of developing embryos is established, full-sized ears can be produced by polar transport of sugars from leaves of sucker branches 8–10 ft from the ear.

In mature sugar beet plants in the dark, sugars in leaves move to the root, and leaves die from lack of sugar while roots contain 15–20% sucrose (fresh weight basis). Young leaves on sugar beets continue to grow rapidly and vigor-ously on root-stored sucrose, which indicates that sugars move in polar fashion *to* young leaves at the same time that sugars are moving *from* older leaves.

Movement of auxin to the morphological base of a stem, even when an ex-cised stem portion is inverted, constitutes another striking example of polar transport.

There are other examples of polar transport involving inorganic ions. Appli-cation of root-forming hormones to the decapitated stem of a bean plant ulti-mately results in the initiation of and growth of roots. These events are preceded by a mobilization of carbohydrates (Alexander, 1938) and of various inorganic ions except Ca^{2+}.

In extreme Ca^{2+} deficiency in bean plants, the root system dies and the plant starts to initiate and grow new roots near the base of the hypocotyl. Total ash determinations revealed that there is a pronounced accumulation of ash constit-uents prior to the initiation and development of lateral roots (Gauch, 1940).

Dual Transport

This type of translocation concerns the transport of two substances simulta-neously. For example, although sugars may be translocated independently, there is considerable evidence that translocation of applied plant growth regu-lators (from leaves) is dependent on simultaneous translocation of photosyn-thate—presumably sugars (Mitchell and Brown, 1946; Weaver and DeRose, 1946; Rice, 1948; Linder et al., 1949; Rohrbaugh and Rice, 1949; Jaworski et al., 1955; Hay and Thimann, 1956; Barrier and Loomis, 1957; Clor, 1959; Mitchell et al., 1960). Phloem transport of 2,4-D, urea, and 3-amino-1,2,4-tria-zole was greatly reduced in carbohydrate-starved cotton leaves (Chlor et al., 1963).

Linder et al. (1949) observed that synthetic growth regulators were readily translocated from bean leaves when the latter contained 2.5–3.5% total sugar, but not when they contained only about 2% total sugar. These percentages were determined by methods that depended on reducing properties of sugars, but certain substances other than sugars have a reducing action—the so-called "noncarbohydrate reducing substances." Apparently, these substances are pres-ent in relatively high concentration in bean leaves, since Mathes (1961) found by paper chromatography that bean leaves with 2% sugar were in fact essentially devoid of sugars. The report by Mathes thus clarifies why 2% sugar was ineffective in the translocation of synthetic plant growth regulators.

According to Crafts (1961), there is increasing evidence indicating a simultaneous correlated flow of many substances (in the phloem) from source to sink. Viruses, growth regulators, hormones, and photosynthate apparently move together.

Gauch and Dugger (1953, 1954) provided an example of dual translocation involving an inorganic ion and an organic molecule. They obtained evidence which they interpreted as indicating that B was involved in sugar translocation. One of the two mechanisms of action for B which they proposed (Gauch and Dugger, 1953) involved the translocation of sugar–borate complexes.

In another example, 2,4-D was not readily translocated in P-deficient tomato plants (Rohrbaugh and Rice, 1956). Inasmuch as sugars and growth regulators have been shown to be translocated at the same time, it was suggested that P might play its role in 2,4-D transport by being a requirement for the release of respiratory energy or by esterification between the sugar and the growth regulator.

Rice and Rohrbaugh (1958) provided yet another example of a relationship between an inorganic ion and the translocation of an organic molecule. They observed that in tomato plants 2,4-D was not readily translocated in K-deficient plants, and that translocation was affected by the K^+ level in plants.

Superfluidity

Superfluidity offers an additional possibility (Mendelssohn, 1958) to account for rapid movement of solution in sieve tubes of the phloem. If molecules exist in a highly ordered state, they might move in a way that is unaccountable by classic physical concepts. Cells may contain little or no random water; most of the water in the closely packed protoplasm of cells may be in an ordered state (Szent-Gyorgyi, 1956). Jacobson (1955) showed that electropolar groups on surfaces may induce order in adjacent water. In a highly ordered state, it is possible that water might move between peptide chains, ordered in the longitudinal direction, by a catenary process if sufficient energy were applied at the upper end (Crafts, 1961).

Electroosmosis

Spanner (1958) proposed that bulk flow of sieve tube contents could be achieved by electroosmosis. According to Crafts (1961), the starting device for this process would be (1) discharge of sugar into the sieve tubes, (2) uptake of water osmotically leading to a vertical surge in cells adjacent to the sieve tube, and (3) flow of K^+ along the direction of flow in the sieve tubes and counter to this direction in neighboring cells. Once established, this circulation would electrically polarize the sieve plate and cause rapid electroosmosis through the plate. Attractive as this hypothesis may be, Crafts (1961) notes that experimental evidence for it is lacking, and he discusses experimental evidence inconsistent with the hypothesis.

TISSUES INVOLVED IN TRANSLOCATION

Upward Movement

For many years most plant physiologists agreed that upward translocation of water and inorganic ions or molecules occurred in the xylem. Curtis (1925, 1935) contradicted this viewpoint and claimed that phloem was the major pathway for inorganic ions. His research (Curtis, 1925) with privet and other stems, in which some were ringed and others were not, indicated that N and ash constituents moved primarily in the phloem. Curtis' experiments were conducted before the interrelationship between salt absorption and respiration was established. Ringing of stems undoubtedly interfered with translocation of sugars to roots; the normal pathway for translocation could thus have been altered. Clements and Engard (1938) voiced strong opposition to Curtis' conclusions. With the advent of isotopes or tracers, overwhelming evidence was obtained which implicated xylem as the normal pathway for inorganic ions. Leaving the tissues intact, Stout and Hoagland (1939) separated phloem and xylem with a layer of waxed paper and showed clearly that xylem was the major pathway for the upward conduction of inorganic nutrients.

The much higher water solubility of ^{45}Ca in xylem and the known immobility of Ca^{2+} in phloem make it likely that xylem provides the radiocalcium for translocation to dogwood foliage (Thomas, 1970). It was concluded that upward movement of ^{45}Ca by exchange reactions in the stems and branches could account for the movement to the tops—a point of agreement with conclusions by Bell and Biddulph (1963).

^{45}Ca applied to roots of apple seedlings moved readily to the developing leaves with kinetin, benzyladenine, and B sprays increasing the movement (Shear and Faust, 1970). NO_3^- increased movement and accumulation of Ca^{2+} in mature leaves; NH_4^+ increased movement into new leaves. It was also concluded that translocation in the stem is effected by a nonspecific ion exchange; any divalent cation could free Ca^{2+} for ascent. Further, it was suggested that this exchange may be a property of lignin, since Ca^{2+} combines with lignin and it easily exchanges from sites it occupies on the lignin molecule.

During senescence of older leaves or during a deficiency of a mobile element, the element may move from older to younger leaves in an upward direction through the phloem.

Downward Movement

When inorganic constituents move out of leaves, it is generally agreed that they do so via the phloem and, if these constituents are moving in a downward direction, they continue to move in the phloem. Biddulph and Markle (1944) observed downward movement of radioactive phosphate through phloem at a rate of 21 cm/hr. Huber (1941) reported that downward movement may be as high as 100 cm/hr in broad-leaved forest trees. In general, velocities of translocation in phloem are of the order of 10–100 cm/hr with extremes ranging up to 300 or even 3000 cm/hr (Zimmermann, 1960).

When two incisions were made in the conducting phloem of white ash (*Fraxinus americana* L.), one above the other, exudation occurred from both incisions (Zimmermann, 1962). *Upward* translocation toward the lower incision indicated a reversal of the normal direction of translocation in the phloem.

Collander (1941a) suggested that in certain species one or more ions might be recycled from the tops back to the roots. Cooil et al. (1965) observed a downward flow of Na^+ from the tops of squash plants to the roots. This downward flow was assumed to be in the phloem inasmuch as Na^+ accumulated above a steamed girdle. They concluded that downward movement of Na^+ might partially account for the impression that upward movement of Na^+ is restricted in most plants. Na^+ has been shown to move downward in *P. vulgaris* but not in *Raphanus sativa* (Wallace and Hemaidan, 1963).

From a sunflower leaf downward translocation of $[^{14}C]$ photosynthate was stimulated by ATP and inhibited by DNP, whereas upward movement was unaffected by ATP or DNP (Shiroya, 1968).

Movement of materials out of leaves occurs through phloem—as amply demonstrated in the case of SO_4^{2-} and phosphate (Biddulph, 1951, 1959; Biddulph and Markle, 1944). It was reported that Ca^{2+} is not transported in phloem, hence does not move out of leaves prior to leaf fall—as is the case with other ions (Sutcliffe, 1962). However, Ca^{2+} may move in the phloem, but at a very slow rate. Foliar-applied Ca^{2+} was observed by Biddulph et al. (1959) to move outward in phloem, but at a rate approximately 1/100 that of P. These investigators also noted that Ca^{2+} was translocated from cotyledons of bean plants during early growth of the seedling, but that later growth depended on external Ca^{2+}, since Ca^{2+} in the cotyledons later became immobile. When Ca^{2+} injected into the stem of a white pine moved into the phloem, the element became immobilized in the form of Ca oxalate crystals (Thomas, 1967).

Bukovac et al. (1960) concluded that Mg^{2+} is immobile in the phloem of young bean plants. On the contrary, when ^{28}Mg was applied to specific leaves of bean *(P. vulgaris)* and barley *(Hordeum vulgare)*, as much as 7% of the absorbed Mg^{2+} was exported from treated bean leaves and 11% was transported basipetally from the treated zone of barley leaves—in 24 hr (Steucek and Koontz, 1970). Transport did not occur past a heat-killed section of treated leaf—indicating that translocation occurred in the phloem. Mg^{2+} movement in phloem was also evident in autoradiograms of bean stem segments in which xylem was separated from phloem by a thin sheet of plastic.

Certain unnatural S amino acids were detected in the phloem of vascular bundles of *P. vulgaris* following the application of $H_2^{35}S$ to the primary leaves (Brandle and Schnyder, 1970).

In studies of the movement of labeled SO_4^{2-} and phosphate from a leaf, it was noted that phloem of a vascular bundle did not behave as a unit, since some of the sieve tubes were radioactive (hence conducting), whereas others were not (Biddulph, 1959).

Although there has been no report with regard to inorganic nutrients, it is interesting in this connection to note that Trip (1969) found that in minor veins of sugar beet leaves sugar was translocated in the companion cells rather than in sieve tubes; in major veins translocation occurred in sieve tubes.

Lateral Movement

In stems in which there are xylem and phloem rays composed of parenchymatous cells, there may be lateral movement of inorganic constituents from xylem to phloem, and even to tissues outward from the phloem.

Lateral movement of inorganic solutes is clearly shown in the case of spiral ringing of a tree (apple, for example) in that upwardly translocated ions or salts move laterally around a spirally ringed trunk, emerge at the end of the spiral, and affect the growth or appearance of branches in line with the open end of the spiral.

Ca^{2+} appears to enter bean stem sections in two ways: a reversible exchange phase, and an irreversible accumulation phase in which Ca oxalate is formed. Biddulph et al. (1961) noted that the xylem cylinder operated essentially as an exchange column for Ca^{2+}. Lateral movement was indicated by the fact that Ca^{2+} was transferred from the xylem of leaf traces to cells of the adjacent phloem.

Simultaneous Upward and Downward Movements in Xylem and Phloem

Zimmerman and Connard (1934) studied in detail the reversal of direction of translocation of solutes in stems. They reported that the orientation of the (intervening) stem was of no consequence in determining the direction of translocation of water, inorganics, and organics, and that *activity* at some distant point was the determining factor affecting the *direction* of translocation of each substance. They therefore concluded that each of these classes of translocatable substances could be moving upward in a given tissue in one part of the stem and downward in this same tissue in another part.

In cotton plants, organic N compounds may be traveling upward at the same time that carbohydrates are moving downward in the phloem (Phillis and Mason, 1936).

In this connection, it has been reported (Mitchell and Worley, 1964) that in immature phloem fibers of bean plants cytoplasmic currents within a fiber usually appeared to move in opposite directions past a point of observation, because of the rotational course of the currents. Streaming in the xylem fibers was also observed, and in what appeared to be fibers in the stems of young cucumber, tomato, sunflower, and flax plants. This bidirectional movement of cytoplasm within a cell is of interest in the movement of substances within a cell and in turn within a plant.

FACTORS THAT MAY ALTER THE PATHWAY OF TRANSLOCATION

Form in Which Element Is Absorbed

The form in which an element is absorbed could possibly determine whether it moves upward in the xylem or in the phloem. In the case of N, NO_3^- would be expected to move upward in xylem; when the element is absorbed as an amino acid, it might very well be carried by the phloem.

Alteration or Lack of Alteration of Form of Element in Roots

In some species, such as apple, absorbed NO_3^- is usually converted into amino acids in the roots (Thomas, 1927; Bollard, 1956, 1957a). In fact, in many species of plants NO_3^- reduction occurs in the roots (Bollard, 1956). In this case, then, N in the organic form may move upward through the phloem.

Interestingly, Klepper and Hageman (1969) noted that leaves of apple seedlings had the highest concentration of nitrate reductase—followed by the stems and petioles and, last, the roots. Roots had by far the highest concentration of NO_3^- of any part of the plant. By supplying higher NO_3^- concentrations, nitrate reductase activity in leaves could be induced.

In apple tracheal sap, organic N (aspartic acid, asparagine, glutamine, glutamic acid, other amino acids, and peptidelike substances) accounted for most of the N present (Bollard, 1957a). There was a trace of NH_4^+ nitrogen, but no detectable NO_3^-.

Similarly, the tracheal sap of a range of woody species of the Rosaceae contained primarily organic N—notably, aspartic acid, asparagine, and glutamine (Bollard, 1957b).

In a more extensive survey, including xylem sap from a range of dicotyledons, monocotyledons, and gymnosperms, Bollard (1957c) reported that NO_3^- occurred in a few species in appreciable amounts, but that in most species it was either absent or present only as a trace. All saps contained a range of amino acids and amides; glutamine and asparagine were quantitatively the most important in nearly all species. Some saps contained either citrulline or allantoic acid as a major constituent.

In cotton (Mason and Maskell, 1931) and sunflower (Leonard, 1936), both NO_3^- and reduced N compounds move in the transpiration stream.

Although P moves in the phloem primarily as P_i (Bieleski, 1969), two unidentified organic P compounds were found in xylem sap—indicating that partial assimilation of P may occur in the root system (Tolbert and Wiebe, 1955). For S, however, the latter authors found only inorganic sulfate—a finding also reported by Thomas et al. (1944). In fact, the latter authors found that when organic S is to be moved out of leaves that the S is apparently changed back to SO_4^{2-} for translocation, and that the S is then resynthesized into organic combination when it arrives at its destination.

In those plants in which NO_3^- reduction normally occurs in the tops, absorbed NO_3^- would not be altered in the roots and would move upward as NO_3^- in the xylem. Whether or not absorbed SO_4^{2-} and phosphate are converted into amino acids in the roots could also determine whether these constituents move upward in the xylem or the phloem.

Carbohydrate Supply to Roots

In those plants, such as apple trees, in which NO_3^- is ordinarily reduced in the roots and incorporated into amino acids that are translocated in the phloem, an insufficiency of carbohydrate translocation to roots could alter the normal events of NO_3^- reduction in roots. If roots were deficient in carbohydrates, N would move upward as NO_3^- in the xylem rather than as amino acids in the phloem.

Species of Plant

The route of translocation of inorganic solutes may vary with the species of plant, or it may be different for herbaceous and woody species. It is possible that in many plants, such as cotton, salts may move principally in the xylem; in woody species they may move primarily in the phloem.

Translocation of Ca^{2+} in the peanut plant is especially interesting in this regard, since Ca^{2+} plays an important role in the formation of peanut fruits at the ends of gynophores or "pegs." Nutrients are absorbed by both roots and pegs. When Ca^{2+} is made available to roots, but not to pegs, fruit formation is very poor. This clearly indicates that translocation of Ca^{2+} from roots to pegs is minimal or essentially nonexistent, and that Ca^{2+} must be directly available to the developing fruit (Brady, 1947). By using ^{45}Ca to study Ca^{2+} translocation in peanut, Bledsoe et al. (1949) obtained convincing evidence to support such a conclusion.

Although K^+, Mg^{2+}, N, and P were translocated from soybean cotyledons, Ohlrogge (1963) reported that Ca^{2+} was not. Gauch (1940) observed that Ca^{2+} did not move out of the cotyledons of Dwarf Red Kidney bean plants raised in a solution to which no Ca^{2+} had been added.

Direction of Flow

Although, in general, it appears that inorganic constituents move upward primarily in the xylem, their downward movement appears to be primarily in the phloem. As indicated earlier, however, if there were a deficiency of a mobile element in the plant, that element would pass out of an older leaf via the phloem and move *upward* through the phloem to a younger leaf.

Amount of an Element Being Absorbed

Even in those cases in which it is believed that the normal pathway for a given element may be the xylem, the absorption of a great amount of that element may cause some spillover into other tissues, for example, the phloem.

FACTORS AFFECTING TRANSLOCATION

pH

The pH of the medium, as well as the concentration of P, may affect the absorption and translocation of Fe^{3+} in plants. For example, according to Biddulph (1951), at pH 7 and high external P (0.001 M) Fe is precipitated on the root surfaces and there is very little Fe in the xylem or in the veins of the leaves. At pH 7 and medium P (0.0001 M), it is found that Fe accumulates in veins but does not enter the mesophyll. At pH 4 and medium P (0.0001 M), there is rapid uptake of Fe and an even distribution throughout the plant—including the leaves.

Transpiration

Inasmuch as water and salts are carried by the xylem, one would expect a positive correlation between rate of transpiration and amount of salts translocated to the tops via the "transpiration stream." In general, such a relationship is found (Pettersson, 1960), but there is generally no linearity between rate of transpiration and amount of salt translocated. This lack of correspondence of rates for the two processes highlights the fact that water and salt absorption are separate, independent processes. Hylmo (1955) may have provided a reconciliation for the conflicting evidence concerning the relationship between transpiration and translocation of salts. He stated that there is a correlation between transpiration and the translocation of certain ions that are not particularly accumulated in the vacuoles of root cells. However, for those ions that are preferentially accumulated in the vacuoles of root cells, there is no relationship between transpiration and either absorption or translocation.

There is a logical explanation for the fact that there may be no relationship (Koontz and Foote, 1966) between the rates of transpiration and of translocation of salts to tops. When the rate of transpiration is high, the concentration of ions in xylem sap is low, and vice versa (Russell and Shorrocks, 1959). With a relatively higher concentration of salts in the xylem sap, under a low rate of transpiration, as much total salt may be translocated to tops as occurs with a relatively low concentration of salts in the xylem sap during a high rate of transpiration.

In order to study the effect of rate of water flow in the xylem on ion transport, Pettersson (1966) applied various pressure deficits to the stems of decapitated sunflower plants and studied the movement of labeled sulfate. A rectilinear correlation existed between the rate of SO_4^{2-} transport in the sap and the water flow at sap flow velocities comparable with transpiration rates. When the transport of water was very slow, the rate of SO_4^{2-} transport became constant and independent of the water stream.

An excellent review on transpiration and on the uptake and translocation of water has been provided by Kozlowski (1964).

Species of Plant

In most plants Ca^{2+} is generally immobile (Bollard, 1960), and Sr^{2+} usually behaves much as Ca^{2+} (Rediske and Selders, 1953); under some conditions Ca^{2+} may be mobile in plants (Biddulph et al., 1959). Ca^{2+} is considered to be highly immobile in phloem (Zimmermann, 1960). Strontium was mobile in bean *(P. vulgaris)* and cotton *(Gossypium hirsutum)* plants; in the latter Sr^{2+} moved from the first two leaves and stems into the third and fourth leaves when they began to develop (Creger and Allen, 1969).

In most species Na^+ may be accumulated by roots, with a lesser concentration in stems and a still lesser concentration in leaves (Collander, 1941a, b; Gauch and Wadleigh, 1945; Truog et al., 1953; McLean, 1956; Berstein et al., 1956; Jacoby, 1964; Cooil et al., 1965). In general, Na^+ is rather uniformly distributed throughout halophytic species. Although it has been suggested (Gauch and Wadleigh, 1945) that cellular membranes of certain extrastelar tissues might retain Na^+ in the root, Jacoby (1964) concluded that Na^+ retention by roots resulted from competition between transport and binding by sites in the path to and in the stele. Cooil et al. (1965) noted a downward flow of Na^+ from the tops of squash plants via the phloem; when such downward movement to the roots occurs, it *could* appear that Na^+ was not readily translocated to the tops, since the concentration in the tops would tend to be and to remain low. After a 4-hr absorption period, approximately 95% of the Na^+ absorbed by bean plants was held by secondary roots; Na^+ translocated to the tops was held by the stem (Pearson, 1967). DNP reduced the amount of Na^+ retained by secondary roots of bean plants and of cotton and, in beans but not in cotton, increased the amount of Na^+ translocated to shoots.

Aeration of Roots

For tomato, soybean, and tobacco subjected to various concentrations of O_2 in the gas mixture around roots, Hopkins et al. (1950) reported that the log transport rate was a function of time and independent of O_2 supply to the roots.

Respiration (Aerobic)

Hopkins et al. (1950) studied the effect of various partial pressures of O_2 around roots and concluded that the rate of transfer of ions from the root to the shoot was independent of aerobic mechanisms in the root.

With regard to the mechanism of salt accumulation and then retention of ions by the stele, the Crafts and Broyer (1938) concept is that ions are accumulated in the cortex prior to their movement through the symplast into the stele where salts leak into the xylem (Laties and Budd, 1964; Lüttge and Laties, 1966, 1967). According to this concept, living cells of the stele would have to have a lower capacity to accumulate and to retain ions than that of the cortical cells. This appears to be true for freshly isolated stelar tissue, although the ion accumulation capacity increases several hours after stelar tissue is separated from the cortex.

However, Yu and Kramer (1967) found that separated stele and stele of intact roots accumulated ^{32}P as effectively as the cortex. Biddulph (1967) and Branton and Jacobson (1962) reported similar findings. Living cells adjacent to xylem vessels may secrete salt into them (Andel, 1953; Ariz, 1956; Sutcliffe, 1962).

In a study of the root cortex and the stele of primary roots of Z. *mays*, it was observed that the stele has a relatively high O_2 consumption, ^{32}P uptake, and strong dehydrogenase activity (Yu and Kramer, 1967). In view of these findings it was concluded that cells of the stele play a direct role in radial translocation of salt across the root and that the stele is able to accumulate salt vigorously. Inasmuch as their findings related only to P, Yu and Kramer (1969) then studied the entry of ^{86}Rb, ^{36}Cl, and ^{32}P into attached and excised corn roots. They found no differences between the stele and cortex of intact roots with respect to the accumulation of these several ions.

Reduction in translocation, following inhibition of respiration, does not necessarily imply an active process; it might reflect association with the movement of another entity or even a change in the structure of the system (Milthorpe and Moorby, 1969).

Temperature

Low temperature (0.5–4.5°C) around petioles greatly reduced the transport of carbohydrates from Red Kidney bean leaves (Curtis and Herty, 1936).

Element Deficiencies

According to Nightingale et al. (1931), translocation of sugars and the digestion of starch proceeded freely in tomato plants extremely deficient in Ca.

In attached leaves of sugarcane, N deficiency decreased the percentage of total ^{14}C activity translocated and the velocity of transport (Hartt, 1970b). In N deficiency, translocation decreased not for lack of sucrose but for some other reason. N deficiency had no effect on transport in detached leaves. It was concluded that the effect of N deficiency upon translocation may be indirect and secondary to the effect on growth of the plant as a whole.

In detached blades of sugarcane grown in nutrient solution with and without K^+, basipetal translocation of K^+ decreased when plants were deprived of K^+ —even when deficiency symptoms were not visible (Hartt, 1970a). Translocation of K^+ was decreased at deficiency concentrations of K^+ that had no effect on photosynthesis. An experiment employing ^{32}P indicated an upset in phosphorylation in stems of K-deficient plants.

Age of Leaves

Although P, for example, moves readily into older leaves on a plant, the concentration of P is highest in younger leaves (Fig. 8.1). Obviously, P is not transferred from the xylem to mesophyll cells in older leaves, and the excess of P entering older leaves is reexported via the phloem to younger leaves (Biddulph, 1951).

Maximal accumulation of ^{45}Ca by bean leaves occurred as each given leaf approached full expansion. Al-Ani and Koontz (1969) noted that when ^{45}Ca was applied to terminal 5-cm portions of specific lateral roots, specific areas of the shoot were preferentially supplied with ^{45}Ca.

Fig. 8.1

Autoradiograph of parts of a bean plant receiving radiophosphorus via the nutrient solution. From O. Biddulph (1951) The translocation of minerals in plants, pp. 261–275. In E. Truog (ed.) Mineral Nutrition of Plants, Univ. Wisconsin Press, Madison. © 1951 by the Regents of the University of Wisconsin—by permission.

In attached and detached bean *(P. vulgaris* L.) leaves, transition from import to export of ^{14}C-labeled assimilates occurred at the same stage, that is, when cotyledons were 63–85% depleted (Kocher and Leonard, 1971). Excision of cotyledons did not delay transition of leaves from import to export. It was suggested that isolated leaves are source–sink systems which could be used successfully to study many problems of transport.

Root Tip Segment Where Absorbed

Although approximately the first millimeter of a root behind the root tip may absorb large amounts of salt, owing to the high metabolic rate of this region, there is little or no transfer of absorbed salt away from this region (Brouwer, 1954; Wiebe and Kramer, 1954; Sutcliffe, 1959). This may be in part related to the fact that there is not a complete conductive tissue at the tip. Differentiated phloem elements occur much nearer the tip of the root than do xylem elements. By using ^{32}P, ^{86}Rb, ^{131}I, and ^{35}S, it was shown that the highest percentage of translocation for barley roots occurs when the isotopes are supplied approximately 27–30 mm from the tip (Wiebe and Kramer, 1954) (Table 8.2).

Table 8.2
Translocation of Salts from Different Regions of the Barley Root[a]

Distance from root tip at which isotope was supplied (mm)	Percentage of absorbed isotope translocated elsewhere			
	^{32}P	^{86}Rb	^{131}I	^{35}S
0–4	1.3	4.2	1.0	1.7
7–10	8.5	14.3	28.3	5.2
27–30	34.4	14.7	28.9	11.8
57–60	24.9	9.4	22.7	9.2

[a] Wiebe and Kramer (1954).

Anatomy

The importance of anatomical considerations in connection with absorption and translocation studies cannot be overemphasized. Translocation of salt from root tips as related to root anatomy was just discussed and is a good case in point.

Nature of Element

In a study of the distribution of B in lily leaves, the concentration of B increased hyperbolically from the leaf base to the tip (Kohl and Oertli, 1961). It was concluded that B moved passively in the transpiration stream and that lack of transport from the leaves was probably a result of the inability of phloem to transport B. Eaton (1944) attributed the lack of mobility of B, from leaves, to the formation of high-molecular-weight B compounds which could not pass through cell membranes.

Light

In detached blades of sugarcane, the initiation of translocation of sugar from the leaf appeared to be under photocontrol. Hartt (1965) found that the effect of light on translocation could not be assigned to its effect on photosynthesis, hence the production of photosynthate. Light apparently affected two aspects of translocation—the polarity or direction of movement and the percentage of basipetal transport of sugar.

Growth Regulators

Triiodobenzoic acid (TIBA) applied to leaves along with Ca^{2+} greatly increased the downward translocation of Ca^{2+} in apple seedlings and tomato plants. Kessler and Moscicki (1958) concluded that the effect of TIBA on Ca^{2+} translocation might be related to its suggested effect in suspending polarity.

Plasmodesmata

In a study of translocation of ^{32}P and ^{14}C assimilates by bean leaves, Leonard and Glenn (1968) believed that translocation was being powered by forces in the plasmodesmata between the border parenchyma and the phloem.

Organic Molecules

In soybeans there was a striking correlation between the concentration of Fe and of citrate in stem exudate (Brown and Tiffin, 1965). When Fe was increased, citrate increased; a decrease in Fe was paralleled by a decrease in citrate. Similarly, in studies with soybean plants and ferric ethylenediamine di-(o-hydroxyphenylacetate) (or FeEDDHA), it was observed that the major form of Fe in exudates was ferric citrate. Tiffin (1970) concluded that Fe is probably translocated as ferric citrate in plants.

Although Fe is apparently complexed with citric acid (Tiffin, 1966, 1970) during translocation, Mn^{2+}, Co^{2+}, and Zn^{2+} appear to be transported predominatly as free inorganic cations (Tiffin, 1967).

Transport of organic substances in phloem has been reviewed in detail by Preston (1963).

Zinc

Ambler (1969) and Ambler et al. (1970) noted that Zn^{2+} interfered with the translocation of Fe in soybean plants. They concluded that Zn^{2+} suppressed Fe transport (1) by suppressing the efflux of H^+ from the roots and (2) by suppressing the production of some substance (Ambler et al., 1971) by the root which is responsible for reducing Fe^{3+} to Fe^{2+}.

Ion Accumulation

In vascular bundles of celery, the young secondary phloem at the sides of the bundles was the most active tissue in accumulating SO_4^{2-} and phosphate, and this tissue was also the most active in the translocation of these two anions (Bieleski, 1966). It was concluded that accumulation plays a primary role in translocation.

Lipids

In grape rootstocks there was a direct correlation between monogalactose diglyceride concentrations in roots and Cl^- transport to leaves (Kuiper, 1968). Mono- and digalactose diglyceride (in the nutrient solution) strongly increased Cl^- transport to all parts of bean and cotton plants—probably by transport of these glycolipids further into the plant (Kuiper, 1969). Kuiper concluded that glycolipids greatly increase Cl^- transport across cell membranes by virtue of H bond formation between galactose groups of the lipid and those of the protein.

RECYCLING

Using Ca^{2+}, phosphate, and SO_4^{2-} labeled with isotopes, Biddulph et al. (1958) reported that phosphate moves to the tops of bean *(P. vulgaris)* plants, then to the roots, and later back to the tops again. Inasmuch as SO_4^{2-} was rapidly incorporated in the young leaves, there was essentially no recycling of SO_4^{2-}; Ca^{2+} did not circulate at all. With the exception of Ca^{2+}, which was not retranslocated, appreciable amounts of other ions reaching a leaf were retranslocated downward via the phloem (Milthorpe and Moorby, 1969).

The mobility of B, and its recycling within a limited portion of the plant, has recently received intensive study by Oertli and Richardson (1970). They found that B readily enters the bark and is translocated within it. They reasoned that since it has also been shown that B remains water soluble in plants, immobility could not be explained by chemical fixation, lack of entry into phloem, or absence of phloem transport. Interestingly, they suggested that B enters the phloem in leaf margins where the concentrations of B are high, is transported in these conduits, is lost therefrom where the concentration in the xylem is low, that is, in basal areas of the leaf and petiole, is transported back in the xylem, and accumulates in terminal places of the transpiration stream. A high mobility of B, together with the essentially unidirectional flow of the transpiration stream, thus causes a cyclic movement of B and prevents the flux of this nutrient. Further, this would explain the immobility over long distances.

LITERATURE CITED

Al-Ani, T. A., and H. V. Koontz. 1969. Distribution of calcium absorbed by all or part of the root system of beans. Plant Physiol. 44:711−716.

Alexander, T. R. 1938. Carbohydrates of bean plants after treatment with indole-3-acetic acid. Plant Physiol. 13:845−858.

Ambler, J. E. 1969. Effect of zinc on the translocation of iron in soybean plants. Ph.D. Thesis, Univ. Maryland, College Park.

Ambler, J. E., J. C. Brown, and H. G. Gauch. 1970. Effect of zinc on translocation of iron in soybean plants. Plant Physiol. 46:320−323.

Ambler, J. E., J. C. Brown, and H. G. Gauch. 1971. Sites of iron reduction in soybean plants. Agron. J. 63:95−97.

Andel, O. M. van. 1953. The influence of salts on the exudation of tomato plants. Acta Bot. Neerl. 2:445−521.

Ariz, W. H. 1956. Significance of the symplasm theory for transport across the root. Protoplasma 46:1−62.

Arnon, D. I., P. R. Stout, and F. Sipos. 1940. Radioactive phorphorus as an indicator of phosphorus absorption of tomato plants at various stages of development. Amer. J. Bot. 27:791−798.

Barr, C. E., and T. C. Broyer. 1964. Effect of light on sodium influx, membrane potential, and protoplasmic streaming in *Nitella*. Plant Physiol. 39:48—52.

Barrier, G. E., and W. E. Loomis. 1957. Absorption and translocation of 2,4-dichloro-phenoxyacetic acid and P^{32} by leaves. Plant Physiol. 32:225—231.

Bell, C. W., and O. Biddulph. 1963. Translocation of calcium. Exchange versus mass flow. Plant Physiol. 38:610—614.

Bernstein, L., J. W. Brown, and H. E. Hayward. 1956. The influence of rootstock on growth and salt accumulation in stone-fruit trees and almonds. Proc. Amer. Soc. Hort. Sci. 68:86—95.

Biddulph, O., and J. Markle. 1944. Translocation of radiophosphorus in the phloem of the cotton plant. Amer. J. Bot. 31:65—71.

Biddulph, O. 1951. The translocation of minerals in plants, pp. 261—275. *In* E. Truog (ed.) Mineral nutrition of plants. Univ. Wisconsin Press, Madison.

Biddulph, O. 1959. Translocation of inorganic solutes, pp. 553—603. *In* F. C. Steward (ed.) Plant physiology — A treatise, Vol. II. Academic, New York.

Biddulph, O., and R. Cory. 1960. Demonstration of two translocation mechanisms in studies of bidirectional movement. Plant Physiol. 35:689—695.

Biddulph, O., and R. Cory. 1965. Translocation of ^{14}C metabolites in the phloem of the bean plant. Plant Physiol. 40:119—129.

Biddulph, O., S. Biddulph, R. Cory, and H. Koontz. 1958. Circulation patterns for phosphorus, sulfur and calcium in the bean plant. Plant Physiol. 33:293—300.

Biddulph, O., R. Cory, and S. Biddulph. 1959. Translocation of calcium in the bean plant. Plant Physiol. 34:512—516.

Biddulph, O., F. S. Nakayama, and R. Cory. 1961. Transpiration stream and ascension of calcium. Plant Physiol. 36:429—436.

Biddulph, S. F. 1967. A microautoradiographic study of ^{45}Ca and ^{35}S distribution in the intact bean root. Planta 74:350—367.

Bieleski, R. L. 1966. Sites of accumulation in excised phloem and vascular tissues. Plant Physiol. 41:455—466.

Bieleski, R. L. 1969. Phosphorus compounds in translocating phloem. Plant Physiol. 44:497—502.

Bledsoe, R. W., C. L. Comar, and H. C. Harris. 1949. Absorption of radioactive calcium by the peanut fruit. Science 109:329—330.

Bollard, E. G. 1956. Nitrogenous compounds in plant xylem sap. Nature 178:1189—1190.

Bollard, E. G. 1957a. Composition of the nitrogen fraction of apple tracheal sap. Australian J. Biol. Sci. 10:279—287.

Bollard, E. G. 1957b. Nitrogenous compounds in tracheal sap of woody members of the family Rosaceae. Australian J. Biol. Sci. 10:288—291.

Bollard, E. G. 1957c. Translocation of organic nitrogen in the xylem. Australian J. Biol. Sci. 10:292—301.

Bollard, E. G. 1960. Transport in the xylem. Annu. Rev. Plant Physiol. 11:141—166.

Brady, N. C. 1947. The effect of period of calcium supply and mobility of calcium in the plant on peanut fruit filling. Proc. Soil Sci. Soc. Amer. 12:336—341.

Brändle, R., and J. Schnyder. 1970. Downward transport of sulfur compounds from primary leaves (*Phaseolus vulgaris*) following gasing with sulfur-35 labeled labeled H_2S. Experientia 26:1395—1396.

Branton, D., and L. Jacobson. 1962. Ion transport in pea plants. Plant Physiol. 37:539—545.

Brouwer, R. 1954. The regulating influence of transpiration and suction tension on the water and salt uptake by roots of intact *Vicia faba* plants. Acta Bot. Neerl. 3:264—312.

Brown, J. C., and L. O. Tiffin. 1965. Iron stress as related to the iron and citrate occurring in stem exudate. Plant Physiol. 40:395–400.

Bukovac, M. J., F. G. Teubner, and S. H. Wittwer. 1960. Absorption and mobility of magnesium-28 in the bean (*Phaseolus vulgaris* L.). Proc. Amer. Soc. Hort. Sci. 75:429–434.

Choi, I. C., and S. Aronoff. 1966. Photosynthate transport using tritiated water. Plant Physiol. 41:1119–1129.

Clements, H. F., and C. J. Engard. 1938. Upward movement of inorganic solutes as affected by a girdle. Plant Physiol. 13:103–122.

Clor, M. A. 1959. Comparative studies on translocation of C^{14}-labeled 2,4-D, urea, and amino triazole in cotton and oaks. Ph.D. Thesis, Univ. California, Davis.

Clor, M. A., A. S. Crafts, and S. Yamaguchi. 1963. Effects of high humidity on translocation of foliar-applied labeled compounds in plants. II. Translocation from starved leaves. Plant Physiol. 38:501–507.

Collander, R. 1941a. Selective absorption of cations by higher plants. Plant Physiol. 16:691–720.

Collander, R. 1941b. The distribution of different cations between root and shoot. Acta Bot. Fennica 29:1–12.

Cooil, B. J., R. K. de la Fuente, and R. S. de la Pena. 1965. Absorption and transport of sodium and potassium in squash. Plant Physiol. 40:625–632.

Crafts, A. S. 1931. Movement of organic materials in plants. Plant Physiol. 6:1–41.

Crafts, A. S. 1932. Phloem anatomy, exudation, and transport of organic nutrients in cucurbits. Plant Physiol. 7:183–225.

Crafts, A. S. 1933. Sieve-tube structure and translocation in the potato. Plant Physiol. 8:81–104.

Crafts, A. S. 1938. Translocation in plants. Plant Physiol. 13:791–814.

Crafts, A. S. 1961. Translocation in plants. Holt, New York. 182 pp.

Crafts, A. S., and T. C. Broyer. 1938. Migration of salts and water into xylem of the roots of higher plants. Amer. J. Bot. 25:529–535.

Creger, C. R., and W. S. Allen. 1969. Strontium mobility in germinating seeds and plants. Plant Physiol. 44:439–441.

Curtis, O. F. 1925. Studies on the tissues concerned in the transfer of solutes in plants. The effect on the upward transfer of solutes of cutting the xylem as compared with that of cutting the phloem. Ann. Bot. 39:573–585.

Curtis, O. F. 1935. The translocation of solutes in plants. McGraw-Hill, New York. 273 pp.

Curtis, O. F., and S. D. Herty. 1936. The effect of temperature on translocation from leaves. Amer. J. Bot. 23:528–532.

Devlin, R. M. 1969. Plant physiology. 2nd ed. Van Nostrand Reinhold. New York, 446 pp.

Eaton, F. M. 1944. Deficiency, toxicity, and accumulation of boron in plants. J. Agr. Res. 69:237–277.

Eschrich, W. 1967. Bidirektionelle Translokation in Siebröhren. Planta 73:37–49.

Fife, J. M., C. Price, and D. C. Fife. 1962. Some properties of phloem exudate collected from root of sugar beet. Plant Physiol. 37:791–792.

Gage, R. S., and S. Aronoff. 1960. Translocation. III. Experiments with carbon 14, chlorine 36, and hydrogen 3. Plant Physiol. 35:53–64.

Gauch, H. G. 1940. Responses of the bean plant to calcium deficiency. Plant Physiol. 15:1–21.

Gauch, H. G., and W. M. Dugger, Jr. 1953. The role of boron in the translocation of sucrose. Plant Physiol. 28:457–467.

Gauch, H. G., and W. M. Dugger. 1954. The physiological action of boron in higher plants: A review and interpretation. Maryland Agr. Exp. Sta. Tech. Bull. A-80.

Gauch, H. G., and C. H. Wadleigh. 1945. Effect of high concentrations of sodium, calcium, chloride, and sulfate on ionic absorption by bean plants. Soil Sci. 59: 139–153.

Hartt, C. E. 1965. Light and translocation of C^{14} in detached blades of sugarcane. Plant Physiol. 40:718–724.

Hartt, C. E. 1970a. Effect of potassium deficiency upon translocation of ^{14}C in detached blades of sugar cane. Plant Physiol. 45:183–187.

Hartt, C. E. 1970b. Effect of nitrogen deficiency upon translocation of ^{14}C in sugarcane. Plant Physiol. 46:419–422.

Hay, J. R., and K. V. Thimann. 1956. The fate of 2,4-dichlorophenoxyacetic acid in bean seedlings. Plant Physiol. 31:446–451.

Honert, T. H. van den. 1932. On the mechanism of transport of organic materials in plants. Proc. Kon. Akad. Wetensch. (Amsterdam) 35:1104–1112.

Hopkins, H. T., A. W. Specht, and S. B. Hendricks. 1950. Growth and nutrient accumulation as controlled by oxygen supply to plant roots. Plant Physiol. 25:193–209.

Huber, B. 1941. Gesichertes und Problematisches in der Wanderung der Assimilate. Ber. Deut. Bot. Ges. 59:181–194.

Hylmö, B. 1955. Passive components in the ion absorption of the plant. I. The zonal ion and water absorption in Brouwer's experiments. Physiol. Plantarum 8:433–449.

Ingelsten, B. 1966. Absorption and transport of sulfate by wheat at varying mannitol concentration in the medium. Physiol. Plantarum 19:563–579.

Jacobson, B. 1955. On the interpretation of dielectric constants of aqueous macromolecular solutions. Hydration of macromolecules. J. Amer. Chem. Soc. 77: 2919–2926.

Jacoby, B. 1964. Function of bean roots and stems in sodium retention. Plant Physiol. 39:445–449.

Jaworski, E. G., S. C. Fang, and V. H. Freed. 1955. Studies in plant metabolism. V. The metabolism of radioactive 2,4-D in etiolated bean stems. Plant Physiol. 30: 272–275.

Kamiya, N. 1960. Physics and chemistry of protoplasmic streaming. Annu. Rev. Plant Physiol. 11:323–340.

Kaufmann, M. R., and P. J. Kramer. 1967. Phloem water relations and translocation. Plant Physiol. 42:191–194.

Kessler, B., and Z. W. Moscicki. 1958. Effect of triiodobenzoic acid and maleic hydrazide upon the transport of foliar applied calcium and iron. Plant Physiol. 33:70–72.

Klepper, L., and R. H. Hageman. 1969. The occurrence of nitrate reductase in apple leaves. Plant Physiol. 44:110–114.

Köcher, H., and O. A. Leonard. 1971. Translocation and metabolic conversion of ^{14}-C-labeled assimilates in detached and attached leaves of *Phaseolus vulgaris* L. in different phases of leaf expansion. Plant Physiol. 47:212–216.

Kohl, H. C., Jr., and J. J. Oertli. 1961. Distribution of boron in leaves. Plant Physiol. 36: 420–424.

Koontz, H. V., and R. E. Foote. 1966. Transpiration and calcium deposition by unifoliate leaves of *Phaseolus vulgaris* differing in maturity. Physiol. Plantarum 19:313–321.

Kozlowski, T. T. 1964. Water metabolism in plants. Harper, New York. 227 pp.

Kuiper, P. J. C. 1968. Ion transport characteristics of grape root lipids in relation to chloride transport. Plant Physiol. 43:1372–1374.

Kuiper, P. J. C. 1969. Effect of lipids on chloride and sodium transport in bean and cotton plants. Plant Physiol. 44:968–972.

Kursanov, A. L. 1963. Metabolism and the transport of organic substances in the phloem. Advance. Bot. Res. 1:209–278.

Laties, G. G., and K. Budd. 1964. The development of differential permeability in isolated steles of corn roots. Proc. Nat. Acad. Sci. 52:462–469.

Leonard, O. A. 1936. Seasonal study of tissue function and organic solute movement in the sunflower. Plant Physiol. 11:25–61.

Leonard, O. A., and R. K. Glenn. 1968. Translocation of assimilates and phosphate in detached bean leaves. Plant Physiol. 43:1380–1388.

Linder, P. J., J. W. Brown, and J. W. Mitchell. 1949. Movement of externally applied phenoxy compounds in bean plants in relation to conditions favoring carbohydrate translocation. Bot. Gaz. 110:628–632.

Loomis, W. E. 1945. Translocation of carbohydrates in maize. Science 101:398–400.

Lüttge, U., and G. G. Laties. 1966. Dual mechanism of ion absorption in relation to long-distance transport in plants. Plant Physiol. 41:1531–1539.

Lüttge, U., and G. G. Laties. 1967. Absorption and long-distance transport by isolated stele of maize roots in relation to the dual mechanism of ion absorption. Planta 74:173–187.

McLean, E. O. 1956. Uptake of sodium and other cations by five crop species. Soil Sci. 82:21–28.

Mason, T. G., and E. J. Maskell. 1928. Studies on the transport of carbohydrates in the cotton plant. II. The factors determining the rate and the direction of movement of sugars. Ann. Bot. 42:571–636.

Mason, T. G., and E. J. Maskell. 1931. Further studies on transport in the cotton plant. I. Preliminary observations on the transport of phosphorus, potassium and calcium. Ann. Bot. 45:125–173.

Mason, T. G., and E. Phillis. 1937. The migration of solutes. Bot. Rev. 3:47–71.

Mathes, M. C. 1961. Factors affecting the translocation of 2,4-D. Ph.D. Thesis, Univ. Maryland, College Park.

Mendelssohn, K. 1958. Superfluids. Science 127:215–221.

Milthorpe, F. L., and J. Moorby. 1969. Vascular transport and its significance in plant growth. Annu. Rev. Plant Physiol. 20:117–138.

Mitchell, J. W., and J. W. Brown. 1946. Movement of 2,4-dichlorophenoxyacetic acid stimulus and its relation to the translocation of organic food materials in plants. Bot. Gaz. 107:393–407.

Mitchell, J. W., and J. F. Worley. 1964. Intracellular transport apparatus of phloem fibers. Science 145:409–410.

Mitchell, J. W., I. R. Schneider, and H. G. Gauch. 1960. Translocation of particles within plants. Science 131:1863–1870.

Münch, E. 1927. Versuche über den Saftkreislauf. Ber. Deut. Bot. Ges. 45:340–356.

Nelson, C. D. 1963. Effect of climate on the distribution and translocation of assimilates, pp. 149–174. *In* L. T. Evans (ed.) Environmental control of plant growth. Academic, New York.

Nightingale, G. T., R. M. Addoms, W. R. Robbins, and L. G. Schermerhorn. 1931. Effects of calcium deficiency on nitrate absorption and on metabolism in tomato. Plant Physiol. 6:605–631.

Oertli, J. J., and W. F. Richardson. 1970. The mechanism of boron immobility in plants. Physiol. Plantarum 23:108–117.

Ohlrogge, A. J. 1963. Mineral nutrition of soybeans, pp. 125–160. *In* A. G. Norman (ed.) The soybean. Academic Press, New York.

Pearson, G. A. 1967. Absorption and translocation of sodium in beans and cotton. Plant Physiol. 42:1171–1175.

Peel, A. J., R. J. Field, C. L. Coulson, and D. C. J. Gardner. 1969. Movement of water and solutes in sieve tubes of willow in response to puncture by aphid stylets. Evidence against a mass flow of solution. Physiol. Plantarum 22:768–775.

Peterson, C. A., and H. B. Currier. 1969. An investigation of bidirectional translocation in the phloem. Physiol. Plantarum 22:1238–1250.

Pettersson, S. 1960. Ion absorption in young sunflower plants. I. Uptake and transport mechanisms for sulphate. Physiol. Plantarum 13:133–147.

Pettersson, S. 1966. Artificially induced water and sulfate transport through sunflower roots. Physiol. Plantarum 19:581–601.

Phillis, E., and T. G. Mason. 1936. Further studies on transport in the cotton plant. IV. On the simultaneous movement of solutes in opposite directions through the phloem. Ann. Bot. 50:161–174.

Pitman, M. G. 1965. Transpiration and the selective uptake of potassium by barley seedlings (*Hordeum vulgare* cv. Bolivia). Australian J. Biol. Sci. 18:987–998.

Preston, R. D. (ed.). 1963. Advances in botanical research, Vol. I. Academic, New York, 396 pp.

Price, C. A. 1970. Molecular approaches to plant physiology. McGraw-Hill, New York. 398 pp.

Rediske, J. H., and A. A. Selders. 1953. The absorption and translocation of strontium by plants. Plant Physiol. 28:594–605.

Rice, E. L. 1948. Absorption and translocation of ammonium 2,4-dichlorophenoxyacetate by bean plants. Bot. Gaz. 109:301–314.

Rice, E. L., and L. M. Rohrbaugh. 1958. Relation of potassium nutrition to the translocation of 2,4-dichlorophenoxyacetic acid in tomato plants. Plant Physiol. 33:300–303.

Richards, L. A., and G. Ogata. 1958. Thermocouple for vapor pressure measurement in biological and soil systems at high humidity. Science 128:1089–1090.

Richardson, M. 1968. Translocation in plants. St. Martin's, New York. 59 pp.

Rohrbaugh, L. M., and E. L. Rice. 1949. Effect of application of sugar on the translocation of sodium 2,4-dichlorophenoxyacetate by bean plants in the dark. Bot. Gaz. 111:85–89.

Rohrbaugh, L. M., and E. L. Rice. 1956. Relation of phosphorus nutrition to the translocation of 2,4-dichlorophenoxyacetic acid in tomato plants. Plant Physiol. 31:196–199.

Russell, R. S., and V. M. Shorrocks. 1959. The relationship between transpiration and the absorption of inorganic ions by intact plants. J. Exp. Bot. 10:301–316.

Salisbury, F. B., and C. Ross. 1969. Plant physiology. Wadsworth, Belmont, Calif. 747 pp.

Shear, C. B., and M. Faust. 1970. Calcium transport in apple trees. Plant Physiol. 45:670–674.

Shiroya, M. 1968. Comparison of upward and downward translocation of ^{14}C from a single leaf of sunflower. Plant Physiol. 43:1605–1610.

Slatyer, R. O. 1958. The measurement of diffusion pressure deficit in plants by a method of vapor equilibration. Australian J. Biol. Sci. 11:349–365.

Soran, V., and G. Lazar. 1969. The relationship between *myo*-inositol, ATP and rotational streaming. Physiol. Plantarum 22:560–569.

Spanner, D. C. 1958. The translocation of sugar in sieve tubes. J. Exp. Bot. 9:332–342.

Steucek, G. L., and H. V. Koontz. 1970. Phloem mobility of magnesium. Plant Physiol. 46:50–52.

Stout, P. R., and D. R. Hoagland. 1930. Upward and lateral movement of salts in certain plants as indicated by radioactive isotopes of potassium, sodium, and phosphorus absorbed by roots. Amer. J. Bot. 26:320–324.

Stout, P. R., R. Overstreet, L. Jacobson, and A. Ulrich. 1947. The use of radioactive tracers in plant nutrition studies. Proc. Soil Sci. Soc. Amer. 12:91–97.

Sutcliffe, J. F. 1959. Salt uptake in plants. Biol. Rev. 34:159–218.

Sutcliffe, J. F. 1960. New evidence for a relationship between ion absorption and protein turnover in plant cells. Nature 188:294–297.

Sutcliffe, J. F. 1962. Mineral salts absorption in plants. Pergamon Press, Oxford. 194 pp.

Szent-Györgyi, A. 1956. Bioenergetics. Science 124:873–875.

Thomas, M. D., R. H. Hendricks, L. C. Bryner, and G. R. Hill. 1944. A study of the sulphur metabolism of wheat, barley and corn using radioactive sulphur. Plant Physiol. 19:227–244.

Thomas, W. 1927. The seat of formation of amino acids in *Pyrus malus* L. Science 66: 115–116.

Thomas, W. A. 1967. Dye and calcium ascent in dogwood trees. Plant Physiol. 42: 1800–1802.

Thomas, W. A. 1970. Retention of calcium-45 by dogwood trees. Plant Physiol. 45: 510–511.

Tiffin, L. O. 1966. Iron translocation. I. Plant culture, exudate sampling, iron-citrate analysis. Plant Physiol. 41:510–514.

Tiffin, L. O. 1967. Translocation of manganese, iron, cobalt, and zinc in tomato. Plant Physiol. 42:1427–1432.

Tiffin, L. O. 1970. Translocation of iron citrate and Phosphorus in xylem exudate of soybean. Plant Physiol. 45:280–283.

Tolbert, N. E., and H. Wiebe. 1955. Phosphorus and sulphur compounds in plant xylem sap. Plant Physiol. 30:499–504.

Trip, P. 1969. Sugar transport in conducting elements of sugar beet leaves. Plant Physiol. 44:717–725.

Trip, P., and P. R. Gorham. 1968a. Bidirectional translocation of sugars in sieve tubes of squash plants. Plant Physiol. 43:877–882.

Trip, P., and P. R. Gorham. 1968b. Translocation of sugar and tritiated water in squash plants. Plant Physiol. 43:1845–1849.

Truog, E., K. C. Berger, and O. J. Attoe. 1953. Response of nine economic plants to fertilization with sodium. Soil Sci. 76:41–50.

Vernon, L. P., and S. Aronoff. 1952. Metabolism of soybean leaves. IV. Translocation from soybean leaves. Arch. Biochem. Biophys. 36:383–398.

Vries, H. de. 1885. Ueber die Bedeutung der Circulation und der Rotation des Protoplasma für den Stofftransport in der Pflanze. Bot. Zeit. 43:1–6, 18–26.

Wallace, A., and N. Hemaidan. 1963. Sodium transport from roots to shoots in bush bean and radish. Plant Physiol. 38:viii.

Weaver, R. J., and H. R. DeRose. 1946. Absorption and translocation of 2,4-dichlorophenoxyacetic acid. Bot. Gaz. 107:509–521.

Webb, J. A., and P. R. Gorham. 1964. Translocation of photosynthetically assimilated ^{14}C in straight-necked squash. Plant Physiol. 39:663–672.

Wiebe, H. H., and P. J. Kramer. 1954. Translocation of radioactive isotopes from various regions of roots of barley seedlings. Plant Physiol. 29:342–348.

Worley, J. F. 1968. Rotational streaming in fiber cells and its role in translocation. Plant Physiol. 43:1648–1655.

Yu, G., and P. J. Kramer. 1967. Radial salt transport in corn roots. Plant Physiol. 42:985–990.

Yu, G., and P. J. Kramer. 1969. Radial transport of ions in roots. Plant Physiol. 44: 1095–1100.

Zimmerman, P. W., and M. H. Connard. 1934. Reversal of direction of translocation of solutes in stems. Contrib. Boyce Thompson Inst. 6:297–302.

Zimmermann, M. H. 1957. Translocation of organic substances in trees. I. The nature of the sugars in the sieve tube exudate of trees. Plant Physiol. 32:288–291.

Zimmermann, M. H. 1960. Transport in the phloem. Annu. Rev. Plant Physiol. 11: 167–190.

Zimmermann, M. H. 1961. Movement of organic substances in trees. Science 133: 73–79.

Zimmermann, M. H. 1962. Translocation of organic substances in trees. V. Experimental double interpretation of phloem in white ash (*Fraxinus americana* L.). Plant Physiol. 37:527–530.

CHAPTER 9

Roles of Macronutrients in Higher Plants

INTRODUCTION

Atomic Structure

Although atomic structure has received considerable study over the years, it has not been possible to predict the essentiality of a single element. Essentiality was determined in each case by the exclusion of a particular element from the medium in which plants were being grown.

Criteria of Essentiality

It is generally agreed that for an element to be regarded as essential (Arnon and Stout, 1939; Arnon, 1950a, 1951) it must fulfill the following requirements. (1) The plant cannot complete its life cycle in the absence of the element. (2) Action of the element must be specific; no other element can wholly substitute for it. (3) The element must be shown to be directly involved in the nutrition of the plant, that is, to be a constituent of an essential metabolite or, at least, required for the action of an essential enzyme. (4) Although sometimes not listed as one of the criteria, it is generally agreed that the element must be shown to be required for higher green plants in general, since the current 16 essential elements pertain to higher plants.

It has proved difficult to apply the first criterion rigidly, since Broyer et al. (1954) showed, for example, that Br^- could at least partially substitute for

Cl^-. These investigators noted, however, that Cl^- is the normal, functional, essential halogen.

The second criterion has been compromised by evidence that essentiality of a given element may be negated when the metabolic reaction in which it normally participates is bypassed. For example, Ichioka and Arnon (1955) showed that *Scenedesmus obliquus* failed to grow with NO_3^- unless 1 ppb Mo was present in the solution; in the presence of NH_4^+ ion or urea, the alga grew equally well with or without Mo. That is, the role of Mo in NO_3^- reduction was bypassed by supplying a reduced form of N. A similar report was made for *Chlorella pyrenoidosa* (Fogg and Wolfe, 1954).

Knowledge of Roles

With regard to elements generally required in large amounts by plants, it is interesting that the roles of some of them are still not clearly delineated, for example, the roles of Ca^{2+} and K^+. Inasmuch as the essentialities of the major (interms of amount) or macro elements were determined before that of the trace or microelements, it would seem logical to expect that roles of macroelements would long since have been rather fully elucidated. Incidentally, it has become increasingly difficult to separate essential elements into macro- and microcategories, since Ca^{2+} and S, for example, are generally required in macroamounts by higher plants, but only in microamounts by most microorganisms (Nicholas, 1961).

The history of micronutrients reveals that for every element the biochemical elucidation of its function *invariably* followed, rather than preceded, acceptance of its essentiality (Arnon, 1950, 1951).

Although there are some conflicting theories concerning the roles of the trace elements, specific roles have at least been assigned to each of them, that is, B, Mn^{2+}, Cu^{2+}, Zn^{2+}, Fe^{2+}, Cl^-, and Mo. The advent of radioisotopes was a great boon to the elucidation of these roles. For example, the role of Mo in NO_3^- reduction was demonstrated (Arnon, 1954) despite the fact that 0.1 ppb Mo in the nutrient solution represented sufficiency (Arnon, 1958).

Additional information on micronutrients has been presented by Lal and Rao (1954), Nicholas (1961), and Stiles (1961). *Diagnostic Criteria for Plants and Soils*, edited by Chapman (1966), contains chapters on the essential elements and on many elements not currently regarded as essential. Considerable information is presented on the roles of essential elements, concentrations of elements in plants and soils, and so on.

GENERAL ROLES OR FUNCTIONS

Soon after the original 10 essential elements (C, H, O, P, K, N, S, Ca, Fe, and Mg) were determined around 1900, various general physiological roles were suggested for these elements. It may be noted that some of the suggested roles might more appropriately have been called effects, since one of them, namely, toxic effect, can hardly be regarded as a role. The commonly listed general roles are:

(1) *Building materials for protoplasm, cell walls, enzymes, and so on.* As examples, one may cite just a few of many possible examples—S, as a constituent of proteins; P, of nucleoproteins, ADP, and ATP; Mg, of chlorophyll; and, C, H, and O of such compounds as carbohydrates, fats, and proteins.

(2) *Development of osmotic pressure in plant cells.* Inasmuch as any ion or molecule contributes to osmotic pressure when present in a solution, inorganic constituents obviously contribute to osmotic pressure. Effects of inorganic constituents in plants, in this regard are usually small compared with those of organic constituents such as sugars, organic acids, and other compounds.

(3) *pH and buffer action.* Inorganic constituents have relatively little influence on pH, although certain ones, for example, phosphate, bicarbonate, and carbonate, may act as buffers and thus prevent marked changes in H^+ concentration. Degree of acidity and buffer action of plant tissues are usually determined primarily by organic acids.

(4) *Maintenance of a desirable degree of hydration of cell colloids.* Physical chemists have been primarily concerned with the effects of various ions on the degree of hydration of colloids. Series have been established that arrange the ions in order of increasing or decreasing effectiveness with regard to hydration. In general, monovalent cations increase hydration, whereas divalent, particularly polyvalent, cations decrease it. It is commonly assumed that degree of hydration of cell colloids must fall within certain limits. It is possible that at least a portion of the deleterious effect resulting from a predominance of one ion or salt in a solution may be caused by too radical a shift in hydration of colloids in one direction or the other.

(5) *Regulation of permeability characteristics of membranes.* Beginning in 1911 (Osterhout, 1911) and continuing into the 1930s, Osterhout (1936) studied the effects of various ions on membrane permeability and electrical potential of cells. Many of the data are conflicting and contradictory, and thus there is confusion with regard to the effects of ions on permeability. Even with our present knowledge, it is impossible to resolve all the apparent discrepancies that have been reported. There remains, however, no doubt concerning the fact that different ions affect permeability differently.

(6) *Toxic effects of elements.* It is not clear why this general role has been listed, inasmuch as toxicity cannot be regarded as a role. Toxicity can be an effect of any element, essential or otherwise, present in sufficient concentration. For example, heavy metals such as Zn^{2+} or Cu^{2+} are known to combine readily with proteins and cause them to be precipitated. Even such ions as Ca^{2+}, Mg^{2+}, K^+, and NO^{3-} can be toxic if present in relatively high concentration.

(7) *Antagonistic effects.* Antagonism pertains to those interactions in which the normal effect of one ion is counteracted or negated by that of another ion.

(8) *Catalytic effects.* Starting around 1930, much of the ambiguity involving catalytic effects of elements was removed. Painstaking research demonstrated specifically how certain elements, particularly micronutrients, bring about catalytic effects. Some of them have been shown to be constituents of or activators of certain enzyme systems.

(9) *Maintenance of electrostatic neutrality.* Within cells electrostatic neutrality exists, positive and negative charges balancing each other. For example, DeKock (1958) suggested that Na^+, K^+, Ca^{2+}, and Mg^{2+}, in the high concentration in which they usually occur in cells, are present in large measure to balance the organic acid and inorganic acid anions. He noted that of the total Mg^{2+} in plants generally less than 0.1% is involved as a constituent of chlorophyll and only extremely minute amounts are required for the action of enzymes for which Mg^{2+} is specifically required.

Cations may, however, have certain specific effects, as reported by Jagendorf and Smith (1962). They noted that removal of cations from chloroplasts led to a decreased ability to phosphorylate and to an increased rate of reduction of ferricyanide.

In a study of various divalent cations, the order of effectiveness suggested that the rate of water substitution from the cation inner coordination hydration sphere may be a rate-limiting step in certain mitochondrial reactions involving electron transport and phosphorylation (Miller et al., 1970).

ROLES OF ESSENTIAL ELEMENTS

Many enzymes require one or more metallic ions as activators. Inasmuch as the metal binds the substrate to the enzyme (Hellerman and Stock, 1938; Klotz, 1954; Smith et al., 1954), the enzyme–metal ion must be able to form coordinate bonds with appropriate groups of the substrate molecule (Hellerman and Stock, 1938).

Although the significance, role, and effect are still to be resolved, increasing attention is being given to the association of various metals with viruses, DNA, and RNA (Steffensen, 1961). It was noted that stability of chromosomes toward various physical and chemical treatments was affected by the type of and presence or absence of various metals. It was concluded that metal ions form some kind of stabilizing bond in the chromosome.

Carbon

C is a constituent of all organic compounds, hence it would be impossible to list all the C-containing compounds that have been found in plants. C can be said to be important in each and every way that organic compounds are important, that is, as constituents of the plant structure, protoplasm, enzymes, and so on. In addition, C is a constituent of HCO_3^- which appears to be important in the exchange of anions by the plant for nutritive anions surrounding the roots (Overstreet and Dean, 1951). This anion is reported to stimulate photophosphorylation and the Hill reaction (Batra and Jagendorf, 1965). At a concentration sufficient to stimulate photophosphorylation, HCO_3^- inhibited the formation of a nonphosphorylated, high-energy condition of chloroplasts, (X_E). It was reasoned that if X_E represented a side pathway, draining energy from the phosphorylation mechanism, this could account for the increase in phosphorylation efficiency associated with high levels of bicarbonate.

Hydrogen

H is a constituent of all organic compounds of which C is a constituent. H^+ is the usual cation exchanged by the root for a cation (nutrient or otherwise) from the soil (Ulrich, 1941; Overstreet and Dean, 1951). Similar to C, H is a component of HCO_3^- which, along with OH^-, may be exchanged by the root for an anion from the soil (Ulrich, 1941).

Young seedlings of six species, and excised embryos from two species of higher plants, evolved H_2 in the complete absence of bacteria (Renwick et al., 1964), indicating that hydrogenase systems are present in seeds and seedlings of higher plants.

Oxygen

O is a constituent of many organic compounds in plants, and only a few organic compounds, such as carotene, do not contain oxygen. O is also a constituent of HCO_3^- and OH^- and as such may also be said to be involved in anion exchange between roots and the external medium (Overstreet and Dean, 1951). It is a constituent of protoplasm, enzymes, and many important compounds in plant structure (such as cellulose). It is a terminal acceptor of H^+ in aerobic respiration.

Nitrogen

N is a constituent of many compounds in plants including amino acids—the subunits of proteins. All enzymes that have been isolated, purified, and characterized are protein in nature—at least the apoenzyme portion. There are four N atoms in each chlorophyll molecule. N-containing compounds of plants are much too numerous to list, but N can also be said to be important in whatever way each N-containing compound is important. As a constituent of protein, enzymes, and chlorophyll, N is involved in all processes associated with protoplasm, enzymic reactions, and photosynthesis. With regard to protein synthesis, Mans (1967) discussed cell-free amino acid-incorporating systems. Such systems will surely prove useful in elucidating certain aspects of protein synthesis in higher plants. To date, however, unequivocal demonstration of protein synthesis by such systems is lacking.

As first shown by Hewitt et al (1956), N (as NO_3^-) is involved in the initiation of nitrate reductase (NR) activity (Ferrari and Varner, 1969), and NH_4^+ ions may repress induction of NR (Morton, 1956). These reports are cited here as examples of the involvement of N in the induction or repression of NR but, owing to the involvement of Mo in N metabolism, this topic is discussed in detail under the role of Mo for higher plants. Fixation of N by higher plants other than legumes has recently been reviewed (Bond, 1967).

It was suggested that NH_4^+, and other substances that inhibit flowering of *Lemna perpusilla*, may act by modifying the leaching of active materials from growing plants through effects on permeability and integrity of membranes (Hillman and Posner, 1971).

A suspension of soybean *(Glycine max* L.) cells did not grow on NO_3^- (25 m*M*) unless the medium was supplemented with NH_4^+ (2 m*M*) or glutamine (Gamborg, 1970). The L and D isomers of 12 amino acids tested singly could not replace NH_4^+. Cells of five other species of plants grew on defined medium with NO_3^- as the sole source of N. Later, it was found that soybean cells grew on defined media with NH_4^+ as the sole N source if Krebs cycle acids were added (Gamborg and Shyluk, 1970). Cells of soybean, wheat, and flax were cultured for extended periods on ammonium citrate medium.

Phosphorus

P is a constituent of many compounds in plants, and its involvement in metabolism is covered in detail by McElroy and Glass (1951, 1952). As a constituent of nucleoproteins, P is involved in that unique portion of protoplasm concerned with cell division and the transfer of hereditary characteristics by the chromosomes. It is a constituent of phospholipids and one of these, lecithin, is believed to be present in cell membranes and to be of universal occurrence in all living cells. Special H^+ carriers, diphosphopyridine nucleotide (DPN or NAD) and triphosphopyridine nucleotide (TPN or NADP) are concerned with H^+ transfers that occur as steps in the Krebs cycle, glycolysis, and the pentose cycle. It is also a constituent of the unique, high-energy compounds ADP and ATP. A considerable portion of the energy liberated by respiration is stored within cells as high-energy phosphate bonds (ADP and ATP) and as reduced coenzymes, NADH and NADPH. High-energy phosphate bonds provide energy for synthesis of such compounds as sucrose, starch, and proteins. Prior to cleavage of a sugar such as glucose, during glycolysis, the molecule becomes

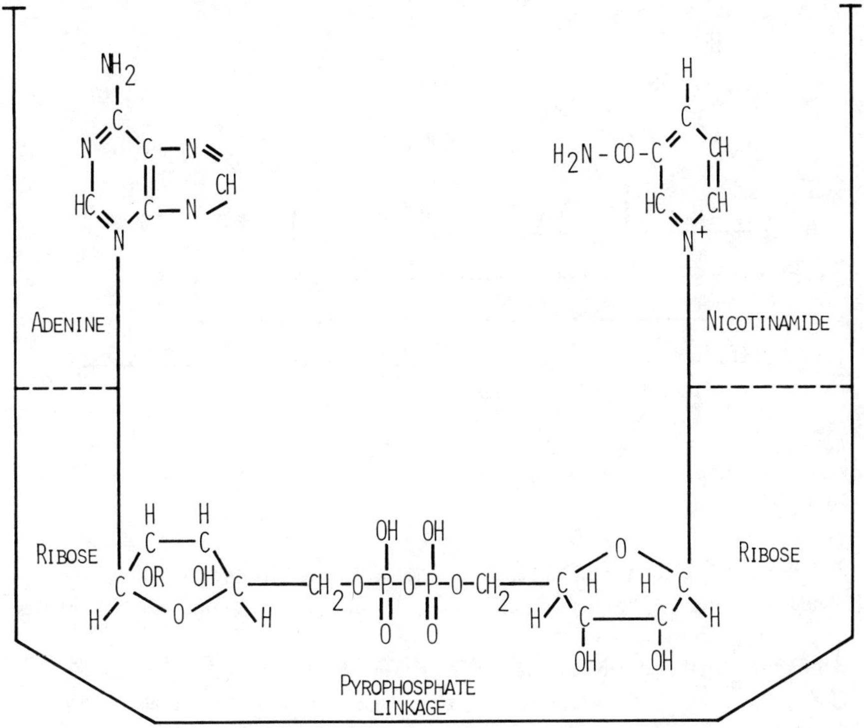

Fig. 9.1

P as a constituent of NAD and NADP. In NAD, R = H; in NADP, R = —PO(OH)$_2$. From D. I. Arnon (1959) Agrochimica, 3:108–139.

phosphorylated at the terminal ends. Starch hydrolysis involves phosphorylation. It has also been postulated that sucrose must be phosphorylated, by sucrose phosphorylase, prior to passage of sucrose from cell to cell in the form of fructose 1,6-diphosphate (Street and Lowe, 1950).

P has long been known to be involved in photosynthesis in connection with phosphorylation of various intermediates in CO_2 assimilation. In the two photochemical reactions occurring during photosynthesis, P is involved in the conversion of light into physiologically useful chemical energy by the formation of NADPH (Fig. 9.1) and ATP (Fig. 9.2).

Phosphate participates more directly in the true photochemical events of photosynthesis than does CO_2; in fact, CO_2 assimilation is dependent on a preceding phosphate assimilation resulting in ATP formation at the expense of light energy (Arnon, 1959).

P is a constituent of pyridoxal phosphate—a coenzyme for transamination systems (Green et al., 1945; Lichstein et al., 1945) and for glutamic acid de-

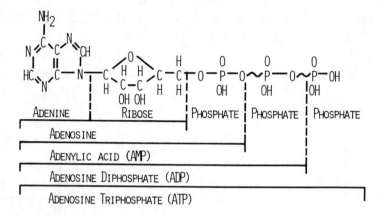

Fig. 9.2

P as a constituent of adenosine, AMP, ADP, and ATP. From D. I. Arnon (1959) Agrochimica, 3:108–139.

carboxylase (Baddiley and Gale, 1945; Schales et al., 1946). P is also a component of sugar phosphates, phytic acid, and other compounds in plants (Evans and Sorger, 1966).

Low external concentrations of P increased the concentrations of Zn^{2+}, Cu^{2+}, Fe^{2+}, and Mn^{2+} in leaves of *Franseria dumosa* (Wallace et al., 1969). Inasmuch as high external concentrations of P may induce Zn^{2+} deficiency, it was suggested that the concept could be generalized to include Zn^{2+} toxicity resulting from low external concentrations of P.

The greater portion of inorganic phosphate (P_i) in cells appears to be stored in the vacuole and to take no part in steady-state metabolism (Loughman, 1960). In this connection, Bieleski (1968) noted that in normal tissue of *Spirodela oligorrhiza* (Kurz) Hegelm, P_i was present in metabolic (12%) and nonmetabolic (88%) pools. In P-deficient tissues approximately 90% of the P_i was in the metabolic pool. This investigator reported that metabolic P_i was located in the cytoplasm and could be used for growth; nonmetabolic P_i was presumably located in the vacuole. During P deficiency, growth was limited by the rate at which P_i was transported out of the vacuole.

Potassium

Although most plants require relatively large amounts of K^+ (Ulrich and Ohki, 1966; Evans and Sorger, 1966; Kilmer et al., 1968), no one has isolated a K-containing compound from plants. K^+ appears to be completely water soluble in plants. There are a few reports, however, indicating that K^+ may be combined with some organic compound(s) (Stout et al., 1947). Olsen (1948) reported that during summer approximately 30% of the K^+ in beech leaves was adsorptively bound. Shortly prior to leaf fall, all the K^+ became soluble. He concluded that K^+ was bound to proteins rather than to cell walls, since the decrease in adsorptively bound K^+ occurred when proteins were hydrolyzed prior to leaf fall. *Chlorella* has also been reported to contain bound K^+ (Scott, 1944).

Along with Mg^{2+}, K^+ is required for the action of fructokinase and other enzymes (Table 9.1). Extracts of K-deficient pea roots showed considerably less pyruvate kinase activity than did those of K-sufficient plants (Evans, 1963). NH_4^+ or Na^+ could only partially substitute for K. Pyruvate kinase from rabbit muscle required K^+ (or Rb^+ or NH_4^+) as well as Mg^{2+}; Ca^{2+} inhibited action of the enzyme (Kachmar and Boyer, 1953).

Although a relationship of K^+ to starch formation was suggested in the earlier literature, recent investigations have more firmly established the involvement of K^+ in starch synthesis (Greenberg and Preiss, 1965; Evans and Sorger, 1966; Akatsuka and Nelson, 1966; Preiss and Greenberg, 1967; Nitsos and Evans, 1968, 1969; Murata and Akazawa, 1968). Nitsos and Evans (1969) showed that the particulate starch synthetase from sweet corn had an absolute requirement for K^+ with an optimal activation occurring at 0.05 *M* KCl. During K^+ deficiency lack of starch synthesis could be the result of reduced *energy supply* (even though sugars accumulate), since K^+ is necessary for glycolysis, oxidative phosphorylation, photophosphorylation, and for adenine synthesis (Evans and Sorger, 1966).

Although K^+ is intimately connected with certain enzyme systems (Ulrich and Ohki, 1966; Evans and Sorger, 1966), it is not clear why plants should require the relatively large amounts and high concentrations (1% or more dry weight basis) that they do for best growth. In this connection, however, additional K^+ may be required for NO_3^- absorption over and above the amount required for other functions that K^+ performs (Nightingale, 1942). Relative K^+ concentration has been reported to condition the rate of energy release through hexose sugar oxidation which in turn conditions the rate of N assimilation (Beckenbach et al., 1940).

It has been suggested that a relatively high concentration of K^+, and other univalent cations, may be required for cation-induced conformational changes of enzymes, since these cations do not form highly stable chelate complexes (Evans and Sorger, 1966). Thus a high concentration of univalent cations would be required for maintenance of these cations with charged groups on protein surfaces. K therefore appears to be the only univalent cation in nature available in sufficient quantity and with appropriate chemical properties to satisfy the requirements of the majority of univalent cation-activated enzymes according to Evans and Sorger (1966). These investigators indicate that K^+ and other univalent cations function as cofactors for approximately 46 known enzymes from animals, higher plants, and microorganisms.

Table 9.1
Roles or Functions of Elements

Element	Constituent of and/or function	Source	Investigator[a]
Fe	Ferredoxin; involved in energy transfer in photosynthesis and in cyclic and non-cyclic photophosphorylation; NO_2 reduction in chloroplasts	—	Mortenson et al., 1962; Tagawa and Arnon, 1962; Arnon, 1965; Joy and Hageman, 1966
Fe	Activator of aconitase	—	Dickman and Cloutier, 1951
Cu	Prosthetic group for polyphenol oxidases	—	Keilin and Mann, 1938; Kubowitz, 1938
Cu	Prosthetic group of laccase	—	Keilin and Mann, 1939a,b
Ca	Activator of phosphatases	Potato tuber	Kalckar, 1944; Krishnan, 1949
Mn	Activator of arginase	Plants and animals	Folley and Greenbaum, 1948
Mn or Mg	Activator of glutamotransferase	Peas	Elliott, 1953
K, NH_4, Na	Activator of pyruvate kinase	Pea roots and leaves	Evans, 1963
K, Rb, NH_4	Activator of pyruvate kinase	Rabbit muscle	Kachmar and Boyer, 1953
Mg	Activator, ATP sulfurylase	Soybean leaves	Adams and Johnson, 1968
Mg, Mn	Activator, glutamine synthetase	Peas	Elliott, 1953
Mg	Activator, succinyl CoA synthetase	Spinach	Kaufman and Alivisatos, 1955
Mg, Mn, Co, and K, Rb, Na, NH_4	Activator, succinyl CoA synthetase	Tobacco	Bush, 1969
Mg	Activator, pyrophosphatase	Yeast	Bailey and Webb, 1944
Mg	Required for oxidative decarboxylation of pyruvic acid to form acetyl CoA	Heart muscle	Korkes et al., 1951, 1952
Mg, Mn, Co	Phosphoglucomutase	Yeast	Cori et al., 1938
Mg, Mn	Activator, enolase	Escherichia coli	Utter and Werkman, 1942

Element(s)	Role	Organism/System	Reference
Mg, Mn, Fe, Co, Ni, Zn	Activator, pyruvic carboxylase	*Proteus vulgaris*	Stumpf, 1945
Mg	Activator, hexokinase	Yeast	Berger et al., 1946; Bailey and Webb, 1948
Mg, Mn	Activator, carboxylase	Soybean	Mee, 1949
Zn	Prosthetic group of carbonic anhydrase	Blood cells	Keilin and Mann, 1939b
Zn	Constituent, alcohol dehydrogenase, alkaline phosphatase, carboxypeptidase B, and other enzymes	—	Evans and Sorger, 1966
Zn	Carbonic anhydrase	Higher plants	Bradfield, 1947
Co	Constituent of B_{12} required by nodule bacteria	Root nodules of alder (*Alnus glutinosa*) and beefwood (*Casuarina cunninghamiana*)	Bond and Hewitt, 1962
Co	N fixation	*Azotobacter chroococcum*	Iswaran and Sundara Rao, 1964
Co	N fixation	*Azobacter vinelandii*	Evans and Kliewer, 1964
Co	N fixation by nodule bacteria	*Phizobium japonicum*	Ahmed and Evans, 1959
Co	N fixation by nodule bacteria	*Rhizobium meliloti*	Reisenauer, 1960
Co	N fixation	Blue-green algae	Holm-Hansen et al., 1954
Co	N fixation	Algae	Hutner and Provasoli, 1964; Evans and Kliewer, 1964
V	Dark enzymatic reactions of photosynthesis	*Scenedesmus obliquus*	Arnon, 1954
P	Constituent of pyridoxal phosphate — coenzyme for glutamic acid decarboxylase and transaminases	Higher plants	Baddiley and Gale, 1945; Green et al., 1945; Lichstein et al., 1945; Schales et al., 1946
Na	Unknown	*Atriplex vesicaria*	Brownell and Wood, 1957; Brownell, 1965
NO_3	Inducer of nitrate reductase	Higher plants	Hewitt and Afridi, 1959
NO_3	Inducer of nitrate reductase	*Neurospora crassa*	Morton, 1956; Kinsky, 1961
B	Unknown	*Nostoc muscorum*	Eyster, 1952

Table 9.1 (continued)

Element	Constituent of and/or function	Source	Investigator[a]
Mo	Along with NO_3^-, an inducer of nitrate reductase	Higher plants	Hewitt and Afridi, 1959
Mo	Constituent of nitrate reductase, with the Mo functioning as electron carrier	*Neurospora crassa*	Nicholas and Nason, 1954a, b
Mn	Activator, fixation of CO_2 by oxalacetate carboxylase	Parsley root	Vennesland and Felsher, 1946; Gollub and Vennesland, 1947
Mn	Involved in O_2 evolution	Chloroplasts	Pirson, 1937; Kessler, 1955; Brown et al., 1958; Eyster et al., 1958; Spencer and Possingham, 1961; Bachofen, 1966; Cheniae and Martin, 1966; Gerhardt, 1966; McKenna and Bishop, 1967; Homann, 1967; Haberman et al., 1968; Cheniae and Martin, 1969
Mn	Activator, oxalacetic carboxylase	Dicots	Vennesland and Felsher, 1946
Cu	Constituent, ascorbic acid, oxidase	Squash	Powers et al., 1944

[a] This list is to be regarded as representative and is not meant to determine the first investigator or to exhaust the list of workers or functions of each element listed.

Na$^+$ and K$^+$ form coordinated complexes with ATP and ADP as reported by Melchior (1954). Chemical models of Na–ATP and K–ATP indicate that Na$^+$ is completely buried in the complex molecule, but that K$^+$ is almost completely exposed. It was concluded that certain enzymes may catalyze reactions involving the K–ATP complex but fail to catalyze those involving the Na–ATP because of steric hindrance.

K$^+$ deficiency in corn plants did not change the base composition of RNA; ribonuclease level in shoots was increased as much as threefold (Hsiao et al., 1968). In some higher plants Rb$^+$ partially but not completely replaced K$^+$ (Richards, 1941, 1944). Na$^+$, Li$^+$, or Cs$^+$ could not even partially substitute for K$^+$ (Richards, 1941).

K$^+$ appears to play a role in translocation, since a deficiency of K$^+$ decreased translocation of labeled photosynthate from leaves to other portions of sugarcane plants (Hartt, 1969). Translocation was inhibited from leaves that showed no visible symptoms of K$^+$ deficiency or any decrease in rate of photosynthesis. The decrease in translocation was considered a primary effect of K$^+$ deficiency.

Working with isolated epidermal strips floated on KCl solutions and receiving CO$_2$-free air, Fischer (1968) and Fischer and Hsiao (1968) observed that uptake of K$^+$ closely paralleled the concomitant increase in stomatal aperture. Although closely linked to uptake of K$^+$, they considered the changes in starch secondary as far as opening of stomata was concerned. It was suggested that absorption of extracellular solutes, such as K$^+$, may be the primary mechanism of stomatal opening (Fujino, 1967; Fischer and Hsiao, 1968).

More recently, Humble and Hsiao (1969a) reported that stomatal opening is a light-activated, highly specific effect of K$^+$ (and Rb$^+$), since light lowered more than 100-fold the concentration of K$^+$ required for maximal opening. Although high concentrations (50 or 100 meq/liter) of Li$^+$, Na$^+$, or Cs$^+$ caused stomata to open, only the effect of K$^+$ (and Rb$^+$) was greatly accentuated by light. NH$_4$$^+$ and Mg^{2+} did not cause opening. They concluded that their work demonstrated, for the first time, a physiological process in plants specifically requiring K$^+$. In recent work, Humble and Hsiao (1969b) concluded that the energy for K$^+$ uptake and stomatal opening could be provided by cyclic photophosphorylation.

Solute accumulation in guard cells is in the range 0.1–0.5 M at the time stomata open (Meidner and Mansfield, 1968). Sawhney and Zelitch (1969) determined that the concentration of K$^+$ in closed stomata was 0.21 M, in fully open stomata 0.50 M, and in adjacent epidermal cells 0.19 M. The increase, associated with a concentration of K$^+$ of 0.29 M, is actually 0.58 M including the anion accompanying the K. This total, 0.58 M, is sufficient for the largest solute accumulation (in guard cells) reported in the literature (Meidner and Mansfield, 1968). The K$^+$ ions are probably supplied from adjacent epidermal cells (Sawhney and Zelitch, 1969). Stomatal opening in light is temperature dependent, requires O$_2$, and is inhibited when ATP formation is repressed

(Zelitch, 1965, 1969). Therefore, it was concluded that any factor that affects ATP synthesis, by noncyclic photophosphorylation in guard cells (including oxidation of glycolate), influences stomatal opening through an effect on the K^+ pump (Zelitch, 1969).

Thomas (1970a) concurred that a supply of ions, particularly K^+, is necessary to initiate and maintain the opening of stomata in light. In the presence of K^+, stomata could be opened in the light and closed in the dark, and this cycle could be repeated. Ca^{2+} or Mg^{2+} caused reductions in the apertures of the stomata. Half-maximal opening of stomata occurred at 0.32 mM K^+—with concentrations of 10 mM resulting in reduction of aperture. The K^+-dependent, light-stimulated opening of stomata of tobacco and *Vicia faba* epidermal strips was rapidly reduced by low concentrations of ouabain according to Thomas (1970b). He therefore concluded that the influx of K^+ into guard cells is associated with a membrane-bound transport ATPase. In *V. faba* opening and closing of stomata, respectively, paralleled the fluxes of labeled K^+ into and out of the strips (Humble and Hsiao, 1970). Gain and loss of K^+ by the strips were shown by cobaltinitrite reaction to be centered in the guard cells. There was little or no light stimulation of opening in strips on Na^+, nor was there stimulation of Na^+ uptake. Inhibition of opening of stomata was generally correlated with inhibition of K^+ uptake. It was suggested that photosystem I and cyclic electron flow could supply the necessary energy for K^+ uptake and stomatal opening. Stomata in strips opened easily under far-red light (>700 nm). In corroborating this work, it was shown that the K^+ staining (with cobaltinitrite) of guard cells of *V. faba* correlated well with absorption of radioactively labeled K^+ (Fischer, 1971). The results indicated that in leaves, as well as in isolated epidermal strips, K^+ and an accompanying anion comprise the major osmotically active solutes in guard cells of open stomata.

Contrary to the preceding evidence of certain workers, it was found that Na^+ was as effective as K^+ for the light-stimulated opening of stomata, and opening in NaCl appeared to proceed by the same mechanism as when stomata were in KCl (Pallaghy, 1970). A similar observation was made earlier (Willmer and Mansfield, 1969). Pallaghy (1970) concluded that the specificity of light-dependent uptake of cations by guard cells is controlled by Ca^{2+} and leads to stomatal opening. It was suggested that Ca^{2+} alters the specificity of ion binding on a site (e.g., a carrier enzyme) responsible for light-stimulated ion uptake.

This is an interesting development concerning causative factors in the opening and closing of stomata—a phenomenon that has received attention since the early work by Lloyd (1908) and Loftfield (1921). Working with *Rumex patientia*, Sayre (1926) noted that opening and closing of stomata were related to pH changes—high pH favoring opening and low pH favoring closing. Later, it was reported (Scarth, 1932; Small et al., 1942) that the pH changed appropriately during opening and closing. In contrast with other leaf cells, starch concentration in guard cells was high in the dark and low in light (Lloyd, 1908; Loftfield, 1921; Sayre, 1926). Concomitant with the higher pH, associated with light, starch was observed to change to sugar which was osmotically active; darkness lowered pH and sugar changed back to starch. Yin and Tung (1948) found phosphorylase in chloroplasts and, since guard cells contain chloroplasts, it was possible that starch and P_i were converted to glucose 1-phosphate in the light and caused a higher osmotic pressure with concomitant opening of stomata. Steward (1964), however, noted that unless glucose 1-phosphate was changed to glucose *and* P_i there would be no change in osmotic pressure. That is, P_i would be active osmotically along with glucose 1-phosphate. Steward therefore proposed a scheme in which glucose and P_i were formed and resulted in opening; conversion of glucose and P_i to glucose 1-phosphate (involving aerobic respiration and ATP) would mediate closure.

The lower axillary buds of intact plants of *Solanum sisymbrifolium* are released from complete inhibition by high concentrations of K^+ in the soil, while complete apical dominance is shown by plants grown at low, but not deficient, concentrations of K (Wakhloo, 1970). The completely inhibited buds of low-K^+ plants could be released from inhibition by supplying kinetin. Partially inhibited buds of the high-K plants, however, were not affected by exogenous kinetin.

In a study of the effect of an antitranspirant, phenylmercuric acetate (PMA), Waisel et al. (1969) noted that the larger stomata of *Betula papyrifera* were more affected by PMA and by environmental changes than were the smaller stomata. They cautioned that stomata of *B. papyrifera*, at least, could not be considered a homogeneous population with respect to their responses to moisture stress, light, and chemical treatment. A similar indication of differences among stomata is indicated by the fact that stomata of cereal plants open and close in "patches" (Miller, 1938).

In a study of seedling root tips of loblolly pine, evidence was obtained concerning possible specificity in the activation of certain enzymes (McClurkin and McClurkin, 1967). The ATPase of nuclei of meristematic cells was stimulated only by Na^+, whereas K^+ stimulated the ATPase of mycorrhizae between root cap cells. This distinction between the two ATPases suggests that certain enzyme systems can be activated only by a specific cation.

Sulfur

Apparently, all plant proteins have S-containing amino acids (cysteine, cystine, and methionine), but their significance in plants has not been fully elucidated. S is also a constituent of the tripeptide, glutathione, since cysteine is one of the three constituent amino acids. This tripeptide may be an important H carrier in animals, but a similar role in plants has not been established. S is a component of lipoic acid, coenzyme A, thiamine pyrophosphate, biotin, adenosine-5′-phosphosulfate and 3-phosphoadenosine-5′-phosphosulfate, and other compounds (Evans and Sorger, 1966). The mustard oil glycosides, of which sinigrin is perhaps the best known, also contain S.

In his excellent review of S metabolism, Thompson (1967) stated that from a quantitative standpoint the most important function of S metabolism in plants is to produce cysteine and methionine. Sulfate must be reduced (Hart and Filner, 1969). Further, SO_4^{2-} reduction (i.e., conversion of SO_4^{2-} to sulfide) serves two purposes. Assimilatory SO_4^{2-} reduction furnishes sulfide for the formation of reduced-S compounds. Dissimilatory reduction provides a flow of electrons for ATP generation; the generated sulfide may also be used metabolically.

Calcium

There are numerous reports to the effect that the middle lamella is composed of Ca pectate. However, Conrad (1926) showed that by placing fresh plant material directly into hot, boiling ethanol, most plant materials did not contain Ca pectate (Table 9.2). Ca pectate apparently formed only when the same plant materials were preserved by oven drying and other relatively slow techniques of inactivating enzymes (Table 9.3). It was concluded that Ca pectate does not exist in the living plant but rather is formed during most conventional procedures for preserving plant material for analysis.

In contrast with Conrad's (1926) findings, Abdel-Halim (1964), using more refined methods, found pectic acid in all plant materials regardless of how they were preserved prior to analysis. However, Ca^{2+} determinations based on extracts for pectates showed no consistent relationship with pectic acid concentration, and Ca^{2+} was present in concentrations far below the levels required for the pectic acid to have existed as Ca pectate. In some tissues Ca^{2+} concentration was below the limit of detection, and yet appreciable quantities of pectic acid were present.

Inasmuch as the evidence had been based on nonspecific extraction and staining procedures, Setterfield and Bayley (1961) indicated that evidence for pectic acid in the middle lamella needed to be reexamined. They concluded that the "middle lamella" may be "a region of the polysaccharide matrix where cellulose is absent, rather than a region of pectate localization." Klein and Ginzburg (1960) also concluded there is no specialized cementing layer between cell walls. In some species pectin rather than pectic acid is concentrated in the middle lamella, for example *Avena* coleoptiles (Albersheim and Bonner, 1959) and onion root tips (Albersheim et al., 1960).

Ginzburg (1958b) noted that conditions required to separate root tip cells

Table 9.2
Protopectin (Including Pectin) and Pectic Acid in Different Plant Tissues[a,b]

| | Percent | |
| | Protopectin and pectin | Pectic acid |
Tissue		
Strawberries, green	8.82	0.00
Strawberries, ripe	6.16	0.00
Tomatoes, green	5.45	0.00
Tomatoes, ripe	2.92	0.00
Potato	2.00	0.00
Beet, red	3.82	0.00
Carrot	10.04	0.00
Turnip	11.93	0.00
Radish, pithy	26.87	15.37
Parsnips	10.68	0.00
Orange	0.10[c]	Trace
Bananas	2.24	0.00
Cherries, green	11.42	0.00
Cherries, ripe	4.32	0.00

[a] Conrad (1926).
[b] Analyses based on dry weight except as otherwise noted.
[c] Based on fresh weight.

Table 9.3
Comparison of Treatments of Parsnip Tissue for Pectic Analysis[a,b]

| | Percent | |
Treatment of tissue	Protopectin	Pectic acid
Dried at 98°C in air oven	6.84	3.95
Dried at 70°C in vacuum	3.83	9.61
Boiling alcohol	10.69	0.00
Sun-dried	7.36	5.26

[a] Conrad (1926).
[b] Values listed are based on dry weight in terms of Ca pectate determined as pectic acid.

with EDTA were far more drastic than would have been expected if simple chelation of Ca^{2+} and solubilization of pectic acid were involved.

Binding of tobacco leaf mesophyll cells involves pectic acid; Ca^{2+} associated with the middle lamella is rather transitory as reported by Zaitlin and Coltrin (1964). They also noted that the middle lamella of tobacco leaf cells is apparently not a region of constant composition. Small amounts of Ca^{2+}, but surprisingly not Mg^{2+}, interfered with the cell separation reaction. The enzyme they used did not, however, separate cells of onion root tips or *Avena* coleoptiles. Zaitlin (1959) reported that the middle lamella of leaves of several species were not attacked by NBC pectinase.

In studying the terminal millimeter of root tips of Alaska pea seedlings,

Ginzburg (1958a) reported that the intercellular cement also contains protein and two types of metal ions, one of which is probably K^+.

Either Ca^{2+} or Sr^{2+} in a mineral nutrient medium prevented toxic effects of other nutrient ions on aerated primary roots of maize; other monovalent or divalent cations substituted for Ca^{2+} did not (Bonds and O'Kelley, 1969).

Marinos (1962) concluded that Ca^{2+} is essential for the maintenance and probably for the formation of cell membrane systems on which functional integrity and cellular metabolism are dependent. Effects of Ca^{2+} on cell walls were regarded as secondary, that is, the deeper staining and gaps in walls of Ca-deficient cells indicated a weakening of wall structure.

In this connection, Steveninck (1965) considered that Ca^{2+} plays an important role in bonding negative charges of the plasma surface and cell wall.

In a study of the effects of auxin and Ca^{2+} on growth and elasticity of sunflower hypocotyls, Uhrstrom (1969) concluded that Ca^{2+} imparts rigidity to the cell wall and that it is necessary for growth. However, at superoptimal concentrations of Ca^{2+} the cell wall becomes too rigid and cell elongation is inhibited.

There has been considerable interest in Ca^{2+} as a cell elongation factor. Root growth is generally promoted by low concentrations of Ca^{2+}, whereas shoot growth may be retarded. Burstrom (1964) noted, however, that inhibitory concentrations for shoot growth are 100 to 1000 times higher than those that promote root growth. Working with Ca-deficient pea stems, he concluded that response of stems of Ca^{2+} did not differ basically from that known for roots. In addition, Burstrom (1963, 1964) decided that the Ca bridge theory advanced by Bennet-Clark (1956) to explain Ca inhibition of cell elongation could be extended to explain Ca promotion of elongation under Ca-deficiency conditions.

Inasmuch as a chelating agent, EDTA, promoted growth of oat coleoptiles, Bennet-Clark (1956) reasoned that the agent removed Ca^{2+} bridges from the cell walls. Thus arose the Ca^{2+} bridge hypothesis. He believed that the growth-promoting effect of IAA was also related to the removal of Ca^{2+} bridges. However, Cleland (1960) found that IAA caused neither a loss of Ca^{2+} from the cell walls nor a redistribution of Ca^{2+} between pectin and protopectin fractions of the walls. Burling and Jackson (1965) also reported that auxin did not affect the Ca^{2+} levels in cell walls.

Ca^{2+} has been reported to be associated (adsorbed) primarily by the cell walls of bean and barley roots (Waisel et al., 1970). Inhibition of elongation of oat coleoptile sections by La^{3+}, Pr^{3+}, and Nd^{3+} is similar to or greater than that by Ca^{2+} at low and intermediate concentrations (Pickard, 1970). It was suggested that the lanthanons might serve as probes in learning more about the manner of action of Ca^{2+}—a normal growth-regulating agent.

Nightingale et al. (1931) reported that in normal tomato plants approximately one-half of the Ca^{2+} was water insoluble and indicated that a portion of the Ca^{2+} was associated with some organic fraction(s). In Ca-deficient plants, he noted that nearly all the Ca^{2+} was water-insoluble. He also observed that tomato plants could not absorb NO_3^- when grown in a solution devoid of added Ca. However, by formulating a minus-Ca solution different from the one employed by Nightingale, Gauch (1940) found that minus-Ca plants could absorb NO_3^-. Apparently, the Mg^{2+} concentration was high enough in Nightingale's minus-Ca solution to interfere with NO_3^- absorption.

Legumes have been reported to be "heavy Ca^{2+} feeders." Until the significance of organic acids in the Krebs cycle was established, organic acids were often regarded as by-products of protein metabolism, and it was assumed that Ca^{2+} precipitated some of the excess of organic acids formed by a high rate of protein synthesis in legumes. Ca^{2+} is no longer regarded primarily as a detoxifier of excess organic acids.

Nodulation of legumes is very poor at pH 5 or lower but, even when the pH is 5.5 or higher, nodulation is very sparse unless there is adequate Ca^{2+} (Albrecht, 1933). In the presence of fixed N in the nutrient solution, Ca^{2+} was required for nodule infection or initiation and, once initiated, nodule development in subterranean clover *(Trifolium subterraneum* L.) proceeded at Ca^{2+} concentrations too low for plant growth (Lowther and Loneragan, 1968). Practically no nodules formed at 100 μM Ca^{2+}, and the number of nodules increased progressively from just above that concentration up to 720 μM Ca^{2+}.

Many species of higher plants form solid deposits of Ca carbonate (Pobeguin, 1954); Ca oxalate is also common, but Ca phosphate and sulfate are somewhat less so.

Ca^{2+} is of utmost importance for the growth and development of the gynophore of the peanut plant, and it must be present in relative abundance in soil immediately surrounding the gynophore (Brady, 1947; Bledsoe et al., 1949). The significance of Ca^{2+} in this regard has not been elucidated.

Mascarenhas and Machlis (1962a) obtained evidence that Ca^{2+} is the chemotropic factor for pollen of *Antirrhinum majus*. No response was obtained with the chlorides of Mg^{2+}, Ba^{2+}, Sr^{2+}, Na^+, or K^+, or sugars, amino acids, organic acids, plant growth regulators, and some other compounds. As a percentage of the dry weight, the upper one-third, middle one-third, and lower one-third of the style, ovary wall, and ovules-plus-placenta contained: 0.51, 0.50, 1.23, 1.3, and 2.17% Ca^{2+}, respectively. Pollen of *Narcissus pseudonarcissus* and *Clivia miniata* also responsed chemotropically to Ca^{2+}. Additional evidence was obtained that Ca^{2+} is a chemotropic agent by Mascarenhas and Machlis (1964). They analyzed various floral tissues of snapdragon and found a correlation between the total concentration of Ca^{2+} and the chemotropic activity of a given tissue. Optimal response to Ca^{2+} was conditioned by B—

boric acid enhancing the chemotropic response to Ca^{2+}. Johri and Vasil (1961) claimed that B enhanced the growth of pollen tubes. Diastase has been reported to have a chemotropic effect on pollen (Mascarenhas and Machlis, 1962b). Its effect may have been caused by Ca^{2+} (Mascarenhas and Machlis, 1964), since diastase has been reported to contain 1 gram-atom or more of Ca^{2+} per mole of enzyme (Fischer et al., 1960; Hsiu et al., 1964). Removal of Ca^{2+} destroys activity of α-amylases, and addition of Ca^{2+} (Mn^{2+}, Ba^{2+}, or Mg^{2+}) restores stability to the protein. Ca^{2+} was reported to be essential for germination and growth of pollen (Brewbaker and Kwack, 1963). For optimal growth pollen required 300–5000 ppm Ca^{2+} (as $Ca(NO_3)_2$), and yet most pollens have a relatively low concentration of internal Ca^{2+}. It was concluded that this low internal level of Ca^{2+} might explain the often observed poor growth of pollen *in vitro* and *in situ*. Recently, Rosen (1968) reviewed the role of Ca^{2+} and other ions on the growth of pollen.

Levels of Ca^{2+}, typical of the concentration of Ca^{2+} in tracheal sap of tomato stems, inhibited *Fusarium* polygalacturonase, and thus may have reduced this wilt disease by interfering with the decomposition of pectic substances within the host (Corden, 1965).

Ca^{2+} is a direct activator of phosphatases in potato tubers, which catalyze the dismutation of two molecules of ADP to yield ATP and adenylic acid, and the removal of two phosphate radicals from ATP to yield adenylic acid (Kalckar, 1944; Krishnan, 1949).

Some workers believe that Ca^{2+} stimulates α-amylase but does not take part in the catalytic action of the enzyme. Others believe that Ca^{2+} stabilizes α-amylase and is also required for its activity. The α-amylases of Ca-deficient pumpkin plants were completely inactive or of only slight activity according to Dvořák and Radotinská-Ledinská (1970). They found that inactive amylases could be reactivated by Ca^{2+} during preparation. They reported that it was impossible to determine whether α-amylases from Ca-deficient plants are inactive *in vivo*, and therefore that it was impossible to judge whether or not α-amylase activity is subject to feedback regulation.

In corn seedlings Sr^{2+} replaced part of the Ca^{2+} ordinarily required (Queen et al., 1963). For enolase of *Escherichia coli*, for which Mg^{2+} or Mn^{2+} serve as activators, Ca^{2+} (and Ni^{2+}) was inhibitory (Utter and Werkman, 1942). Ca^{2+} was also inhibitory for pyruvate kinase from rabbit muscle (Kachmar and Boyer, 1953).

Interestingly, Ca^{2+} may affect fluorescence (Homann, 1969). There is an overriding influence of ionic environment on the fluorescence yield of isolated chloroplasts; the yield can be changed at will simply by varying the composition of the suspending medium. Various ions have different effects on fluorescence. With Ca^{2+}, for example, low concentrations increased fluorescence, whereas higher concentrations decreased it.

A definitive role of Ca^{2+} in N metabolism is indicated by the work of Paulsen and Harper (1968). Severely Ca-deficient *Triticum aestivum* L. seedlings accumulated high concentrations of NO_2^- and moderate levels of NO_3^-. Synthesis of NR was reduced by Ca^{2+} deficiency and NO_2^- accumulation, but activity of NR was unaffected by this deficiency. It was concluded therefore that Ca^{2+} is involved in intracellular transport of NO_2^- and not in the induction or activity of enzymes (i.e., NR and/or NiR).

Magnesium

Each chlorophyll molecule contains one atom of Mg, that is, 2.7% of the weight of the chlorophyll molecule. Mg may thus be said to be involved in the most important synthetic reaction on earth, namely, photosynthesis. Of the total Mg^{2+} in plants, generally less than 0.1% occurs in chlorophyll; much of the Mg^{2+} is paired off with organic acid and inorganic anions (DeKock, 1958).

Mg^{2+} is an activator for enolase (Utter and Werkman, 1942), yeast phosphoglucomutase (Cori, *et al.*, 1938), hexokinase (Berger et al., 1946; Bailey and Webb, 1948), yeast pyrophosphatase (Bailey and Webb, 1944), and carboxylase (Stumpf, 1945; Mee, 1949). Pyruvate kinase was shown to require Mg^{2+} and K^+ (Kachmar and Boyer, 1953). Later, Miller and Evans (1957) reported that the enzyme required a monovalent cation (K^+, Rb^+, or NH_4^+) and a divalent cation (Mg^{2+} or Mn^{2+}); Ni^{2+} or Ca^{2+} was ineffective. All phosphokinases depend on $-SH$ groups and are activated by Mg^{2+} (Dixon, 1949).

Mg^{2+} was required for the activity of ATP sulfurylase which mediates the reaction between ATP and SO_4^{2-} to yield adenosine-5'-phosphosulfate (Adams and Johnson, 1968). It was also required for glutamine synthetase activity (Elliott, 1953). Succinyl-CoA synthetase was first isolated from plant tissues (spinach) by Kaufman and Alivisatos (1955) and shown to require Mg^{2+} and ADP or ATP. The enzyme from tobacco showed a requirement for a univalent and a divalent cation—K^+, Rb^+, NH_4^+, or Na^+ and Mg^{2+}, Mn^{2+}, Ca^{2+}, or Co^{2+} (Bush, 1969). One ATPase that appears to be involved in transport processes (Robertson, 1960) requires Mg^{2+} (or Mn^{2+}), but it is further stimulated by K^+ and other monovalent ions (Fisher and Hodges, 1969). Along with at least four other cofactors, Mg^{2+} was required for oxidative decarboxylation of pyruvic acid to form acetyl-CoA (Korkes et al., 1951, 1952; Gunsalus, 1954).

There was a higher Mg^{2+} requirement at higher temperatures than at lower ones for a reaction sequence in Co_2 fixation in photosynthesis (Wallace and Mueller, 1962). Additionally, when ATP was equal to or in excess of Mg^{2+}, there was decreased $^{14}CO_2$ fixation with ribulose-5-phosphate as substrate.

High Mg^{2+} levels resulted in more effective use of high ATP levels than did low Mg^{2+} levels.

Mg^{2+} is necessary for the action of hexokinase (Berger et al., 1946) and presumably is also required for the action of one or more enzymes responsible for lipid synthesis. Seeds high in oils also tend to be high in Mg^{2+}, and *Vaucheria* (Miller, 1938) does not form oil droplets (in its cytoplasm) when grown in a nutrient medium devoid of Mg^{2+} but does so in its presence. Mg^{2+} was required for citric acid synthesis in *Aspergillus niger* (Bernhauer et al., 1940).

Mg^{2+}, replaceable by Co^{2+} but not Mn^{2+}, is required in all photosynthetic phosphorylations (Arnon, 1959).

The enzyme (−) S-adenosyl-L-methionine-Mg protoporphyrin methyltransferase, which catalyzes transfer of a methyl group from (−) S-adenosyl-L-methionine to Mg protoporphyrin to form Mg protoporphyrinmonomethyl ester, has been detected in isolated chloroplasts from Z. *mays* (Radmer and Bogorad, 1967).

Binding of 30S and 50S subunits to form a 70S ribosome is dependent on Mg^{2+} concentration according to Watson (1965). When the Mg^{2+} concentration was reduced below that found normally in growing E. *coli* cells, the 70S ribosome was reduced to 30S and 50S subunits. This investigator noted that in cell-free systems the concentration of Mg^{2+} can determine whether or not there will be "reading mistakes" in protein synthesis. Anomalous leucine incorporation resulted when excessive concentrations of Mg^{2+} were used in the incorporation experiments.

NONESSENTIAL ELEMENTS

Almost all of the naturally occurring elements have been detected in plants. It is possible that with further refinements in technique and in the purity of salts, additional elements may be shown to be essential—at least for certain organisms. Infrequent claims for the essentiality of some of these elements have been unsubstantiated, hence their inclusions at this time would not be appropriate.

The literature on Al^{3+} justifies discussion at this time. Although Al^{3+} is not regarded as an essential element for plants (Woolley, 1957), it is involved in an interesting biochemical reaction. When pink-flowered hydrangea plants are transferred to an acid soil or the plants are treated with Al^{3+} salts, the color of the flowers changes from pink to blue. Apparently a blue-colored, acid-stable, Al^{3+} lake is formed which may be regarded as a colloidal complex or a loose combination of the delphinidine pigment with Al^{3+} (Chenery, 1948). The pH of the sap of hydrangea flowers is approximately 4.4.

In lemon and orange cuttings, there appeared to be marked increases in top and root growth from the addition of 0.1 ppm Al^{3+} to solution cultures, but it was discovered that Al^{2+} was negating an effect of copper. With the Cu^{2+} concentration of the solution was decreased, addition of Al^{3+} was without effect on the growth of the plants (Liebig et al., 1942). This research is particularly noteworthy inasmuch as it indicates how without careful study a given element might be listed as "stimulatory" or "essential" when in reality it merely affects the action of some other element—often an essential element.

Although not regarded as essential, 8.7 μg Al^{3+}/liter stimulated growth of *Chlorella*, as reported by Bertrand and Wolf (1966). They concluded that Al^{3+} was a "dynamic" microelement for plants, although the amount required was very small. Al^{3+} is known to stimulate reduction of cytochrome (Randall and Vose, 1963). Al^{3+} is accumulated in relatively high concentrations in *Lycopodium, Hicoria,* and *Symplocus* (Hutchinson, 1945).

Si had no observable effect on growth of tomato plants (Woolley, 1957). I^- was capable of inducing premature leaf abscission in plants as observed by Crosby and Vlitos (1962). They hypothesized that I^- induces abscission by increasing, either directly or indirectly, the decomposition of IAA within the leaf or by interfering with its normal biosynthesis.

Although there are some reports that at low concentrations Se is stimulative for plant growth, it is not an essential element according to Ganje (1966). He noted that it may prove to be essential in animal nutrition and that there is good evidence that Se prevents certain animal diseases such as muscular dystrophy (white muscle disease) in lambs and calves, exudative diathesis in chicks, and liver necrosis in rats.

There was no evidence that Se is required by alfalfa or subterranean clover (Broyer et al., 1966). Peterson and Butler (1962), in a study of various species of plants, reported on the incorporation of Se into various Se amino acids. Shrift (1961) and Shrift et al. (1961a, b) reported that the Se analog of methionine temporarily blocked cell division of *Chlorella vulgaris,* but it eventually resumed; growth continued during the time division was blocked and giant cells resulted. Presumably, Se methionine-altered proteins could function in growth but not in unadapted cells in cell division.

Several Se-accumulating plants, *Astragalus crotalariae* and *Oonopsis condensata* (Shrift and Virupaksha, 1963), and *Astragalus bisulcatus* (Trelease et al., 1960), convert selenite into Se methylselenocysteine. In *Stanleya pinnata,* selenate is incorporated into the Se analog of cystathionine (Virupaksha and Shrift, 1963).

Although there has not been much research on Ga, it appears to be nonessential for higher plants and seldom occurs in concentrations over 1.0 ppm (Liebig, 1966).

There is no evidence that Cr is an essential element. Although the element is widely distributed, with usually less than 1000 ppm in the soil, the concentration in plants is usually lesss than 1.0 ppm (Pratt, 1966a). It was reported that only serpentine soils derived from ultrabasic or serpentine rocks have concentrations of Cr as high as several percent.

Anderson (1951) mentioned a possible relationship between Ti and N fixation by legumes. Ti is nonessential, and there is no evidence that it is phytotoxic, as reported by Pratt (1966b). He added that toxicity from Ti would not be expected, since the element is of extremely low solubility in the pH range 4–8.

Zr is not essential for plants and even with a spectograph the element is generally not detectable in plants (Pratt, 1966c). The solubility of Zr is so low in soils that little of the element would be expected in plants, and certainly phytotoxicity would not be expected.

In most plants the concentration of Sn is below 2.0 ppm on a dry weight basis as observed by Wallihan (1966). He noted that Sn in plant tissues is not related to the concentration of Sn in the soil. He speculated that Sn did not appear in plants possibly because of the lack of an accumulation mechanism for the element.

Ba is neither essential for the growth of plants (or animals), nor of benefit to plant growth according to Vanselow (1966a). He reported that Ba^{2+} may, however, in part replace Ca^{2+}; the amount of Ba^{2+} absorbed by plants is roughly dependent on the exchangeable Ba^{2+} in the soil. Clear-cut cases of Ba^{2+} excess in plants have not been reported.

Ag occurs in seawater, rocks, soils, plants, and animals but generally in rather minute amounts. Vanselow (1966b) noted that although the element is almost universally found in soils and plants, it is seldom present in excess of 1 ppm in the dry matter; in fact, its concentration is generally below 0.01 ppm. Ag has not been shown to be essential.

Although Vanselow (1966c) showed Sr^{2+} values as high as 17,500 ppm (dry weight basis) in elm leaves and twigs, he stressed that Sr^{2+} was not essential for the growth of any plant, and that none had been shown to benefit from its addition to the medium. Owing to the similarity of Sr^{2+} to Ca^{2+}, he suggested that Sr^{2+} may replace Ca^{2+} at least partially. From nutrient solution Ca^{2+} and SR^{2+} are absorbed generally in proportion to their respective concentrations in the solution (Vanselow, 1966c).

Interestingly, Hewitt (1951) suggested how nonessential elements might after all be involved in metabolism. He noted that certain enzymes may be activated not only by some essential element but also by apparently nonessential elements such as Co^{2+} or Ni. He therefore suggested that stimulatory effects of nonessential elements reported from time to time might have been based on their playing a beneficial role as an enzyme activator when the normal essential element was deficient in the plant. It has also been suggested by Bear and

Prince (1945) that even for essential nutrient ions there are at least two functions. They speculated that at least one of these roles for each nutrient, such as Ca^{2+}, Mg^{2+}, and K^+, was *specific,* whereas the other role(s) could be satisfied interchangeably by either of the other two cations. Once the supply of each cation was adequate to meet the specific requirement, there could be a wide range in acceptable concentrations of the other two cations.

LITERATURE CITED

Abdel-Halim, M. A. 1964. Presence and composition of the middle lamella in higher plants. Ph. D. Thesis, Univ. Maryland, College Park.

Adams, C. A., and R. E. Johnson. 1968. ATP sulfurylase activity in the soybean [*Glycine max* (L.) Merr.] Plant Physiol. 43:2041–2044.

Ahmed, S., and H. J. Evans. 1959. Effect of cobalt on the growth of soybeans in the absence of supplied nitrogen. Biochem. Biophys. Res. Commun. 1:271–275.

Akatsuka, T., and O. E. Nelson. 1966. Granule-bound adenosine diphosphate glucose-starch glucosyl transferase of maize seeds. J. Biol. Chem. 241:2280–2286.

Albersheim, P., and J. Bonner. 1959. Metabolism and hormonal control of pectic substances. J. Biol. Chem. 234:3105–3108.

Albersheim, P., K. Mühlethaler, and A. Frey-Wyssling. 1960. Stained pectin as seen in the electron microscope. J. Biophys. Biochem. Cytol. 8:501–506.

Albrecht, W. A. 1933. Inoculation of legumes as related to soil acidity. J. Amer. Soc. Agron. 25:512–522.

Anderson, A. J. 1951. The influence of plant nutrients on symbiotic nitrogen fixation. Proc. Spec. Conf. Plant and Animal Nutrition (Australia) 1949:190–199.

Arnon, D. I. 1950a. Inorganic micronutrient requirements of higher plants. Proc. 7th Int. Bot. Congr. (Stockholm), 263 pp.

Arnon, D. I. 1950b. Criteria of essentiality of inorganic micronutrients for plants, with special reference to molybdenum, pp. 31–39. *In* Trace elements in plant physiology. Chronica Botanica, Waltham, Mass. 144 pp.

Arnon, D. I. 1951. Growth and function as criteria in determining the essential nature of inorganic nutrients, pp. 313–341. *In* E. Truog (ed.) Mineral nutrition of plants. Univ. Wisconsin Press, Madison.

Arnon, D. I. 1954. Some recent advances in the study of essential micronutrients for green plants. Huitieme Congr. Int. Bot. (Paris) 1954:73–80.

Arnon, D. I. 1958. The role of micronutrients in plant nutrition with special reference to photosynthesis and nitrogen assimilation, pp. 1–32. *In* C. A. Lamb, O. G. Bentley, and J. M. Beattie (eds.) Trace elements (Proc. Conf. Ohio Agr. Exp. Sta., Wooster, Ohio, Oct. 14–16, 1957). Academic, New York.

Arnon, D. I. 1959. Phosphorus and the biochemistry of photosynthesis. Agrochimica 3:108–139.

Arnon, D. I. 1965. Ferredoxin and photosynthesis. Science 149:1460–1470.

Arnon, D. I., and P. R. Stout. 1939. The essentiality of certain elements in minute quantity for plants with special reference to copper. Plant Physiol. 14:371–375.

Bachofen, R. 1966. Die Oxydation von Mangan durch Chloroplasten im Licht. Z. Naturforsch. 216:278–284.

Baddiley, J., and E. F. Gale. 1945. Codecarboxylase function of "pyridoxal phosphate." Nature 155:727–728.

Bailey, K., and E. C. Webb. 1944. Purification and properties of yeast pyrophosphatase. Biochem. J. 38:394–398.

Bailey, K., and E. C. Webb. 1948. Purification of yeast hexokinase and its reaction with $\beta\beta'$-dichlorodiethyl sulphide. Biochem. J. 42:60–68.

Batra, P. P., and A. T. Jagendorf. 1965. Bicarbonate effects on the Hill reaction and photophosphorylation. Plant Physiol. 40:1074–1079.

Bear, F. E., and A. L. Prince. 1945. Cation-equivalent constancy in alfalfa. J. Amer. Soc. Agron. 37:217–222.

Beckenbach, J. R., W. R. Robbins, and J. W. Shive. 1940. Nutrition studies with corn. III. A statistical interpretation of the relation between nutrient ion concentration and the carbohydrate and nitrogenous content of the tissue. Soil Sci. 49:219–238.

Bennet-Clark, T. A. 1956. Salt accumulation and mode of action of auxin, pp. 284–291. *In* R. L. Wain and F. Wightman (eds.) The chemistry and mode of action of plant growth substances. Butterworths, London.

Berger, L., M. W. Slein, S. P. Colowick, and C. F. Cori. 1946. Isolation of hexokinase from baker's yeast. J. Gen. Physiol. 29:379–391.

Bernhauer, K., A. Iglauer, H. Knoblock, and O. Zippelius. 1940. Über die Säurebildung aus Zucker durch *Aspergillus niger*. VIII. Mitteilung: Der Einfluss von Magnesium auf die Säurebildung. Biochem. Z. 303:300–307.

Bertrand, D., and A. de Wolf. 1966. L'aluminium, oligo-element dynamique pour les vegetaux superieurs. Compt. Rend. Acad. Sci. (Paris) 262:479–481.

Bieleski, R. L. 1968. Effect of phosphorus deficiency on levels of phosphorus compounds in *Spirodela*. Plant Physiol. 43:1309–1316.

Bledsoe, R. W., C. L. Comar, and H. C. Harris. 1949. Absorption of radioactive calcium by the peanut fruit. Science 109:329–330.

Bond, G. 1967. Fixation of nitrogen by higher plants other than legumes. Annu. Rev. Plant Physiol. 18:107–126.

Bond, G., and E. J. Hewitt. 1962. Cobalt and the fixation of nitrogen by root nodules of *Alnus* and *Casuarina*. Nature 195:94–95.

Bonds, E., and J. C. O'Kelley. 1969. Effects of Ca and Sr on *Zea mays* seedling primary root growth. Amer. J. Bot. 56:271–274.

Bradfield, J. R. G. 1947. Plant carbonic anhydrase. Nature 159:467–468.

Brady, N. C. 1947. The effect of period of calcium supply and mobility of calcium in the plant on peanut fruit filling. Proc. Soil Sci. Soc. Amer. 12:336–341.

Brewbaker, J. L., and B. H. Kwack. 1963. The essential role of calcium ion in pollen germination and pollen tube growth. Amer. J. Bot. 50:859–865.

Brown, T. E., H. C. Eyster, and H. A. Tanner. 1958. Physiological effects of manganese deficiency, pp. 135–155. *In* C. A. Lamb, O. G. Bentley, and J. M. Beattie (eds.) Trace elements (Proc. Conf. Ohio Agr. Exp. Sta., Wooster, Ohio, Oct. 14–16, 1957). Academic, New York.

Brownell, P. F. 1965. Sodium as an essential micronutrient element for a higher plant (*Atriplex vesicaria*). Plant Physiol. 40:460–468.

Brownell, P. F., and J. G. Wood. 1957. Sodium as an essential micronutrient element for *Atriplex vesicaria* Heward. Nature 179:635–636.

Broyer, T. C., A. B. Carlton. C. M. Johnson, and P. R. Stout. 1954. Chlorine — A micronutrient element for higher plants. Plant Physiol. 29:526–532.

Broyer, T. C., D. C. Lee, and C. J. Asher. 1966. Selenium nutrition of green plants. Effects of selenite supply on growth and selenium content of alfalfa and subterranean clover. Plant Physiol. 41:1425–1428.

Burling, E., and W. T. Jackson. 1965. Changes in calcium levels in cell walls during elongation of oat coleoptile sections. Plant Physiol. 40:138–141.

Burström, H. 1963. Growth regulation by metals and chelates. Advance. Bot. Res. 1:73–100.

Burström, H. 1964. Calcium, water conditions, and growth of pea seedling stems. Physiol. Plantarum 17:207–219.

Bush, L. P. 1969. Influence of certain cations on activity of succinyl CoA synthetase from tobacco. Plant Physiol. 44:347–350.

Chapman, H. D. (ed.). 1966. Diagnostic criteria for plants and soils. Div. Agr. Sci., Univ. California, Berkeley. 793 pp.

Chenery, E. M. 1948. Aluminum in plants and its relation to plant pigments. Ann. Bot. 12:121–136.

Cheniae, G. M., and I. F. Martin. 1966. Studies on the function of manganese in photosynthesis. Brookhaven Symp. Biol. 19:406–417.

Cheniae, G. M., and I. F. Martin. 1969. Photoreactivation of manganese catalyst in photosynthetic oxygen evolution. Plant Physiol. 44:351–360.

Cleland, R. 1960. Effect of auxin upon loss of calcium from cell walls. Plant Physiol. 35:581–584.

Conrad, C. M. 1926. A biochemical study of the insoluble pectic substances in vegetables. Amer. J. Bot. 13:531–547.

Corden, M. C. 1965. Influence of calcium nutrition on *Fusarium* wilt of tomato and polygalacturonase activity. Phytopathology 55:222–224.

Cori, G. T., S. P. Colowick, and C. F. Cori. 1938. The enzymatic conversion of glucose-1-phosphoric ester to 6-ester in tissue extracts. J. Biol. Chem. 124:543–555.

Crosby, D. G., and A. J. Vlitos. 1962. Leaf abscission induced by the iodide ion. Plant Physiol. 37:358–363.

DeKock, P. C. 1958. The nutrient balance in plant leaves. Agr. Progress 33:88–95.

Dickman, S. R., and A. A. Cloutier. 1951. Factors affecting the activity of aconitase. J. Biol. Chem. 188:379–388.

Dixon, M. 1949. Multienzyme systems. Cambridge Univ. Press, Cambridge. 100 pp.

Dvořák, M., and V. Radotínská-Ledinská. 1970. The amylolytic activity of Ca-deficient pumpkin plants (*Cucurbita pepo* L.). Biol. Plantarum 12:117–124.

Elliott, W. H. 1953. Isolation of glutamine synthetase and glutamotransferase from green peas. J. Biol. Chem. 201:661–672.

Evans, H. J. 1963. Effect of potassium and other univalent cations on activity of pyruvate kinase in *Pisum sativum*. Plant Physiol. 38:397–402.

Evans, H. J., and M. Kleiwer. 1964. Vitamin B_{12} compounds in relation to the requirements of cobalt for higher plants and nitrogen-fixing organisms. Ann. N.Y. Acad. Sci. 112:735–755.

Evans, H. J., and G. J. Sorger. 1966. Role of mineral elements with emphasis on the univalent cations. Annu. Rev. Plant Physiol. 17:47–76.

Eyster, C. 1952. Necessity of boron for *Nostoc muscorum*. Nature 170:755.

Eyster, C., T. E. Brown, H. A. Tanner, and S. L. Hood. 1958. Manganese requirement with respect to growth, Hill reaction and photosynthesis. Plant Physiol. 33:235–241.

Ferrari, T. E., and J. E. Varner. 1969. Substrate induction of nitrate reductase in barley aleurone layers. Plant Physiol. 44:85–88.

Fischer, E. H., W. N. Sumerwell, J. Junge, and E. A. Stein. 1960. Calcium and the molecular structure of α-amylases. Proc. 4th Int. Congr. Biochem. (Vienna, 1958) 8:124–137.

Fischer, R. A. 1968. Stomatal opening: Role of potassium uptake by guard cells. Science 160:784–785.

Fischer, R. A. 1971. Role of potassium in stomatal opening in the leaf of *Vicia faba*. Plant Physiol. 47:555–558.

Fischer, R. A. and T. C. Hsiao. 1968. Stomatal opening in isolated epidermal strips of

Vicia faba. II. Responses to KCl concentration and the role of potassium absorption. Plant Physiol. 43:1953—1958.

Fisher, J., and T. K. Hodges. 1969. Monovalent ion stimulated adenosine triphosphatase from oat roots. Plant Physiol. 44:385—395.

Fogg, G. E., and M. Wolfe. 1954. The nitrogen metabolism of the blue-green algae (Myxophyceae), pp. 99—125. *In* B. A. Fry and J. L. Peel (eds.) Autotrophic microorganisms (4th Symp. Soc. Gen. Microbiol.). Cambridge Univ. Press, Cambridge.

Folley, S. J., and A. L. Greenbaum. 1948. Determination of the arginase activities of homogenates of liver and mammary gland: Effects of pH and substrate concentration and especially of activation by divalent metal ions. Biochem. J. 43:537—549.

Fujino, M. 1967. Role of adenosinetriphosphate and adenosinetriphosphatase in stomatal movement. Sci. Bull. Fac. Educ. Nagasaki Univ. 18:1—47.

Gamborg, O. L. 1970. The effects of amino acids and ammonium on the growth of plant cells in suspension culture. Plant Physiol. 45:372—375.

Gamborg, O. L., and J. P. Shyluk. 1970. The culture of plant cells with ammonium salts as the sole nitrogen source. Plant Physiol. 45:598—600.

Ganje, T. J. 1966. Selenium, pp. 394—404. *In* H. D. Chapman (ed.) Diagnostic criteria for plants and soils. Div. Agr. Sci., Univ. California, Berkeley.

Gauch, H. G. 1940. Responses of the bean plant to calcium deficiency. Plant Physiol. 15:1—21.

Gerhardt, B. 1966. Manganeffekte in photosynthetischen Reaktionen von *Anacystis.* Ber. Deut. Bot. Ges. 79:63—68.

Ginzburg, B. S. 1958a. Evidence for a protein gel structure cross-linked by metal cations in the intercellular cement of plant tissue. J. Exp. Bot. 12:85—107.

Ginzburg, B. S. 1958b. Evidence for a protein component in the middle lamella of plant tissue: A possible site for indolylacetic acid action. Nature 181:398—400.

Gollub, M., and B. Vennesland. 1947. Fixation of carbon dioxide by a plant oxalacetate carboxylase. J. Biol. Chem. 169:233—234.

Green, D. E., L. F. Leloir, and V. Nocito. 1945. Transaminases. J. Biol. Chem. 161: 559—582.

Greenberg, E., and J. Preiss. 1965. Biosynthesis of bacterial glycogen. II. Purification and properties of the adenosine diphosphoglucose-glycogen transflucosylase of *Arthrobacter* species NRRL B 1973. J. Biol. Chem. 240:2341—2348.

Gunsalus, I. C. 1954. Group transfer and acyl-generating functions of lipoic acid derivatives, pp. 599—604. In W. D. McElroy and B. Glass (eds.) Mechanism of enzyme action. Johns Hopkins Press, Baltimore.

Habermann, H. M., M. A. Handel, and P. McKellar. 1968. Kinetics of chloroplast-mediated photooxidation of diketogulonate. Photochem. Photobiol. 7:211—224.

Hart, J. W., and P. Filner. 1969. Regulation of sulfate uptake by amino acids in cultured tobacco cells. Plant Physiol. 44:1253—1259.

Hartt, C. E. 1969. Effect of potassium deficiency upon translocation of ^{14}C in attached blades and entire plants of sugarcane. Plant Physiol. 44:1461—1469.

Hellerman, L., and C. C. Stock. 1938. Activation of enzymes. V. The specificity of arginase and the nonenzymatic hydrolysis of guanidino compounds. Activating metal ions and liver arginase. J. Biol. Chem. 125:771—793.

Hewitt, E. J. 1951. The role of the mineral elements in plant nutrition. Annu. Rev. Plant Physiol. 2:25—52.

Hewitt, E. J. and M. M. R. K. Afridi. 1959. Adaptive synthesis of nitrate reductase in higher plants. Nature 183:57—58.

Hewitt, E. J., E. G. Fisher, and M. C. Candela. 1956. Factors affecting the activity of nitrate reductase in cauliflower plants. Annu. Rep. Long Ashton Res. Sta. 1955:202.

Hillman, W. A., and H. B. Posner. 1971. Ammonium ion and the flowering of *Lemna perpusilla*. Plant Physiol. 47:586–587.

Holm-Hansen, O., G. C. Gerloff, and F. Skoog. 1954. Cobalt as an essential element for blue-green algae. Physiol. Plantarum 7:665–675.

Homann, P. 1967. Studies on the manganese of the chloroplast. Plant Physiol. 42:997–1007.

Homann, P. 1969. Cation effects on the fluorescence of isolated chloroplasts. Plant Physiol. 44:932–936.

Hsiao, T. C., R. H. Hageman, and E. H. Tyner. 1968. Effects of potassium nutrition on ribonucleic acid and ribonuclease in *Zea mays* L. Plant Physiol. 43:1941–1946.

Hsiu, J., E. H. Fischer, and E. A. Stein. 1964. Alpha-amylases as calcium-metalloenzymes. II. Calcium and the catalytic activity. Biochemistry 3:61.

Humble, G. D., and T. C. Hsiao. 1969a. Specific requirement of potassium for light-activated opening of stomata in epidermal strips. Plant Physiol. 44:230–234.

Humble, G. D., and T. C. Hsiao. 1969b. Light-dependent influx and efflux of guard cell potassium during stomatal opening and closing. Plant Physiol. 44(Suppl.):no. 97.

Humble, G. D., and T. C. Hsiao. 1970. Light-dependent influx and efflux of potassium of guard cells during stomatal opening and closing. Plant Physiol. 46:483–487.

Hutchinson, E. 1945. Aluminum in soils, plants, and animals. Soil Sci. 60:29–40.

Hutner, S. H., and L. Provasoli. 1964. Nutrition of algae. Annu. Rev. Plant Physiol. 15:37–56.

Ichioka, P. S., and D. I. Arnon. 1955. Molybdenum in relation to nitrogen metabolism. II. Assimilation of ammonia and urea without molybdenum by *Scenedesmus*. Physiol. Plantarum 8:552–560.

Iswaran, V., and W. V. B. Sundara Rao. 1964. Role of cobalt in nitrogen fixation by *Azotobacter chroococcum*. Nature 203:549.

Jagendorf, A. T., and M. Smith. 1962. Uncoupling phosphorylation in spinach chloroplasts by absence of cations. Plant Physiol. 37:135–141.

Johri, B. M., and I. K. Vasil. 1961. Physiology of pollen. Bot. Rev. 27:325–381.

Joy, K. W., and R. H. Hageman. 1966. The purification and properties of nitrite reductase from higher plants and its dependence on ferredoxin. Biochem. J. 100:263–273.

Kachmar, J. F., and P. D. Boyer. 1953. Kinetic analysis of enzyme reactions. II. The potassium activation and calcium inhibition of pyruvic phosphoferase. J. Biol. Chem. 200:669–682.

Kalckar, H. M. 1944. Adenylpyrophosphatase and myokinase. J. Biol. Chem. 153:355–367.

Kaufman, S., and S. G. A. Alivasatos. 1955. Purification and properties of the phosphorylating enzyme from spinach. J. Biol. Chem. 216:141–152.

Keilin, D., and T. Mann. 1938. Polyphenol oxidase. Purification, nature and properties. Proc. Roy. Soc. (London) B125:187–204.

Keilin, D., and T. Mann. 1939a. Laccase, a blue copper-protein oxidase from the latex of *Rhus succedanea*. Nature 143:23–24.

Keilin, D., and T. Mann. 1939b. Carbonic anhydrase. Nature 144:442–443.

Kessler, E. 1955. On the role of manganese in the oxygen-evolving system of photosynthesis. Arch. Biochem. Biophys. 59:527–529.

Kilmer, V. J., S. E. Younts, and N. C. Brady (eds.). 1968. The role of potassium in agriculture. Amer. Soc. Agron., Madison, Wisc. 509 pp.

Kinsky, S. C. 1961. Induction and repression of nitrate reductase in *Neurospora crassa*. J. Bacteriol. 82:898–904.

Klein, S., and B. Ginzburg. 1960. Effects of ethylenedinitrilotetraacetic acid on plant cell walls. J. Biophys. Biochem. Cytol. 7:335–338.

Klotz, I. M. 1954. Thermodynamics and molecular properties of some metal-protein complexes, pp. 257–290. *In* W. D. McElroy, and H. B. Glass (eds.) The mechanism of enzyme action. Johns Hopkins Press, Baltimore.

Korkes, S., A. Campillo, I. C. Gunsalus, and S. Ochoa. 1951. Enzymatic synthesis of citric acid. IV. Pyruvate as acetyl donor. J. Biol. Chem. 193:721–735.

Korkes, S., A. Campillo, and S. Ochoa. 1952. Pyruvate oxidation system of heart muscle. J. Biol. Chem. 195:541–547.

Krishnan, P. S. 1949. Studies of apyrases. II. Some properties of potato apyrase. Arch. Biochem. 20:272–283.

Kubowitz, F. 1938. Spaltung und Resynthese der Polyphenoloxydase und des Hamocyanins. Biochem. Z. 299:32–57.

Lal, K. N., and M. S. Rao. 1954. Micro-element nutrition of plants. Banaras Hindu Univ. Press, Banaras, India. 246 pp.

Lichstein, H. C., I. C. Gunsalus, and W. W. Umbreit. 1945. Function of the vitamin B$_6$ group: Pyridoxal phosphate (codecarboxylase) in transamination. J. Biol. Chem. 161:311–320.

Liebig, G. F., Jr. 1966. Gallium, pp. 197–199. *In* H. D. Chapman (ed.) Diagnostic criteria for plants and soils. Div. Agri. Sci., Univ. California, Berkeley.

Liebig, G. F., Jr., A. P. Vanselow, and H. D. Chapman. 1942. Effects of aluminum on copper toxicity, as revealed by solution-culture and spectrographic studies of citrus. Soil Sci. 53:341–351.

Lloyd, F. E. 1908. The physiology of stomata. Carnegie Inst. Washington Pub. 82.

Loftfield, J. V. G. 1921. The behavior of stomata. Carnegie Inst. Washington Pub. 314.

Loughman, B. C. 1960. Uptake and utilization of phosphate associated with respiratory changes in potato tuber slices. Plant Physiol. 35:418–424.

Lowther, W. L., and J. F. Loneragan. 1968. Calcium and nodulation in subterranean clover (*Trifolium subterraneum* L.). Plant Physiol. 43:1362–1366.

McClurkin, I. T., and D. C. McClurkin. 1967. Cytochemical demonstration of a sodium-activated and a potassium-activated adenosine triphosphatase in loblolly pine seedling root tips. Plant Physiol. 42:1103–1110.

McElroy, W. D., and B. Glass (eds.). 1951. Phosphorus metabolism – A symposium on the role of phosphorus in the metabolism of plants and animals, Vol. I. Johns Hopkins Press, Baltimore. 762 pp.

McElroy, W. D., and B. Glass (eds.). 1952. Phosphorus metabolism – A symposium on the role of phosphorus in the metabolism of plants and animals, Vol. II. Johns Hopkins Press, Baltimore. 930 pp.

McKenna, J. M., and N. I. Bishop. 1967. Studies on the photooxidation of manganese by isolated chloroplasts. Biochim. Biophys. Acta 131:339–349.

Mans, R. J. 1967. Protein synthesis in higher plants. Annu. Rev. Plant Physiol. 18:127–146.

Marinos, N. G. 1962. Studies on submicroscopic aspects of mineral deficiencies. I. Calcium deficiency in the shoot apex of barley. Amer. J. Bot. 49:834–841.

Mascarenhas, J. P., and L. Machlis. 1962a. Chemotropic response of *Antirrhinum majus* pollen to calcium. Nature 196:292–293.

Mascarenhas, J. P., and L. Machlis. 1962b. The normonal control of the directional growth of pollen tubes. Vitamins Hormones 20:347–372.

Mascarenhas, J. P., and L. Machlis. 1964. Chemotropic response of the pollen of *Antirrhinum majus* to calcium. Plant Physiol. 39:70–77.

Mee, S. 1949. A study of carboxylase in soybean sprouts. Arch. Biochem. 22:139–148.

Meidner, H., and T. A. Mansfield. 1968. Physiology of stomata. McGraw, New York.

Melchior, N. C. 1954. Sodium and potassium complexes of adenosinetriphosphate: equilibrium studies. J. Biol. Chem. 208:615–627.

Miller, E. C. 1938. Plant physiology (with reference to the green plant), 2nd ed. McGraw-Hill, New York, 1201 pp.

Miller, G., and H. J. Evans. 1957. The influence of salts on pyruvate kinase from tissue of higher plants. Plant Physiol. 32:346–354.

Miller, R. J., S. W. Dumford, D. E. Koeppe, and J. B. Hanson. 1970. Divalent cation stimulation of substrate oxidation by corn mitochondria. Plant Physiol. 45:649–653.

Mortenson, L. E., R. C. Valentine, and J. E. Carnahan. 1962. An electron transport factor from *Clostridium pasteurianum*. Biochem. Biophys. Res. Commun. 7:448–452.

Morton, A. G. 1956. A study of nitrate reduction in mould fungi. J. Exp. Bot. 7:97–112.

Murata, T., and T. Akazawa. 1968. Enzymic mechanism of starch synthesis in sweet potato root. I. Requirement of potassium ions for starch synthetase. Arch. Biochem. Biophys. 126:873–879.

Nicholas, D. J. D. 1961. Minor element nutrients. Annu. Rev. Plant Physiol. 12:63–90.

Nicholas, D. J. D., and A. Nason. 1954a. Molybdenum and nitrate reductase. II. Molybdenum as a constituent of nitrate reductase. J. Biol. Chem. 207:353–360.

Nicholas, D. J. D., and A. Nason. 1954b. Mechanism of action of nitrate reductase from *Neurospora*. J. Biol. Chem. 211:183–197.

Nightingale, G. T. 1942. Potassium and phosphate nutrition of pineapple in relation to nitrate and carbohydrate reserves. Bot. Gaz. 104:191–223.

Nightingale, G. T., R. M. Addoms, W. R. Robbins, and L. G. Schermerhorn. 1931. Effects of calcium deficiency on nitrate absorption and on metabolism in tomato. Plant Physiol. 6:605–631.

Nitsos, R. E., and H. J. Evans. 1968. The effects of univalent cations on particulate starch synthetase. Abstr. Amer. Soc. Plant Physiol., Western Section, Program of Meetings, Utah State Univ., Logan, Utah. June 25–27.

Nitsos, R. E., and H. J. Evans. 1969. Effects of univalent cations on the activity of particulate starch synthetase. Plant Physiol. 44:1260–1266.

Olsen, C. 1948. Adsorptively bound potassium in beech leaf cells. Physiol. Plantarum 1:136–141.

Osterhout, W. J. V. 1911. The permeability of living cells to salts in pure and balanced solutions. Science 34:187–189.

Osterhout, W. J. V. 1936. Electrical phenomena in large plant cells. Physiol. Rev. 16:216–237.

Overstreet, R., and L. A. Dean. 1951. The availability of soil anions, pp. 79–105. *In* E. Truog (ed.) Mineral nutrition of plants. Univ. Wisconsin Press, Madison.

Pallaghy, C. K. 1970. The effect of Ca^{++} on the ion specificity of stomatal opening in epidermal strips of *Vicia faba*. Z. Pflanzenphysiol. 62:58–63.

Paulsen, G. M., and J. E. Harper. 1968. Evidence for a role of calcium in nitrate assimilation in wheat seedlings. Plant Physiol. 43:775–780.

Peterson, P. J., and G. W. Butler. 1962. The uptake and assimilation of selenite by higher plants. Australian J. Biol. Sci. 15:126.

Pickard, B. G. 1970. Comparison of calcium and lanthanon ions in the *Avena*-coleoptile growth test. Planta 91:314–320.

Pirson, A. 1937. Ernährungs- und Stoffwechselphysiologische Untersuchungen an *Fontinalis* und *Chlorella*. Z. Bot. 31:193–267.

Pobeguin, T. 1954. Contribution à l'étude des carbonates de calcium, précipitation du calcaire par les végétaux, comparison avec le monde animal. Ann. Sci. Nat. Bot. 15:29–109.

Powers, W. H., S. Lewis, and C. R. Dawson. 1944. The preparation and properties of highly purified ascorbic acid oxidase. J. Gen. Physiol. 27:167—180.

Pratt, P. F. 1966a. Chromium, pp. 136—141. *In* H. D. Chapman (ed.). Diagnostic criteria for plants and soils. Div. Agr. Sci., Univ. California, Berkeley.

Pratt, P. F. 1966b. Titanium, pp. 478—479. *In* H. D. Chapman (ed.) Diagnostic criteria for plants and soils. Div. Agr. Sci., Univ. California, Berkeley.

Pratt, P. F. 1966c. Zirconium, p. 500. *In* H. D. Chapman (ed.) Diagnostic criteria for plants and soils. Div. Agr. Sci., Univ. California, Berkeley.

Preiss, J., and E. Greenberg. 1967. Biosynthesis of starch in *Chlorella pyrenoidosa*. I. Purification and properties of the adenosine diphosphoglucose:α-1,4-glucan, α-4-glucosyl transferase from *Chlorella*. Arch. Biochem. Biophys. 118:702—708.

Queen, W. H., H. W. Fleming, and J. C. O'Kelley. 1963. Effects on *Zea mays* seedlings of a strontium replacement for calcium in nutrient media. Plant Physiol. 38:410—413.

Radmer, R. J., and L. Bogorad. 1967. (—)S-adenosyl-L-methionine-magnesium proto-porphyrin methyltransferase, an enzyme in the biosynthetic pathway of chlorophyll in *Zea mays*. Plant Physiol. 42:463—465.

Randall, P. J., and P. B. Vose. 1963. Effect of aluminum on uptake and translocation of phosphorus by perennial ryegrass. Plant Physiol. 38:403—409.

Reisenauer, H. M. 1960. Cobalt in nitrogen fixation by a legume. Nature 186:375—376.

Renwick, G. M., C. Giumarro, and S. M. Siegel. 1964. Hydrogen metabolism in higher plants. Plant Physiol. 39:303—306.

Richards, F. J. 1941. Physiological studies in plant nutrition. XI. The effect on growth of rubidium with low potassium supply, and modification of this effect by other nutrients. Part I. The effect on total dry weight. Ann. Bot. 5:263—296.

Richards, F. J. 1944. Physiological studies in plant nutrition. XI. The effect on growth of rubidium with low potassium supply, and modification of this effect by other nutrients. Part II. The effect on dry weight distribution, net assimilation rate, tillering, fertility, etc. Ann. Bot. 8:323—356.

Robertson, R. N. 1960. Ion transport and respiration. Biol. Rev. 35:231—264.

Rosen, W. G. 1968. Ultrastructure and physiology of pollen. Annu. Rev. Plant Physiol. 19:435—462.

Sawhney, B. L., and I. Zelitch. 1969. Direct determination of potassium ion accumulation in guard cells in relation to stomatal opening in light. Plant Physiol. 44:1350—1354.

Sayre, J. D. 1926. Physiology of the stomata of *Rumex patientia*. Ohio J. Sci. 26: 233—266.

Scarth, G. W. 1932. Mechanism of the action of light and other factors on stomatal movement. Plant Physiol. 7:481—504.

Schales, O., V. Mims, and S. S. Schales. 1946. Glutamic acid decarboxylase of higher plants. I. Distribution; preparation of clear solutions; nature of prosthetic group. Arch. Biochem. 10:455—465.

Scott, G. T. 1944. Cation exchanges in *Chlorella pyrenoidosa*. J. Cellular Comp. Physiol. 23:47—58.

Setterfield, G., and S. T. Bayley. 1961. Structure and physiology of cell walls. Annu. Rev. Plant Physiol. 12:35—62.

Shrift, A. 1961. Biochemical interrelations between selenium and sulfur in plants and microorganisms. Fed. Proc. 20:695—702.

Shrift, A., and T. K. Virupaksha. 1963. Biosynthesis of Se-methylselenocysteine from selenite in selenium-accumulating plants. Biochim. Biophys. Acta 71:483.

Shrift, A., J. Nevyas, and S. Turndorf. 1961a. Mass adaptation to selenomethionine in populations of *Chlorella vulgaris*. Plant Physiol. 36:502—509.

Shrift, A., J. Nevyas, and S. Turndorf. 1961b. Stability and reversibility of adaptation to selenomethionine in *Chlorella vulgaris*. Plant Physiol. 36:509—519.

Small, J., M. I. Clarke, and J. Crosbie-Baird. 1942. pH phenomena in relation to stomatal opening. II—V. Proc. Roy. Soc. (Edinburgh) B61:233—266.

Smith, E. L., N. C. Davis, E. Adams, and D. H. Spackman. 1954. The specificity and mode of action of two metal-peptidases, pp. 291—318. *In* W. D. McElroy and H. B. Glass (eds.) The mechanism of enzyme action. Johns Hopkins Press, Baltimore.

Spencer, D., and J. V. Possingham. 1961. The effect of manganese deficiency on photophosphorylation and the oxygen-evolving sequence in spinach chloroplasts. Biochim. Biophys. Acta 52:379—381.

Steffensen, D. M. 1961. Chromosome structure with special reference to the role of metal ions. Int. Rev. Cytol. 12:163—197.

Steveninck, R. F. M. van. 1965. The significance of calcium on the apparent permeability of cell membranes and the effects of substitution with other divalent ions. Physiol. Plantarum 18:54—69.

Steward, F. C. 1964. Plants at work. Addison-Wesley, Reading, Mass.

Stiles, W. 1961. Trace elements in plants, 3rd ed. Cambridge Univ. Press, Cambridge. 249 pp.

Stout, P. R., R. Overstreet, L. Jacobson, and A. Ulrich. 1947. The use of radioactive tracers in plant nutrition studies. Proc. Soil Sci. Soc. Amer. 12:91—97.

Street, H. E., and J. S. Lowe. 1950. The carbohydrate nutrition of tomato roots. II. The mechanism of sucrose absorption by excised roots. Ann. Bot. 14:307—329.

Stumpf, P. K. 1945. Pyruvic oxidase of *Proteus vulgaris*. J. Biol. Chem. 159:529—544.

Tagawa, K., and D. I. Arnon. 1962. Ferredoxins as electron carriers in photosynthesis and in the biological production and consumption of hydrogen gas. Nature 195: 537—543.

Thomas, D. A. 1970a. The regulation of stomatal aperture in tobacco leaf epidermal strips. I. The effect of ions. Australian J. Biol. Sci. 23:961—979.

Thomas, D. A. 1970b. The regulation of stomatal aperture in tobacco leaf epidermal strips. II. The effect of ouabain. Australian J. Biol. Sci. 23:981—989.

Thompson, J. F. 1967. Sulfur metabolism in plants. Annu. Rev. Plant Physiol. 18: 59—84.

Trelease, S. F., A. A. Di Somma, and A. L. Jacobs. 1960. Seleno-amino acids found in *Astragalus bisulcatus*. Science 132:618.

Uhrström, I. 1969. The time effect of auxin and calcium on growth and elastic modulus in hypocotyls. Physiol. Plantarum 22:271—287.

Ulrich, A. 1941. Metabolism of non-volatile organic acids in excised barley roots as related to cation-anion balance during salt accumulation. Amer. J. Bot. 28:526—537.

Ulrich, A., and K. Ohki. 1966. Potassium, pp. 362—393. *In* H. D. Chapman (ed.) Diagnostic criteria for plants and soils. Div. Agr. Sci., Univ. California, Berkeley.

Utter, M. F., and C. H. Werkman. 1942. Effect of metal ions on the reactions of phosphopyruvate by *Escherichia coli*. J. Biol. Chem. 146:289—300.

Vanselow, A. P. 1966a. Barium, pp. 24—32. *In* H. D. Chapman (ed.) Diagnostic criteria for plants and soils. Div. Agr. Sci., Univ. California, Berkeley.

Vanselow, A. P. 1966b. Silver, pp. 405—408. *In* H. D. Chapman (ed.) Diagnostic criteria for plants and soils. Div. Agr. Sci., Univ. California, Berkeley.

Vanselow, A. P. 1966c. Strontium, pp. 433—443. *In* H. D. Chapman (ed.) Diagnostic criteria for plants and soils. Div. Agr. Sci., Univ. California, Berkeley.

Vennesland, B. and R. Z. Felsher. 1946. Oxalacetic and pyruvic carboxylases in some dicotyledonous plants. Arch. Biochem. 11:279—306.

Virupaksha, T. K., and A. Shrift. 1963. Biosynthesis of selenocystathionine from selenate in *Stanleya pinnata*. Biochim. Biophys. Acta 74:791.

Waisel, Y., G. A. Borger, and T. T. Kozlowski. 1969. Effects of phenylmercuric acetate on stomatal movement and transpiration of excised *Betula papyrifera* Marsh. leaves. Plant Physiol. 44:685–690.

Waisel, Y., A. Hoffen, and A. Eshel. 1970. The localization of aluminum in the cortex cells of bean and barley roots by x-ray microanalysis. Physiol. Plantarum 23:75–79.

Wakhloo, J. L. 1970. Role of mineral nutrients and growth regulators in the apical dominance in *Solanum sisymbrifolium*. Planta 91:190–194.

Wallace, A., and R. T. Mueller. 1962. Effects of magnesium concentrations on *in vitro* CO_2 fixing reaction. Plant Physiol. 37:154–157.

Wallace, A., E. M. Romney, and R. T. Mueller. 1969. Effect of the phosphorus level on the micronutrient content of *Franseria dumosa*. Phyton 26:151–154.

Wallihan, E. F. 1966. Tin, pp. 476–477. *In* H. D. Chapman (ed.) Diagnostic criteria for plants and soils. Div. Agr. Sci., Univ. California, Berkeley.

Watson, J. D. 1965. Molecular biology of the gene. Benjamin, New York. 494 pp.

Willmer, C. M., and T. A. Mansfield, 1969. A critical examination of the use of detached epidermis in studies of stomatal physiology. New Phytol. 68:363.

Woolley, J. T. 1957. Sodium and silicon as nutrients for the tomato plant. Plant Physiol. 32:317–321.

Yin, H. C., and Y. T. Tung. 1948. Phosphorylase in guard cells. Science 108:87–88.

Zaitlin, M. 1959. Isolation of tobacco leaf cells capable of supporting virus multiplication. Nature 184:1002–1003.

Zaitlin, M., and D. Coltrin. 1964. Use of pectic enzymes in a study of the nature of intercellular cement of tobacco leaf cells. Plant Physiol. 39:91–95.

Zelitch, I. 1965. Environmental and biochemical control of stomatal movement in leaves. Biol. Rev. 40:463–482.

Zelitch, I. 1969. Stomatal control. Annu. Rev. Plant Physiol. 20:329–348.

CHAPTER 10

Roles of Micronutrients
in Higher Plants

The distinction between macro- and micronutrients for higher plants is an arbitrary one—at least for certain elements. An element required in large amounts by one higher plant may be required only in small amounts by another. However, the elements discussed in this chapter are generally required in much smaller amounts than those discussed in the preceding chapter.

ROLES

Iron

It has been established that Fe is a constituent of peroxidase, catalase, and cytochrome oxidase. Significant roles of the first two enzymes in plant nutrition have not been established, but cytochrome oxidase (Fritz and Beevers, 1955) is involved in the terminal step of oxidation in the Krebs cycle. Fe also appears to be essential for chlorophyll synthesis, but it is not known whether it is directly or indirectly concerned—possibly as a constituent of or activator of some enzyme system.

Fe is a component of ferredoxin—an electron-transferring protein (Burris, 1966) discovered by Mortenson et al. (1962) and associated with chloroplasts (Smillie, 1963). It functions in photosynthesis (Fig. 10.1), NO_3^- and NO_2^- reductions (Losada et al., 1965; Betts and Hewitt, 1966; Joy and Hageman, 1966), and in N fixation. It has a lower oxidation–reduction potential than any

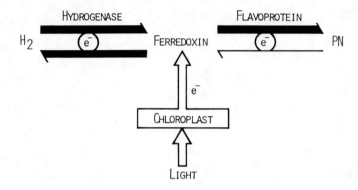

Fig. 10.1
Diagrammatic representation of the role of ferredoxin in photochemical pyridine nucleotide reduction and in the biological production and consumption of H_2 gas. The flavoprotein is normally bound in chloroplasts. (Tagawa and Arnon, 1962.)

other carrier that has been isolated. The potential of ferredoxin, about −420 mV, is approximately that of the hydrogen electrode (Burris, 1966).

For about 10 yr prior to the discovery of ferredoxin, proteins had been isolated from chloroplasts and assigned various functions and names—methemoglobin-reducing factor (Davenport et al., 1952), TPN-reducing factor (Arnon et al., 1957), photosynthetic pyridine nucleotide reductase (PPNR) (San Pietro and Lang, 1958), heme-reducing factor (Davenport and Hill, 1960), and the "red" enzyme (with 1% Fe) (Gewitz and Volker, 1962). All of these terms are synonymous (Arnon, 1965) and have been replaced by ferredoxin.

According to Arnon (1965), "the family of ferredoxins would include those non-heme, non-flavin proteins that transfer to appropriate enzyme systems some of the most 'reducing' electrons in cellular metabolism—electrons released by the photochemical apparatus of photosynthesis or by the H_2–hydrogenase system." In order to distinguish ferredoxins from heme and nonheme proteins (including flavoproteins) with more electropositive oxidation–reduction potentials, Arnon (1965) favored retaining the tentative definition of ferredoxins as "iron proteins which function as electron carriers on the 'hydrogen side' of pyridine nucleotides."

Ferrous Fe is required for the action of aconitase (Dickman and Cloutier, 1951). When Fe is deficient, most of it is combined with phosphoproteins (Liebich, 1941), and citric acid accumulates (Iljin, 1951), presumably because aconitase activity is low (DeKock, 1958). In a cell with adequate Fe, there is some unbound Fe in the ferrous state (DeKock, 1958). Fe deficiency markedly reduces nitrate reductase activity and hemoglobin concentration in soybean nodules (Cheniae and Evans, 1960).

Plant ferritin (phytoferritin) is a protein–Fe complex located within plastids and chloroplasts in regions not occupied by the thylakoid system (Seckbach, 1969). There are no reports on the nature of the Fe core of phytoferritin. The chief function of ferritin is believed to be one of Fe storage (Hyde et al., 1963; Catesson, 1966; Marinos, 1967).

A positive correlation has been reported between Fe concentration of the medium and concentrations of chlorophyll and heme in leaves (Evans, 1959; DeKock et al., 1960). A similar relationship was observed in cowpea leaves by Marsh et al. (1963a). They reported that in Fe deficiency the concentration of heme enzymes, such as catalase, was reduced; activity of photosynthetic pyridine nucleotide reductase in leaves was strikingly decreased. In cowpeas *(Vigna sinensis* L.), Fe deficiency reduced the synthesis of δ-aminolevulinic acid—the precursor of porphyrins (Bogorad, 1960)—and this reduction was associated with a retardation of chlorophyll synthesis (Marsh et al., 1963b). Fe therefore appears to be involved in synthesis of δ-aminolevulinic acid (Marsh et al., 1963a).

Fe played the key role and acted as a "trigger" of the action of coconut milk in stimulating cell division of cultured explants of *Daucus carota* (Neumann and Steward, 1968). Neither Mo nor Mn^{2+} could replace Fe in this respect. Fe emerged as the key trace element that interacted with factors present in coconut milk to induce growth in the otherwise quiescent carrot tissue. Steward et al. (1968) stated that an outstanding effect of Fe was its determination of the level of protein synthesized in cultured explants of *D. carota*. Fe promoted the use of ^{14}C from fructose and directed it into protein; neither Mo nor Mn^{2+} achieved this effect either separately or in combination.

Price (1968) reviewed the numerous effects of Fe deficiencies on lower and higher plants, including reported changes in activities of various enzymes. He also discussed the roles and functioning of natural and synthetic chelates with respect to Fe.

Sodium

Working with a halophyte, Brownell and Wood (1957) demonstrated a Na requirement for bladder saltbush *(Atriplex vesicaria)*. When grown in the absence of Na^+, yields were approximately 10% of those of the control, and Na-deficient plants eventually died. One-tenth meq Na_2SO_4/liter gave significantly better growth than 0.02; there was no difference in growth between 0.1 and 0.6 meq/liter. Leaves of plants receiving no (added) Na^+ contained 350 ppm Na^+, whereas those from plants receiving 0.1 meq Na_2SO_4/liter in the culture solution contained 9100 ppm of Na^+. Later, Brownell (1965) reported that Li^+, K^+, or Rb^+ could not substitute for Na^+ in *A. vesicaria* Heward. In the absence of Na^+, some of the plants died. Upon the addition of Na^+, some of the plants died. Upon the addition of Na^+ to Na-deficient *Atriplex nummu-laria,* the rate of anaerobic CO_2 production was increased, suggesting that the effect of Na^+ was in the glycolytic sequence (Brownell and Jackman, 1966). No other univalent cation was able to substitute for Na^+ in this regard.

The effect of Rb^+ on growth and development of sugar beet plants depended on Rb^+ concentration, K^+ supply, and relative abundance of Na^+ (El-Sheikh and Ulrich, 1970). Rb^+ added either to a low- or high-K^+ solution, with or without added Na^+, increased leaf blade size greatly—possibly through an effect on phytohormones or through a "partitioning effect" on the distribution of carbohydrates—top growth being favored over storage roots. Na^+ increased the growth of the plants whether K^+ supply was low or high. In addition, Na^+ or Rb^+ added to a high-K^+ solution increased the sucrose percentage and total sucrose of the storage roots. It was suggested that sugar beet plants may possibly require Rb^+ and Na^+ per se for growth and development, but that in such small amounts needs are generally satisfied by traces of Na^+ or Rb^+ present as contaminants in nutrient salts.

Halogeton glomeratus (M. Bieb.) C. A. Mey is a poisonous annual weed on approximately 10,500,000 acres of western range (Williams, 1960). The poison is oxalic acid, and leaves usually contain in excess of 30% soluble oxalates (dry weight basis) and 3 or 4% insoluble oxalates (presumably of Ca^{2+} and/or Mg^{2+}). Na^+ was essential for vigorous growth; 0.1 N NaCl in the nutrient solution was particularly effective in promoting growth, vigor, size, and oxalate formation. Na^+ was the chief cation associated with oxalate (i.e., soluble oxalate). This species also apparently required 5–10 ppm Cl^- in the nutrient solution to prevent Cl^--deficiency symptoms.

Na^+ may partially substitute for K^+ in Italian ryegrass (Hylton et al., 1967) and other higher plants (Harmer and Benne, 1945).

A concentration of 1.0 mM NaCl in the nutrient solution effected a 12% increase in dry weight of tomato plants, as reported by Woolley (1957). He concluded, however, that larger effects would have to be observed before Na^+ could be considered an essential element. Similarly, El-Sheikh et al. (1967) stated that essentiality of Na^+ and/or Rb^+ for growth of sugar beets might be inferred, but that other criteria would have to be fulfilled for conclusive proof.

In a review of the literature, Lunt (1966) reported that in various species of plants Na^+ may be required for certain enzymes, activate certain enzyme systems, be required for photosynthesis *(Synechoccus cedorum),* or increase CO_2 assimilation. He added that it is perhaps not surprising that the role of Na^+ is still obscure, since even the function(s) of K^+ is not well understood (Lunt, 1966; Ulrich and Ohki, 1966).

Chlorine

The earliest indication of the essentiality of Cl^- for plants came from research by Lipman (1938) for various plants, and by Eaton (1942), which demonstrated significant increases in growth of tomatoes and cotton with low concentration of Cl^-. Raleigh (1948) obtained significant increases in the growth of table beets by the addition of Cl^-. For sugar beets, Cl^- was necessary for top and root growth, and for sugar formation rather than for utilization of sugar (Ulrich and Ohki, 1956).

Although a possible role was not suggested, Stout et al. (1956) demonstrated that Cl^- is an essential micronutrient for tomatoes, lettuce, and cabbage. Without Cl^- leaves develop chlorosis, followed by necrosis, and plants do not set fruit. Other researchers have also demonstrated the essentiality of Cl^- for higher plants (Broyer et al., 1954; Johnson et al., 1957; Martin and Lavollay, 1958).

Low internal concentrations of Cl^- markedly increased the percentage of water in tobacco leaves throughout the season and tended to protect leaves against "drought spot" (Garner et al., 1930). Slightly higher concentrations of Cl^- very markedly reduced combustibility of the leaf and thus lowered the quality.

Haas (1945) reviewed the literature on effects of Cl^- on plants and provided additional evidence of the effect of Cl^- on increasing the hydration of plant tissues.

Warburg and Lüttgens (1946) concluded that Cl^- was a coenzyme essential for photochemical reactions in photosynthesis. Arnon and Whatley (1949), however, postulated that Cl^- or Br^- prevent a rapid light-induced deterioration of some substance essential for O_2 evolution. This preventive action was believed to be required for and to occur only in isolated chloroplasts—not in intact cells.

Boron

B has been regarded as an essential element for higher plants since around 1910 (Agulhon, 1910a, b). Essentiality was also demonstrated early by Warington (1923) and by Sommer and Lipman (1926).

Before discussing "complexing," it is pertinent to recall the types of complexes that B may form with polyhydroxyl compounds (Isbell et al., 1948):

Type A

Type BD

$$\left[\begin{array}{c} -\overset{|}{C}-O \\ -\overset{|}{C}-O \end{array} \!\! \overset{}{\searrow}\!\! B \!\! \overset{O-\overset{|}{C}-}{\underset{O-\overset{|}{C}-}{\diagup}} \right] H^+$$

Type BD_2

Type BD_2 is the one that would be involved in an interaction of B with chains of cellulose or long-chain compounds of other polyhydroxyl molecules. For example, a given concentration of agar results in a much firmer gel when traces of B are added, presumably as a result of cross-binding between the chains.

Although compounds containing B have not been isolated from plants, B (as borate) would be expected to combine with certain OH-rich compounds, such as sugars, to form complexes. For example, mannitol reacts with boric acid so that boric acid then responds as a strong acid and may therefore be readily titrated. This is the basis for one of the early methods of determining B. There is no reason to question that borate and sugars react in plants in the same way that they do *in vitro*. The question is rather one of interpreting the location of and the significance of borate–carbohydrate complexing.

Owing to the negative charge of sugar–borate complexes, it is conceivable that they might pass through negatively charged cell surfaces or membranes more readily than neutral sugar molecules. However, owing to a possible reaction of borate with membrane constituents, it is conceivable that borate loci or borate reaction centers on membranes might facilitate the passage of sugar through membranes. In this connection, Brown (1952) presented evidence he interpreted as indicating that sugars pass over specific reaction centers or sites in their movement through membranes.

In order to review the literature on B, particularly the more recent literature, it is pertinent to review briefly a report on the role of B in plants which appeared in 1953.

Gauch and Dugger (1953) obtained evidence interpreted as indicating a role of B in the translocation of sugars in plants. It was suggested that B might perform this role either (1) through sugar–borate complexes which traversed membranes more readily than nonborated sugar molecules or (2) through the association of B with membranes which, in that location, facilitated the passage of sugar. The second suggestion was favored for the mode of action of B. A year later, Gauch and Dugger (1954) listed approximately 15 roles postulated for B and attempted to explain these various suggested roles in the light of their theory, that is, that B is concerned with sugar translocation.

Beginning in 1956 an alternative theory for the role of B in plants was advanced. In essence, this theory is based on the fact that a deficiency of B causes a reduction in cellular activities (division, differentiation, maturation, respiration, growth and so on) in the growing points of tops and roots—the very regions were B-deficiency symptoms first appear (Skok, 1956, 1957a, 1958). Inasmuch as sugars are not being utilized in these regions of B-deficient plants, as compared with B-sufficient plants, a gradient in favor of sugar translocation to the growing points no longer exists, and sugar translocation is reduced or essentially nil. Conversely, when a plant receives adequate B, the growing points are active, there is a steep gradient in the concentration of sugar from leaves to growing points, and translocation is favored. The growing points of B-sufficient plants may be regarded as physiological "sinks"—with regard to sugar.

This second concept is clearly as logical an explanation as the first one, and is reasonable from a physiological point of view. The burden of proof rests on those who champion the sugar–borate concept.

It is not easy to design experiments that might serve to support or to refute one or the other of these hypotheses. Certainly much of the research of the past 15 years can be "explained" by invoking either hypothesis. In the future the most critical need will be for experiments designed in such a way that they clearly support or refute one of the two opposing hypotheses.

Research has been reported which appears to circumvent the physiological sinks. Turnowska-Starck (1960a) immersed the cut ends of triofoliolate leaves of bean plants in a solution of sugar or in a solution of sugar plus B. The presence of B increased the absorption and translocation of sugar. The leaves were not B-deficient, and there were no active growing points in these mature leaves, thus sinks were not involved.

Mitchell et al. (1953a, b) presented evidence in favor of a role of B in sugar translocation, which appears not to be explainable on the basis of physiological sinks in B-sufficient plants. As background, it should be recalled that various researchers demonstrated that growth regulators do not move from carbohydrate-depleted leaves unless sugar is also supplied along with a given growth regulator. Mitchell et al. (1953a, b) reasoned that, if sugar were required for the translocation of growth regulators and B-facilitated sugar translocation, the addition of B to sugar-plus-growth regulator mixtures should enhance translocation of the growth regulator (Figs. 10.2 and 10.3). Inasmuch as the bean plants were grown in a fertile soil and the experiments were generally concluded within 2–4 hr after initiation, "physiological sinks" would not be expected to develop for plants receiving B in the mixture and not in those without B. The addition of only B with the growth regulator increased the movement of growth regulators as compared with that occurring in the absence of added B. Presumably B as an additive increased the movement of some native sugar and, in addition, increased the translocation of growth regulator. Perhaps sugar translocation can be increased, even in B-sufficient plants, by additional B. Eaton's (1944) report that maximal growth of plants and the first signs of B toxicity may overlap could be pertinent. In an experiment by McIlrath and Palser (1956) on the role of B in plants, it was noted that plus-B tomato plants

Fig. 10.2
Stem curvatures of bean plants resulting from application of 0.9 μg of 2,4–D(NH$_4$), 2000 μg of sucrose, and 160 μg of B to the right leaf of each plant (right), compared with that induced by an equal amount of 2,4–D(NH$_4$) and sugar (left). Photographed 3.5 hr after treatment. From J. W. Mitchell et al. (1953a): Science, 118:354–355. American Association for the Advancement of Science—by permission. Photo—U.S. Dept. of Agriculture.

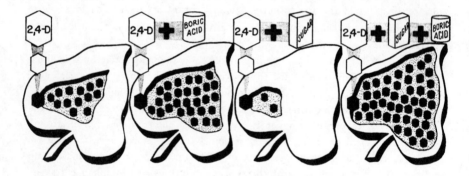

Fig. 10.3
Diagrammatic representation of the effect of B on the translocation of sugar and in turn of 2,4–D from bean leaves. (Mitchell et al., 1953b.)

showed moderate B-toxicity symptoms, and yet the plants were apparently otherwise very normal in appearance. With a limiting concentration of growth regulator applied on a primary leaf, Mitchell et al. (1953a, b) noted that addition of sugar actually decreased the translocation of growth regulator to the hypocotyl, presumably by interfering with the uptake of growth regulator by the leaf. However, with the same concentration of growth regulator and of sugar, the addition of B resulted in the greatest translocation of growth regulator. This was interpreted as an indication that B had more than offset the competition between growth regulator and sugar for entry into the leaf. Presumably, B enhanced the entry of sugar into the leaf and in turn the extra, absorbed sugar increased the translocation of growth regulator to the hypocotyl. This effect of B was shown with various growth regulators and various sugars. In fact, the degree of effect of added B was generally related to its known degree of reaction in complexing with the various sugars. This latter point seems to highlight the significance of complexing in the role that B plays in plants.

There is considerable evidence and support for the concept (Skok, 1958) that B is required for active growing points which constitute sinks for sugar, and that the alleged effect of B on sugar translocation is strictly an indirect effect. In a study of x-irradiated plus-B and minus-B sunflower plants, x-ray damage was positively correlated with the degree of cellular and metabolic activity of plants at the time they were irradiated (Skok, 1956, 1957a). B deficiency reduces cellular activity, and B-deficient plants came through the radiation with less damage than B-sufficient ones. It was concluded that the results emphasize the relationship between B and cellular activity. Removal of the terminal bud of a sunflower plant reduced sugar translocation toward the tip more than a deficiency of B in similar plants did (Skok, 1957a). Whittington (1957) also concluded that a greater utilization of sugar occurs in the presence of B than in its absence, and that utilization leads to a higher rate of translocation. In an interesting approach Dyar and Webb (1961) showed that naphthalene acetic acid applied to meristems of B-deficient bean plants increased translocation of ^{14}C-labeled photosynthates to the tips. They added that B-deficient plants have a functional conducting system, but that it does not conduct because substrates are not being utilized. Accordingly, B-deficient plants would not be limited in growth by a sugar deficiency as claimed by Gauch and Dugger (1953, 1954); B would not be necessary per se for sugar translocation. They concluded that B is essential for auxin metabolism, possibly synthesis. [It should be added that although B deficiency apparently had not progressed enough to result in necrosis and destruction of the phloem in the plants used by Dyar and Webb (1961), *prolonged* B deficiency clearly results in destruction of the phloem (Palser and McIlrath, 1956).] Without doubt numerous studies on the role of B in plants have involved plants well beyond incipient B deficiency which have had necrotic and nonfunctional phloem.

In a review of the biochemistry and fine structure of phloem in relation to transport, Eschrich (1970) concluded that the literature he examined gave no indication that sinks are actively involved in the process of long-distance transport. More recently, others have disagreed with that conclusion. Translocation rate, ATP concentration, and CO_2 production of a developing leaf (sink leaf) were studied in sugar beet plants prior to and during anaerobic treatment of the sink leaf (Geiger and Christy, 1971). Within 3–5 min after onset of treatment with N_2, translocation into the sink leaf decreased to near zero and then recovered to a level of approximately 50% of the control over the next 2 hr. Decline in CO_2 output and ATP levels coincided with the attainment of the new translocation rate. It was concluded that the correlation between ATP concentration and translocation rate indicated that ATP-dependent active transport in the sink leaf augments the driving force for translocation.

Edwards (1971) studied translocation in healthy and powdery mildew-infected (*Erysiphe graminis* f. sp. *hordei,* race CR3) barley. The sinklike properties of the powdery mildew infection were used to determine what effect imposing a sink in the midst of normal source tissue (mature primary leaf) would have on the translocation process. In the healthy primary leaf of barley, ^{14}C fixed in the tip section of the blade was preferentially translocated to the root, whereas ^{14}C fixed in the basal section was primarily translocated to the shoot. The mildew infection did not alter this basic pattern.

When $^{14}CO_2$ was given to leaves of B-deficient and B-sufficient Hawkeye soybeans, the B-deficient plants contained ^{14}C concentrations in the stem exudate and root sap, and citrate in the root sap, that were as high as or higher than those in B-sufficient plants (Brown and Ambler, 1969). Therefore, in soybeans B deficiency did not interfere with transport of photosynthate or precursors required for citrate synthesis.

Skok (1957b) was unable to overcome B deficiency by application of a solution of sugar directly to plant tips. Weiser et al. (1964) found that B enhanced the uptake of sucrose applied to leaves and concluded that this phenomenon of increased uptake had given rise to the erroneous conclusion by other workers that B enhances translocation of sugar. In a study with tomato, turnip, and cotton, McIlrath and Palser (1956) suggested that accumulation of carbohydrates in B-deficient leaves, as compared with B-sufficient leaves, could be the result of destruction of phloem. In a companion study (Palser and McIlrath, 1956), it was shown that necrosis of the phloem occurred in tomato and turnip, but not in cotton. Sugar–B sprays on B-deficient tomato and turnip plants, but not cotton, resulted in highly significant increases in growth as reported by McIlrath and Palser (1956). They noted that the growth of tomato and turnip plants sprayed with a sugar–B solution was very similar to that of plants grown on nutrient solution containing B. The plus-B series of plants received 5 ppm B, and tomato plants showed moderate B toxicity (i.e., burning around the margins of the leaves). These investigators did not indicate why they used 5.0 ppm B instead of the more customary 0.5 ppm B for the plus-B series, but the concentration used could possibly be justified based on Eaton's (1944) observation that maximal growth of plants and incipient B toxicity may overlap. McIlrath and Palser (1956) concluded that "although the data presented do not particularly support the hypothesis that one of the primary roles of boron is to facilitate sugar translocation, neither do they completely refute it."

In studies on peanut leaves *(Arachis hypogaea* L.) from plants grown in two levels of B, B-deficient leaves had higher concentrations of amino acids and N but the same concentration of protein as B-sufficient leaves (Shiralipour et al., 1969). Results were expressed on a per cell (milligrams of DNA) basis. It was concluded that the increase in concentration of amino acids could be partially or completely explained by the higher concentration of N in B-deficient plants. The possibiliy that a deficiency of carbohydrates in B-deficient leaves might have led to the accumulation of N and amino acids was not considered. Inasmuch as B may alter the amount of DNA per cell and cell counts were not made, it is possible that milligrams of DNA may not have constituted a per cell basis.

Inasmuch as other complex-forming elements, for example, Sr^{2+}, Al^{3+}, and Ge^{6+}, could partially substitute for B in sunflower plants, Skok (1957b) concluded that the physiological role of B very likely involved its ability to form complexes with OH-rich compounds. Thus he has clearly and consistently supported the complexing of B with OH-rich compounds but has with good reasons placed a different interpretation on how and why this complexing is significant in plants. Later, McIlrath and Skok (1966) reported that Ge^{6+} delayed B-deficiency symptoms of sunflower plants. They concluded that Ge^{6+} had not substituted for B in metabolic functions, but that it had increased the mobility of B and, by binding monmetabolic polyhydroxyl sites, played a sparing role for the limited quantity of available, soluble B.

In regard to the Gauch and Duggar (1953, 1954) hypothesis, certain researchers supported this concept, and the work of others could possibly be interpreted as supporting it. For bean plants grown at a relatively low level of B, additional B increased the translocation of [^{14}C] sucrose applied to leaves as observed by Turnowska-Starck (1960b). She reported that the radioactivity of plants on a low level of B was several times lower than that of plants with a high concentration of this element, although the former plants manifested no external symptoms of B deficiency. Phloem tissues showed no necrosis in the B-deficient plants; both types of plants had the same concentration of soluble sugars and the same degree of hydration of tissues. It was concluded that the "inhibition of ^{14}C migration in plants with a low B concentration could have been caused neither by anatomical nor morphological changes of the phloem." In a study with maize, Shkolnik and Abdurashitov (1958) observed that B increased the translocation of sucrose and monsaccharides. They stated that, "the assumption is made and substantiated that the cause of the positive influence of boron on translocation of carbohydrates can be found in its ability to form complex compounds with carbohydrates which are translocated with greater ease." Garman et al. (1953) obtained a positive correlation between the concentration of B and of sucrose in apple fruits; high B was associated with high sucrose concentrations. The main effect of B, in the nutrient solution, was on increasing the percent of soluble solids in cantaloupes (Stark and Haut, 1958).

Working with isolated cereal roots, Almestrand (1957) reported that 0.0035 and 5.0 ppm B had the same effects and that neither concentration increased sugar uptake or respiration of excised roots. Turnowska-Starck (1959) reported that B added to sucrose solutions did not increase the absorption or desorption of sucrose from onion epidermis. It should be noted, however, that translocation per se was not being studied in this case. Three reports can be reconciled by either the sink or the sugar–borate complexing concepts. Sisler et al. (1956) reported increased translocation of sugars in B-sufficient plants as compared with B-deficient plants. Following exposure of single, intact sunflower leaves to $^{14}CO_2$, the velocity of translocation of ^{14}C was greater in B-sufficient than in B-deficient plants (Lee et al., 1966). Nelson and Gorham (1957) found that 5 ppm B enhanced the uptake of glucose applied to primary soybean leaves (in the presence of a detergent), but they did not regard this phenomenon as a function of the complexing ability of B since the process was relatively insensitive to the pH range 4.8–8.4.

The remaining portion of the section on B covers research that has appeared since the publication of a review article on B (Gauch and Dugger, 1954), or was overlooked at the time the review article was published.

In contrast with several studies reported earlier by Gauch and Dugger (1954), a 200-fold variation in external B concentration did not alter the concentration of water-soluble Ca^{2+} in bean roots (Neales and Hinde, 1962).

In a study of the translocation of dialyzable (soluble) B from mature leaves (bean, chard, cocklebur, cotton, cucumber, sunflower, tobacco, tomato, and turnip) deprived of an external supply of B, only cotton and turnip showed a high degree of B mobility (McIlrath, 1965). A high mobility of B in only these two species is very interesting in view of the fact that B has generally been regarded as highly immobile.

Stability constants for various borate–carbohydrate complexes have been determined (Malcolm et al., 1964). B in organic complexes has been reported for seawater (Noakes and Hood, 1961). By electrophoresis, Linskens (1955) concluded that sugar–borate complexes were present in pollinated styles of petunia. He stated that he was unable to determine whether sugar–B complexes operated *in vivo* in pollen, played a role in intermediate metabolism, or played a role in the passage of sugar through membranes in the form of ionized, sugar–B complexes. Skok (1958) concluded that the capacity of borate to complex with polyhydroxy and related compounds was probably related to the physiological function of B but added that the relationship is not clear. He stated, however, that the effect on translocation was probably an indirect one related to the effect of B on growth and cellular activity. Inasmuch as Sr^{2+}, Al^{3+} and particularly Ge^{6+}, all of which complex with diols, alleviated B-deficiency symptoms temporarily, Skok (1957b) concluded that the physiological role of B is likely related at least in part to the complexing property of the borate ion. Inasmuch as B is not reutilizable and must be continually supplied to plants, he suggested that the complexing reaction is related to the formation of a structural unit or "building block" rather than to a metabolic step reaction. Corroborating a report by Sisler et al. (1956), he noted that spraying sugar or citric acid on B-deficient stem tips did not alleviate B deficiency. McIlrath and Palser (1956) "detected no, or very slight, increases in growth of B-deficient tomato, turnip and cotton plants after spraying with sugar, and the B deficiency symptoms were not alleviated by the treatment." B was required by a diatom, *Cylindrotheca fusiformis*, but germanium dioxide, although it complexes with diols, did not suffice for this diatom (Neales, 1967). He concluded that the solid geometry of the OH groups of $B(OH)_4$ and $Ge(OH)_6$ may be different, and although Ge^{6+} partially substituted for B in sunflower (Skok, 1957b), the diatom would not accept Ge^{6+} as a substitute. He noted that an alternative to the complexing hypothesis is one that relates the biological activity of borate to the fact that boric and phenylboric acids are Lewis acids. Therefore boric acid does not act as a proton donor but, as a Lewis acid, accepts the electron pair of the base (for example, OH) to form the tetrahedral anion $B^-(OH)_4$. Inasmuch as plants require a continuous supply of B, he hypothesized that each borate ion undergoes a biologically important reaction only once and does not participate in a reversible cyclic change as do the transition metals, for example, Fe and Mo. He proposed that this once-only reaction is:

$$B(OH)_3 + OH^- = [B(OH)_4]^-$$
(Triangular)   (Tetrahedral)

Ranganathan and Rengassamy (1965) suggested that the biological necessity of B may be partially explained on the basis of its electron deficiency which makes possible its participation in electron transfer reactions.

With regard to cell walls, Spurr (1957a) noted that the walls of collenchyma of celery petioles of B-deficient plants failed to develop normal thickening and attributed this effect to a deficiency of carbohydrates. He reasoned that since the ground parenchyma and phloem parenchyma become much thicker-walled, carbohydrates from the major source, that is, the phloem, were condensed in these intervening tissues and therefore did not reach the more peripheral collenchyma. He concluded that histological evidence favored the hypothesis that B is involved in sugar translocation. In another paper Spurr (1957b) reported that in B deficiency the fine structure of cell walls became coarser and that there were fewer lamellae. He suggested that B acts as a morphogenic agent affecting the development of specific form in cell walls. In contrast with other reports, Whittington (1959) claimed that the primary effect of B deficiency in bean radicles is cessation of cell division. His results indicated that division did not cease because of lack of sugar or lack of synthesis of nucleic acids and proteins. Rather, he considered that abnormalities in the formation of the cell wall prevented the cell from "becoming organized for mitosis." He concluded that B is concerned with the formation of pectin from uridine diphosphate-D-glucose. Odhnoff (1961) determined that boric acid or phenylboric acid reacts with carbohydrates of the cell wall, and she concluded that B is involved in the synthesis of cell walls. Cell walls of tobacco pith tissue showed a reduction of galactan in B deficiency (normal galactan, 8%), but no significant changes in the relative amounts of pectic substances or cellulose as reported by Wilson (1961). He observed that B deficiency caused a marked increase in the total amount of cell wall material. Using material from B-sufficient and B-deficient plants, Starck (1963) examined stem tip material with an electron microscope. Cell walls of B-deficient plants did not reveal microfibrillar structure but only a disorganized, scarce mesh of fibrils. Whittington (1959) concluded that B is concerned with cell wall metabolism and, in a later paper Slack and Whittington (1964) indicated that B is specifically involved in cell wall bonding. They cautioned that apparent differences in pectins in B-deficient and normal root tips were a result of differences in ages of cells in the two kinds of root tips, and that B had no true effect on pectin synthesis. Odhnoff (1957) reported that a deficiency of B did not inhibit cell division and reasoned that B deficiency would interfere first with the growth phase most closely connected with carbohydrate metabolism, that is, the intussusception phase of cell wall growth. In support of this idea, she noted that the first and most severe B-deficiency symptoms occur in roots and that in roots intussusception presumably dominates.

Skok (1961) noted that gibberellic acid speeded up the growth of sunflowers and the appearance of B-deficiency symptoms. Red light depressed development and maturation and slowed down the appearance of B-deficiency symptoms. Such results were interpreted as evidence of a role of B in differentiation and maturation.

There appears to be some lack of agreement on the possible involvement of B in lignin synthesis, since B deficiency has been reported to cause both a reduction (Palser and McIlrath, 1956; Dutta and McIlrath, 1964) and an increase in lignin (Neales, 1960; Odhnoff, 1961). Inasmuch as there is some indication that lignin precursors (i.e., phenolic compounds) may be converted to lignin only if H_2O_2 is available, Presley and McIlrath (1956) added H_2O_2 to the substrate in an attempt to improve the growth of B-deficient plants. They observed no effect with tomato and turnip plants, and only a slight improvement in B-deficient cotton plants. Neales (1960) considered B to be involved in lignin synthesis and noted that maize radicles grew in the absence of B for 4 days, whereas bean radicles showed a decreased rate of growth in the absence of B after 5 hr. As a possible explanation for these differences, he regarded as germain the report by Brown et al. (1959) that monocots have a pathway for lignin synthesis that dicots do not. Watanabe et al. (1961) also suggested that B is involved in the polymerization of precursors of lignin, since a phenolic compound, scopoletin (6-methoxy-7-hydroxycoumarin), was 20 times as concentrated in leaves of B-deficient as in the leaves of B-sufficient tobacco plants. Following this report, B-deficient sunflower plants were shown to accumulate two, major, blue-fluorescing compounds—scopolin (7-gluosyloxy-6-methoxy-coumarin) and a glucose derivative of gentisic acid (Watanabe et al., 1964). Esculin, isoquercitrin, and scopoletin were also identified for the first time in sunflowers. McIlrath and Skok (1964b) concluded that B had some function in lignin production, but they could not establish whether or not the role was a direct one. They very critically discussed the literature on B and lignin and, significantly, noted that sampling procedures and methods of expressing results have much to do with the conclusions that are drawn. For example, a low concentration of lignin in normal plants could result from growth and dilution of lignin; conversely, lack of growth in B-deficient plants would tend to cause a high concentration of lignin. Their critical analysis of the relationship of B to lignin could very well explain the divergent results indicated earlier in this section. For sunflower, Dutta and McIlrath (1964) proposed that the lower peroxidase activity of B-deficient tissues, as compared with that of B-sufficient tissues, could in part account for the lower degree of lignification in B-deficient plants. Lee and Aronoff (1967) proposed that B is required by organisms in which there is an appreciable synthesis of phenolic ligands that can complex with borate. Further, they reported that borate partitions metabolism between the glycolytic and pentose shunt pathways. Accordingly, borate reacts with 6-phosphogluconic acid to form a substrate which inhibits the action of 6-phosphogluconate dehydrogenase. In the absence of B, inhibition of the enzyme is released, and excess phenolic compounds are formed. These acids in turn associate strongly with borate and thus develop an autocatalytic system for the formation of excess phenolic acids that result in necrosis of tissue and eventual death of the plant. Inasmuch as B appears to function in xylem formation and lignification (Skok, 1958), the gibberellic acid-induced stem proliferation of debudded tobacco stems was regarded as an appropriate phenomenon for studying directly the effect of B on xylem formation and lignification. The effect of B would not then be confounded by its effect on elongation growth (Skok, 1968). Gibberellic acid-induced proliferation consists primarily of xylem tissue, and B-deficient tobacco plants showed less proliferation. The reduction in proliferation and formation of xylem was regarded as additional evidence for a role of B in xylem formation.

A possible relation between B and polyphenols has recently been suggested. Oil palm *(Elaeis guineensis)* seedlings were produced in nutrient solutions with and without B (Rajaratnam et al., 1971). Leaves were analyzed for leucocyanidin (detected as cyanidin) several months prior to the time that B-deficiency symptoms were expected. Plants receiving B had high concentrations of leuco-cyanidin in the leaves, whereas those not receiving B had no detectable leuco-cyanidin. It was concluded that if B is required only or mainly for flavonoid synthesis, an explanation had been provided as to why higher plants require B.

Shkolnik et al. (1961) credited M. A. Belousov with first advancing the hypothesis that B is directly responsible for sugar translocation (as later suggested by Gauch and Duggar (1953, 1954)) but they considered the hypothesis to be "poorly grounded" and thus turned to a study of RNA and B. In this and in later work (Shkolnik and Soloviyova-Troitzkaya, 1962; Shkolnik and Mayevskaya, 1962), it was claimed that B-deficiency symptoms could be alleviated by the addition of RNA to B-free nutrient solution. Cessation of root growth within 6 hr (Albert and Wilson, 1961) and a decrease in RNA concentration have also been reported for B-deficient tomato root tips (Albert, 1965). Neither of these effects occurred when thymine, guanine, or cytosine was added to minus-B nutrient solution (Johnson and Albert, 1967). Addition of 6-azauracil or barbituric acid to plus-B nutrient solutions decreased the elongation of roots and the concentration of RNA—coincident with the onset of B-deficiency symptoms. It was concluded therefore that B may play a role in some aspect of nitrogen base synthesis or utilization and thereby in RNA metabolism.

With regard to enzymes and B, Dugger et al. (1957) postulated that B inhibits conversion of sugars to starch, hence the higher concentration of sugars in B-sufficient plants, as compared with B-deficient ones, would favor sugar translocation in the former group. They concluded that B inhibits starch formation by forming an unreactive glucose-1-phosphate–B complex. However, Khym and Cohn (1953) reported that glucose-1-phosphate does not complex with borate and, since there are no adjacent *cis*-hydroxyl groups, one would not expect complexing—according to Zittle (1951). Scott (1960) concluded that B does not inhibit starch formation by forming a complex with glucose-1-phosphate, but that instead there is truly competitive inhibition with B complexing at the active site of starch phosphorylase. Complexes between B and protein have been reported (Winfield, 1945; Zittle, 1951). Scott (1960) noted that high-B leaves had higher concentrations of sugars than leaves of plants on low B; higher sugar concentrations in the high-B leaves was ascribed to B complexing with the active site of starch phosphorylase rather than to impeded translocation. B deficiency caused an extreme accumulation of starch in *Lemna minor* (Scholz, 1962). Apparently, B inhibits the action of phosphogluco-mutase, so that glucose-6-phosphate, hence fructose-6-phosphate and fructose would be reduced; thus sucrose synthesis would be restricted (Loughman, 1961).

Excised flax roots grown in sterile culture required B, and 0.05 ppm B was sufficient (Neales, 1959). Root growth stopped after 48 hr in media lacking B.

The B requirement of excised roots was similar to that of intact roots. Inasmuch as roots in B-deficient media were in sugar solution and yet died, it was concluded that B was not involved in the movement of sugars into flax root cells, that is, B deficiency was not a matter of sugar deficiency. In a later paper, Neales (1964) confirmed his earlier observation that excised and intact tomato roots both required B and at about the same concentration. By contrast, Albert and Wilson (1961) reported that growth of intact tomato root tips ceased within 6 hr when B was withheld, but that excised tomato roots grew in the absence of B for 17 days. Incidentally, they reported that increased light intensity was more causally related to speeding up the appearance of B-deficiency symptoms that was lengthening of the photoperiod. Although cessation of root growth was the first symptom of B deficiency in tomato plants, Yih and Clark (1965) concluded that cessation was not attributable to a deficiency of carbohydrates. In fact, B-deficient root tips had a higher dry weight and greater concentrations of carbohydrate and protein than did B-sufficient control plants. They reasoned that higher carbohydrate and protein concentrations in B-deficient plants reflected enhanced maturation of tissues closer to the root tips in the case of B-deficient plants.

With regard to other work on B and pollen, Visser (1955) covers in detail the significance of B in the germination and growth of pollen. For trumpet vine [*Tecoma radicans* (L.) Juss.], B did not increase the absorption of or respiration of fructose, but did so with glucose and sucrose (O'Kelley, 1957). It was concluded therefore that the specific effect of B on pollen may concern protein synthesis or synthesis of pectic compounds. Tributyl borate, Na tetraborate (borax), boric acid, K metaborate, and K tetraborate stimulated the germination of *Amaryllis hybrida* pollen at a concentration of 5 ppm (Stanley, 1961). In a later paper (Stanley and Lichtenberg, 1963), the importance of B for pollen is discussed, and the various compounds of B that are suitable are listed. Myo-inositol-2-[H^3] was shown to be readily converted into both D-galacturonosyl and L-arabinosyl units of pectin in pear pollen tube membranes (Stanley and Loewus, 1964). Incidentally, while most hexoses and related sugars chelate with borate through two OH^- functions, *myo*-inositol and certain related naturally occurring cyclitols can form a tridentate complex with boron, that is, three OH^- groups of the cyclitol participate. Vasil (1964) also discussed the importance of B for the germination and growth of pollen. B was reported to have a beneficial effect on the germination of pollen grains of 56 species of fruit, vegetables, and ornamental and wild plants (Daniel and Varoczy, 1957). In their review of the physiology of pollen, Johri and Vasil (1961) reported that B enhanced growth of pollen tubes and that the effect of B on pollen germination and tube growth was much more pronounced than the effect of any known hormone, vitamin, or other chemical. They reported further that B promoted sugar absorption and metabolism by forming sugar–borate complexes, increased the O_2 uptake, and was involved in synthesis of pectic materials required for the walls of actively elongating pollen tubes. They added the further note that pollen of most species is naturally deficient in B. Pollen-derived tissue of *Ginkgo biloba* L. showed a requirement for B with an optimal level of 0.1 ppm according to Roy et al. (1966). In its absence there was (1) a reduced rate of cell division and (2) an alteration in the composition of the cell walls (i.e., there was less arabinose). These investigators concluded that their results did not support the Gauch–Dugger theory that B is involved in sugar uptake and/or translocation. However, they (Roy et al., 1966) studied a tissue in which "translocation over long distances through specialized tissues . . . " was not involved. In view of the type of tissue with which they worked, then, it appears that their data are not appropriate to corroborate or refute the Gauch–Dugger theory.

Phenylboric acid has been employed in an attempt to elucidate the role of B in plants. Torssell (1956) noted that phenylboric acid strongly complexes with sugars and greatly influences the growth of roots. He claimed that boric acid and phenylboric acid form complexes with cellulose and pectin of the cell wall and thus prevent bonding of the chains by van der Waal forces or by hydrogen bonds because of steric hindrance and partly because of repulsion caused by the ionized complexes. This prevention of bonding was believed to retard the crystallization of cellulose, so that the cell wall remains stretchable longer. Interestingly, he suggested that phenylboric acid may transport sugar across membranes. He stated that phenylboric acid derivatives with a lipophilic group could be incorporated more easily than boric acid in the plasma membrane and thus take part in sugar translocation to a greater extent than boric acid. In a study with roots, Burstrom (1957) stated that there are no structural reasons to include phenylboric acid among the auxins and, further, that phenylboric "probably has nothing to do with the genuine boron activity either."

Several workers have attempted to gain information on the role of B by observing compounds that induce symptoms similar to those characteristic of B deficiency. Symptoms of 2,6-dichlorobenzonitrile poisoning in plants were very similar to the symptoms of B deficiency according to Milborrow (1964). He concluded that the process affected by this compound was fairly closely related to that requiring B, but he was unable to determine whether or not the two reactions were identical. The onset of necrosis in the leaves and growing point of sunflower as a result of B deficiency was characterized by an increase in the ratio of caffeic to chlorogenic acid (Dear and Aronoff, 1965). It was suggested that the B-deficiency syndrome may be the result of excessive, free caffeic acid.

With regard to B and phloem, Ziegler (1956) reported that the concentration of B in sieve tube sap was 5 ppm. Later, Ziegler (1958) presented data on the O_2 consumption (in cubic millimeters of O_2 per hour at 25°C per milligram of protein N) of parenchyma, phloem, and xylem of *Heracleum mante-*

gazzianum Somm. et Lev., as: 33 ± 3.5, 115 ± 33, and 95 ± 23, respectively. In other words, he found that the respiration rate for phloem is about three to four times that of the parenchyma in which the bundles lie. However, Duloy and Mercer (1961) caution that there is no "peculiar metabolism" in the phloem. On a fresh weight basis, considering O_2 uptake, respiration of phloem is 5 to 40 times higher than that of ground parenchyma. They noted, however, that this apparent difference could be explained on the basis of a high concentration of protoplasm in phloem—a concentration difference which just about exactly offsets the extent to which, on a fresh weight basis, the respiration of phloem appears to be higher than that of ground parenthyma.

With regard to B and photosynthesis, B has been reported to be present in a relatively high concentration in chloroplasts (Otting, 1951; Skok and McIlrath, 1958). Leaves accumulate B in excess of their requirements, while accumulation in stems is more closely correlated with metabolic requirement (McIlrath and Skok, 1964a). In view of the accumulation of B in chloroplasts, and Skok's (1958) suggestion that B is required only by autotrophic plants, McIlrath and Skok (1964a) concluded that such evidence was "indicative of a role for boron in photosynthesis (or some initial step in carbohydrate metabolism)." Electron microscopy of mesophyll cells of sunflower leaves showed that during B deficiency chloroplasts degenerate and cell walls undergo profound structural changes before any visual deficiency symptoms are apparent (Lee and Aronoff, 1966). Mitochondria increase in number as B deficiency develops and frequently shows myeline figures, while some of the nuclei develop dense rhombohedral structures in the absence of B.

With regard to B and auxins, naphthalene acetic acid applications to meristems of B-deficient bean plants, *Phaseolus vulgaris* var. Black Valentine, increased the translocation of ^{14}C-labeled photosynthates to the tips (Dyar and Webb, 1961). It was suggested that B-deficient plants may have a functional conducting system, but that it does not conduct because substrates are not being utilized; thus B-deficient plants are not limited in growth by sugar deficiency. It was concluded that B is unnecessary for sugar translocation per se but rather for auxin metabolism—possibly auxin synthesis. B increased the production of roots on *Phaseolus* cuttings in association with indole and IAA (Gorter, 1958). Natural plant auxins can move in polar fashion and so can sugar in certain plants. Sugars move from leaves of beet plants (sugar beets in particular) to the root against a sugar gradient, and carrot tissues, rich in sugar, go on to absorb all the glucose from dilute glucose solutions (Grant and Beevers, 1961). Such movements of sugar appear to be unexplainable on the basis of sugar moving to regions where the concentration of sugar is lower, that is, to physiological sinks.

Bitter pit of apples appears somehow to be related to B, and incidence of the disorder was effectively reduced by B sprays—particularly when applied during blossoming (Dunlap and Thompson, 1959). It is possible that bitter pit might be caused by localized deficiencies of carbohydrates in the localized areas where bitter pit occurs in the fruit, since B sprays tend to prevent the disorder.

Evidence by Baker et al. (1956) indicated that there is an increase in hydrophilic colloids in B-deficient plants as compared with B-sufficient plants. Decrease in rate of water loss by B-deficient plants, as compared with B-sufficient ones, was believed to be related to the higher concentration of hydrophilic colloids in B-deficient plants.

With regard to B and algae, boric acid can be a sex-determining factor for the bisexual green alga *Chlamydomonas* which secretes both male and female sex-determining substances—the latter a methyl ether of quercetin (Kuhn et al., 1942). By combining with the quercetin derivative, boric acid permits the excess of male factor to determine the sex. Working with the Columbia strain of *Chlorella vulgaris*, 0.5 ppm B, as compared with zero B, increased cell number and dry weight per cell (Skok and McIlrath, 1957; McIlrath and Skok, 1958). Despite the apparent effect of B on growth, Skok and McIlrath (1957) clearly indicated that their results did not unequivocably establish a B requirement since no attempt was made to substitute other elements. Bowen et al. (1965) were unable to demonstrate a B requirement for *Chlorella* and later, McBride et al. (1967) also reported that they were unable to demonstrate a requirement. *Chlorella* and two other green algae did not require B and did not absorb the element in appreciable amounts as observed by Gerloff (1968). He found, however, that three species of blue-green algae readily absorbed B and showed a marked growth response to B with N-fixing species when NO_3^- was omitted from the culture medium. *Cylindrotheca fusiformis* Reinmann and Lewin required B for growth both in the light and in the dark (Lewin, 1965, 1966a). It was found that 0.5 ppm B supported maximal growth rate of this diatom. Growth rate was also affected by the initial Si/B ratio in the medium and this indicated a possible interaction between Si and B. B was required for 12 species of marine pennate diatoms, four species of marine centric diatoms, and eight freshwater species of diatoms as reported by Lewin (1966b). She concluded that B is essential for the growth of most, probably all, diatoms. She showed that some species of marine algal flagellates also required B for growth; others apparently did not. As indicated by Lewin (1965), the wall of the diatom is unique as compared with higher plants because of the absence of cellulose and the presence of silica.

For fungi, 3.2 ppm B more than doubled the number of perithecia produced by *Sordaria macrospora* Auersw., but 8 ppm or higher concentrations were inhibitory for the production of ascospores, perithecia, and mycelia (Turian, 1955). When dipped in phosphate-buffered solutions of boric acid (25–500 ppm B), male and female gametangia of *Allomyces javanicus* liberated giant male and female gametes (Turian, 1956). It was concluded that boric acid may have interferred with RNA—an essential constituent of the nuclear cap —during cleavage of gamete units in the gametangia. Zittle (1951) reported that B reacts both with RNA and pyridoxine. B inhibited certain oxidative enzymes of which riboflavin is the coenzyme; the inhibition apparently involved the pyridoxine portion of the riboflavin molecule (Adrian, 1962). Bowen and Gauch (1966) showed that the following fungi did not have a B requirement: *Saccharomyces cerevisiae, Aspergillus niger, Neurospora crassa,* and *Penicillium chrysogenum.* Gerloff (1968) showed that *P. chrysogenum* neither required nor absorbed B.

In a study involving various fungi, liverworts, *Selaginella,* and ferns, Thellier (1963) found that B was stimulatory for most of the organisms but could not conclude that B was essential for any one of these plants. However, Bowen and Gauch (1965) showed that the strobilus was apparently aborted in B-deficient *Selaginella,* and that very few sori were formed on B-deficient plants of *Dryopteris dentata;* the B-deficient fern plants died.

Manganese

According to Gerretsen (1950), after photoinduced electron transfer, Mn^{2+} may be the active agent in the induction of photooxidation of water. At the same time Mn^{2+}, as such or adsorbed on a carrier, might also easily combine with the OH^- radical with the result being the primary photochemical oxidation product. This product would give rise to H_2O_2 which is readily decomposed into water and free oxygen.

The generally accepted hypothesis concerning the role of Mn^{2+} in photosynthesis is that it is a catalyst within the O_2-evolving reactions of photosystem II (Pirson, 1937; Pirson et al., 1952; Kessler, 1955, 1957; Brown et al., 1958; Eyster et al., 1958; Spencer and Possingham, 1961; Possingham and Spencer, 1962; Bachofen, 1966; Cheniae and Martin, 1966; Gerhardt, 1966; McKenna and Bishop, 1967; Homann, 1967; Habermann et al., 1968; Cheniae and Martin, 1969). The Mehler reaction (a Hill reaction) was stimulated by Mn^{2+}, and the reaction and whole-cell photosynthesis were inhibited by Cu^{2+} (Habermann, 1969).

In extremely Mn-deficient cells of *Anacystis nidulans,* light was required for cells supplied with Mn^{2+} to form active O_2-evolving centers, as reported by Cheniae and Martin (1969). They concluded that there are six managanese sites on the oxidizing (O_2-evolving) side of each trapping center of photosystem II.

Mn was required for maintenance of chloroplast structure (Teichler-Zallen, 1969) (Fig. 10.4). Mn-deficient cotton leaves showed abnormally high IAA oxidase activity (Taylor et al., 1968).

Mn was an activator of arginase which converts arginine into urea and orni-

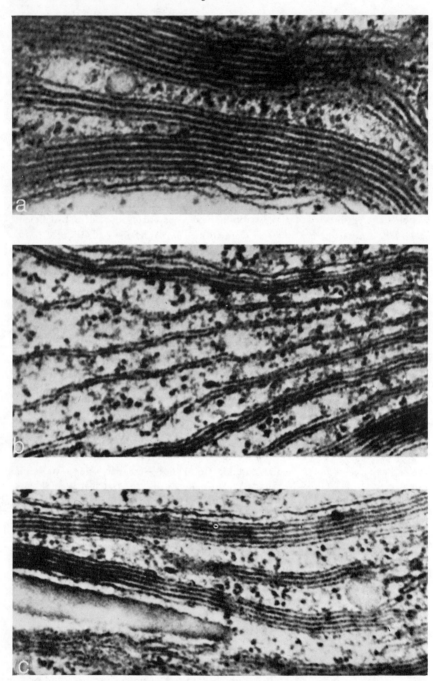

Fig. 10.4
Section of *Chalamydomonas reinhardi* chloroplast from a normal cell (a) ($\times$121,000), manganese-deficient cell (b) ($\times$88,500), and recovered cell (c) ($\times$88,500). (D. Teichler-Zallen, 1969.)

thine (Folley and Greenbaum, 1948), and it was the most effective of various metals tested. Mn is the activator of β-carboxylases which catalyze the assimilation of CO_2 and lead to the formation of di- and tricarboxylic acids. Mn was required for fixation of CO_2 by parsley root oxalacetate carboxylase (Vennesland and Felscher, 1946; Gollub and Vennesland, 1947; Speck, 1949). It was also required for glutamotransferase in peas, Mg^{2+} being less effective than Mn^{2+} (Elliott, 1953).

Zinc

Sommer and Lipman (1926) demonstrated the essentiality of Zn^{2+} for higher plants. A relationship between Zn^{2+} and auxin was first suggested by Skoog (1940). Working with tomato and sunflower, he noted that Zn-deficient plants contained little or no auxin, and that auxin decreased before Zn-deficiency symptoms appeared. One or two days after adding Zn^{2+}, the auxin concentration in previously Zn-deficient plants increased. He concluded that Zn^{2+} was not required for auxin synthesis but for its maintenance in an active state; lack of Zn^{2+} led to destruction (presumably oxidation) of auxin. He further noted that Zn-deficiency symptoms were more acute when plants were grown in blue light than when grown in red light. Ultraviolet radiation was reported to cause oxidation of auxin (Burkholder and Johnston, 1937; Popp and McIlvaine, 1937; Galston and Hillman, 1961).

Although the enzyme system(s) has not been isolated, there is good evidence that Zn^{2+} is required for synthesis of tryptophan (Tsui, 1948a) which in turn may be the precursor (Wildman et al., 1947; Sherwin and Purves, 1969) of IAA, the principal hormone of higher plants. Therefore Zn^{2+} may be said to be indirectly required for the formation of this important natural plant growth regulator (Wildman et al., 1947; Tsui, 1948a).

Zn-deficient tomato plants have a significantly lower percentage of water (Tsui, 1948b) than Zn-sufficient plants, and this may be related to the role of auxin in water absorption.

Carbonic anhydrase contains Zn^{2+} as the prosthetic group (Keilin and Mann, 1939b). Although the enzyme was initially studied in red blood corpuscles, carbonic anhydrase also occurs in higher plants (Bradfield, 1947). Zn may also be a constituent of lactic dehydrogenase (Price, 1962), and it is an essential component of many respiratory enzymes (Vallee, 1959). Zn is an essential constituent of alcohol dehydrogenase, glutamic dehydrogenase, alkaline phosphatase, carboxypeptidase B, and other enzymes (Evans and Sorger, 1966). Price (1966) discusses various processes sensitive to Zn in plants and microorganisms.

Copper

The essentiality of Cu^{2+} for higher plants was established in the early 1930s. Sommer (1931) showed that Cu^{2+} was essential for tomato, flax, and sunflower plants. The evidence for the essentiality of Cu^{2+} and its known functions by 1950 were reviewed by Arnon (1950b).

Based on a symposium, a comprehensive review was published (McElroy and Glass, 1950) discussing the occurrence and roles of Cu^{2+} in animal, plant, and soil relationships. More recently, Reuther and Labanauskas (1966) reviewed the status of Cu^{2+} in plants.

Cu^{2+} is a constituent of polyphenoloxidase (Keilin and Mann, 1938; Kubowitz, 1938; Bonner, 1950), ascorbic acid oxidase (Powers et al., 1944), and laccase (Keilin and Mann, 1939a; Tissieres, 1948). Polyphenoloxidase was reported to occur in beet *(Beta vulgaris)* chloroplasts (Arnon, 1949). As one of the polyphenoloxidases, tyrosinase may be involved in the terminal step of aerobic respiration. Laccase is of limited distribution in higher plants. Ascorbic acid oxidase might also theoretically serve in the terminal step of aerobic respiration, but its role in this regard has not been established. Cu^{2+} is also a constituent of monoamine oxidase, uricase, cytochrome oxidase, and galactose oxidase, as well as plastocyanin from *Spinacia oleracea* (Evans and Sorger, 1966).

Cu^{2+} along with Fe, or Fe and Mn, toughens and thickens the skin of Irish potato tubers (McLéan et al., 1944).

Molybdenum

The biological importance of Mo was not realized until it was shown that it was highly beneficial in the fixation of gaseous N by *Azotobacter chroococcum* (Bortels, 1930). Arnon (1938) obtained the first evidence that Mo might be essential for higher plants when he noted that the addition of a group of seven elements (Mo, V, Ti, W, Cr, Ni, and Co) caused a significant increase in the growth of lettuce and asparagus. With the exception of Mo, the other six elements have not been shown to be essential for higher green plants.

Mo was first clearly demonstrated to be essential for higher plants by Arnon and Stout (1939). Essentiality of Mo for higher plants was confirmed by Piper (1940), Hoagland (1941), Hewitt and Jones (1947), and Mulder (1948). Hoagland (1945) reviewed the literature up to 1945 which indicated the essentiality of Mo.

Mo-sufficient tomato plants contain approximately 10 ppb Mo on a fresh weight basis (Stout and Meagher, 1948). Mo-deficient tomato plants growing on sepentine barren soils contained less than 0.1 ppm Mo on a dry weight basis (Walker, 1948).

Mo-deficient cauliflower and broccoli plants exhibit a modification in growth known as "whip-tail." The appearance of such plants strongly suggests that in Mo deficiency there is an alteration in the production, translocation, or action of hormones responsible for growth and differentiation of various plant organs.

In an outstanding series of papers, the role(s) of Mo in tomato (Hewitt and McCready, 1956) and *Brassica* crops was reported (Hewitt and Bolle-Jones, 1952a, b; Agarawala and Hewitt, 1954a, b, 1955a, b). Mo deficiency in cauliflower was produced with a nutrient concentration of 0.000005 ppm Mo, and also of 0.00005 ppm Mo with a high concentration of NO_3^- present (Agarawala and Hewitt, 1954a). It was noted that growth diminished sharply with more than 0.005 ppm of Mo. Whip-tail occurred at low levels of Mo regardless of the form of N supplied (Agarawala and Hewitt, 1954b, 1955b). This indicated that Mo plays a role other than its role in NO_3^- reduction. Mo was also required for tomato plants regardless of the form of N supplied (Hewitt and McCready, 1956). Evans (1956) concluded that Mo is required for the reduction of NO_3^- and the fixation of N, but that Mo plays some unknown role in addition to those two roles.

Mo was reported to be a nondialyzable component of xanthine oxidase (Totter et al., 1953).

Owing to the involvement of Mo in N metabolism, literature on NR and nitrite reductase (NiR) and also on N fixation have been assembled here even though involvement of elements other than Mo may have been reported. For detailed coverage of the literature, excellent reviews are available on the subjects of biological reduction of SO_4^{2-} and NO_3^- (Bandurski, 1965) and of N fixation (Burris, 1965, 1966).

Approximately 0.01 ppm Mo was sufficient in the nutrient solution for tomato plants (Arnon and Stout, 1939). Despite the low concentration of Mo that suffices for the growth of most plants much has been learned about its role(s). In many plants, NO_3^- accumulates in Mo-deficient plants and disappears when Mo is supplied (Hewitt and Jones, 1947; Mulder, 1948; Meagher, 1952; Spencer and Wood, 1954). Mo-deficient tomato plants may accumulate NO_3^- to the extent of 12% of the dry weight of the plants. This value was observed to drop to about 1% within 48 hr after Mo was supplied (Stout and Meagher, 1948). Its role in NO_3^- reduction was confirmed when Nicholas and Nason (1954a, b) and Evans and Hall (1955) demonstrated that Mo is the metal component of NR obtained from *N. crassa* and soybean leaves, respectively.

NR can utilize either NADH or $FMNH_2$ as cofactors but *in vivo* (for higher plants) NADH is presumably the electron donor, inasmuch as $FMNH_2$ does not exist in high enough concentration (Schrader et al., 1968). It was noted that the half-life of NR was approximately 4 hr.

The suggested reaction sequence mediated by NR from *Neurospora* (Nicholas and Nason, 1954b) is:

$$TPNH \longrightarrow FAD \ (or \ FMN) \longrightarrow Mo \longrightarrow NO_3^-$$

Mo functions as an electron carrier in this sequence.

According to Beevers and Hageman (1969), NR most commonly found in plant leaves is depicted as a molybdoflavoprotein capable of accepting electrons directly from NADH, or $FMNH_2$ ($FADH_2$), but not from NADPH. These investigators discuss the evidence indicating that 3-PGAld and the cytoplasmically located NAD-dependent PGD (phosphoglyceraldehyde dehydrogenase) is the electron-generating system for NR in both light and dark. For roots and nongreen tissues, the reducing potential is presumed to be furnished by respiration of carbohydrates. They state that the electron donor for NO_2^- reduction in nonphotosynthetic tissue is unknown, and that the mechanism by which NO_3^- is reduced to NH_3 in nongreen tissues is not well established. With regard to NO_2^- reduction, not only is the precise mechanism unsettled, but it appears that NiR affects the transfer of six electrons at one enzyme surface in converting NO_2^- and NH_3—"a disturbing phenomenon on the basis of thermodynamics and kinetics."

NR was not present in or associated with partially purified, intact chloroplasts aqueously isolated from *Wolffia arrhiza* (Swader and Stocking, 1971). The chloroplasts were capable of using NO_2^- but not NO_3^- as an electron acceptor during light-stimulated electron transport in the absence of additional cytoplasmic components.

Addition of Ni (4×10^{-3} M) to the extracting buffer enhanced NR activity in preparations of young grain sorghum [*Sorghum bicolor* L. (Moench)] leaf tissue by as much as sixfold (Maranville, 1970). Ni also increased the activity of NR in Sudan grass [*Sorghum sudanense* P. (Stapf)].

NR has been at least partially characterized from soybean leaves (Evans and Nason, 1953), cauliflower (Candela et al., 1957), wheat embryos (Spencer, 1959), corn leaves (Hageman and Flesher, 1960), and marrow leaves (Cresswell, 1961). It has been found in tomato leaves and roots and characterized as a sulfhydryl-containing enzyme inasmuch as cysteine was required for its extraction in an active state as observed by Sanderson and Cocking (1964a). These investigators also found NR in tomato plants but were unable to identify the natural electron donor. The latter enzyme reduced NO_3^- to NH_3 in the absence of grana with benzyl viologen as an electron donor but did not do so with reduced pyridine nucleotides as donors (Sanderson and Cocking, 1964b).

In a study of the NR activity of various inbred parent corn lines and hybrids, two hybrids of the "low × low" category exhibited heterosis; the level of NR activity of the hybrids was significantly higher than the level of either inbred parent (Schrader et al., 1966). However, this effect did not occur with certain other hybrids.

In higher plants the Mo requirement can be greatly reduced when NH_4^+ rather than NO_3^- is supplied, but the requirement cannot be completely abolished (Mulder, 1948; Meagher, 1952; Agarawala, 1952; Hewitt and McCready, 1954). Presumably, in higher plants at least, Mo plays a role in addition to NO_3^- reduction.

It has been reported that isolated spinach chloroplasts contain all enzymes required for reduction of NO_3^- to NH_3 (Losada et al., 1965). However, in corn seedlings, Ritenour et al. (1967) deduced that NiR was localized in chloroplasts but that NR and glutamic acid dehydrogenase were not. Although their report concerning the location of NR is at variance with the report by Losada et al. (1965), Ritenour et al. (1967) recognized that NR might be localized on the external chloroplast membrane. According to Ritenour et al. (1967), NR is located in the cytoplasm and NADH is involved in its reduction to NO_2^-. Sugars come from the chloroplast, NADH is formed, and the latter is used during NO_3^- reduction to NO_2^-. NO_2^- then moves into the chloroplast where, mediated by ferredoxin, it is reduced to NH_3. Intermediates in this process have not been isolated. NiR may be near the inner surface of the chloroplast membrane but in any event the reductase must be located sufficiently close to ferredoxin to mediate the reduction of NO_2^- to NH_3. NR could be on the outer surface of the chloroplast membrane, rather than in the cytoplasm, but attempts to associate NR with the chloroplast membrane have been unsuccessful.

There was no NR in dry barley *(Hordeum vulgare* L. 'Himalaya') seed, and it did not appear during incubation of the half-seed or isolated aleurone layer. NR activity was initiated in the aleurone layer only after the addition of NO_3^- (Ferrari and Varner, 1969). NO_3^- has also been reported to be required for the induction of NR in mold fungi (Morton, 1956) and cauliflower (Afridi and Hewitt, 1964). For cauliflower, Mo was also required. For mold fungi, presence of the end product (NH_4^+ ions) prevented the induction of NR (Morton, 1956). In a study of the role of Mo in the utilization of NH_4^+ and NO_3 by *A. niger,* Steinberg (1937) suggested that Mo was essential for the activation of enzymes involved in NO_3^- reduction.

NR activity decreased during the dark (Candela et al., 1957; Hageman and Flesher, 1960; Hageman, et al., 1961; Sanderson and Cocking, 1964b), and protein synthesis was apparently involved in its disappearance (Travis et al., 1969).

In *Lemna minor*, NR and NiR were low when plants were grown on amino acids or NH_3 but both enzymes were rapidly induced by NO_3^- (Joy, 1969). NO_2^- induces NiR only. NH_3 represses the NO_3-induced synthesis of both NR and NiR. NH_3 has also been reported to repress induction of NR in *Chlorella* (Syrett and Morris, 1963).

NO_2^- seldom occurs in appreciable concentration in plants but does so for a few hours when Mo-deficient plants high in NO_3^- are given Mo as observed by Hageman et al. (1962). Their work with marrow leaf NiR indicated that free hydroxylamine is not a natural intermediate in the pathway of NO_3^- and NO_2^- reduction. They concluded that NiR probably depends on TPNH as a primary electron donor with an unidentified cofactor which can be replaced by benzyl viologen *in vitro.*

In contrast with the NH_3 repression of NR in algae and fungi, NO_2^-, nitropropionic acid, NH_4^+ ions, and amino acids did not act as inhibitors of NR activity or synthesis in corn seedlings (Schrader and Hageman, 1967). It was found that certain secondary metabolites, for example, coumarin, *trans*-cinnamic, and *trans-o*-hydroxycinnamic acid, inhibited induction of NR.

Working with cauliflower, Hewitt et al. (1956) were the first investigators to demonstrate that NR is inducible by NO_3^-. In corn (Champigny, 1963), cauliflower (Candela, et al., 1957; Afridi and Hewitt, 1964), and other higher plants (Hewitt and Afridi, 1959), both NO_3^- and Mo appeared to be required for induction of NR. For radish *(Raphanus sativus* L., var. Cherry Belle) cotyledons and corn *(Zea mays* L., var. Hy2 $\times$ Oh7) seedlings, NO_3^- was required for induction of NR as reported by Beevers et al. (1965). They found that, in contrast with fungi (Morton, 1956; Kinsky, 1961) and *Chlorella* (Morris and Syrett, 1963), NH_4^+ ions did not prevent induction of NR. With respect to the effect of light, Beevers et al. (1965) reported that the effect was indirect inasmuch as light increased the permeability of tissue to NO_3^- uptake; induction of NR was proportional to the amount of NO_3^- available.

K^+ was about twice as effective as Rb^+ for the induction of synthesis of NR in *N. crassa;* Na^+, Li^+, or NH_4^+ either failed to stimulate induction or completely inhibited it (Nitsos and Evans, 1966).

In alfalfa, Evans et al. (1950) concluded that Mo functions as a catalyst in symbiotic N fixation and in the reduction of NO_3^-. Mo is apparently required for all plants that reduce NO_3^- or fix free N (Nicholas, 1963).

Mo deficiency markedly reduced NR activity and hemoglobin concentration of soybean nodules (Cheniae and Evans, 1960). Similar effects were noted following application of NO_3^- and, sometimes, NH_4^+ nitrogen.

The nitrogenase system of cell-free extracts of soybean nodules was separated into two fractions with Mo and Fe concentrated in one fraction and Fe in the other (Klucas et al., 1968). Combination of the fractions produced a striking stimulation in activity as compared with those of the individual fractions.

The hydrogenase of soybean nodules releases readily measurable quantities of H (Burris, 1965). Incidentally, cells in nodules are polyploid, but the polyploidy is not believed to be induced by the bacteria; rather, only the spontaneously polyploid root cells are invaded and proliferate to form nodules (Wipf and Cooper, 1940).

It has been suggested the inhibition of modulation in the presence of combined N may be explained by the following events. Rhizobia reduce NO_3^- to NO_2^-; and, NO_2^- catalytically destroys IAA (Tanner and Anderson, 1964). It was noted that ordinarily rhizobia make IAA from tryptophan, and then root infection (i.e., development of nodules) occurs. However, rhizobia synthesize IAA from tryptophan better in the absence than in the presence of NO_3^-.

Mo was required for fixation of atmospheric N by *Azotobacter*, and by root nodule bacteria associated with roots of pea plants (Mulder, 1948).

For some years it was thought that all N-fixing blue-green algae belonged to the *Nostocaceae*, but now fixation has been demonstrated in *Tolypothrix* sp., a representative of the *Scytonemataceae*, in *Calothrix parietina* (family *Rivulariaceae*), and in several other genera in these families (Burris, 1965). Singh (1961) reported that *Aulosira fertilissima* (family *Microchaetaceae*) is the principal organism for N fixation in Indian rice paddies. All of the photosynthetic bacteria examined have proved capable of fixing N (Burris, 1965).

Cobalt

Although Co is not essential for higher plants, with the exception of legumes having N-fixing bacteria, CO^{2+} has certain effects on the metabolism of higher plants. Working with pea roots, Galston and Siegel (1954) discovered that Co^{2+} decreases the apparent rate of synthesis of peroxides and also prevents peroxidative destruction of IAA. Decreasing the rate of destruction of IAA could lead to an increase in growth. Co^{2+}-initiated chain decompositions of hydroperoxides are well known (Tobolsky and Mesrobian, 1954).

In a study of the effect of Co^{2+} on bean leaf expansion and oxidative phosphorylation, it was noted that Co^{2+} counteracts the inhibition by DNP in oxidative phosphorylation and reduces the activity of ATPase (Loercher and Liverman, 1964).

Co^{2+}, but not Mn^{2+}, replaces Mg^{2+} in photosynthetic phosphorylations (Arnon, 1959).

Co^{2+} occurs in high concentrations in the stigma and style of *Lilium longiflorum* (Yamada, 1969).

LITERATURE CITED

Adrian, J. 1962. Antagonisme nutritionnel entre l'acide borique et la riboflavine. Arch. Sci. Physiol. 16:139–166.

Afridi, M. M. R. K., and E. J. Hewitt. 1964. The inducible formation and stability of nitrate reductase in higher plants. I. Effects of nitrate and molybdenum on enzyme activity in cauliflower (*Brassica oleracea* var. *botrytis*). J. Exp. Bot. 15:251–261.

Agarwala, S. C. 1952. Relation of nitrogen supply to the molybdenum requirement of cauliflower grown in sand culture. Nature 169:1099.

Agarwala, S. C., and E. J. Hewitt. 1954a. Molybdenum as a plant nutrient. III. The interrelationships of molybdenum and nitrate supply in the growth and molybdenum content of cauliflower plants grown in sand culture. J. Hort. Sci. 29:278–290.

Agarwala, S. C., and E. J. Hewitt. 1954b. Molybdenum as a plant nutrient. IV. The interrelationships of molybdenum and nitrate supply in chlorophyll and ascorbic acid fractions in cauliflower plants grown in sand culture. J. Hort. Sci. 29:291.

Agarwala, S. C., and E. J. Hewitt. 1955a. Molybdenum as a plant nutrient. V. The interrelationships of molybdenum and nitrate supply in the concentration of sugars, nitrate and organic nitrogen in cauliflower plants grown in sand cultures. J. Hort. Sci. 30:151.

Agarwala, S. C., and E. J. Hewitt. 1955b. Molybdenum as a plant nutrient. VI. Effects of molybdenum supply on the growth and composition of cauliflower plants given different sources of nitrogen supply in sand culture. J. Hort. Sci. 30:163—180.

Agulhon, H. 1910a. Recherches sur la presence et le role du bore chez les vegetaux. Ph.D. Thesis, Univ. Paris, Paris.

Agulhon, H. 1910b. Emploi du bore comme engrais catalytique (The use of boron as a catalytic fertilizer). Compt. Rend. Acad. Sci. (Paris) 150:288—291.

Albert, L. S. 1965. Ribonucleic acid content, boron deficiency symptoms, and elongation of tomato root tips. Plant Physiol. 40:649—654.

Albert, L. S., and C. M. Wilson. 1961. Effect of boron on elongation of tomato root tips. Plant Physiol. 36:244—251.

Almestrand, A. 1957. Growth and metabolism of isolated cereal roots. Physiol. Plantarum 10:521—620.

Arnon, D. I. 1938. Microelements in culture-solution experiments with higher plants. Amer. J. Bot. 25:322—325.

Arnon, D. I. 1949. Copper enzymes in isolated chloroplasts. Polyphenoloxidase in *Beta vulgaris*. Plant Physiol. 24:1—15.

Arnon, D. I. 1950. Criteria of essentiality of inorganic micronutrients for plants, with special reference to molybdenum, pp. 31—39. *In* Trace elements in plant physiology. Chronica Botanica, Waltham, Mass.

Arnon, D. I. 1959. Phosphorus and the biochemistry of photosynthesis. Agrochimica 3:108—139.

Arnon, D. I. 1965. Ferredoxin and photosynthesis. Science 149:1460—1470.

Arnon, D. I., and P. R. Stout. 1939. Molybdenum as an essential element for higher plants. Plant Physiol. 14:599—602.

Arnon, D. I., and F. R. Whatley. 1949. Is chloride a coenzyme of photosynthesis? Science 110:554—556.

Arnon, D. I., F. R. Whatley, and M. B. Allen. 1957. Triphosphopyridine nucleotide as a catalyst of photosynthetic phosphorylation. Nature 180:182—185.

Bachofen, R. 1966. Die Oxydation von Mangan durch Chloroplasten im Licht. Z. Naturforsch. 216:278—284.

Baker, J. E., H. G. Gauch, and W. M. Dugger, Jr. 1956. Effects of boron on the water relations of higher plants. Plant Physiol. 31:89—94.

Bandurski, R. S. 1965. Biological reduction of sulfate and nitrate, pp. 467—490. *In* J. Bonner and J. E. Warner (eds.) Plant biochemistry. Academic, New York.

Beevers, L., and R. H. Hageman. 1969. Nitrate reduction in higher plants. Annu. Rev. Plant Physiol. 20:495—522.

Beevers, L., L. E. Schrader, D. Flesher, and R. H. Hageman. 1965. The role of light and nitrate in the induction of nitrate reductase in radish cotyledons and maize seedlings. Plant Physiol. 40:691—698.

Betts, G. F., and E. J. Hewitt. 1966. Photosynthetic nitrite reductase and the significance of hydroxylamine in nitrite reduction in plants. Nature 210:1327—1329.

Bogorad, L. 1960. The biosynthesis of protochlorophyll, pp. 227—256. *In* M. B. Allen (ed.) Comparative biochemistry of photoreactive systems. Academic, New York.

Bonner, J. 1950. Plant biochemistry. Academic, New York. 537 pp.

Bortels, H. 1930. Molyldän als Katalysator bei der biologisehen Stickstoffbindung. Arch. Mikrobiol. 1:333—342.

Bowen, J. E., and H. G. Gauch. 1965. The essentiality of boron for *Dryopteris dentata* and *Selaginella apoda*. Amer. Fern J. 55:67—73.

Bowen, J. E., and H. G. Gauch. 1966. Nonessentiality of boron in fungi and the nature of its toxicity. Plant Physiol. 41:319—324.

Bowen, J. E., H. G. Gauch, R. W. Krauss, and R. A. Galloway. 1965. The nonessentiality of boron for *Chlorella*. J. Phycol. 1:151—154.

Bradfield, J. R. G. 1947. Plant carbonic anhydrase. Nature 159:467—468.

Brown, J. C., and J. E. Ambler. 1969. Characterization of boron deficiency in soybeans. Physiol. Plantarum 22:177—185.

Brown, R. 1952. Protoplast surface enzymes and absorption of sugar, Int. Rev. Cytol. 1:107.

Brown, S. A., D. Wright, and A. C. Neish. 1959. Studies in lignin biosynthesis using isotopic carbon. VII. Can. J. Biochem. Physiol. 37:25.

Brown, T. E., H. C. Eyster, and H. A. Tanner. 1958. Physiological effects of manganese deficiency, pp. 135—155. *In* C. A. Lamb, O. G. Bentley, and J. M. Beattie (eds.) Trace elements (Proc. Conf. Ohio Agr. Exp. Sta., Wooster, Ohio. Oct. 14—16, 1957). Academic, New York, 410 pp.

Brownell, P. F. 1965. Sodium as an essential micronutrient element for a higher plant (*Atriplex vesicaria*). Plant Physiol. 40:460—468.

Brownell, P. F., and M. E. Jackman. 1966. Changes during recovery from sodium deficiency in *Atriplex*. Plant Physiol. 41:617—622.

Brownell, P. F., and J. G. Wood. 1957. Sodium as an essential micronutrient element for *Atriplex vesicaria* Heward. Nature 179:635—636.

Broyer, T. C., A. B. Carlton, C. M. Johnson, and P. R. Stout. 1954. Chlorine — A micronutrient element for higher plants. Plant Physiol. 29:526—532.

Burkholder, P. R., and E. S. Johnston. 1937. Inactivation of plant growth substances by light. Smithsonian Inst. Misc. Collections, Vol. 95, No. 20.

Burris, R. H. 1965. Nitrogen fixation, pp. 961—979. *In* J. Bonner and J. E. Varner (eds.) Plant biochemistry. Academic, New York.

Burris, R. H. 1966. Biological nitrogen fixation. Annu. Rev. Plant Physiol. 17:155—184.

Burström, H. 1957. Root surface development, sucrose inversion, and free space. Physiol. Plantarum 10:741—751.

Candela, M. I., E. G. Fisher, and E. J. Hewitt. 1957. Molybdenum as a plant nutrient. X. Some factors affecting the activity of nitrate reductase in cauliflower plants grown with different nitrogen sources and molybdenum levels in sand culture. Plant Physiol. 32:280—288.

Catesson, A. M. 1966. Presence de phytoferritine dans le cambium et les tissus conducteurs de la tige de sycomore "*Acer pseudoplatanus.*" Compt. Rend. Acad. Sci. (Paris) 262:1070—1073.

Champigny, M. L. 1963. Sur l'activite et l'induction de la nitrate reductase dans les plantules de Mais. Physiol. Veg. 1:139—169.

Cheniae, G., and H. J. Evans. 1960. Physiological studies on nodule-nitrate reductase. Plant Physiol. 35:454—462.

Cheniae, G., and I. F. Martin. 1966. Studies on the function of manganese in photosynthesis. Brookhaven Symp. Biol. 19:406—417.

Cheniae, G., and I. F. Martin. 1969. Photoreactivation of manganese catalyst in photosynthetic oxygen evolution. Plant Physiol. 44:351—360.

Cresswell, C. F. 1961. An investigation into the nitrate, nitrite and hydroxylamine metabolism in higher plants. Ph.D. Thesis, Univ. Bristol, Bristol.

Daniel, L., and E. Varoczy. 1957. Physiological studies of pollen. II. The effect of boron on the germination of pollen grains and on the growth of pollen tubes. Novenyuedelem 6:309–330.

Davenport, H. E., and R. Hill. 1960. A protein from leaves catalysing the reduction of haem-protein compounds by illuminated chloroplasts. Biochem. J. 74:493–501.

Davenport, H. E., R. Hill, and F. R. Whatley. 1952. A natural factor catalyzing reduction of methaemoglobin by isolated chloroplasts. Proc. Roy. Soc. (London) B139: 346–358.

Dear, J., and S. Aronoff. 1965. Relative kinetics of chlorogenic and caffeic acids during the onset of boron deficiency in sunflower. Plant Physiol. 40:458–459.

DeKock, P. C. 1958. The nutrient balance in plant leaves. Agr. Progress 33:88–95.

DeKock, P. C., K. Commisiong, V. C. Farmer, and R. H. E. Inkson. 1960. Inter-relationships of catalase, peroxidase, hematin, and chlorophyll. Plant Physiol. 35:599–604.

Dickman, S. R., and A. A. Cloutier. 1951. Factors affecting the activity of aconitase. J. Biol. Chem. 188:379–388.

Dugger, W. M., Jr., T. E. Humphreys, and B. Calhoun. 1957. The influence of boron on starch phosphorylase and its significance in translocation of sugars in plants. Plant Physiol. 32:364–370.

Duloy, M. D., and V. F. Mercer. 1961. Studies in translocation. I. The respiration of the phloem. Australian.J. Biol. Sci. 14:391–401.

Dunlap, D. B., and A. H. Thompson. 1959. Effect of boron sprays on the development of bitter-pit in the York Imperial apple. Maryland Agr. Exp. Sta. Bull. A-102.

Dutta, T. R., and W. J. McIlrath. 1964. Effects of boron on growth and lignification in sunflower tissue and organ cultures. Bot. Gaz. 125:89–96.

Dyar, J. J., and K. L. Webb. 1961. A relationship between boron and auxin in C^{14} translocation in bean plants. Plant Physiol. 36:672–676.

Eaton, F. M. 1942. Toxicity and accumulation of chloride and sulfate salts in plants. J. Agr. Res. 64:357–399.

Eaton, F. M. 1944. Deficiency, toxicity, and accumulation of boron in plants. J. Agr. Res. 69:237–277.

Edwards, H. H. 1971. Translocation of carbon in powdery mildewed barley. Plant Physiol. 47:324–328.

Elliott, W. H. 1953. Isolation of glutamine synthetase and glutamotransferase from green peas. J. Biol. Chem. 201:661–672.

El-Sheikh, A. M., and A. Ulrich. 1970. Interactions of rubidium, sodium, and potassium on the nutrition of sugar beet plants. Plant Physiol. 46:645–649.

El-Sheikh, A. M., A. Ulrich, and T. C. Broyer. 1967. Sodium and rubidium as possible nutrients for sugar beet plants. Plant Physiol. 42:1202–1208.

Eschrich, W. 1970. Biochemistry and fine structure of phloem in relation to transport. Annu. Rev. Plant Physiol. 21:193–214.

Evans, H. J. 1956. Role of molybdenum in plant nutrition. Soil Sci. 81:199–208.

Evans, H. J. 1959. The biochemical role of iron in plant metabolism. School Forest. Bull. 15, Duke Univ., Durham, N.C.

Evans, H. J., and N. S. Hall. 1955. Association of molybdenum with nitrate reductase from soybean leaves. Science 122:922–923.

Evans, H. J., and A. Nason. 1953. Pyridine nucleotide-nitrate reductase from extracts of higher plants. Plant Physiol. 28:233–254.

Evans, H. J., and G. J. Sorger. 1966. Role of mineral elements with emphasis on the univalent cations. Annu. Rev. Plant Physiol. 17:47–76.

Evans, H. J., E. R. Purvis, and F. E. Bear. 1950. Molybdenum nutrition of alfalfa. Plant Physiol. 25:555–566.

Eyster, C., T. E. Brown, H. A. Tanner, and S. L. Hood. 1958. Manganese requirement with respect to growth, Hill reaction and photosynthesis. Plant Physiol. 33:235—241.

Ferrari, T. E., and J. E. Varner. 1969. Substrate induction of nitrate reductase in barley aleurone layers. Plant Physiol. 44:85—88.

Folley, S. J., and A. L. Greenbaum. 1948. Determination of the arginase activities of homogenates of liver and mammary gland: Effects of pH and substrate concentration and especially of activation of divalent metal ions. Biochem. J. 43:537—549.

Fritz, G., and H. Beevers. 1955. Cytochrome oxidase content and respiratory rates of etiolated wheat and barley seedlings. Plant Physiol. 30:309—317.

Galston, A. W., and W. S. Hillman. 1961. The degradation of auxin, pp. 647—670. *In* W. Ruhland (ed.) Encyclopedia of plant physiology. Vol. XIV. Growth and growth substances. Springer, Berlin.

Galston, A. W., and S. M. Siegel. 1954. Antiperoxidative action of the cobaltous ion and its consequences for plant growth. Science 120:1070—1071.

Garman, P., L. G. Keirstead, and W. T. Mathis. 1953. Quality of apples as affected by sprays. Connecticut Agr. Exp. Sta. Bull. 576.

Garner, W. W., J. E. McMurtrey, Jr., J. D. Bowling, and E. G. Moss. 1930. Role of chlorine in nutrition and growth of the tobacco plant and its effect on the quality of the cured leaf. J. Agr. Res. 40:627—648.

Gauch, H. G., and W. M. Dugger, Jr. 1953. The role of boron in the translocation of sucrose. Plant Physiol. 28:457—467.

Gauch, H. G., and W. M. Dugger, Jr. 1954. The physiological action of boron in higher plants: A review and interpretation. Maryland Agr. Exp. Sta. Tech. Bull. A-80.

Geiger, D. R., and A. L. Christy. 1971. Effect of sink region anoxia on translocation rate. Plant Physiol. 47:172—174.

Gerhardt, B. 1966. Manganeffekte in photosynthetischen Reaktionen von *Anacytis*. Ber. Deut. Bot. Ges. 79:63—68.

Gerloff, G. C. 1968. The comparative boron nutrition of several green and blue-green algae. Physiol. Plantarum 21:369—377.

Gerretsen, F. C. 1950. Manganese in relation to photosynthesis. II. Redox potentials of illuminated crude chloroplast suspensions. Plant and Soil 2:159—193.

Gewitz, H., and W. Völker. 1962. Über die Atmungsfermente der *Chlorella*. Hoppe-Seyler's Z. Physiol. Chemie 330:124—131.

Gollub, M., and B. Vennesland. 1947. Fixation of carbon dioxide by a plant oxalacetate carboxylase. J. Biol. Chem. 169:233—234.

Gorter, C. J. 1958. Synergism of indole and indole-3-acetic acid in the root production of *Phaseolus* cuttings. Physiol. Plantarum 11:1—9.

Grant, B. R., and H. Beevers. 1961. Hexose absorption by carrot discs. Plant Physiol. 36:xxxi.

Haas, A. R. C. 1945. Influence of chlorine on plants. Soil Sci. 60:53—61.

Habermann, H. 1969. Reversal of copper inhibition in chloroplast reactions by manganese. Plant Physiol. 44:331—336.

Habermann, H. M., M. A. Handel, and P. McKellar. 1968. Kinetics of chloroplast-mediated photooxidation of diketogulonate. Photochem. Photobiol. 7:211—224.

Hageman, R. H., and D. Flesher. 1960. Nitrate reductase activity in corn seedlings as affected by light and nitrate content of nutrient media. Plant Physiol. 35:700—708.

Hageman, R. H., D. Flesher, and A. Gitter. 1961. Diurnal variation and other light effects influencing the activity of nitrate reductase and nitrogen metabolism in corn. Crop Sci. 1:201—204.

Hageman, R. H., C. F. Cresswell, and E. J. Hewitt. 1962. Reduction of nitrate, nitrite and hydroxylamine to ammonia by enzymes extracted from higher plants. Nature 193:247—250.

Harmer, P. M., and E. J. Benne. 1945. Sodium as a crop nutrient. Soil Sci. 60:137–148.

Hewitt, E. J., and M. M. R. K. Afridi. 1959. Adaptive synthesis of nitrate reductase in higher plants. Nature 183:57–58.

Hewitt, E. J., and E. W. Bolle-Jones. 1952a. Molybdenum as a plant nutrient. I. The influence of molybdenum on the growth of some *Brassica* crops in sand culture. J. Hort. Sci. 27:245–256.

Hewitt, E. J., and E. W. Bolle-Jones. 1952b. Molybdenum as a plant nutrient. II. The effect of molybdenum deficiency on some horticultural and agricultural crop plants in sand culture. J. Hort. Sci. 27:257–265.

Hewitt, E. J., and E. W. Jones. 1947. The production of molybdenum deficiency in plants in sand culture with special reference to tomato and *Brassica* crops. J. Pom. Hort. Sci. 23:254–262.

Hewitt, E. J., and C. C. McCready. 1954. Relation of nitrogen supply to the molybdenum requirement of tomato plants grown in sand culture. Nature 174:186–187.

Hewitt, E. J., and C. C. McCready. 1956. Molybdenum as a plant nutrient. VII. The effects of different molybdenum and nitrogen supplies on yield and composition of tomato plants grown in sand culture. J. Hort. Sci. 31:284–290.

Hewitt, E. J., E. G. Fisher, and M. C. Candela. 1956. Factors affecting the activity of nitrate reductase in cauliflower plants. Annu. Rep. Long Ashton Res. Sta. 1955:202.

Hoagland, D. R. 1941. Water culture experiments on molybdenum and copper deficiencies of fruit trees. Proc. Amer. Soc. Hort. Sci. 38:8–12.

Hoagland, D. R. 1945. Molybdenum in relation to plant growth. Soil Sci. 60:119–123.

Homann, P. 1967. Studies on the manganese of the chloroplast. Plant Physiol. 42:997–1007.

Hyde, B. B., A. J. Hodge, A. Kahn, and M. L. Birnstiel. 1963. Studies on phytoferritin. I. Identification and localization. J. Ultrastruct. Res. 9:248–258.

Hylton, L. O., A. Ulrich, and D. R. Cornelius. 1967. Potassium and sodium inter-relations in growth and mineral content of Italian ryegrass. Agron. J. 59:311–314.

Iljin, W. S. 1951. Metabolism of plants affected with lime-induced chlorosis (calciose). II. Organic acids and carbohydrates. Plant and Soil 3:339–351.

Isbell, H. S., J. F. Brewster, N. B. Holt, and H. L. Frush. 1948. Behavior of certain sugars and sugar alcohols in the presence of tetraborates — Correlation of optical rotation and compound formation. Nat. Bur. Standards Res. Paper RP 1862.

Johnson, C. M., P. R. Stout, T. C. Broyer, and A. B. Carlton. 1957. Comparative chlorine requirements of different plant species. Plant and Soil 8:337–353.

Johnson, D. L., and L. S. Albert. 1967. Effect of selected nitrogen-bases and boron on the ribonucleic acid content, elongation, and visible deficiency symptoms of tomato root tips. Plant Physiol. 42:1307–1309.

Johri, B. M., and I. K. Vasil. 1961. Physiology of phloem. Bot. Rev. 27:325–381.

Joy, K. W. 1969. Nitrogen metabolism of *Lemna minor*. II. Enzymes of nitrate assimilation and some aspects of their regulation. Plant Physiol. 44:849–853.

Joy, K. W., and R. H. Hageman. 1966. The purification and properties of nitrite reductase from higher plants and its dependence on ferredoxin. Biochem. J. 100:263–273.

Keilin, D., and T. Mann. 1938. Polyphenol oxidase. Purification, nature and properties. Proc. Roy. Soc. (London) B125:187–204.

Keilin, D., and T. Mann. 1939a. Laccase, a blue copper-protein oxidase from the latex of *Rhus succedanea*. Nature 143:23–24.

Keilin, D., and T. Mann. 1939b. Carbonic anhydrase. Nature 144:442–443.

Kessler, E. 1955. On the role of manganese in the oxygen-evolving system of photo-synthesis. Arch. Biochem. Biophys. 59:527–529.

Kessler, E. 1957. Stoffwechselphysiologische Untersuchungen an Hydrogenase enhaltenden Grunalgen. I. Über die Rolle des Mangans bei Photoreduktion und Photosynthese. Planta 49:435−454.

Khym, J. X., and W. E. Cohn. 1953. The separation of sugar phosphates by ion exchange with the use of the borate complex. J. Amer. Chem. Soc. 75:1153−1156.

Kinsky, S. C. 1961. Induction and repression of nitrate reductase in *Neurospora crassa*. J. Bacteriol. 82:898−904.

Klucas, R. B., B. Koch, S. A. Russell, and H. J. Evans. 1968. Purification and some properties of the nitrogenase from soybean (*Glycine max* Merr.) nodules. Plant Physiol. 43:1906−1912.

Kubowitz, F. 1938. Spaltung und Resynthese der Polyphenoloxydase und des Hämocyanins. Biochem. Z. 299:32−57.

Kuhn, R., I. Löw, and F. Moewus. 1942. Kurze Originalmitteilung. Naturwissenschaften 30:407.

Lee, K., C. M. Whittle, and H. J. Dyer. 1966. Boron deficiency and translocation profiles in sunflower. Physiol. Plantarum 19:919−924.

Lee, S. G., and S. Aronoff. 1966. Investigations on the role of boron in plants. III. Anatomical observations. Plant Physiol. 41:1570−1577.

Lee, S. G., and S. Aronoff. 1967. Boron in plants: A biochemical role. Science 158:798−799.

Lewin, J. C. 1965. The boron requirement of a marine diatom. Naturwissenschaften 70:1−2.

Lewin, J. C. 1966a. Physiological studies of the boron requirement of the diatom *Cylindrotheca fusiformis* Reinmann and Lewin. J. Exp. Bot. 17:473−479.

Lewin, J. C. 1966b. Boron as a growth requirement for diatoms. J. Phycol. 2:160−163.

Liebich, H. 1941. Quantitativ-chemische Untersuchungen über das Eisen in den Chloroplasten und übrigen Zellestandteilen von *Spinacia oleracea*. Z. Bot. 37:129.

Linskens, H. 1955. Physiologische Untersuchungen der Pollenschlauch-Hemmung selbststeriler Petunien. Z. Bot. 43:1−44.

Lipman, C. B. 1938. Importance of silicon, aluminum, and chlorine for higher plants. Soil Sci. 45:189−198.

Loercher, L., and J. L. Liverman. 1964. Influence of cobalt on leaf expansion and oxidative phosphorylation. Plant Physiol. 39:720−725.

Losada, M., J. M. Ramirez, A. Paneque, and F. F. Del Campo. 1965. Light and dark reduction of nitrate in a reconstituted chloroplast system. Biochim. Biophys. Acta 109:86−96.

Loughman, B. C. 1961. Effect of boric acid on the phosphoglucomutase of pea seeds. Nature 191:1399−1400.

Lunt, O. R. 1966. Sodium, pp. 409−432. *In* H. D. Chapman (ed.) Diagnostic criteria for plants and soils. Div. Agr. Sci., Univ. California, Berkeley.

McBride, L., W. Chorney, and J. Skok. 1967. The boron requirement of *Chlorella*. Plant Physiol. 42(Suppl.):S-5.

McElroy, W. D., and B. Glass (eds.). 1950. Copper metabolism − A symposium on animal, plant and soil relationships. Johns Hopkins Press, Baltimore. 443 pp.

McIlrath, W. J. 1965. Mobility of boron in several dicotyledonous species. Bot. Gaz. 126:27−30.

McIlrath, W. J., and B. F. Palser. 1956. Responses of tomato, turnip, and cotton to variations in boron nutrition. I. Physiological responses. Bot. Gaz. 118:43−52.

McIlrath, W. J., and J. Skok. 1958. Boron requirement of *Chlorella vulgaris*. Bot. Gaz. 119:231−233.

McIlrath, W. J., and J. Skok. 1964a. Distribution of boron in the tobacco plant. Physiol.

Plantarum 17:839—845.

McIlrath, W. J., and J. Skok. 1964b. Boron nutrition and lignification in sunflower and tobacco stems. Bot. Gaz. 125:268—271.

McIlrath, W. J., and J. Skok. 1966. Substitution of germanium for boron in plant growth. Plant Physiol. 41:1209—1212.

McKenna, J. M., and N. I. Bishop. 1967. Studies on the photooxidation of manganese by isolated chloroplasts. Biochim. Biophys. Acta 131:339—349.

McLean, J. G., W. C. Sparks, and A. M. Binkley. 1944. The effect of certain minor elements on yield, size, and skin thickness of potato tubers. Proc. Amer. Soc. Hort. Sci. 44:362—368.

Malcolm, E. W., J. W. Green, and H. A. Swenson. 1964. A study of the borate-carbohydrate complex. J. Chem. Soc. 903:4669—4676.

Maranville, J. W. 1970. Influence of nickel on the detection of nitrate reductase in sorghum extracts. Plant Physiol. 45:591—593.

Marinos, N. G. 1967. Multifunctional plastids in the meristematic region of potato ruber buds. J. Ultrastruct. Res. 17:91—113.

Marsh, H. V., Jr., H. J. Evans, and G. Matrone. 1963a. Investigations on the role of iron in chlorophyll metabolism. I. Effect of iron deficiency on chlorophyll and heme content and on the activities of certain enzymes in leaves. Plant Physiol. 38:632—638.

Marsh, H. V., Jr., H. J. Evans, and G. Matrone. 1963b. Investigations on the role of iron in chlorophyll metabolism. II. Effect of iron deficiency on chlorophyll synthesis. Plant Physiol. 38:638—642.

Martin, G., and J. Lavollay. 1958. Le chlore, oligo-élément indispensable pour *Lemna minor*. Experientia 14:333—334.

Meagher, W. R. 1952. Molybdenum nutrition of the higher plant as revealed by nitrogen metabolites. Ph.D. Thesis, Univ. of California, Berkeley.

Milborrow, B. V. 1964. 2,6-Dichlorobenzonitrile and boron deficiency. J. Exp. Bot. 15:515—524.

Mitchell, J. W., W. M. Dugger, Jr., and H. G. Gauch. 1953a. Increased translocation of plant-growth-modifying substances due to application of boron. Science 118: 354—355.

Mitchell, J. W., W. M. Dugger, Jr., and H. G. Gauch. 1953b. Greater effects from growth modifiers. Agr. Res. 2:15.

Morris, I., and P. J. Syrett. 1963. The development of nitrate reductase in *Chlorella* and its repression by ammonium. Arch. Mikrobiol. 47:32—41.

Mortenson, L. E., R. C. Valentine, and J. E. Carnahan. 1962. An electron transport factor from *Clostridium pasteurianum*. Biochem. Biophys. Res. Commun. 7:448—452.

Morton, A. G. 1956. A study of nitrate reduction in mould fungi. J. Exp. Bot. 7:97—112.

Mulder, E. G. 1948. Importance of molybdenum in the nitrogen metabolism of microorganisms and higher plants. Plant and Soil 1:94—19

Neales, T. F. 1959. The boron requirement of flax roots grown in sterile culture. J. Exp. Bot. 10:426—436.

Neales, T. F. 1960. Some effects of boron on root growth. Australian J. Biol. Sci. 13:232—248.

Neales, T. F. 1964. A comparison of the boron requirement of intact tomato plants and excised tomato roots grown in sterile culture. J. Exp. Bot. 15:647—653.

Neales, T. F. 1967. The boron nutrition of the diatom, *Cylindrotheca fusiformis*, grown on agar, and the biological activity of some substituted phenylboric acids. Australian J. Biol. Sci. 20:67—76.

Neales, T. F., and R. W. Hinde. 1962. A test of the calcium-boron interaction hypothesis using the growth of the bean radicle. Physiol. Plantarum 15:217—228.

Nelson, C. D., and P. R. Gorham. 1957. Uptake and translocation of C^{14}-labeled sugars applied to primary leaves of soybean seedlings. Can. J. Bot. 35:340–347.

Neumann, K. H., and F. C. Steward. 1968. Investigations on the growth and metabolism of culture explants of *Daucus carota*. I. Effects of iron, molybdenum and manganese on growth. Planta 81:333–350.

Nicholas, D. J. D. 1963. Inorganic nutrient nutrition of microorganisms, pp. 363–447. *In* F. C. Steward (ed.) Plant physiology, Vol. III. Academic, New York.

Nicholas, D. J. D., and A. Nason. 1954a. Molybdenum and nitrate reductase. II. Molybdenum as a constituent of nitrate reductase. J. Biol. Chem. 207:353–360.

Nicholas, D. J. D., and A. Nason. 1954b. Mechanism of action of nitrate reductase from *Neurospora*. J. Biol. Chem. 211:183–197.

Nitsos, R. E., and H. J. Evans. 1966. Effects of univalent cations on the inductive formation of nitrate reductase. Plant Physiol. 41:1499–1504.

Noakes, J. E., and D. W. Hood. 1961. Boron-boric acid complexes in sea-water. Deep-Sea Res. 8:121–129.

Odhnoff, C. 1957. Boron deficiency and growth. Physiol. Plantarum 10:984–1000.

Odhnoff, C. 1961. The influence of boric acid and phenylboric acid on the root growth of bean (*Phaseolus vulgaris*). Physiol. Plantarum 14:187–220.

O'Kelley, J. C. 1957. Boron effects on growth, oxygen uptake and sugar absorption by germinating pollen. Amer. J. Bot. 44:239–244.

Otting, W. 1951. Borgehalt und Verteilung des Bors in verschiedenen Pflanzen. Z. Pflanzenernähr. Düng. Bodenk. 55:235–247.

Palser, B. F., and W. J. McIlrath. 1956. Responses of tomato, turnip, and cotton to variations in boron nutrition. II. Anatomical responses. Bot. Gaz. 118:53–71.

Piper, C. S. 1940. Molybdenum as an essential element for plant growth. J. Australian Inst. Agr. Sci. 6:162–164.

Pirson, A. 1937. Ernährungs- und Stoffwechselphysiologische Untersuchungen an *Fontinalis* und *Chlorella*. Z. Bot. 31:193–267.

Pirson, A., C. Tichy, and G. Wilhelmi. 1952. Stoffwechsel und Mineralsalzernahrung einzelliger Grunalgen. Mitteilung I. Vergleichende Untersuchungen an Mangel-kulturen von *Ankistrodesmus*. Planta 40:199–253.

Popp, H. W., and H. R. C. McIlvaine. 1937. Growth substances in relation to the mechanism of the action of radiation on plants. J. Agr. Res. 55:931–936.

Possingham, J. V., and D. Spencer. 1962. Manganese as a functional component of chloroplasts. Australian J. Biol. Sci. 15:58–68.

Powers, W. H., S. Lewis, and C. R. Dawson. 1944. The preparation and properties of highly purified ascorbic acid oxidase. J. Gen. Physiol. 27:167–180.

Presley, H. J., and W. J. McIlrath. 1956. Effect of substrate hydrogen peroxide on growth of boron deficient plants. Trans. Illinois State Acad. Sci. 49:36–42.

Price, C. A. 1962. Zinc-dependent lactate dehydrogenase in *Euglena gracilis*. Biochem. J. 82:61–66.

Price, C. A. 1966. Control of processes sensitive to zinc in plants and microorganisms, pp. 1–20. *In* A. S. Prasad (ed.), Zinc metabolism. C. C. Thomas, Springfield, Ill.

Price, C. A. 1968. Iron compounds and plant nutrition. Annu. Rev. Plant Physiol. 19:239–248.

Rajaratnam, J. A., J. B. Lowry, P. N. Avadhani, and R. H. V. Corley. 1971. Boron: Possible role in plant metabolism. Science 172:1142.

Raleigh, G. J. 1948. Effects of the sodium and of the chloride ion in the nutrition of the table beet in culture solutions. Proc. Amer. Soc. Hort. Sci. 51:433–436.

Ranganathan, V., and P. Rengasamy. 1965. Periodic position of major and minor elements that occur in plants. J. Proc. Inst. Chem. (India) 37:193.

Reuther, W., and C. K. Labanauskas. 1966. Copper, pp. 157—179. *In* H. D. Chapman (ed.) Diagnostic criteria for plants and soils. Div. Agr. Sci., Univ. California, Berkeley.

Ritenour, G. L., K. W. Joy, J. Bunning, and R. H. Hageman. 1967. Intracellular localization of nitrate reductase, nitrite reductase, and glutamic acid dehydrogenase in green leaf tissue. Plant Physiol. 42:233—237.

Roy, T. T., F. K. Hille, and H. E. Clark. 1966. Requirement of Ginkgo pollen-derived tissue cultures for boron and effects of boron deficiency. Plant Physiol. 41:815—820.

Sanderson, G. W., and E. C. Cocking. 1964a. Enzymic assimilation of nitrate in tomato plants. I. Reduction of nitrate to nitrite. Plant Physiol. 39:416—422.

Sanderson, G. W., and E. C. Cocking. 1964b. Enzymic assimilation of nitrate in tomato plants. II. Reduction of nitrite to ammonia. Plant Physiol. 39:423—431.

San Pietro, A., and H. M. Lang. 1958. Photosynthetic pyridine nucleotide reductase. I. Partial purification and properties of the enzyme from spinach. J. Biol. Chem. 231:211—229.

Scholz, G. 1962. Versuche mit Bor an Lemnaceen. Die Wirkung des Bors auf den Kohlenhydratgehalt. Die Kulturpflanze 10:63—71.

Schrader, L. E., and R. H. Hageman. 1967. Regulation of nitrate reductase activity in corn (*Zea mays* L.) seedlings by endogenous metabolites. Plant Physiol. 42: 1750—1756.

Schrader, L. E., D. M. Peterson, E. R. Leng, and R. H. Hageman. 1966. Nitrate reductase activity of maize hybrids and their parental inbred. Crop Sci. 6:169—173.

Schrader, L. E., G. L. Ritenour, G. L. Eilrich, and R. H. Hageman. 1968. Some characteristics of nitrate reductase from higher plants. Plant Physiol. 43:930—940.

Scott, E. G. 1960. Effect of supra-optimal boron levels on respiration and carbohydrate metabolism of *Helianthus annuus*. Plant Physiol. 35:653—661.

Seckbach, J. 1969. Iron content and ferritin in leaves of iron treated *Xanthium pennsylvanicum* plants. Plant Physiol. 44:816—820.

Sherwin, J. E., and W. K. Purves. 1969. Tryptophan as an auxin precursor in cucumber seedlings. Plant Physiol. 44:1303—1309.

Shiralipour, A., H. C. Harris, and S. H. West. 1969. Boron deficiency and amino acid and protein contents of peanut leaves. Crop Sci. 9:455—456.

Shkolnik, M. Y., and S. A. Abdurashitov. 1958. [Influence of microelements on synthesis and translocation of carbohydrates]. Plant Physiol. (USSR) 5(5):393—399.

Shkolnik, M. Y., and A. N. Mayevskaya. 1962. Significance of boron in nucleic acid metabolism. Plant Physiol. (USSR) 9:270—278.

Shkolnick, M. Y., and E. A. Soloviyova-Troitzkaya. 1962. On the physiological role of boron. 2. The significance of temperature in the elimination of boron deficiency by supplying RNA. Bot. J. 47:626—635.

Shkolnik, M. Y., A. N. Mayevskaya, and E. A. Soloviyova. 1961. The significance of boron in the nucleate metabolism. 5th Int. Congr. Biochem.

Singh, R. N. 1961. Role of blue-green algae in nitrogen economy of Indian agriculture. Indian Council Agr. Res:, New Delhi, India.

Sisler, E. C., W. M. Dugger, Jr., and H. G. Gauch. 1956. The role of boron in the translocation of organic compounds in plants. Plant Physiol. 31:11—17.

Skok, J. 1956. Relationship of boron nutrition to radiosensitivity of sunflower plants. Plant Physiol. 31(Suppl.):xii.

Skok, J. 1957a. Relationship of boron nutrition to radiosensitivity of sunflower plants. Plant Physiol. 32:648—658.

Skok, J. 1957b. The substitution of complexing substances for boron in plant growth. Plant Physiol. 32:308—312.

Skok, J. 1958. The role of boron in the plant cell, pp. 227—243. *In* C. A. Lamb,

O. G. Bentley, and J. M. Beattie (eds.) Trace elements. Academic, New York.

Skok, J. 1961. Effect of GA and light on the utilization of boron and the development of boron deficiency symptoms. Plant Physiol. 36(Suppl.):xv–xvi.

Skok, J. 1968. Relationship of boron to gibberellic acid-induced proliferation in debudded tobacco plants. Plant Physiol. 43:384–388.

Skok, J., and W. J. McIlrath. 1957. Influence of boron on the growth of *Chlorella*. Plant Physiol. 32(Suppl.):xxiii.

Skok, J., and W. J. McIlrath. 1958. Distribution of boron in cells of dicotyledonous plants in relation to growth. Plant Physiol. 33:428–431.

Skoog, F. 1940. Relationships between zinc and auxin in the growth of higher plants. Amer. J. Bot. 27:939–951.

Slack, C. R., and W. J. Whittington. 1964. The role of boron in plant growth. III. The effects of differentiation and deficiency on radicle metabolism. J. Exp. Bot. 15:495–514.

Smillie, R. M. 1963. Formation and function of soluble proteins in chloroplasts. Can. J. Bot. 41:123–154.

Sommer, A. L. 1931. Copper as an essential element for plant growth. Plant Physiol. 6:339–345.

Sommer, A. L., and C. B. Lipman. 1926. Evidence of the indispensable nature of zinc and boron for higher green plants. Plant Physiol. 1:231–249.

Speck, J. F. 1949. The effect of cations on the decarboxylation of oxalacetic acid. J. Biol. Chem. 178:315–324.

Spencer, D. 1959. A diphosphopyridine nucleotide-specific nitrate reductase from germinating wheat. Australian J. Biol. Sci. 12:181–192.

Spencer, D., and J. V. Possingham. 1961. The effect of manganese deficiency on photophosphorylation and the oxygen evolving-sequence in spinach chloroplasts. Biochim. Biophys. Acta 52:379–381.

Spencer, D., and J. G. Wood. 1954. The role of molybdenum in nitrate reduction in higher plants. Australian J. Biol. Sci. 7:425–434.

Spurr, A. R. 1957a. Boron in morphogenesis of plant cell walls. Science 126:78–80.

Spurr, A. R. 1957b. The effect of boron on cell wall structure in celery. Amer. J. Bot. 44:637–651.

Stanley, R. C. 1961. Effects of various forms of boron on germinating pollen. Plant Physiol. 36(Suppl.):xvi.

Stanley, R. C., and E. A. Lichtenberg. 1963. The effect of various boron compounds on *in vitro* germination of pollen. Physiol. Plantarum 16:337–346.

Stanley, R. C., and F. A. Loewus. 1964. Boron and myo-inositol in pollen pectin biosynthesis, pp. 1–9. *In* H. F. Linskens, (ed.) Int. Symp. Pollen Physiol. and Fertilization.

Starck, J. R. 1963. Effect of boron on the cell wall structure of sunflower. Acta Soc. Bot. Pol. 32:619–623.

Stark, F. C., and I. C. Haut. 1958. Mineral nutrient requirements of cantaloupes. Maryland Agr. Exp. Sta. Bull. A-93.

Steinberg, R. A. 1937. Role of molybdenum in the utilization of ammonium and nitrate nitrogen by *Aspergillus niger*. J. Agr. Res. 55:891–902.

Steward, F. C., K. H. Neumann, and K. V. N. Rao. 1968. Investigations on the growth and metabolism of cultured explants of *Daucus carota*. II. Effects of iron, molybdenum and manganese on metabolism. Planta 81:351–371.

Stout, P. R., and W. R. Meagher. 1948. Studies of the molybdenum nutrition of plants with radioactive molybdenum. Science 108:471–473.

Stout, P. R., C. M. Johnson, and T. C. Broyer. 1956. Chlorine in plant nutrition –

Experiments with plants in nutrient solutions establish chlorine as a micronutrient essential to plant growth. California Agr. 10:10.

Swader, J. A., and C. R. Stocking. 1971. Nitrate and nitrite reduction by *Wolffia arrhiza*. Plant Physiol. 47:189–191.

Syrett, P. J., and I. Morris. 1963. Inhibition of nitrate assimilation by ammonia in *Chlorella*. Biochim. Biophys. Acta 67:566–577.

Tagawa, K., and D. I. Arnon. 1962. Ferredoxins as electron carriers in photosynthesis and in the biological production of hydrogen gas. Nature 195:537–543.

Tanner, J. W., and I. C. Anderson. 1964. External effect of combined nitrogen on nodulation. Plant Physiol. 39:1039–1043.

Taylor, D. M., P. W. Morgan, H. E. Joham, and J. V. Amin. 1968. Influence of substrate and tissue manganese on the IAA-oxidase system in cotton. Plant Physiol. 43: 243–247.

Teichler-Zallen, D. 1969. The effect of manganese on chloroplast structure and photosynthetic ability of *Chlamydomonas reinhardi*. Plant Physiol. 44:701–710.

Thellier, M. 1963. Contribution a l'etude de la nutrition en bore des vegetaux. Ph.D. Thesis, Univ. Paris, Paris.

Tissières, A. 1948. Reconstruction of laccase from its protein and copper. Nature 162: 340–341.

Tobolsky, A. V., and R. B. Mesrobian. 1954. Organic peroxides. Interscience, New York.

Torssell, K. 1956. Chemistry of arylboric acids. VI. Effects of arylboric acids on wheat roots and the role of boron in plants. Physiol. Plantarum 9:652–664.

Totter, J. R., W. T. Burnett, Jr., R. A. Monroe, I. B. Whitney, and C. L. Comar. 1953. Evidence that molybdenum is a nondialyzable component of xanthine oxidase. Science 118:555.

Travis, R. L., W. R. Jordan, and R. C. Huffaker. 1969. Evidence for an inactivating system of nitrate reductase in *Hordeum vulgare* L. during darkness that requires protein synthesis. Plant Physiol. 44:1150–1156.

Tsui, C. 1948a. The role of zinc in auxin synthesis in the tomato plant. Amer. J. Bot. 35:172–179.

Tsui, C. 1948b. The effect of zinc on water relation and osmotic pressure of the tomato plant. Amer. J. Bot. 35:309–311.

Turian, G. 1955. Recherches sur l'action de l'acide borique sur la fructification des *Sordaria*. Phytopath. Z. 25:181–189.

Turian, G. 1956. Le corps paranucleaire des gametes geants d'*Allomyces javanicus* traite a l'acide borique. Protoplasma 47:135–138.

Turnowska-Starck, Z. 1959. [Researches on the absorption of sucrose by the epidermis of *Allium cepa*.] Bot. Pol. Acta Soc. 28:409–424.

Turnowska-Starck, Z. 1960a. [The influence of boron and phosphorus on the absorption and translocation of sucrose.] Acta Soc. Bot. Pol. 29:219–247.

Turnowska-Starck, Z. 1960b. [The influence of boron on the translocation of sucrose in bean seedlings.] Acta Soc. Bot. Pol. 29:533–552.

Ulrich, A., and K. Ohki. 1956. Chlorine, bromine and sodium as nutrients for sugar beet plants. Plant Physiol. 31:171–181.

Ulrich, A., and K. Ohki. 1966. Potassium, pp. 362–393. *In* H. D. Chapman (ed.) Diagnostic criteria for plants and soils. Div. Agr. Sci., Univ. California, Berkeley.

Vallee, B. L. 1959. Biochemistry, physiology, and pathology of zinc. Physiol. Rev. 39:443–490.

Vasil, I. K. 1964. Effect of boron on pollen germination and pollen tube growth. *In* H. F. Linskens (ed.), Pollen physiology and fertilization. North-Holland, Amsterdam.

Vennesland, B., and R. Z. Felsher. 1946. Oxalacetic and pyruvic carboxylases in some dicotyledonous plants. Arch. Biochem. 11:279–306.

Visser, T. 1955. Germination and storage of pollen. Lab. voor Tuinbouwplantenteelt, Landbouwhogeschool, Wageningen. Pub. 134.

Walker, R. B. 1948. Molybdenum deficiency in serpentine barren soils. Science 108: 473—475.

Warburg, O., and W. Lüttgens. 1946. Photochemical reduction of quinone in green cells and granules. Biochimia 11:303—322.

Warrington, K. 1923. The effect of boric acid and borax on the broad bean and certain other plants. Ann. Bot. 37:629—672.

Watanabe, R., W. J. McIlrath, J. Skok, W. Chorney, and S. H. Wender. 1961. Accumulation of scopoletin glucoside in boron deficient tobacco leaves. Arch. Biochem. Biophys. 94:241—243.

Watanabe, R., W. Chorney, J. Skok, and S. H. Wender. 1964. Effect of boron deficiency on polyphenol production in the sunflower. Phytochemistry 3:391—393.

Weiser, C. J., L. T. Blaney, and P. Li. 1964. The question of boron and sugar translocation in plants. Physiol. Plantarum 17:589—599.

Whittington, W. J. 1957. The role of boron in plant growth. I. The effect on general growth, seed production and cytological behaviour. J. Exp. Bot. 8:353—367.

Whittington, W. J. 1959. The role of boron in plant growth. II. The effect on growth of the radicle. J. Exp. Bot. 10:93—103.

Wildman, S. G., M. G. Ferri, and J. Bonner. 1947. The enzymatic conversion of tryptophan to auxin by spinach leaves. Arch. Biochem. 13:131—144.

Williams, M. C. 1960. Effect of sodium and potassium salts on growth and oxalate content of *Halogeton.* Plant Physiol. 35:500—505.

Wilson, C. M. 1961. Cell wall carbohydrates in tobacco pith parenchyma as affected by boron deficiency and by growth in tissue culture. Plant Physiol. 36:336—341.

Winfield, M. E. 1945. The role of boron in plant metabolism. 3. The influence of boron on certain enzyme systems. Australian J. Exp. Biol. 23:267—272.

Wipf, L., and D. C. Cooper. 1940. Somatic doubling of chromosomes and nodular infection in certain Leguminosae. Amer. J. Bot. 27:821—824.

Woolley, J. T. 1957. Sodium and silicon as nutrients for the tomato plant. Plant Physiol. 32:317—321.

Yamada, Y. 1969. The effect of cobalt on the growth of pollen. IV. Intracellular localization of labelled cobalt determined with autoradiographs. Bot. Mag. (Tokyo) 82:488—490.

Yih, R. Y., and H. E. Clark. 1965. Carbohydrate and protein content of boron-deficient tomato root tips in relation to anatomy and growth. Plant Physiol. 40:312—315.

Ziegler, H. 1956. Untersuchungen uber die Leitung und Sekretion der Assimilate. Planta 47:447—500.

Zeigler, H. 1958. Über die Atmung und den Stofftransport in den Isolierten Leitbündeln der Blattstiele von *Heracleum Mantegazzianum* Somm. et Lev. Planta 51:186—200.

Zittle, C. A. 1951. Reaction of borate with substances of biological interest, pp. 493—527. *In* F. F. Nord (ed.) Advances in Enzymology and Related Subjects of Biochemistry, Vol. XII. Interscience, New York.

CHAPTER 11

Roles of Elements in Lower Plants

INTRODUCTION

In this section the roles of various elements are discussed primarily with regard to their requirements or roles in algae, fungi, or bacteria. Certain of these elements may be nonessential for some of these organisms.

Ketchum (1954), Krauss (1958), Hutner and Provasoli (1964), and O'Kelley (1968) reviewed the requirements and roles for various elements in algae. Ketchum (1954) summarized the compositions of nutrient solutions used by various researchers for the growth of algae (Table 11.1).

Krauss and Thomas (1954) described an apparatus for the mass culture of *Scenedesmus obliquus*. They stressed the role of a continuous CO_2 supply in maintaining a desirable pH and in preventing a pH rise to an inhibitory level. Daily utilization of nutrients was followed in order to determine at what rates the nutrients would have to be replenished for continuous culture of algae.

In continuous pure culture, Eyster (1967) demonstrated that *Chlorella sorokiniana* required, in addition to C, H, and O, the following: N, P, Mg, S, K, Fe, Ca, Mn, Zn, and Cu, but not B, Na, Co, or Cl. He was unable to demonstrate a requirement for Mo but assumed that cells received enough as a contaminant to satisfy the role of this element in NO_3^- reduction. He noted that the comparative molar content of harvested cells for each nutrient, except C, H, and O, could be expressed as: $N_{24,000}K_{3,000}P_{1,400}S_{500}Mg_{400}Ca_{150}Mn_{20}Fe_{10}Zn$ and Cu. Optimal nutrient concentrations were: Mg, 1×10^{-4} to 2×10^{-2} M; P, 5×10^{-4} to 2×10^{-2} M; K, 5×10^{-4} to 1×10^{-1} M; Fe, 2×10^{-5} to 1×10^{-3} M; Ca, 5×10^{-4} to 3×10^{-3} M; Mn, 1×10^{-5} to 1×10^{-2} M; Zn, 1×10^{-5} to 1×10^{-2} M; and, Cu, 10^{-8} to 10^{-3} M.

Table 11.1
Concentration of Major Nutrients in Various Culture Media (for growing algae) [a,b]

Investigator	Ca	Mg	K	Na	S	N	P	Fe	Mn	Si
Craig and										
Trelease (1937)	—	486.0	1680.0	2.0	640.0	350.0	558.0	0.6	0.4	0.1
Chu (1942)	2.0	2.4	24.0	2.0	—	5.0	18.0	0.8	—	1.8
Chu (1942)	91.0	7.6	18.7	—	—	17.0	9.0	—	—	9.1
Scott (1943)	244.0	50.0	202.0	—	66.7	171.0	57.0	4.4	1.0	—
Myers and										
Clark (1944)	—	240.0	820.0	—	323.3	168.0	279.0	7.0	0.5	—
Rodhe (1948)	14.7	1.0	2.2	7.5	1.3	10.2	0.9	0.2	0.01	4.6
Witsch (1948)	12.2	9.8	100.0	—	13.0	106.0	34.0	3.5	—	—
Gerloff et al.,										
(1950)	—	0.1	2.3	—	0.8	13.6	0.5	0.03	—	—
Tanada (1951)	0.3	6.1	399.0	1.8	8.4	87.0	64.0	0.2	24.0	—
Geohegan (1953)	—	486.0	1320.0	2.0	1240.0	305.0	558.0	0.6	0.4	0.1

[a] Ketchum (1954).
[b] Values are given in parts per million.

In order to determine the nutrient requirements of algae, Gerloff and Fishbeck (1969) argued in favor of determining critical concentrations within algal cells rather than of varying the concentrations of nutrients in the medium. Critical concentration was the concentration that permitted maximal or near-maximal growth. They studied four green algae, *Chlorella pyrenoidosa*, *Scenedesmus quadricauda*, *Draparnaldia plumosa*, and *Stigeoclonium tenue*, and two blue-green algae, *Microcystis aeruginosa* and *Nostoc muscorum*. For these six algae requirements for Ca^{2+} were extremely low (0.06% or less dry weight): for Mg^{2+}, equal to or only slightly less than those for higher plants (0.15–0.30%), with the exception of *S. quadricauda* (0.05%); requirements for K^+ varied greatly from values less than those for higher plants (0.25–0.50%) to values equal to or in excess of those of higher plants (0.80–2.40%). *Scenedesmus quadricauda* had extremely low requirements for Ca^{2+}, Mg^{2+}, and K^+ (Table 11.2).

ROLES OR EFFECTS

Sodium

As early as 1906–1912, it was demonstrated that Na^+ was essential for certain marine algae, including, among others, *Ulva lactuca*, *Porphyra perforata*, *P. naiadum*, and *Chondrus crispus* (Osterhout, 1912). More recently, several blue-green algae were shown to require Na^+ (Allen and Arnon, 1955b; Kratz and Myers, 1955). For both higher plants (Harmer and Benne, 1945) and lower plants (Allen, 1952), Na^+ may partially substitute for K^+.

Table 11.2
The Critical Concentration and the Range of Concentration of Ca, Mg, and K in Various Species of Green and Blue-Green Algae[a]

Species	Critical concentration (%)			Range of concentration (%)		
	Ca	Mg	K	Ca	Mg	K
Chlorella pyrenoidosa	0.00	0.15	0.40	0.00–0.01	0.03–0.37	0.11–1.44
Scenedesmus quadricauda	0.06	0.05	0.25	0.01–0.66	0.03–0.08	0.15–0.38
Draparnaldia plumosa	0.03	0.20	2.40	0.01–0.26	0.02–0.52	0.52–3.91
Stigeoclonium tenue	0.03	0.25	1.90	0.03–0.25	0.09–0.47	0.44–2.57
Microcystis aeruginosa	0.04	0.30	0.50	0.03–0.95	0.17–1.56	0.21–0.55
Nostoc muscorum	0.02	0.25	0.80	0.01–0.44	0.18–0.46	0.51–0.99

[a] Gerloff and Fishbeck (1969).

Of the many species and strains of blue-green algae studied (Allen, 1952), *N. muscorum* grew with K^+ but not with Na^+, whereas the following species grew with Na^+ but not with K^+: *Synechoccus cedorum, Chroococcus turgidus, Chroococcus* sp., and *Oscillatoria* sp. strain E-2. Many of the species were indifferent as to whether Na^+ or K^+ was the major alkali metal in the medium. One of the photosynthetic bacteria, *Rhodopseudomonas spheroides,* required Na^+ when grown aerobically in the dark (Grosse, 1963). Most, or perhaps all, moderate halophytic bacteria, similar to the marine species, have specific Na^+ requirements (MacLeod and Onofrey, 1956).

Optimal growth of a blue-green alga, *Anabaena cylindrica,* required 5 ppm Na^+ in the culture medium (Allen and Arnon, 1955b). K^+, Li^+, Rb^+, or Cs^+ could not substitute for Na^+. This alga required a higher level of Na^+ (0.4 meq/liter) to prevent chlorosis when grown with NO_3^- than when grown without combined N (0.004 meq Na Cl/liter) (Brownell and Nicholas, 1967). Na^+ deficiency resulted in increased incorporation of $^{15}NO_3$, $^{15}NO_2$, $^{15}NH_3$, or [^{14}C]-glutamate into protein—as compared with normal cells. However, Na^+ deficiency reduced N fixation and it was therefore concluded that Na^+ was very likely required for conversion of N_2 to NH_3 in this species.

Sulfur

Hase et al. (1960) reported that S-deficient *Chlorella ellipsoidea* cells did not divide, but did so when SO_4^{2-} was supplied. They noted further that at the onset of cell division a S-containing peptide nucleotide appeared in high concentration; its concentration decreased during cell division.

In studies employing x-irradiation, mutants of *Ophiostoma multiannulatum* (Hedgc. and Davids) (Fries, 1945, 1946) and *Escherichia coli* (Lampen and Jones, 1947; Lampen et al., 1947a, b) were produced that were unable to utilize SO_4^{2-}. These forms, called "parathiotrophic" in contrast with normal forms called "euthiotrophic," could assimilate cysteine and cystine but could not reduce SO_4^{2-}.

Chlorella pyrenoidosa Chick (Emerson strain 3) utilized thiosulfate for growth as effectively as SO_4^{2-} (Hodson et al., 1968b), and crude enzyme preparations from the alga were shown to be capable of reducing SO_4^{2-} to thiosulfate (Levinthal and Schiff, 1968) with adenosine-3′-phosphate-5′-phosphosulfate (PAPS) as an intermediate (Hodson et al., 1968a). During sulfate reduction by this organism, cysteine, glutathione, and homocysteine were formed (Schiff, 1964).

Working with cell-free extracts from several microorganisms, Hodson and Schiff (1971a) came to the conclusion that thiosulfate-forming activity appears to be a common feature of SO_4^{2-}-reducing systems, and that this activity may be present in enzymic systems previously thought to be forming sulfite. They described additional properties of crude extracts from *C. pyrenoidosa* (Emerson strain 3) which reduce adenosine-3′-phosphate-5′-phosphosulfate (PAPS) to thiosulfate, and separation of the extracts into two components both of which must be present to obtain maximal formation of acid-volatile radioactivity (Hodson and Schiff, 1971b). Seven mutants of *C. pyrenoidosa* (Emerson strain 3), impaired for SO_4^{2-} utilization, were isolated after treatment of the wild-type organism with nitrosoguanidine by replica plating on media containing thiosulfate and L-methionine (Hodson et al., 1971). All mutants grew on thiosulfate and all possessed the activating enzymes that convert SO_4^{2-} to PAPS.

SO_4^{2-} and various reduced forms of SO_4^{2-} satisfy the S requirements of bacteria (Roberts et al., 1955; Postgate, 1965; Villarejo and Westley, 1966), fungi (Maw, 1960; Leinweber and Monty, 1964), algae (Hodson et al., 1968b), and higher plants (Thomas, 1958). In higher plants, too, SO_4^{2-} must be reduced. Although there is no evidence with regard to higher plants, SO_4^{2-}-reducing enzymes in bacteria (Dreyfuss and Monty, 1963; Pasternak et al., 1965; Wheldrake and Pasternak, 1965) are repressed—apparently by cysteine (Ellis et al., 1964; Wheldrake, 1967).

It has recently been reported that SO_4^{2-} reduction in higher plants may proceed via adenylylsulfate instead of phosphoadenylylsulfate as in microorganisms (Ellis, 1969).

Cobalt

Co received little attention agriculturally until 1935 when it was shown to be of great importance in the diet of ruminants and, in 1948, was shown to be a constituent of vitamin B_{12}—apparently required by all animals (Vanselow, 1966). In plants, only certain blue-green algae (*Nostoc muscorum, Calothrix parietina, Coccochloris peniocystis*, and *Diplocystis aeruginosa*) have been shown to require Co^{2+} or vitamin B_{12} (Holm-Hansen et al., 1954). A concentration of 0.4 μg Co^{2+}/liter in the medium permitted near-optimal growth. Several other groups of algae also require either Co^{2+} or vitamin B_{12} (Hutner and Provasoli, 1964).

Legumes do not require Co^{2+}, but bacteria in nodules of legumes require it in order to fix atmospheric nitrogen. This requirement has been demonstrated for soybean plants *(Glycine max)* grown in symbiosis with *Rhizobium japonicum* (Ahmed and Evans, 1959) and for alfalfa *(Medicago sativa)* grown in symbiosis with *Rhizobium meliloti* (Reisenauer, 1960). In the latter case 0.1 μmole of Co^{2+}/liter was present in the medium for alfalfa to satisfy the Co^{2+} requirement of the nodule bacteria. Further evidence of a Co^{2+} requirement for legumes was presented by Ahmed and Evans (1960, 1961) and Delwiche et al. (1961). The cobamide coenzyme from *R. meliloti* is 5,6-dimethylbenzimidazolylcobamide—vitamin B_{12} coenzyme (Kliewer and Evans, 1963a).

Even nonleguminous plants with symbiotic N-fixing bacteria require Co^{2+} at 0.01 ppm in the nutrient solution—for example alder *(Alnus glutinosa)* and beefwood *(Casuarina cunninghamiana)* (Bond and Hewitt, 1962). In nutrient solution soybeans received adequate Co^{2+} at a concentration of 0.1–1.0 ppb (Ahmed and Evans, 1960, 1961) and alfalfa at 2×10^{-8} M (Delwiche et al., 1961). Co^{2+} was required by the free-living, N-fixing bacteria *Azotobacter vinelandii* (Evans and Kliewer, 1964) and *A. chroococcum* (Iswaran and Sundara Rao, 1964). For the latter organism, 0.1 ppm Co^{2+} in the medium was optimal.

Co^{2+} was essential for the symbiotic growth of *Azolla filiculoides* and *Anabaena azollae* in the absence of combined N, but the Co^{2+} requirement appeared to be associated solely with the *A. azollae* component (Johnson et al., 1966). *Azotobacter vinelandii* required an extremely low concentration (0.001 ppb) of Co^{2+} for normal growth (Kliewer and Evans, 1963b).

As little as 0.1 ppm Co^{2+} in solutions is damaging to many species of plants, as observed by Vanselow (1966). He added that Co^{2+} does not occur in soils in toxic concentration anywhere. In fact, the only problem is a dietary deficiency of Co^{2+}, with forage low in Co^{2+}, which causes various disorders in ruminants.

Vanadium

V is widely distributed in the ocean, soils, and plants; most plants contain approximately 1.0 ppm V on a dry weight basis (Pratt, 1966). V is not essential for higher plants, but it may be essential for certain bacteria and algae. Bertrand (1941a, b, 1942) reported that V, in a low concentration, acted as a growth factor for *Aspergillus niger*. Several researchers reported that V stimulated N fixation by *Azotobacter* (Konishi and Tsuge, 1933; Shibuya and Saeki, 1934; Burk and Horner, 1935; Horner et al., 1942)—possibly by partly replacing Mo in the N-fixation process (Bove et al., 1957). Burk (1934) showed that the Mo requirement of *Azotobacter* for N fixation could be satisfied by V.

To date, V has been reported to be essential for only one organism, *Scenedesmus obliquus* (Arnon and Wessel, 1953). In the nutrient solution 0.01 ppm was sufficient. Later it was observed that at high light intensity the rate of photosynthesis of *S. obliquus*, on a unit chlorophyll basis, was twice as high in the plus-V as in the minus-V cells (Arnon, 1954). It was concluded that V deficiency did not primarily interfere with photochemical events but rather with dark enzymatic reactions that limit the overall rate of photosynthesis at high light intensity. V has been implicated in the reduction of CO_2 (Warburg et al., 1955).

Toxicities from V have not been reported under field conditions; in nutrient solution culture 0.5 ppm or greater is toxic for plants (Pratt, 1966).

Gallium

In earlier experiments, Steinberg appeared to demonstrate the essentiality of Ga (0.01–0.02 ppm in the medium) for *A. niger* (Steinberg, 1938, 1942a, b) and for a higher plant, *Lemna minor* (Steinberg, 1941). Later, Steinberg (1946) was unable to demonstrate essentiality for these plants as strikingly as he had in his earlier experiments.

Ga stimulated cell production in *Nitschia closterium* cultures deficient in P and N but had no effect when these nutrients were added (Riley, 1943).

Rubidium

In some algae and bacteria, Rb^+ appears to substitute completely for K^+. Complete substitution of Rb^+ for K^+ was demonstrated in *Streptococcus faecalis* (MacLeod and Snell, 1950) and in algae adapted to grow in Rb^+— *Chlorella vulgaris* (Pirson, 1939) and *Ankistrodesmus braunii* (Kellner, 1955). For *Staphlococcus pyogenes*, Rb^+ replaced K^+ with an efficiency of 20% to less than 3%—depending on the concentration of the two ions (Wyatt, 1963). However, there was no growth of *Leuconostoc mesenteroides* when Rb^+ was added to a K^+-free medium (MacLeod and Snell, 1950). In *C. pyrenoidosa* Chick., replacing K^+ with Rb^+ may spare or replace K^+ in cell metabolism, but the growth rate of the alga is decreased (Osretkar and Krauss, 1965).

Magnesium

Owing to its involvement in transphosphorylations, Mg^{2+} is•required not only for higher plants but also for algae, fungi, and bacteria (Epstein, 1965).

Trelease and Selsam (1939) reported that *C. vulgaris* tolerated high concentrations of Mg^{2+} salts and grew considerably even in 0.42 M $MgSO_4$. All algae that contain chlorophyll obviously require Mg^{2+} as a constituent of the pigment. An adequate concentration of Mg^{2+} for algae may be quite low. For *Ankistrodesmus falcatus*, 0.1 ppm Mg^{2+} produced optimal growth; there was no inhibition of growth up to 10 ppm (Rodhe, 1948).

For *Chlorella*, the concentration of chlorophyll was proportional to Mg^{2+} concentration from 0.2 to 2.8 ppm (Finkle and Appleman, 1953a). Mg^{2+} was also required for catalase synthesis, but a lower concentration sufficed for this synthesis than for that of chlorophyll. Mg^{2+} deficiency of *Chlorella* occurred with 0.2, 0.5, and 1.0 ppm, but essentially optimal growth was obtained with 2.8 ppm Mg^{2+} (Finkle and Appleman, 1953b). Mg^{2+}-deficient cells had 20 times the cell volume of cells receiving adequate Mg^{2+}.

Calcium

According to some reports (Molisch, 1895; Pringsheim, 1926; Hopkins and Wann, 1926; Trelease and Selsam, 1939; Pratt, 1941; Scott, 1943; Myers, 1944), *Chlorella* can be grown successfully in culture solutions lacking Ca^{2+}. Other workers (Noack et al., 1940; Manuel, 1944) reported that Ca^{2+} is required for *Chlorella*. A marine diatom, *N. closterium*, had a Ca^{2+} requirement of 5 ppm (Hutner, 1948). Using a chelate extinction procedure, Hutner and Provasoli (1951) established a 5 ppm Ca^{2+} requirement for *Chlorella* and *Euglena*. This technique involved successively increasing the concentration of citrate and then determining the increments of Ca^{2+} required in order to restore growth. In the cells 10 ppm Ca^{2+} was sufficient for *Chlorella* on a dry weight basis (Walker, 1954, 1955).

Bacteria and fungi (Trelease and Selsam, 1939) and a blue-green alga, *Coccochloris peniocystis* (Gerloff et al., 1950), either did not require Ca^{2+} or else very minute concentrations were sufficient. For example, when fixing atmospheric N, *A. vinelandii* produced 80% of maximal growth with approximately 0.2 ppm Ca^{2+}; the organism had an absolute Ca^{2+} requirement on both free and fixed N (Jacobsons et al., 1962). However, Yocum (1960) reported that Ca^{2+} was required for *Azotobacter* growth on molecular N but not on combined N. Some of the Cyanophyceae and some of the green algae (*Spirogyra*, *Haematoccus*, and *Vaucheria*) require Ca^{2+} for their development (Trelease and Selsam, 1939).

Ca^{2+} has been reported to be required for the growth of various fungi, for example, *Achlya* spp. (Reischer, 1951; Barksdale, 1962), *Phycomyces* sp.

(Ødegard, 1952), *Allomyces* spp. (Machlis, 1953; Ingraham and Emerson, 1954), *Neurospora* sp. (Castel and Bertrand, 1954), *Coprinus* sp. (Fries, 1956), and *Phytophthora* spp. (Davies, 1959; Erwin and Katznelson, 1961). *Rhizoctonia solani* produced only 14% of its normal growth in the absence of Ca^{2+} (Steinberg, 1948). Production of perithecia of *Chaetomium globosum* (Basu, 1951) and of mature sporocarps of *Coprinus sterquilinus* (Heintz and Jayko, 1965) occur only in the presence of Ca^{2+}.

The absolute requirement for Ca^{2+} and/or sterol for utilization of nitrate, but not asparagine, by *Phytophthora parasitica* var. *nicotianae* appears to be unique (Hendrix and Guttman, 1970).

Ca^{2+} was essential for some strains of *Azotobacter*, both in the presence and absence of fixed N, while five strains did not require Ca^{2+} (Norris and Jensen, 1957). A liverwort was reported to have a very low requirement for Ca^{2+} (Machlis, 1962).

Halimeda opuntia (L.) Lamouroux and *Halimeda discoidea* Decaisne, prominent calcifying algae, were studied *in situ* in coral reef lagoon ecosystems in the Caribbean and also in the laboratory under controlled conditions for light, CO_2, aeration, and other factors. Both algae showed a stimulation of incorporation of Ca^{2+} (as $CaCO_3$) by light and a diurnal rhythm of incorporation under identical conditions of illumination (Stark et al., 1969). Both phenomena paralleled the rhythm of chloroplast migration within the plants. Calcification was stimulated by CO_2, respiration, and aeration. N sources inhibited incorporation of Ca^{2+} during the day—indicating that NH_3 production was not responsible for precipitation of $CaCO_3$. Differential wash-out rates of Ca^{2+} absorbed during the day as compared with that absorbed at night support the concept of a two-step mechanism for calcification.

Numerous photographs showing the manifold forms of fossil $CaCO_3$ depositions laid down by algae can be found in Johnson (1961).

O'Kelley and Herndon (1961) suggested that Ca^{2+} is required in algae for the formation of pectic materials in addition to Ca pectate and that the Ca^{2+} requirement varies quantitatively with the amounts of pectic materials produced by various algae. However, Gerloff and Fishbeck (1969) reasoned that, because of the alleged role of Ca^{2+} as a constituent of the middle lamella, filamentous algae might have a higher Ca^{2+} requirement than that of unicellular species. This was not found to be the case, since the lowest and highest Ca^{2+} requirements were observed in two unicellular algae—*C. pyrenoidosa* and S. *quadricauda*. It was concluded that *C. pyrenoidosa* did not require Ca^{2+}—at least not more than the mere traces in solution as a contaminant. This alga absorbed Ca^{2+} in barely detectable amounts even when the solution contained 5 ppm of Ca^{2+}. Maximal cell concentration of Ca^{2+} at this external level was 0.01%. The other three green algae and the two blue-green algae they studied required Ca^{2+}.

In a recent, thorough coverage of the mineral nutrition of algae, O'Kelley

(1968) stated that it appears probable that all algae have a Ca^{2+} requirement (at least in the absence of Sr^{2+}), but that it is difficult to demonstrate it in some species. For one strain of *Chlorella*, the Ca^{2+} requirement was in the micronutrient range (Walker, 1953) but, for other algae, in the macronutrient range (O'Kelley and Herndon, 1961). Ca^{2+} was required for zoospore release in *Chlorococcum echinozygotum* and *Protosiphon botryoides* (O'Kelley and Herndon, 1959; O'Kelley and Deason, 1962; Gilbert and O'Kelley, 1964) because parent cell wall digestion was activated only by Ca^{2+}.

In a study of growth and N fixation by *Anabaena cylindrica*, Allen and Arnon (1955a) found that Ca^{2+} was required whether NO_3^- or N_2 was supplied; Ca^{2+} could not be replaced by Sr^{2+}. Although the specific role of Ca^{2+} was not determined, it was noted that a deficiency of Ca^{2+} decreased the rate of N fixation by *Rhizobium trifolii* (strain TA-1) in nodules of subterranean clover (Banath et al., 1966). Ca^{2+} deficiency did not interfere, however, with translocation of reduced N from nodules.

Aspergillus niger was reported to have the same mineral requirements as those of higher plants except that it did not require Ca^{2+}, Si, and B (Steinberg, 1945). Ca^{2+} was required for the differentiation of sporangia and gemmae in *Achlya* sp.; it could not be replaced by Mg^{2+}, K^+, or Na^+ (Griffin, 1966).

Potassium

In a study of K^+ nutrition of *C. pyrenoidosa*, it was noted that K^+ was firmly bound within cells and could not be removed by washing with distilled water or with solutions of Na^+, K^+, or Mg^{2+} chlorides (Scott, 1944).

After consecutive transfers of four species of aquatic *Hyphomycetes* to K^+-deficient media, a sudden reduction in growth was observed by Hickman (1969). The concentration of K^+ required to prevent this reduction in growth was so low as to make K^+ a micronutrient for these fungi. He concluded that perhaps no other organisms are known to require such low concentrations of K^+.

Strontium

In *C. pyrenoidosa*, Sr^{2+} can apparently substitute completely for Ca^{2+} (Walker, 1953, 1955). However, for *Selenastrum minutum* and the D-1 and D-3 strains of *S. obliquus*, Sr^{2+} could not replace Ca^{2+} (Walker, 1955). In addition, it was noted that 1 ppm Sr^{2+} strongly inhibited the growth of *Coccomyxa pringsheimii*.

Phosphorus

Polyphosphates are of wide distribution in algae (Krauss, 1958), and they could serve both as a phosphate reservoir and as an effective energy source—as suggested by Stich (1955) and for *Acetabularia* by Thilo et al. (1956).

Nitrogen

Chlorella vulgaris Beijerinck var. viridis (Chodat) did not show a detectable urease activity (Baker and Thompson, 1962). The rest of the literature dealing with N is discussed in the following section.

Molybdenum

Mo was required by N-fixing, blue-green algae (Bortels, 1938, 1940), by a green alga, *C. pyrenoidosa* (Walker, 1953), and *A. niger* (Steinberg, 1945) grown on NO_3^-. Loneragan and Arnon (1954) noted that Mo deficiency in *C. pyrenoidosa* resulted in a reduction in dry weight, chlorophyll, and photosynthesis, and an increase in endogenous respiration.

Mo-deficient *S. obliquus* cells grew poorly (Arnon, 1954) and were unable to assimilate NO_3^-; starch accumulated and cells were unable to grow or divide (Arnon et al., 1955; Arnon, 1958). Chlorophyll was markedly reduced, but increased strikingly upon addition of Mo. The most rapid cell division occurred when the Mo supply was above 3000 Mo atoms per cell; no cell division occurred when the Mo supply dropped to 1500 to 1700 atoms per cell. In order to produce a marked Mo deficiency, the concentration of Mo in the nutrient solution had to be reduced to between 10^{-9} and 10^{-8} g/liter. $\frac{1}{10}$ ppb Mo was sufficient for vigorous growth, 1 ppb caused no increased growth over that occurring with 0.1 ppb, and 10 ppb was only slightly inhibitory.

In additional research with *S. obliquus* (strain D-3), Ichioka and Arnon (1955) and Arnon (1958) reported that the Mo requirement could be abolished when NH_4^+ or urea was substituted for NO_3^+. Chlorophyll synthesis was normal when NH_4^+ or urea, rather than NO_3^-, was supplied; the Mo effect on chlorophyll was considered indirect and to be operative only when NO_3^- reduction was blocked (for lack of Mo). A concentration of 1 ppb Mo was sufficient in the nutrient solution. NR and NiR have been identified in *Scenedesmus* (Omura, 1954).

Anabaena (Arnon, 1958) and pea root nodules and *Azotobacter* (Mulder, 1948) require Mo in order to fix atmospheric N. *Anabaena cylindrica* did not grow with NO_3^- unless Mo was present; with NH_4^- or urea the organism grew equally well with or without Mo (Fogg and Wolfe, 1954). For N fixation, the Mo requirement of *Azotobacter* could be satisfied by adding V (Burk, 1934).

Nicholas and Nason (1954a, b) conclusively identified Mo in *Neurospora crassa* as the specific metal component of NR—the flavoprotein that catalyzes oxidation of reduced pyridine nucleotides by NO_3^- to yield NO_2 and oxidized coenzymes. Mo was associated with hydrogenase in an anaerobic N-fixing bacterium, *Clostridium pasteurianum* (Shug et al., 1954).

In *Chlorella* (Morris and Syrett, 1963), *N. crassa* (Kinsky, 1961), and mold fungi (Morton, 1956), NO_3^- was required for induction of NR, and the presence of the end product, NH_4^- ions, prevented its induction. By contrast, NH_4^- ions did not inhibit induction of NR in radish cotyledons or corn seedlings (Beevers et al., 1965).

Mo appears to be required for all plants that reduce NO_3^- or fix free N (Nicholas, 1963).

Chlorine

In one of the rare reports of Cl^--containing compounds in plants, Haines (1969) found that the sulfolipids of *Ochromonas danica* were polar lipids and that they contained Cl^- as a constituent. The most important Cl^--containing compound was tentatively identified as *threo-(R)-13-chloro-1-(R)-14-docosanol* (L-13,D-14).

Petty (1961) listed 29 chlorometabolites synthesized by microorganisms, but no essential function was indicated for any of them.

Boron

Eyster (1952) claimed that B was essential for *N. muscorum,* and McIlrath and Skok (1958) reported that "it appears safe to assume that B is essential for the normal metabolism of *Chlorella vulgaris* (Columbia strain)." Bowen et al. (1965) were unable to demonstrate a B requirement for several species and strains of *Chlorella* including the Columbia strain. Later, McBride et al. (1967) also reported that *Chlorella* does not have a B requirement.

Suneson (1945) reported a favorable influence of B on development of zyqotes and germlings of *Ulva lactuca* in seawater. Henkel (1951) also studied the effect of added B on marine species and reported an increased number of cells (per filament) developing from spores of *Bangia* and *Porphyra* after the addition of 2 ppm B to seawater. Seawater contains 4.6 ppm of B.

For a marine diatom, *Cylindrotheca fusiformis,* B was required for growth (Lewin, 1965, 1966a), a conclusion later corroborated by Neales (1967).

Gerloff (1968) found that three species of blue-green algae readily absorb B; he also observed a marked growth response to B with N-fixing species when NO_3^- was omitted from the culture medium. *Chlorella* and two other green algae did not require B and did not absorb it in appreciable amounts. *Scenedesmus obliquus* did not show a B requirement; cell number and dry weight were not limited by a B concentration of 0.5 μg/liter (Dear and Aronoff, 1968). Winfield (1945) and Gerloff (1963) reported that fungi do not require B. Bowen and Gauch (1966) showed that the following fungi do not have B requirements: *Saccharomyces cerevisiae*, *A. niger*, *N. crassa*, and *Penicillium chrysogenum*. They reported, in addition, on the nature of B toxicity in fungi. Guirard and Snell (1962) did not list B as essential for bacteria. However, B was shown to be required for N fixation by *Azotobacter chroococcum* and for growth of *A. niger* (Gerretsen and de Hoop, 1954).

When grown on a glucose–starch–yeast extract medium with 8 ppm B, *Sordaria* sp. produced only abortive or abnormal asci without differentiated ascospores (Turian, 1954). When no B was added to the medium, the ascospores were normal in appearance.

Iodine, Bromine, and Fluorine

Although there is no solid evidence that higher plants require I$^-$, there is a possibility that some red and brown algae do (Stosch, 1961, 1962).

I$^-$ was required for growth and reproduction of a marine brown alga, *Petalonia fascia* (Hsiao, 1969). Working with axenic cultures in a completely defined medium, Fries (1964, 1966) studied the I$^-$ requirement of different red algae and showed that *Polysiphonia urceolata* has an absolute requirement for I$^-$. I$^-$ constituted 1% of the dry wright of *Laminaria digitata* (Black, 1948). In some algae, I$_2$ and Br$_2$ are formed by enzymic oxidation of I$^-$ and Br$^-$ (Roche et al., 1963).

I$^-$ is not required for higher plants. As little as 1.0 ppm has been reported to be toxic to plants, but such a concentration does not exist in soils; therefore toxicity from I$^-$ has not been reported in nature (Martin, 1966b).

Several reports were summarized indicating that: seawater contains 10–25 ppb I$^-$; surface and well waters $<$ 10–318 ppb; soils and rocks $<$ 0.1–70 ppm; sea plants $<$ 0.001–12,500 ppm; and land plants $<$ 0.1–20 ppm (Martin, 1966b). In sea plants I$^-$ accumulates especially in the Laminariales (Young and Langille, 1958).

Kendall (1919) established the essentiality of I$^-$ in animal nutrition by isolating the I$^-$-containing compound thyroxine (4,5,6-trihydro-4,5,6-triiodo-2-oxy-B-indolpropionic acid) in cyrstalline form from the thyroid gland.

Organic I$^-$ compounds have been reported in wheat roots (Boszorményi et al., 1959).

Ectocarpus fasiculatus appeared to have an absolute requirement for I$^-$, but was inhibited by 64 μmole KI/liter as observed by Pédersen (1969). She found that *Lithosiphon pusillus* grew best in the highest concentration (64 μmole/liter) of KI tested, but that there was always some growth without I$^-$. These two organisms showed no response to Br$^-$ in media containing KI.

Growth of zoospores of *Pylaiella litoralis* was remarkably increased by addition of KI to a culture medium consisting of vitamin-free Asp6F with B_{12} (1 μg/liter) and kinetin.

Br^- occurs in plants in low concentrations, but it is regarded as nonessential and nontoxic (Martin, 1966a). In fact, the element has often been used in nutrient absorption studies as a nonessential and nontoxic element. Toxicity to Br^- has on a few occasions been reported to follow the use of methyl bromide or ethylene dibromide for soil fumigation.

Acacia georginae was reported to synthesize fluoroacetate (Oelrichs and McEwan, 1961; Murray et al., 1961). Seeds of *Dichapetalum toxicarium* contain ω-fluorooleic acid, $FCH_2(CH_2)_7CH:CH(CH_2)_7COOH$, and smaller concentrations of fluoropalmitic acid (Ward et al., 1964).

Iron, Manganese, Zinc, and Copper

These elements have been reported to be essential for many algae (Wiessner, 1962; Hutner and Provasoli, 1964) and nonphotosynthetic microorganisms (Nicholas, 1963)—as well as for higher plants. Thus the universal requirement for these elements is generally recognized.

Cu has been implicated in photoreduction by S. *obliquus* strain D-3; Cu, as plastocyanin, is an essential component of the electron transport mechanism of pigment system I of photosynthesis (Bishop, 1964).

The requirement for Fe by algae has been extensively studied by many workers—among them Hopkins and Wann (1925, 1927b), Hopkins (1930), and Myers (1951). In cells 30 ppm of Fe was sufficient for *Chlorella* on a dry weight basis (Walker, 1954, 1955). For continuous-flow cultures of *Chlorella*, 8 ppm of Fe was adequate, whereas 0.6 ppm limited growth (Myers and Clark, 1944). In a search for the "alkaline limit" for the growth of *Chlorella*, it was noted that growth fell off rapidly at pH 7; the limit for growth was pH 8.4 (Hopkins and Wann, 1927a). It was concluded that *Chlorella* might have grown satisfactorily at pH 7 and higher if more Fe had been added. In cell-free preparations of *A. cylindrica*, it appeared that ferredoxin was involved in the electron transport chain between NADPH and NO_2^- (Hattori and Myers, 1966). In *Euglena gracilis*, Fe required for growth was located in sites distinct from those required for chlorophyll synthesis (Price and Carell, 1964). In fact, there was a pool of Fe in the cell that was unavailable for chlorophyll synthesis. Therefore at Fe concentrations where growth is normal, chlorophyll synthesis may be as little as one-third that of the control rates.

The Zn^{2+} requirement of *Chlorella* was demonstrated by Stegman (1940) and of *A. niger* by Steinberg (1919). In fact, Fe was also required by *A. niger*, and both Zn^{2+} and Fe had to be added for a marked increase in growth. In cells 4.5 ppm Zn^{2+} was sufficient for *Chlorella* on a dry weight basis (Walker,

1954, 1955). Using protein formation as a measure of growth of *E. gracilis*, Price and Quigley (1966) determined that the relative growth rate was a linear function of internal Zn^{2+} concentration in the alga. They presented a formula for determining the quantitative Zn^{2+} requirement for growth. Zn^{2+} appeared to have some general effect in *E. gracilis* (Klebs) since limitation in growth of the organism could not be directly explained by the reduced rate of oxidation of ethanol (Price and Millar, 1962). Zn^{2+}-deficiency symptoms in *Euglena* suggested that Zn^{2+} was involved in syntheses leading from RNA to protein (Wacker, 1962).

Aspergillus niger required Zn^{2+} for the synthesis of tryptophan, but Cd^{2+} could substitute for it in the enzyme (Bertrand and Wolf, 1959). Zn^{2+} was required in *A. niger* for glucose-6-phosphate and 6-phosphogluconic dehydrogenases (Bertrand and Wolf, 1957). Hexokinase from *N. crassa* required Zn^{2+} for its activity, and a deficiency in the element reduced DPNH and TPNH diaphorase and increased acid phosphatase (Medina and Nicholas, 1957).

Approximately 50% (w/w) of the dry weight of chloroplast lamellae is lipid. Characteristically, the lipid is rich in α-linolenate and other related polyunsaturated fatty acids linked to mono- and digalactosyldiglycerides (Constantopoulos, 1970). It was found that growth of photoautotrophic *E. gracilis* Z. was strongly inhibited by Mn^{2+} deficiency, whereas chlorophyll formation was not appreciably affected. Galactosyldiglyceride concentration of Mn^{2+}-deficient *Euglena* was approximately 40% lower on the basis of either chlorophyll concentration or dry weight than was that of Mn^{2+}-sufficient cells. Dark-grown cells, grown photoheterotrophically in light sufficient for greening or photosynthesis, showed a 40% reduction in both chlorophyll and galactosyldiglycerides under Mn^{2+} deficiency. Chloroplast formation appeared to be affected. The fatty acids of photoheterotrophic Mn-deficient cells were mainly saturated with an unusual accumulation (approximately 45% of the total fatty acids) of myristic acid. In spite of this, the galactosyldiglycerides contain mainly unsaturated fatty acids. The ratio of chlorophyll to galactosyldiglyceride was remarkably constant at all Mn^{2+}-deficiency concentrations.

Euglena gracilis (Klebs) cultures were grown under conditions in which limitation in Mn^{2+} limited chlorophyll concentration much more than growth (Gavalas and Clark, 1971). Although initial rates of photosynthetic O_2 evolution were not affected by the concentration of Mn^{2+}, photoinhibition in high-intensity light was markedly influenced. Mn^{2+}-deficient cells were more sensitive to CMU inhibition of photosynthesis than those receiving adequate Mn^{2+}. It was concluded that the site of action of Mn^{2+} was on the reducing side of photosystem II—close to the CMU-sensitive site. Further, they suggested that this Mn^{2+}-affected site may represent a secondary structural or metabolic consequence of Mn^{2+} deficiency not necessarily involved in quantum yields of O_2.

Additions of Mn^{2+} to Mn^{2+}-deficient *Chlorella* cells immediately increased the rate of photosynthesis (Pirson, 1937). Eyster et al. (1956) reported that Mn^{2+} deficiency resulted in a 75% reduction in photosynthesis and a failure of the Hill reaction in *C. pyrenoidosa*. Mn^{2+} appears to play a role in NO_3^- reduction in *C. vulgaris*—according to Krauss (1958). Similarly, Kylin (1943) noted that Mn^{2+} stimulates the growth of *Ulva* sporlings if NO_3^-, but not NH_4^+, is used as a N source. On a dry weight basis in cells, 2.5 ppm Mn^{2+} is sufficient for *Chlorella* (Walker, 1954, 1955).

Silicon

Inasmuch as diatoms have silicified cell walls, it is perhaps not surprising that this group of plants requires Si (Lewin, 1962). For dense cultures of *Navicula pelliculosa*, Lewin (1955a) reported that 35 ppm Si in the nutrient solution is optimal; growth is proportional to Si concentration at the lower concentrations of Si. When *N. pelliculosa* was grown in media containing 3.5, 8.3, and 43 ppm Si, generation (doubling) times were 20.4, 13.7, and 13.1 hr, respectively, as observed by Lewin (1957). She also reported that SiO_2 varied from 4 to 22% of the dry weight of the cells. There appeared to be a close relationship between Si uptake and aerobic respiration in *N. pelliculosa* (Lewin, 1955b), and Si uptake and sulfhydryl groups in the cell membrane (Lewin, 1954).

It has been reported that low silicate concentrations may limit the rate of growth of diatoms in the Antarctic during summer, inasmuch as unusually thin-walled diatoms have been observed at that time of the year in that region (Hart, 1934; Clowes, 1938).

The role of Si in plant growth has been covered in an excellent, exhaustive review by Lewin and Reimann (1969). They reported that Si in the form of silica gel is deposited in epidermal cells and cell walls of many plant species—particularly, Equisetaceae, Cyperaceae, Gramineae, and Urticaceae. The cell walls of diatoms (Bacillariophyceae) and certain silicified flagellates (Chrysophyceae) also contain Si. Two major groups of animals use Si to form skeletal structures composed of silica gel—the radiolarians (in the Protozoa) and the sponges (Porifera). Silica gel is a form of hydrated amorphous silica, $SiO_2 \cdot nH_2O$, or polymerized silicic acid.

Si in xylem sap is entirely in the form of monosilicic acid or orthosilicic acid, $Si(OH)_4$ (Barber and Shone, 1966; Handreck and Jones, 1967); this is the form in which Si is absorbed by higher plants and diatoms (Lewin, 1962; Jones and Handreck, 1965). Up to 16% of the dry weight of horsetails (*Equistetum arvense*) may be silica (Lewin and Reimann, 1969). Rice and horsetails exhibit a "weeping willow" habit of growth under Si-deficient conditions (Yoshida, et al., 1959). Below pH 9, in soil and natural waters, Si exists as uncharged monosilicic acid; ATP appears to be involved in silicic acid uptake (Lewin, 1955b).

Si can prevent Mn^{2+} toxicity in barley, rice, rye, and ryegrass by decreasing the uptake of Mn^{2+} (Vlamis and Williams, 1967). Si reduces Fe and Mn toxicities in rice plants by increasing the oxidation of these elements and causing a precipitation of the elements on the surface of roots, as observed by Okuda and Takahashi (1965). These investigators also showed that Si-deficient rice plants transpire at a greater rate than Si-sufficient plants.

Silicon may be involved in protein metabolism of *Cyclotella cryptica* (Werner, 1966, 1968). A decrease in the glutamic acid pool precedes the inhibition of total protein synthesis in Si deficiency. The acetyl-CoA pathway is not inhibited. It was concluded that silicic acid is involved with reactions between the condensing enzyme (acetyl-CoA and oxalacetate) and α-ketoglutarate (Werner, 1968). There is no cellulose; the carbohydrate material of the diatom cell wall is glucuronomannan (Ford and Percival, 1965).

It was concluded that although Si was dispensable for most plants it is necessary for healthy development of many plants; it is essential for those plants with high Si concentrations—such as rice and other grasses, horsetails, and diatoms (Lewin and Reimann, 1969).

LITERATURE CITED

Ahmed, S., and H. J. Evans. 1959. Effect of cobalt on the growth of soybeans in the absence of supplied nitrogen. Biochem. Biophys. Res. Commun. 1:271–275.

Ahmed, S., and H. J. Evans. 1960. Cobalt: A micronutrient element for the growth of soybean plants under symbiotic conditions. Soil Sci. 90:205–210.

Ahmed, S., and H. J. Evans. 1961. The essentiality of cobalt for soybean plants grown under symbiotic conditions. Proc. Nat. Acad. Sci. 47:24–36.

Allen, M. B. 1952. The cultivation of Myxophyceae. Arch. Mikrobiol. 17:34–53.

Allen, M. B., and D. I. Arnon. 1955a. Studies on nitrogen-fixing blue-green algae. I. Growth and nitrogen fixation by *Anabaena cylindrica* Lemm. Plant Physiol. 30:366–372.

Allen, M. B., and D. I. Arnon. 1955b. Studies on nitrogen-fixing blue-green algae. II. The sodium requirement of *Anabaena cylindrica*. Physiol. Plantarum 8:653–659.

Arnon, D. I. 1954. Some recent advances in the study of essential micronutrients for green plants. Huitieme Congr. Int. Bot. (Paris) 1954:73–80.

Arnon, D. I. 1958. The role of micronutrients in plant nutrition with special reference to photosynthesis and nitrogen assimilation, pp. 1–32. *In* C. A. Lamb, O. G. Bentley, and J. M. Beattie (eds.) Trace elements in plants (Proc. Conf. Ohio Agr. Exp. Sta., Wooster, Ohio, Oct. 14–16, 1957). Academic, New York.

Arnon, D. I., and G. Wessell. 1953. Vanadium as an essential element for green plants. Nature 172:1039–1040.

Arnon, D. I., P. S. Ichioka, G. Wessel, A. Fujiwara, and J. T. Woolley. 1955. Molybdenum in relation to nitrogen metabolism. I. Assimilation of nitrate nitrogen by *Scenedesmus*. Physiol. Plantarum 8:538–551.

Baker, J. E., and J. F. Thompson. 1962. Metabolism of urea and ornithine cycle intermediates by nitrogen-starved cells of *Chlorella vulgaris*. Plant Physiol. 37:618–624.

Banath, C. L., E. A. N. Greenwood, and J. F. Loneragan. 1966. Effects of calcium
deficiency on symbiotic nitrogen fixation. Plant Physiol. 41:760–763.

Barber, D. A., and M. G. T. Shone. 1966. The absorption of silica from aqueous solutions
by plants. J. Exp. Bot. 17:569–578.

Barksdale, A. W. 1962. Effect of nitritional deficiency on growth and sexual reproduction
of *Achlya ambisexualis*. Amer. J. Bot. 49:633–638.

Basu, S. N. 1951. Significance of calcium in the fruiting of *Chaetomium* species,
particularly *Chaetomium globosum*. J. Gen. Microbiol. 5:231–238.

Beevers, L., L. E. Schrader, D. Flesher, and R. M. Hageman. 1965. The role of light
and nitrate in the induction of nitrate reductase in radish cotyledons and maize
seedlings. Plant Physiol. 40:691–698.

Bertrand, D. 1941a. Importance of the trace element vanadium for *Aspergillus niger*.
Compt. Rend. Acad. Sci. (Paris) 213:254–257.

Bertrand, D. 1941b. Vanadium as a growth factor for *Aspergillus niger*. Bull. Soc. Chim.
Biol. 23:467–471.

Bertrand, D. 1942. Vanadium as an essential trace element for *Aspergillus niger*.
Ann. Inst. Pasteur 68:226–244.

Bertrand, D., and A. De Wolf. 1957. Sur la nécessité du zinc, comme oligoélément, pour
la glucose-6-phosphatedéhydrogénase et la 6-phosphategluconique-dehydrogénase
de l'*Aspergillus niger*. Compt. Rend. Acad. Sci. (Paris) 245:1179.

Bertrand, D., and A. De Wolf. 1959. Sur la nécessité de l'oligoélément zinc pour la synthèse
du tryptophane, chez l'*Aspergillus niger* et son remplacement possible par le cadmium.
Compt. Rend. Acad. Sci. (Paris) 249:2237.

Bishop, N. I. 1964. Site of action of copper in photosynthesis. Nature 204:401–402.

Black, W. A. P. 1948. Seasonal variations in chemical constitution of some of the sub-
littoral seaweeds common to Scotland. II. *Laminaria digitata*. III. *Laminaria
saccharina* and *Saccorhiza bulbosa*. J. Soc. Chem. Ind. (London) 67:169–172.

Bond, G., and E. J. Hewitt. 1962. Cobalt and the fixation of nitrogen by root nodules
of *Alnus* and *Casuarina*. Nature 195:94–95.

Bortels, H. 1938. Entwicklung und Stickstoffbindung bestimmter Mirkoorganismen in
Abhängigkeit von Spurenelemente und vom Wetter. Ber. Deut. Bot. Ges. 56:153–160.

Bortels, H. 1940. Uber die Bedeutung des Molybdäns für stickstoffbindene Nostocaceen.
Arch. Mikrobiol. 11:155–186.

Böszörményi, Z., E. Cseh, and L. Gaspar. 1959. The synthesis of organic iodine compounds
in wheat roots. Naturwissenschaften 46:584.

Bove, J., C. Bove, and D. I. Arnon. 1957. Molybdenum and vanadium requirements of
Azotobacter for growth and nitrogen fixation. Plant Physiol. 32(Suppl.):23.

Bowen, J. E., and H. G. Gauch. 1966. Nonessentiality of boron in fungi and the nature
of its toxicity. Plant Physiol. 41:319–324.

Bowen, J. E., H. G. Gauch, R. W. Krauss, and R. A. Galloway. 1965. The nonessentiality
of boron for *Chlorella*. J. Phycol. 1:151–154.

Brownell, P. F., and D. J. D. Nicholas. 1967. Some effects of sodium on nitrate assimilation
and N_2 fixation in *Anabaena cylindrica*. Plant Physiol. 42:915–921.

Burk, D. 1934. Azotase and nitrogenase in *Azotobacter*. Ergeb. Enzymforsch. 3:23–56.

Burk, D., and C. K. Horner. 1935. The specific catalytic role of molybdenum and vanadium
in nitrogen fixation and amide utilization by *Azotobacter*. Trans. 3rd Int. Congr.
Soil Sci. 1:152–155.

Castel, G., and D. Bertrand. 1954. Calcium, an essential oligoelement of *Neurospora crassa.* Compt. Rend. Acad. Sci. (Paris) 239:1546–1548.

Chu, S. P. 1942. The influence of the mineral composition of the medium on the growth of planktonic algae. I. Methods and culture media. J. Ecol. 30:284–325.

Clowes, A. J. 1938. Phosphate and silicate in the southern ocean. Discovery Rep. 19: 1–120.

Constantopoulos, G. 1970. Lipid metabolism of manganese-deficient algae. I. Effect of manganese deficiency on the greening and the lipid composition of *Euglena gracilis.* Z. Plant Physiol. 45:76–80.

Craig, F. N., and S. F. Trelease. 1937. Photosynthesis of *Chlorella* in heavy water. Amer. J. Bot., 24:232–242.

Davies, M. E. 1959. The nutrition of *Phytophthora fragariae.* Trans. Brit. Mycol. Soc. 42:193–200.

Dear, J. M., and S. Aronoff. 1968. The non-essentiality of boron for *Scenedesmus.* Plant Physiol. 43:997–998.

Delwiche, C. C., C. M. Johnson, and H. M. Reisenauer. 1961. Influence of cobalt on nitrogen fixation by *Medicago.* Plant Physiol. 36:73–78.

Dreyfuss, J., and K. K. Monty. 1963. The biochemical characterization of cysteine-requiring mutants of *Salmonella typhimurium.* J. Biol. Chem. 238:1019–1024.

Ellis, R. J. 1969. Sulphate activation in higher plants. Planta 88:34–42.

Ellis, R. J., S. K. Humphries, and C. A. Pasternak. 1964. Repressors of sulphate activation in *Escherichia coli.* Biochem. J. 92:167–172.

Epstein, E. 1965. Mineral metabolism, pp. 438–466. *In* J. Bonner and J. E. Varner (eds.) Plant biochemistry. Academic, New York.

Erwin, D. C., and H. Katznelson. 1961. Nutrition of *Phytophthora cryptogea.* Can. J. Microbiol. 7:15–25.

Evans, H. J., and M. Kliewer. 1964. Vitamin B_{12} compounds in relation to the requirements of cobalt for higher plants and nitrogen-fixing organisms. Annu. N.Y. Acad. Sci. 112:735–755.

Eyster, C. 1952. Necessity of boron for *Nostoc muscorum.* Nature 170:755.

Eyster, C. 1967. Mineral nutrient requirements of *Chlorella Sorokiniana* in continuous pure culture. USAF School Aerospace Med., Aerospace Med. Div. (AFSC), Brooks Air Force Base, Texas. SAM-TR-67-40. 53 pp.

Eyster, C., T. E. Brown, and H. A. Tanner. 1956. Manganese requirement with respect to respiration and the Hill reaction in *Chlorella pyrenoidosa.* Arch. Biochem. Biophys. 64:240–241.

Finkle, B. J., and D. Appleman. 1953a. The effect of magnesium concentration on chlorophyll and catalase development in *Chlorella.* Plant Physiol. 28:652–663.

Finkle, B. J., and D. Appleman. 1953b. The effect of magnesium concentration on growth of *Chlorella.* Plant Physiol. 28:664–673.

Fogg, G. E., and M. Wolfe. 1954. The nitrogen metabolism of the blue-green algae (Myxophyceae), pp. 99–125. *In* B. A. Fry and J. L. Peel (eds.) Autotrophic microorganisms (4th Symp. Soc. Gen. Microbiol.). Cambridge Univ. Press, Cambridge.

Ford, C. W., and E. Percival. 1965. Carbohydrates of *Phaeodactylum tricornutum.* II. A sulfonated glucuronomannan. J. Chem. Soc. 1965:7042–7046.

Fries, L. 1956. Studies in the physiology of *Coprinus.* II. Influence of pH, metal factors and temperature. Svensk. Bot. Tidskr. 50:47–96.

Fries, L. 1964. *Polysiphonia urceolata* in axenic culture. Nature 202:110.

Fries, L. 1966. Influence of iodine and bromine on growth of some red algae in axenic culture. *Physiol. Plantarum* 19:800–808.

Fries, N. 1945. X-ray induced mutations in the physiology of *Ophiostoma.* Nature 155: 757–758.

Fries, N. 1946. X-ray parathiotrophy in *Ophiostoma*. Svensk. Bot. Tidskr. 40:127—140.

Gavalas, N. A., and H. E. Clark. 1971. On the role of manganese in photosynthesis (Kinetics of photoinhibition in manganese-deficient and 3-(4-chlorophenyl)-1, 1-dimethylurea-inhibited *Euglena gracilis*.). Plant Physiol. 47:139—143.

Geoghegan, M. J. 1953. Experiments with *Chlorella* at Jealott's Hill, pp. 182—189. *In* J. S. Burlew (ed.) Algal culture from laboratory to pilot plant. Carnegie Inst. Washington Pub. 600.

Gerloff, G. C., G. P. Fitzgerald, and F. Skoog. 1950. The mineral nutrition of *Coccochloris peniocystis*. Amer. J. Bot. 37:835—840.

Gerloff, G. C. 1963. Comparative mineral nutrition of plants. Annu. Rev. Plant Physiol. 14:107—124.

Gerloff, G. C. 1968. The comparative boron nutrition of several green and blue-green algae. Physiol. Plantarum 21:369—377.

Gerloff, G. C., and K. A. Fishbeck. 1969. Quantitative cation requirements of several green and blue-green algae. J. Phycol. 5:109—114.

Gerretsen, F. C., and H. de Hoop. 1954. Boron, an essential micro-element for *Azotobacter chroococcum*. Plant and Soil 5:349—367.

Gilbert, W. A., and J. C. O'Kelley. 1964. The effects of replacement of calcium by strontium on the reproduction of *Chlorococcum echinozygotum*. Amer. J. Bot. 51:866—869.

Griffin, D. H. 1966. Effect of electrolytes on differentiation in *Achlya* sp. Plant Physiol. 41:1254—1256.

Grosse, W. 1963. Growth and phosphate metabolism of *Rhodopseudomonas spheroides* as influenced by sodium, potassium, and magnesium nutrition. Flora 153:157—193.

Guirard, B. M., and E. E. Snell. 1962. Nutritional requirements of microorganisms, pp. 33—93. *In* I. C. Gunsalus and R. Y. Stanier (eds.) The bacteria — A treatise on structure and function, Vol. IV. Academic, New York.

Haines, T. H. 1969. Algal sulfolipids. Paper delivered at 158th Amer. Chem. Soc. Nat. Meeting, New York, Sept. 8—12.

Handreck, K. A., and L. H. P. Jones. 1967. Uptake of monosilicic acid by *Trifolium incarnatum* (L.). Australian J. Biol. Sci. 20:483—485.

Harmer, P. M., and E. J. Benne. 1945. Sodium as a crop nutrient. Soil Sci. 60:137—148.

Hart, T. J. 1934. On the phytoplankton of the south-west Atlantic and the Bellingshausen Sea, 1929—31. Discovery Rep. 8:3—268.

Hase, E., S. Mihard, and H. Tamiya. 1960. Role of sulfur in the cell division of *Chlorella*, with special reference to the sulfur compounds appearing during the process of cell division. I. Plant Cell Physiol. 1:131—142.

Hattori, A., and J. Myers. 1966. Reduction of nitrate and nitrite by subcellular preparations of *Anabaena cylindrica*. I. Reduction of nitrite to ammonia. Plant Physiol. 41:1031—1036.

Heintz, C. E., and L. G. Jayko. 1965. The effect of calcium on fruit body production in *Coprinus sterquilinus*. Bacteriol. Proc. 1965:22.

Hendrix, J. W., and S. Guttman. 1970. Sterol or calcium requirement by *Phytophthora parasitica* var. *nicotianae* for growth on nitrate nitrogen. Mycologia 62:195—198.

Henkel, R. 1951. Ernährungsphysiologische Untersuchungen an Meeresalgen, insbesondere an *Bangia pumila*. Kieler Meeresforsch. 8:192—211.

Hickman, D. W. 1969. Potassium as a micronutrient for aquatic Hyphomycetes. Plant Physiol. 44(Suppl.):no. 100.

Hodson, R. C., and J. A. Schiff. 1971a. Studies of sulfate utilization by algae. 8. The ubiquity of sulfate reduction to thiosulfate. Plant Physiol. 47:296—299.

Hodson, R. C., and J. A. Schiff. 1971b. Studies of sulfate utilization by algae. 9.

Fractionation of a cell-free system from *Chlorella* into two activities necessary for the reduction of adenosine 3'-phosphate 5'-phosphosulfate to acid-volatile radio-activity. Plant Physiol. 47:300—305.

Hodson, R. C., J. A. Schiff, and J. P. Mather. 1971. Studies of sulfate utilization by algae. 10. Nutritional and enzymatic characterization of *Chlorella* mutants impaired for sulfate utilization. Plant Physiol. 47:306—311.

Hodson, R. C., J. A. Schiff, A. J. Scarsella, and M. Levinthal. 1968a. Studies of sulfate utilization by algae. 6. Adenosine-3'-phosphate-5'-phosphosulfate (PAPS) as an intermediate in thiosulfate formation from sulfate by cell-free extracts of *Chlorella*. Plant Physiol. 43:563—569.

Hodson, R. C., J. A. Schiff, and A. J. Scarsella. 1968b. Studies of sulfate utilization by algae. 7. *In vivo* metabolism of thiosulfate by *Chlorella*. Plant Physiol. 43:570—577.

Holm-Hansen, O., G. C. Gerloff, and F. Skoog. 1954. Cobalt as an essential element for blue-green algae. Physiol. Plantarum 7:665—675.

Hopkins, E. F. 1930. Iron-ion concentration in relation to growth and other biological processes. Bot. Gaz. 89:209—240.

Hopkins, E. F., and F. B. Wann. 1925. The effect of the H ion concentration on the availability of iron for *Chlorella* sp. J. Gen. Physiol. 9:205—210.

Hopkins, E. F., and F. B. Wann. 1926. Relation of hydrogen-ion concentration to growth of *Chlorella* and to the availability of iron. Bot. Gaz. 81:353—376.

Hopkins, E. F., and F. B. Wann. 1927a. Further studies on growth of *Chlorella* as affected by hydrogen-ion concentration. Bot. Gaz. 83:194—201.

Hopkins, E. F., and F. B. Wann. 1927b. Iron requirement for *Chlorella*. Bot. Gaz. 84:407—427.

Horner, C. K., D. Burk, F. E. Allison, and M. S. Sherman. 1942. Nitrogen fixation by *Azotobacter*, as influenced by molybdenum and vanadium. J. Agr. Res. 65:173—193.

Hsiao, S. I. C. 1969. Life history and iodine nutrition of the marine brown alga, *Petalonia fascia*. Can. J. Bot. 47:1611—1616.

Hutner, S. H. 1948. Essentiality of constituents of sea water for growth of a marine diatom. Trans. New Hampshire Acad. Sci., Ser. 2, 10:136—141.

Hutner, S. H., and L. Provasoli. 1951. The phytoflagellates, pp. 27—128. *In* A. Lwoff (ed.) Biochemistry and physiology of protozoa, Vol. I. Academic, New York.

Hutner, S. H., and L. Provasoli. 1964. Nutrition of algae. Annu. Rev. Plant Physiol. 15:37—56.

Ichioka, P. S., and D. I. Arnon. 1955. Molybdenum in relation to nitrogen metabolism. II. Assimilation of ammonia and urea without molybdenum by *Scenedesmus*. Physiol. Plantarum 8:552—560.

Ingraham, J. L., and R. Emerson. 1954. Studies of the nutrition and metabolism of the aquatic Phycomycete, *Allomyces*. Amer. J. Bot. 41:146—152.

Iswaran, V., and W. V. B. Sundara Rao. 1964. Role of cobalt in nitrogen fixation by *Azotobacter chroococcum*. Nature 203:549.

Jacobsons, A., E. A. Zell, and P. W. Wilson. 1962. A re-investigation of the Ca requirement of *Azotobacter vinelandii* using purified media. Arch. Mikrobiol. 41:1—10.

Johnson, G. V., P. A. Mayeux, and H. J. Evans. 1966. A cobalt requirement for symbiotic growth of *Azolla filiculoides* in the absence of combined nitrogen. Plant Physiol. 41:852—855.

Johnson, J. H. 1961. Limestone-building algae and algal limestones. Colorado School Mines, Dep. Publ., Golden, Colorado.

Jones, L. H. P., and K. A. Handreck. 1965. Studies of silica uptake in the oat plant. III. Uptake of silica from soils by the plant. Plant and Soil 23:79—96.

Kellner, K. 1955. Die Adaptation von *Ankistrodesmus braunii* an Rubidium und Kupfer.

Biol. Zentralbl. 74:662–691.

Kendall, E. C. 1919. Isolation of the iodine compound which occurs in the thyroid. J. Biol. Chem. 39:125–147.

Ketchum, B. H. 1954. Mineral nutrition of phytoplankton. Annu. Rev. Plant Physiol. 5:55–74.

Kinsky, S. C. 1961. Induction and repression of nitrate reductase in *Newrospora crassa*. J. Bacteriol. 82:898–304.

Kliewer, M., and H. J. Evans. 1963a. Identification of cobamide coenzyme in nodules of symbionts and isolation of the B_{12} coenzyme from *Rhizobium meliloti*. Plant Physiol. 38:55–59.

Kliewer, M., and H. J. Evans. 1963b. Cobamide coenzyme contents of soybean nodules and nitrogen fixing bacteria in relation to physiological conditions. Plant Physiol. 38:99–104.

Konishi, K., and T. Tsuge. 1933. Effect of inorganic constituents of soil solution on the growth of *Azotobacter*. J. Agr. Chem. Soc. (Japan) 9:129–144.

Kratz, W. A., and J. Myers. 1955. Nutrition and growth of several blue-green algae. Amer. J. Bot. 42:282–287.

Krauss, R. W. 1958. Physiology of the fresh-water algae. Annu. Rev. Plant Physiol. 9: 207–244.

Krauss, R. W., and W. H. Thomas. 1954. The growth and inorganic nutrition of *Scenedesmus obliquus* in mass culture. Plant Physiol. 29:205–214.

Kylin, A. 1943. The influence of trace elements on the growth of *Ulva lactuca* Kungl. Fysiograf. Sällskap. i Lund, Förh. 13:185–192.

Lampen, J. O., and M. J. Jones. 1947. Studies on the sulfur metabolism of *Escherichia coli*. II. Interrelations of norleucine and methionine in the nutrition of *Escherichia coli* and of a methionine-requiring mutant of *Escherichia coli*. Arch. Biochem. 13: 47–53.

Lampen, J. O., M. J. Jones, and A. B. Perkins. 1947a. Studies on the sulfur metabolism of *Escherichia coli*. I. The growth characteristics and metabolism of a mutant strain requiring methionine. Arch. Biochem. 13:33–45.

Lampen, J. O., R. R. Roepke, and M. J. Jones. 1947b. Studies on the sulfur metabolism of *Escherichia coli*. III. Mutant strains of *Escherichia coli* unable to utilize sulfate for their complete sulfur requirements. Arch. Biochem. 13:55–66.

Leinweber, F. J., and K. J. Monty. 1964. Cysteine biosynthesis in *Neurospora crassa*. I. The metabolism of sulfite, sulfide and cysteinesulfinic acid. J. Biol. Chem. 240: 782–787.

Levinthal, M., and J. A. Schiff. 1968. Studies on sulfate utilization by algae. 5. Identification of thiosulfate as a major acid-volatile product formed by a cell-free sulfate-reducing system from *Chlorella*. Plant Physiol. 43:555–562.

Lewin, J. C. 1954. Silicon metabolism in diatoms. I. Evidence for the role of reduced sulfur compounds in silicon utilization. J. Gen. Physiol. 37:589–599.

Lewin, J. C. 1955a. Silicon metabolism in diatoms. II. Sources of silicon for growth of *Navicula pelliculosa*. Plant Physiol. 30:129–134.

Lewin, J. C. 1955b. Silicon metabolism in diatoms. III. Respiration and silicon uptake in *Navicula pelliculosa*. J. Gen. Physiol. 39:1–10.

Lewin, J. C. 1957. Silicon metabolism in diatoms. IV. Growth and frustule formation in *Navicula pelliculosa*. Can. J. Microbiol. 3:427–433.

Lewin, J. C. 1962. Silicification, pp. 445–455. *In* R. A. Lewin (ed.) Physiology and biochemistry of algae. Academic, New York.

Lewin, J. C. 1965. The boron requirement of a marine diatom. Naturwissenschaften 70:1–2.

Lewin, J. C. 1966. Physiological studies of the boron requirement of the diatom

Cylindrotheca fusiformis   Reinmann and Lewin. J. Exp. Bot. 17:473–479.

Lewin, J. C., and B. E. F. Reimann. 1969. Silicon and plant growth. Annu. Rev. Plant Physiol. 20:289–304.

Loneragan, J. F., and D. I. Arnon. 1954. Molybdenum in the growth and metabolism of *Chlorella*. Nature 174:459.

McBride, L., W. Chorney, and J. Skok. 1967. The boron requirement of *Chlorella*. Plant Physiol. 42(Suppl.):S-5.

Machlis, L. 1953. Growth and nutrition of water molds in the subgenus *Euallomyces*. II. Optimal composition of the minimal medium. Amer. J. Bot. 40:450–460.

Machlis, L. 1962. The effects of mineral salts, glucose, and light on the growth of the liverwort, *Sphaerocarpus donnellii*. Physiol. Plantarum 15:354–362.

McIlrath, W. J., and J. Skok. 1958. Boron requirement of *Chlorella vulgaris*. Bot. Gaz. 119:231–233.

MacLeod, R. A., and E. Onofrey. 1956. Nutrition and metabolism of marine bacteria. II. The relation of sea water to the growth of marine bacteria. J. Bacteriol. 71:661–667.

MacLeod, R. A., and E. E. Snell. 1950. Ion antagonism in bacteria as related to anti-metabolites. Ann. N.Y. Acad. Sci. 52:1249–1259.

Manuel, M. E. 1944. The cultivation of *Chlorella* sp. Plant Physiol. 19:359–369.

Martin, J. P. 1966a. Bromine, pp. 62–64. *In* H. D. Chapman (ed.) *In* Diagnostic criteria for plants and soils. Div. Agr. Sci., Univ. California, Berkeley.

Martin, J. P. 1966b. Iodine, pp. 200–202. *In* H. D. Chapman (ed.) *In* Diagnostic criteria for plants and soils. Div. Agr. Sci., Univ. California, Berkeley.

Maw, G. A. 1960. Utilization of sulfur compounds by a brewer's yeast. J. Inst. Brewing 66:162–167.

Medina, A., and D. J. D. Nicholas. 1957. Properties of a zinc-dependent hexokinase from *Neurospora crassa*. Biochem. J. 66:573.

Molisch, H. 1895. Die Ernährung der Algen. Sitz. Akad. Wiss. Wien Math.-Nat. Kl. 104:783–800.

Morris, I., and P. J. Syrett. 1963. The development of nitrate reductase in *Chlorella* and its repression by ammonium. Arch. Mikrobiol. 47:32–41.

Morton, A. G. 1956. A study of nitrate reduction in mould fungi. J. Exp. Bot. 7:97–112.

Mulder, E. G. 1948. Importance of molybdenum in the nitrogen metabolism of micro-organisms and higher plants. Plant and Soil 1:94–119.

Murray, L. R., J. D. McConnell, and J. H. Whittem. 1961. Suspected presence of fluoro-acetate in *Acacia georginae* F. M. Bailey. Australian J. Sci. 24:41–42.

Myers, J. 1944. The growth of *Chlorella pyrenoidosa* under various culture conditions. Plant Physiol. 19:579–589.

Myers, J. 1951. Physiology of the algae. Annu. Rev. Microbiol. 5:157–180.

Myers, J., and L. B. Clark. 1944. Culture conditions and the development of the photo-synthetic mechanism. II. An apparatus for the continuous culture of *Chlorella*. J. Gen. Physiol. 28:103–112.

Nicholas, D. J. D. 1963. Inorganic nutrient nutrition of microorganisms, pp. 363–447. *In* F. C. Steward (ed.) Plant physiology, Vol. III. Academic, New York.

Nicholas, D. J. D., and A. Nason. 1954a. Molybdenum and nitrate reductase. II. Molybdenum as a constituent of nitrate reductase. J. Biol. Chem. 207:353–360.

Nicholas, D. J. D., and A. Nason. 1954b. Mechanism of action of nitrate reductase from *Neurospora*. J. Biol. Chem. 211:183–197.

Noack, K., A. Pirson, and G. Stegmann. 1940. Der Bedarf an Spurenelementen bei *Chlorella*. Naturwissenschaften 28:172–173.

Norris, J. R., and H. L. Jensen. 1957. Calcium requirements of *Azotobacter*. Nature 180:1493.

Odegård, K. 1952. On the physiology of *Phycomyces blakesleeanus* Burgeff. I. Mineral requirements on a glucose-asparagine medium. Physiol. Plantarum 5:583–609.

Oelrichs, P. B., and T. McEwan. 1961. Isolation of the toxic principle in *Acacia georginae*. Nature 190:808–809.

O'Kelley, J. C. 1968. Mineral nutrition of algae. Annu. Rev. Plant Physiol. 19:89–112.

O'Kelley, J. C., and T. R. Deason. 1962. Effect of nitrogen, sulfur and other factors on zoospore production by *Protosiphon botryoides*. Amer. J. Bot. 49:771–777.

O'Kelley, J. C., and W. R. Herndon. 1959. Effect of strontium replacement for calcium on production of mobile cells in *Protosiphon*. Science 130:718.

O'Kelley, J. C., and W. R. Herndon. 1961. Alkaline earth elements and zoospore release and development in *Protosiphon botryoides*. Amer. J. Bot. 48:796–802.

Okuda, A., and E. Takahashi. 1965. *In* The mineral nutrition of the rice plant, pp. 123–146. Proc. Symp. Int. Rice Res. Inst. Johns Hopkins Press, Baltimore.

Omura, H. 1954. On the nitrate and nitrite reductase in green algae. Enzymologia 17:127–132.

Osretkar, A., and R. W. Krauss. 1965. Growth and metabolism of *Chlorella pyrenoidosa* Chick. during substitution of Rb for K. J. Phycol. 1:22–32.

Osterhout, W. J. V. 1912. Plants which require sodium. Bot. Gaz. 54:532–536.

Pasternak, C. A., R. J. Ellis, M. G. Jones-Mortimer, and C. E. Crichton. 1965. The control of sulphate reduction in bacteria. Biochem. J. 96:270–275.

Pedersén, M. 1969. The demand for iodine and bromine of three marine brown algae grown in bacteria-free cultures. Physiol. Plantarum 22:680–685.

Petty, M. A. 1961. Introduction to the origin and biochemistry of microbial halo-metabolites. Bacteriol. Rev. 25:111–130.

Pirson, A. 1937. Ernährungs- und Stoffwechselphysiologische Untersuchungen an *Fontinalis* und *Chlorella*. Z. Bot. 31:193–267.

Pirson, A. 1939. Über die Wirkung von Alkalionen auf Wachstum und Stoffwechsel von *Chlorella*. Planta 29:231–251.

Postgate, J. R. 1965. Recent advances in the study of sulphate-reducing bacteria. Bacteriol. Rev. 29:425–441.

Pratt, P. F. 1966. Vanadium, pp. 480–483. *In* H. D. Chapman (ed.) Diagnostic criteria for plants and soils. Div. Agr. Sci., Univ. California, Berkeley.

Pratt, R. 1941. Studies on *Chlorella vulgaris*. IV. Influence of the molecular proportions of KNO_3, KH_2PO_4, and $MgSO_4$ in the nutrient solution on the growth of *Chlorella*. Amer. J. Bot. 28:492–497.

Price, C. A., and E. Millar. 1962. Zinc, growth, and respiration in *Euglena*. Plant Physiol. 37:423–427.

Price, C. A., and E. F. Carell. 1964. Control by iron of chlorophyll formation and growth in *Euglena gracilis*. Plant Physiol. 39:862–868.

Price, C. A., and J. W. Quigley. 1966. A method for determining quantitative zinc requirements for growth. Soil Sci. 101:11–16.

Pringsheim, E. G. 1926. Ueber das Ca-Bedürfnis einiger Algen. Planta 2:555–568.

Reischer, H. S. 1951. Growth of Saprolegniaceae in synthetic media. I. Inorganic nutrition. Mycologia 43:142–155.

Reisenauer, H. M. 1960. Cobalt in nitrogen fixation by a legume. Nature 186:375–376.

Riley, G. A. 1943. Physiological aspects of spring diatom flowering. Bull. Bingham Oceanogr. Coll. 8, Art. 4:1–53.

Roberts, R. B., P. H. Abelson, D. B. Cowie, E. T. Bolton, and R. J. Britten. 1955. Studies of biosynthesis in *Escherichia coli*. Carnegie Inst., Washington, D. C.

Roche, J., M. Fontaine, and J. Leloup. 1963. Halides, pp. 493–547. *In* M. Florkin and H. S. Mason (eds.) Comparative biochemistry, Vol. V. Academic, New York.

Rodhe, W. 1948. Environmental requirements of fresh-water plankton algae. Symbolae Bot. Upsalienses 10:1–149.

Schiff, J. A. 1964. Studies of sulfate utilization by algae. II. Further identification of reduced compounds formed from sulfate by *Chlorella*. Plant Physiol. 39:176–179.

Scott, G. T. 1943. The mineral composition of *Chlorella pyrenoidosa* grown in culture media containing various concentrations of calcium, magnesium, potassium, and sodium. J. Cellular Comp. Physiol. 21:327–338.

Scott, G. T. 1944. Cation exchanges in *Chlorella pyrenoidosa*. J. Cellular Comp. Physiol. 23:47–58.

Shibuya, K., and H. Saeki. 1934. The effect of vanadium on the growth of plants. II. J. Soc. Trop. Agr. (Japan) 6:721–729.

Shug, A. L., P. W. Wilson, D. E. Green, and M. R. Mahler. 1954. The role of molybdenum and flavin in hydrogenase. J. Amer. Chem. Soc. 76:3355–3356.

Stark, L. M., L. Almodovar, and R. W. Krauss. 1969. Factors affecting the rate of calcification in *Halimeda opuntia* (L.) Lamouroux and *Halimeda discoidea* Decaisne. J. Phycol. 5:305–312.

Stegmann, G. 1940. Die Bedeutung der Spurenelemente für *Chlorella*. Z. Bot. 35: 385–422.

Steinberg, R. A. 1919. A study of some factors in the chemical stimulation of the growth of *Aspergillus niger*. Amer. J. Bot. 6:330–356; 357–372.

Steinberg, R. A. 1938. The essentiality of gallium to growth and reproduction of *Aspergillus niger*. J. Agr. Res. 57:569–574.

Steinberg, R. A. 1941. Use of *Lemna* for nutrition studies on green plants. J. Agr. Res. 62:423–430.

Steinberg, R. A. 1942a. Effect of trace elements on growth of *Aspergillus niger* with amino acids. J. Agr. Res. 64:455–475.

Steinberg, R. A. 1942b. Influence of carbon dioxide on response of *Aspergillus niger* to trace elements. Plant Physiol. 17:129–132.

Steinberg, R. A. 1945. Use of microorganisms to determine essentiality of minor elements. Soil Sci. 60:185–189.

Steinberg, R. A. 1946. Mineral requirements of *Lemna minor*. Plant Physiol. 21:42–48.

Steinberg, R. A. 1948. Essentiality of calcium in the nutrition of fungi. Science 107:423.

Stich, H. von. 1955. Synthese und Abbau der Polyphosphate von *Acetabularia* nach autoradiographischen Untersuchungen des P-Stoffwechsels. Z. Naturforsch. 10B: 281–284.

Stosch, H. A. von. 1961. Wirkungen von Jud und Arsenit auf Meeresalgen in Kulture, pp. 142–150. *In* D. De Virville and J. Feldmann (eds.), Proc. 4th Int. Seaweed Symp., Pergamon, Oxford.

Stosch, H. A. von. 1962. Jodbedarf bei Meeresalgen. Naturwissenschaften 49:42–43.

Suneson, S. 1945. Einige Versuche über den Einfluss des Bors auf die Entwicklung und Photosynthese der Meeresalgen. Kung. Fysiogr. Sallsk. Forh. Lund 15:185–197.

Tanada, T. 1951. The photosynthetic efficiency of carotenoid pigments in *Navicula minima*. Amer. J. Bot. 38:276–283.

Thilo, E., H. Gruze, J. Hämmerling, and G. Werz. 1956. Über Isolierung und Identifizierung der Polyphosphate aus *Acetabularia mediterranea*. Z. Naturforsch. 11B:266–270.

Thomas, M. D. 1958. Assimilation of sulfur and physiology of essential S-compounds, pp. 37–63. *In* W. Ruhland (ed.), Encyclopedia of Plant Physiology, Vol. IX. Springer, Berlin.

Trelease, S. F., and M. E. Selsam. 1939. Influence of calcium and magnesium on the growth of *Chlorella*. Amer. J. Bot. 26:339—341.

Turian, G. 1954. L'acide borique, inhibiteur de la differenciation des ascospores chez *Sordaria*. Experientia 10:183.

Vanselow, A. P. 1966. Cobalt, pp. 142—156. *In* H. D. Chapman (ed.) Diagnostic criteria for plants and soils. Div. Agr. Sci., Univ. California, Berkeley.

Villarejo, M., and J. Westley. 1966. Sulfur metabolism of *Bacillus subtilis*. Biochim. Biophys. Acta 117:209—216.

Vlamis, J., and D. E. Williams. 1967. Manganese and silicon interaction in the Gramineae. Plant and Soil 27:131—140.

Wacker, W. E. C. 1962. Nucleic acids and metals. III. Changes in nucleic acid, protein, and metal content as a consequence of zinc deficiency in *Euglena gracilis*. Biochemistry 1:859—865.

Walker, J. B. 1953. Inorganic micronutrient requirements of *Chlorella*. I. Requirements for calcium (or strontium), copper, and molybdenum. Arch. Biochem. Biophys. 46:1—11.

Walker, J. B. 1954. Inorganic micronutrient requirements of Chlorella. II. Quantitative requirements for iron, manganese, and zinc. Arch. Biochem. Biophys. 53:1—8.

Walker, J. B. 1955. Strontium inhibition of calcium utilization by a green alga. Arch. Biochem. Biophys. 60:264—265.

Warburg, O., G. Krippahl, and W. Buchholz. 1955. Wirkung von Vanadium auf die Photosynthese. Z. Naturforsch. 10B:422.

Ward, P. F. V., R. J. Hall, and R. A. Peters. 1964. Fluoro-fatty acids in the seeds of *Dichapetalum toxicarium*. Nature 201:611.

Werner, D. 1966. Die Kieselsäure im Stoffwechsel von *Cyclotella cryptica* Reimann, Lewin und Guillard. Arch. Mikrobiol. 55:278—308.

Werner, D. 1968. Stoffwechselregulation durch den Zellwandbaustein Kieselsäure: Polgrössenänderungen von α-Ketoglutarsäure, Aminosäuren und Nucleosidphosphaten. Z. Naturforsch. 23B:268—272.

Wheldrake, J. F. 1967. Intracellular concentration of cysteine in *Escherichia coli* and its relation to repression of the sulphate-activating enzymes. Biochem. J. 105:697—699.

Wheldrake, J. F., and C. A. Pasternak. 1965. The control of sulphate activation in bacteria. Biochem. J. 96:276—280.

Wiessner, W. 1962. Inorganic nutrients, pp. 267—286. *In* R. A. Lewin (ed.) Physiology and biochemistry of algae. Academic, New York.

Winfield, M. E. 1945. The role of boron in plant metabolism. 3, The influence of boron on certain enzyme systems. Australian J. Exp. Biol. 23:267—272.

Witsch, H. von. 1948. Physiologischer Zustand und Wachstumsintensität bei *Chlorella*. Arch. Mikrobiol. 14:128—141.

Wyatt, H. V. 1963. The effect of alkali metals on the growth of *Staphylococcus pyrogenes*. Exp. Cell Res. 30:56—61.

Yocum, C. S. 1960. Nitrogen fixation. Annu. Rev. Plant Physiol. 11:25—36.

Yoshida, S., Y. Ohnishi, and K. Kitagishi. 1959. Soil Sci. Plant Nutrition [(Soil Plant Food (Tokyo)] 5:127—133). (Original not seen; title unavailable).

Young, E. G., and W. M. Langille. 1958. The occurrence of inorganic elements in the marine algae of the Atlantic provinces of Canada. Can. J. Bot. 36:301—310.

CHAPTER 12

Hydroponics

INTRODUCTION

The artificial culture of plants in some medium other than soil has played a vital role in the advancement of knowledge about plants and their requirements (Hoagland and Arnon, 1948). As a matter of fact, this technique is the only one that has enabled scientists to determine which elements are essential for higher plants. With a technique whereby one element can be omitted, while other elements known to be essential are included, it is possible to develop and to describe the deficiency symptoms of the omitted element (if indeed it is an essential element). Such a technique is also required for precise studies involving variations in pH of the medium, variations in proportions of nutrients in the medium, and for many other studies concerning growth requirements of plants.

Synonyms

Various terms have been applied to techniques for growing plants in some medium other than soil, including hydroponics, soilless culture, tank farming, tray agriculture, and nutriculture. Nutriculture is an all-inclusive term for several methods of growing plants in artificial media—water culture, aggregate culture, and adsorbed nutrient culture (Hoagland and Arnon, 1950).

History

In 1679 Mariotte was one of the first investigators to grow plants in water culture. Woodward (1699) published the first definite account of growing

Fig. 12.1
Water culture equipment of the type employed by Sachs around 1860. (Hoagland and Arnon, 1950.)

plants without soil. He grew spearmint, potatoes, and vetch in rain, spring, river, conduit, and distilled water to determine whether water or soil particles nourished the plants. He noted a correlation between plant growth and the amount of "terrestrial matter" these different waters contained. In 1758, Duhamel was the first to grow plants to maturity in water culture.

In the late 1850s, Sachs (1859), Knop (1859), and others began intensive studies of the artificial culture of plants (Fig. 12.1). Although these early workers were unaware of the significance of certain factors, such as aeration for the roots, they nevertheless devised and used, more or less satisfactorily, nutrient solutions very similar to those in use at the present time.

From around 1900 until the 1930s, plant physiologists were primarily concerned with three aspects of plant nutrition: (1) What elements are essential for plant growth? (2) What is the "best" nutrient solution? (3) How important is the proportion of one element to another in the nutrient solution, that is, "balance."

From the 1930s on, research centered largely around: (1) the physiological functions of essential elements; (2) interrelationships among nutrient ions; (3) the mechanism by which ions are absorbed by plants; and (4) the translocation of inorganic constituents—more recently with radioactive forms of the elements.

Requirements for Plant Growth

For the growth of plants, *identical* requirements of the environment must be met whether plants are grown in soil or in artificial culture. Requirements can be met in different ways in soil and in hydroponics. If they are met, equally good growth can be obtained in water, sand, gravel, or soil culture (Hoagland and Arnon, 1938, 1939; Arnon and Hoagland, 1939, 1940; Laurie and Kiplinger, 1940).

Light. Any misconceptions to the contrary, the requirement for light is just as essential for plants grown hydroponically as for those in soil. In artificial culture it is relatively easy to supply all requirements for plant growth except the one for an adequate intensity of light.

When plants are grown hydroponically out-of-doors, one may rely mostly— if not entirely—on solar radiation. Indoors it is necessary to provide artificial illumination.

For many plants increases in plant growth follow increases in light intensity up to that of full sunlight (approximately 10,000 ft-c). This relationship has been determined by studying different degrees of shading of plants exposed to full sunlight. Presumably, increases in growth with increasing light intensity are related to the fact that some leaves on a plant are shaded by others, hence the shaded ones may not attain the maximal rate of photosynthesis. It might also be added that leaves in full sunlight receive more intense light than they can use. For single leaves maximal rates of photosynthesis have been reported for approximately $\frac{1}{10}$ to $\frac{1}{5}$ of full sunlight (i.e., with 1000–2000 ft-c). In artificial culture rooms or chambers, using fluorescent and Mazda lamps, intensities of approximately 2000 ft-c have generally been attained. This intensity of light, if of proper balance, permits a good growth of plants. With the advent of high-intensity light sources, it is now possible to attain 5000 ft-c or more, but such intensities are naturally more expensive to install and to operate than the lesser intensities. Leiser et al. (1960) evaluated various light sources for plant growth. They reported that intensities of 4000 ft-c, or above were superior to those of lower intensities (1200–1500 ft-c).

Soybean plants, grown in controlled environment cabinets under high-intensity (220 W/m^2, 400–700 nm) and low-intensity (90 W/m^2) light, showed differences in growth rate, leaf anatomy, chloroplast ultrastructure, leaf starch, chlorophyll, and chloroplast lipid concentration (Ballentine and Forde, 1970).

Intensity of light is not the only consideration for plant growth so far as light is concerned. Different regions of the spectrum must be represented and be in proper balance with respect to each other. In the past balance has been achieved by using a combination of fluorescent tubes and ordinary Mazda lamps, so that both the blue and red ends of the spectrum, respectively, were present.

For Red Kidney bean plants, Rohrbaugh (1942) reported that leaf area, percent of Ca^{2+} in leaves, and dry weight were higher in plants receiving a

Fig. 12.2
Comparison of the growth of tomato plants in soil (A), solution (B), and sand cultures (C).
(Hoagland and Arnon, 1950.)

balanced radiation—as compared with plants receiving narrower regions of the visible spectrum.

When growth of plants in a well-fertilized and well-watered soil of good physical characteristics is compared with that in an artificial culture, it is found that plants can be spaced just as closely in soil as in an artificial culture (Arnon and Hoagland, 1940; Hoagland and Arnon, 1950) (Fig. 12.2). Even though all other requirements are adequately met except light, shading of one plant by another may be a limiting factor in growth.

Temperature. Contrary to the mystical claims for hydroponics, a favorable temperature is just as important for plants in artificial culture as for those in soil (Hoagland and Arnon, 1950). Corn, soybeans, snapbeans, and lima beans, for example, grow best when day temperature is relatively high; peas and spinach, however, are regarded as cool-season crops. Night temperature profoundly affects stem growth and fruit setting of tomato (Went and Cosper, 1945).

Water. Water is another requirement for plant growth that must be satisfied in hydroponics as well as in soil. Some of the water absorbed is used by the plant, but most of it is lost by transpiration from the leaves; a significant amount may be lost by evaporation.

Aeration for the Roots. In order for roots of most plants to absorb water and nutrients, the roots require a certain amount of O_2. Plants do not grow well in "water-logged" soil devoid of air spaces, and most plants do not grow well in water culture unless provision is made to aerate the solution by circulating it or by bubbling air into it.

In poorly-aerated cultures, Arnon (1937) noted that plants provided with NO_3^- grew much better than those supplied with NH_4^+ nitrogen. This report is pertinent with regard to compacted, poorly aerated soils. Later, Gilbert and Shive (1945) reported that NO_3^- supplied an internal source of O_2, following reduction of the NO_3^- by the plant, and that the amount of NO_3^- absorbed by soybeans, and the extra CO_2 evolved, showed a very close correspondence (Fig. 12.3).

Inorganic Nutrients. In artificial culture nutrients are usually supplied as soluble, freely available salts. As with the other factors, however, the nutrient requirements remain the same regardless of the way plants are grown or the manner in which nutrients are supplied.

In hydroponics, inorganic nutrition may be partially or wholly altered at any time so desired and, from an experimental point of view for the scientist interested in plant nutrition, this is a distinct advantage of hydroponics as compared with soil culture. Furthermore, in hydroponics it is possible to attain and to maintain whatever concentrations of various nutrients are desired.

Anchorage. For plants growing in soil, sand, or gravel cultures, anchorage is no problem. However, when plants are grown in water culture, it is necessary to provide some means of support for the seedlings, and later the plants, above the nutrient solution. Even in sand culture, support can be a problem when conditions are right for phenomenal plant growth, for example, soybean plants approximately 10–12 ft tall (Fig. 12.4) (Gauch and Weigel, 1968).

METHODS OR TECHNIQUES

A detailed presentation and discussion of apparatus, solutions, and techniques can be found in several books on soilless gardening (Connors and Tiedjens, 1940; Turner and Henry, 1939; Phillips, 1940; Laurie, 1940; Gericke, 1942; Ellis, et al., 1947). In outstanding publications, Hewitt (1952, 1966) described sand and water culture methods for studies of plant nutrition.

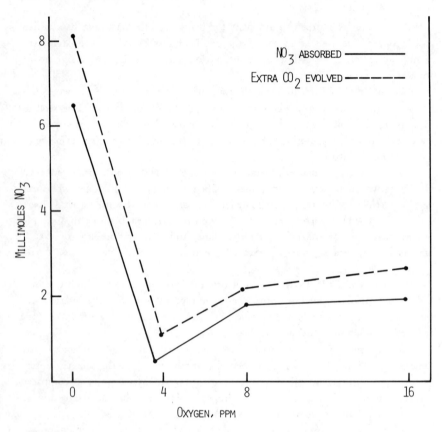

Fig. 12.3

Relation between rates of NO₃⁻ ion absorption and extra CO₂ production by soybean roots after 23 days of treatment at four O₂ levels. From S. G. Gilbert and J. W. Shive (1945) Soil Sci., 59:453–461. © 1945, The Williams and Wilkins Co., Baltimore, Md. 21202, U.S.A.—by permission.

Water Culture

In a water culture setup, a thin layer of cork or a board with holes for the plants may be placed above a container holding the nutrient solution, and the plants may be supported in the holes by wads of nonabsorbent cotton, foam rubber, or some other suitable material (Hoagland and Arnon, 1939). Sometimes vertical supports are provided next to the plants to support the tops and to prevent undue squeezing of bases of stems while providing support. In other types of water culture equipment, there is placed above the solution a shallow tray made of hardware cloth or other suitable material, which is filled with sawdust, wood shavings, rice hulls, or some other relatively inert medium (Fig. 12.5). Seeds or seedlings are placed in this medium which is kept moist until the roots are long enough to reach the nutrient solution in the reservoir below the tray.

Fig. 12.4
Soybean plants, 10 ft tall, growing in automatically irrigated sand cultures in a greenhouse. Providing support for plants of this size can be a problem.

Osmond and Clark (1970) suggested using 18-liter linear polyethylene containers for water culture, with plants held upright by wires extending from a center dowel securely fastened to tight-fitting lids.

Sand Culture

Using sand as an inert medium, many different types of culture equipment have been devised (Robbins, 1928; Offutt et al., 1939; Thomas et al. 1943; Gauch and Wadleigh, 1943; Magistad et al., 1943; Davidson, 1946; Robbins, 1946; Withrow and Withrow, 1948; Maynard et al., 1970). Certain types of sand culture equipment are used primarily because of their simplicity, whereas rather complicated automatically irrigated types have been devised for the more critical nutrition experiments conducted by scientists. Thomas et al. (1943) devised large sand culture units with automatic equipment for subirrigation.

Drip Culture. One of the simplest, semiautomatic types of sand culture involves a container filled with sand. The sand is kept moist by nutrient solution which siphons and drips from a capillary glass tube—one end of which passes into the solution contained in an inverted jar placed next to it (Shive and Robbins, 1937).

Fig. 12.5
General arrangement of tank equipment and method of planting. A frame supporting a wire screen fits over the metal tank filled with nutrient solution; tomato plants are placed with their roots immersed in the nutrient solution; a layer of excelsior is spread over the netting, as shown in the far end of the tank; planting is completed by spreading a layer of rice hulls over the excelsior. (Hoagland and Arnon, 1950.)

"Slop" Culture. Of the various types of sand culture, unquestionably the simplest is the so-called "slop" culture technique. One may use any type of container with a hole in the bottom, and an excess of nutrient solution is applied to the sand so that some solution passes through.

Soxhlet-Type Culture. A soxhlet-type, automatic, sand culture has been developed (Eaton and Bernardin, 1962). Its distinctive feature lies in the fact that the sand is alternately flooded and drained several times during each 15-min "on" period of the time clock.

Automatically Irrigated Culture. Eaton (1936) was one of the first investigators to develop various types of automatically irrigated sand culture equipment, including large-scale cultures (Eaton, 1941a, 1942a).

Fig. 12.6
(A) Cross-sectional view of automatically irrigated sand culture equipment. Insert A, one-way check valve (also 24) which has since been replaced by a commercially available, plastic, one-way valve; 29, polyethylene tubing with holes through which nutrient solution is delivered to sand culture (35); 5, pipe with outlets for source of intermittently supplied, low-pressure air for forcing nutrient solution from the ½-gal jug to the sand culture (35). For other details, see original article. (Gauch and Wadleigh, 1943.) (B) Close-up of sand cultures showing certain modifications from the original design shown in (A).

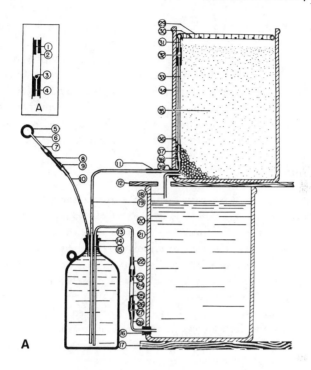

For critical investigations of the nutrition of plants, this type of equipment is highly desirable. Ordinarily, in this type of apparatus each plant container is provided with its own reservoir of nutrient solution (Fig. 12.6A and B) (Gauch and Wadleigh, 1943). The solution, or some portion of it, is automatically delivered to the sand culture at hourly or other desired time intervals. In most types of automatically irrigated sand culture, the excess of nutrient solution is returned to the reservoir and, from time to time, distilled or demineralized water is added to the reservoir to bring the nutrient solution back to volume.

Gravel Culture

In general, when gravel has been used as the inert medium, automatically irrigated systems have been devised (Alexander et al., 1939; Laurie and Kiplinger, 1940; Kiplinger and Laurie, 1948). Ordinarily, the nutrient solution is delivered into the bottom of a crock or greenhouse bench by so-called "subirrigation," and the pumping time is adjusted so that further delivery of nutrient solution ceases when the level of the solution approaches the surface of the gravel (Fig. 12.7). The excess of solution drains by gravity into the reservoir.

Following the lead of Eaton (1939), who developed subirrigation for sand cultures, Withrow and Biebel (1938) developed the mechanics for the subirrigation of gravel cultures; a somewhat similar system was later proposed by Connors and Tiedjens (1940).

Fig. 12.7
Diagrammatic illustration of a design for subirrigated gravel culture. (Alexander, et al., 1939.)

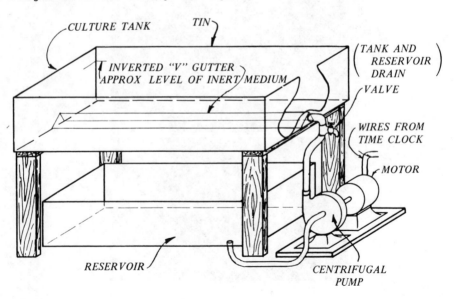

Colloidal Clay Culture

Albrecht and Schroeder (1939) stressed the desirability of using colloidal clay in studies of plant nutrition. They noted that colloidal clay duplicates very closely the salient features of soil, while permitting a complete chemical control of the medium, hence the nutrition of plants. Large amounts of nutrients can be adsorbed onto the surface of clay without causing any appreciable change in OP of the medium; still the ions are freely available to the plants. The influence of such basic ions as Ca^{2+} can be separated from the effects of pH per se (Albrecht, 1933). Another advantage of colloidal clay is exemplified by the fact that at pH 5 it is possible to have 650 times as much H^+ on a 2% colloidal clay suspension as exists in a solution at pH 5 and still have little or no deleterious effect of H^+ ions on plant growth (Albrecht, 1940). In addition, it is possible to make an efficiency evaluation or determination of the percentage of the available nutrient supply absorbed (from the clay) by the plants. Albrecht (1937) suggested that soil colloids may be very important in the delivery of nutrients to plants and, under some conditions, in the loss of nutrients from plants. Jenny and Overstreet (1939) demonstrated that soil colloids are very important in the delivery of nutrients to plants and, on other occasions, in the loss of nutrients from plants.

Albrecht and Schroeder (1939) prepared colloidal clay from a subsoil high in beidellite clay. After total BEC was determined, they added nutrients in the amount and ratio desired and set the pH value. Clay was mixed with leached white sand to furnish a suitable culture medium. The clay was saturated either wholly with one kind of ion or with definite quantities (hence proportions) of several kinds of ions. Later, Albrecht (1946) published procedures for the preparation of acid colloidal clay on a comparatively large scale.

Wittwer et al. (1947) also mentioned that, with colloidal clay colloids carrying nutrient ions in sand culture, ions could be added without changing the osmotic concentration of the substrate and without the addition of soluble ions of the opposite charge.

Synthetic Ion Exchange Materials

Plants have been grown successfully in sand culture (Schlenker, 1942) or gravel culture (Converse et al., 1943) with cations adsorbed on Permutit (Decalso, an artifical zeolite) (Schlenker, 1942) and anions on De-acidite, an aniline material (Converse et al., 1943). A comparison of equal amounts of soluble salts and adsorbed ions showed the adsorbed ions resulted in a greater plant growth (Schlenker, 1942) than that resulting from soluble salts in solution.

Graham and Albrecht (1943) and Jenny (1946) reported that NO_3^- adsorbed on Amberlite anion exchanger was available for plant growth.

Later, Arnon and Grossenbacher (1947) reported on the suitability of synthetic ion exchange materials (Amberlites) for sources of adsorbed cations and anions. It was observed that the percentage saturation had to be relatively high for Ca^{2+} and Mg^{2+} when K^+ was also present, or Ca^{2+} and Mg^{2+} were unavailable for plant growth (Arnon and Meagher, 1947). Similarly, it was found that the percentage saturation with PO_4^{3-} and SO_4^{2-} had to be relatively high when NO_3^- was also adsorbed onto Amberlite in order to prevent a deficiency of P and S.

Amberlite cation exchanger (IR-100) is a phenol formaldehyde resin in which the H^+ of the phenolic group can be exchanged for other cations; Amberlite anion exchanger (IR-4) is an amine formaldehyde resin (Harrisson et al., 1943).

Nutrient Vapor Bath

In this type of artificial culture equipment, nutrient solution bathes the roots in vapor form—supplied by means of an ordinary vaporizer (Klotz, 1944).

Sterile Root Medium

Blanchard and Diller (1950) developed an apparatus for growing a plant with its roots in a sterile medium; contamination occurred in only about 5% of the cultures.

USEFULNESS OF HYDROPONICS

In Research

Determining Which Elements are Essential. One of the very important uses of artificial culture equipment has been that of determining which elements are essential for plant growth. Essentiality of a given element can be studied successfully only by resorting to some artificial culture technique.

Producing and Characterizing a Given Element Deficiency. By means of the artificial culture technique, it has been possible to produce deficiency symptoms for various essential elements and to characterize the symptoms associated with these deficiencies. Inasmuch as a single element is omitted in any one case, the symptoms are specific for a single element deficiency, and they are not complicated by the appearance of deficiency symptoms of one or more other elements—such as often occurs under field conditions.

Studying the Physiological Role of an Essential Element. Obviously, when a scientist wishes to study the role of a given element shown to be essential, it is mandatory that all other essential elements be supplied, and that the solution be free of the omitted element under study or contain known, added amounts of the element.

Determining Interrelationships between or among Elements. In precise studies of the interrelationships among elements involving variations in their concentrations around roots, the use of artificial culture equipment is highly desirable. With such a system it is possible to determine the exact composition of the solution in which the plants are grown. However, when salts or a nutrient solution of known composition are added to the soil, a reaction may occur between added salts and ions adsorbed on the surfaces of the inorganic and organic colloids, so that the composition of the "soil solution" will not be what the investigator sought to achieve.

Commercially

Artificial culture has been used to a limited extent commercially—particularly for the production of high-value floricultural or horticultural crops (Robbins, 1928; Biekart and Connors, 1935). Its applicability to commercial production was suggested by Robbins (1928) and Spencer and Shive (1933), and fully demonstrated by Biekart and Connors (1935) and others. Arnon and Hoagland (1940) also suggested the possibility of producing crops on a large scale by water or sand culture.

Feasibility. Stuart (1948) described in detail the use of sand culture equipment by the United States Army Air Force during World War II. The first installation was established on Ascenstion Island early in 1945, and it consisted of 25 beds each 400 ft long and 3 ft wide. The yield of fresh salad vegetables, cucumbers, tomatoes, radishes, lettuce, and green peppers was 94,000 lb during the first year of operation.

In the summer of 1945, 75 beds were constructed at Atkinson Field, British Guiana, and production of vegetables at this installation in 1946 amounted to 234,337 lb.

A third soilless culture garden was constructed on Iwo Jima late in 1945. About the same time, hydroponic gardens were constructed in Japan where, at Chofu, there were 5 acres under glass and 50 acres out-of-doors; at Otsu there were 25 acres out-of-doors. It was expected that during 1947 more than 10,000,000 servings of fresh vegetables would be produced (Stuart, 1948).

From 1947 to 1957, approximately 100,000,000 lb of fresh vegetables were produced at Chofu and Otsu for the United States Armed Forces in Korea, Japan, and Okinawa (Culbertson et al., 1957). Production costs per pound were about the same in hydroponics as in soil.

EVALUATION OF HYDROPONICS

Advantages

(1) It is possible to *attain* and to *maintain* whatever nutrient environment is desired.

(2) It provides a method of growing plants in an area where soil is lacking (for example, coral reef islands) (Arnon and Hoagland, 1940), or where soil is present and fertile but is contaminated with disease-producing organisms or some toxic principal.

(3) There is general uniformity of plant growth (Laurie and Kiplinger, 1940) and plant products inasmuch as the nutritional factors are at an optimum and are the same from one crop to the next.

(4) The nutrition can be altered at any time to coordinate it with changes in the weather, particularly changes in sunshine and temperature.

(5) The onset of and degree of flowering and fruiting are affected by N supply in particular. Inasmuch as the nutrition can be altered at any time, much can be done to regulate the vegetative or the reproductive phases of plant growth.

(6) The acid–base balance (pH) of the solution can be easily set and maintained in the range most suitable for a given crop.

(7) Less, if any, trouble is encountered with soil-borne fungi, bacteria, and insects (Culbertson et al., 1957). Apparently, Dunlap (1936) was the first investigator to use sand for the growing of seedlings, with the direct purpose of avoiding damping-off of seedlings.

(8) The equipment can be made automatic, hence, the labor (Laurie and Kiplinger, 1940) and expense of watering the plants are avoided. Weeding, mulching, and changing of soil are also eliminated.

(9) With the usual greenhouse watering of plants no longer necessary, there is lower humidity which may mean less trouble from mildew (for those crops susceptible to mildew), black spot (of roses), and rust or other leaf spots of carnations.

(10) Especially with the subirrigation system, the surface of the gravel is either kept dry or is permitted to dry out several times daily. Damping-off of seedlings is largely if not altogether prevented.

(11) Many experienced growers have found this method to be a cheaper and more satisfactory medium than soil for growing plants or, at least, certain of their crops. Owing to the tendency toward uniformity of plant growth (Culbertson et al., 1957), they often report a lower percentage of cull plants or cull fruit.

Disadvantages

(1) Production is limited as compared with field production.

(2) Considerable technical knowledge is necessary to design equipment, to plan routine procedures, and particularly to handle special problems which may arise (Culbertson et al., 1957).

(3) Economically, its use under greenhouse conditions is usually limited to high-value, specialty crops.

(4) Disease, when and if it does appear, may affect all plants in the container or bed, since circulation of the nutrient solution tends to spread disease-producing organisms to the roots of all plants.

(5) Special modifications of equipment and procedure are necessary for certain crops—such as Irish potato. Edible roots of certain crops, potatoes, carrots, and parsnips, may be undesirably altered in shape with coarse gravel but not when sand is used.

NUTRIENT SOLUTIONS

Attempts to Find the "Best" Solution

In the early 1900s, many investigators were interested in the possibility that there was a "best" solution for the culture of plants in general or, at least, for a given species. The "triangle" system, for proportioning the major nutrient salts, was used by Schreiner and Skinner (1910), Tottingham (1914), Shive (1915), and their followers (McCall, 1916; Livingston and Tottingham, 1918) to vary the composition of nutrient solutions and, hopefully, to find the best solution. This search culminated in a combined effort involving many cooperating plant physiologists. It was directed by a special committee of the Division of Biology and Agriculture of the National Research Council (Livingston et al., 1919). The committee established detailed procedures that were to be followed and even provided a common pool of nutrient salts to be used by all cooperating personnel.

It soon became evident that there was no one best solution. Analyses of the results showed that this goal could not be achieved. Even with the same species, different solutions emerged as the best solution for plants grown at different seasons of the year. Clearly, there was a variation in the solution that was best depending on daylength, temperature, amount of sunlight, and so on. As an example, it has been shown that, other factors being at their optima, plants can use and actually require more NO_3^- during periods of sunny than during periods of cloudy weather (Nightingale, 1942); essentially equally good growth occurred over a wide range in the overall concentration of nutrient solutions and in the proportionality among the various salts (Shive, 1914; Johnston, 1932; Stuart, 1948). This finding should perhaps have been anticipated inasmuch as equally good growth occurs in various fertile soils despite the fact that the compositions of the soils and their soil solutions are quite varied. If there were only a very critical composition of soil solution that would support excellent plant growth, there would probably be very few soils, hence very few areas, where good crop production would occur. Some of the discrepancies in early work may have been related to variation in the frequencies with which nutrient solutions were changed or renewed. It soon became obvious that the solution that was best for one species might sometimes not be the best, or even very good, for another one. However, in view of the wide latitude in nutrient composition over which equally good growth is possible many species of plants may actually grow quite well on a given solution. In nature, a good fertile soil is suitable for the growth of many species of plants. Even for a given species, most investigators found that the composition of the best solution varied at different stages of growth. Those findings are consistent with our current knowledge of the seasonal variations in the absorption of different ions. In this regard, the outstanding example is the generally higher requirement for P in the early as compared with the later stages of growth. The solution that proves best may vary depending on the criterion employed, that is, whether top growth, root growth, yield, or some other aspect of growth or yield is considered.

Composition

Nutrient solutions used by Sachs (1860), Knop (1865), Pfeffer (1900), and Crone (1902) are listed in Miller's (1938) textbook of plant physiology. The solutions these early investigators used were quite similar in composition to the ones used today with the exception that trace elements in particular are now usually added in a chelated form.

In 1920, Hoagland reported that the following nutrient solution (in parts per million) was satisfactory for the growth of many plants: K^+, 190; Ca^{2+}, 172; Mg^{2+}, 52; PO_4^{3-}, 117; NO_3^-, 700; and, SO_4^{2-}, 202. Hoagland and Arnon's (1950) formulations for nutrient solutions (Table 12.1) have probably been used more widely than any. In recent years Hoagland's solution has most frequently been used at half the original concentration. In fact, the 0.5 ppm of Mn^{2+} in Hoagland's solution caused Mn^{2+} toxicity in barley leaves; 0.025 ppm was satisfactory (Vlamis and Williams, 1962).

Numerous nutrient solutions have been formulated by different workers. Table 12.2 lists the composition of a nutrient solution we have used for the successful growth of tomato, snapbean, sunflower, impatiens, pea, corn, and many other species.

Table 12.1
Hoagland and Arnon's Nutrient Solutions[a]

Stock solution	Milliliters in a liter of nutrient solution[b]
Solution 1	
1 M KH$_2$PO$_4$	1
1 M KNO$_3$	5
1 M Ca(NO$_3$)$_2$	5
1 M MgSO$_4$	2
Solution 2	
1 M NH$_4$H$_2$PO$_4$	1
1 M KNO$_3$	6
1 M Ca(NO$_3$)$_2$	4
1 M MgSO$_4$	2

[a] Hoagland and Arnon (1950).
[b] To either of these nutrient solutions, trace elements are added to give the following concentrations of B, Mn, Zn, Cu, and Mo: 0.5, 0.5, 0.05, 0.02, and 0.01 ppm, respectively. Fe is added at the rate of 1 ml 0.5% Fe tartrate per liter of nutrient solution.

Table 12.2
Composition of a Nutrient Solution Proven Satisfactory for Many Types of Plants[a]

Milliequivalents/liter						Parts per million				
Ca	Mg	K	NO$_3$	H$_2$PO$_4$	SO$_4$	Cu	Zn	Mn	Fe[b]	B
10	4	4	10	4	4	0.02	0.05	0.5	3	0.5

[a] Solution developed by Gauch.
[b] Fe supplied as Na Fe–EDTA.

Various Methods of Formulating Nutrient Solutions

Triangle System. The triangle method for systematically varying the proportions of various salts in a solution was first proposed by Schreinemakers (1893) and Bancroft (1902) for certain percentage experiments in physical chemistry. Later, the method was adopted by Schreiner and Skinner (1910), Tottingham (1914), Shive (1914, 1915), McCall (1916), Shive (1917), Johnston (1920), and others as a means of varying the composition of nutrient solutions for water and sand cultures. Tottingham (1914) used a four-salt solution made with the following salts: KNO$_3$, Ca(NO$_3$)$_2$, KH$_2$PO$_4$, and MgSO$_4$. Shive (1914, 1915) was the first to suggest a three-salt nutrient solution, because of its greater simplicity, composed of Ca(NO$_3$)$_2$, KH$_2$PO$_4$, and MgSO$_4$. Miller (1938) gives a detailed presentation of the triangle system. For supplying the six major elements (Ca, Mg, K, S, P, N), six types of combinations of salts are possible (see Miller, 1938). These six types or combinations were numbered and cataloged by Livingston and Tottingham (1918), and these designations were universally adopted by all workers.

"Hamner" System. Beckenbach et al. (1936) first published a technique for varying the proportion of cations in a nutrient solution without a concomitant variation in anions, and vice versa. The same procedure was later suggested by Hamner (1940), and it has been employed by others (Hamner et al. 1942; Lyon and Garcia, 1944). Unfortunately, this system or experimental design cannot be submitted to factorial analysis (Richards, 1944).

Concentration Series. Particularly when there is no information on which to base the selection of concentration levels for various elements in an experiment, there is much to be gained from studying plant growth response over a range of concentrations of each element to be studied. Ideally, one or more of the levels should be low enough to induce deficiency, and one or more of them high enough to induce toxicity, that is, to reduce growth. In this way the responses of the plant may be "bracketed" by the wide range in concentrations for a given element.

There are certain advantages in using a concentration series:

(1) With a closely integrated series of concentrations of a given element, there is always the possibility of finding a response in some concentration range that would not have been observed if fewer, widely spaced concentrations had been studied. Two illustrations suffice to demonstrate the significance of this possibility. Eaton (1942b) observed growth responses of tomato at extremely low concentrations of Cl^- that might never have been revealed had he not employed a closely integrated series of small increments in concentration. Thomas (1952) observed a double-peak response of pollen germination and growth in a range of concentrations of B. Other examples of double-peak responses have been reported which might very easily have been missed if widely separated and fewer levels had been studied.

(2) The results of a concentration series experiment are very useful for selecting, for example, the so-called "low", "medium", and "high" levels for a factorial experiment, as discussed in the next section.

Factorial Experiment. A factorial design permits a study of the importance of interaction between factors, as well as a study of the effects of the factors themselves. The basic requirement is that each level of each factor occur in combination with each level of every other factor being studied. In addition to being able to measure interactions, the factorial design is highly efficient because, through repeated stratification of the data, each main effect is measured with the use of all observations (Wadleigh and Tharp, 1940). There need be but few replicates and, in fact, as stressed by Brandt (1937), reliable information can be obtained without the use of any true replications.

The use of statistical measurements is justified and in fact mandatory when certain types of information are being sought. For certain types of inquiry, particularly those designed to discover new phenomena, the use of statistics may not necessarily be justified or even desirable. In a recent highly entertaining but thought-provoking article by an authority in the field of statistics, Feller (1969) has cautioned that "in biological experimental work, a major abuse of statistics has been overemphasis on the role of statistics in evaluating results. The aim of basic research is not to produce statistically valid results, but to study new phenomena." He notes that the results of basic research can rarely be evaluated statistically, and that this point of view was emphasized by the great pioneer of modern statistics, R. A. Fisher. The sole purpose of basic research is simply to discover new things—new basic facts. "Such experiments have little in common with standard routines, and must be considered on their own merits."

pH Tolerance, Adjustment, and Control

Although there is evidence that pH of the soil (Stout and Overstreet, 1950) or a nutrient solution may vary from approximately pH 4 to 8 and still produce good plant growth, if the nutrients are maintained and supplied in the proper forms (Arnon and Johnson, 1942), for any given experiment it is customary to set and to maintain the pH at some selected value.

If a general statement can be made, it is that in most nutrient solutions containing NO_3^- the pH tends to become more basic. NO_3^- is a fairly rapidly absorbed ion and, as it is absorbed, OH^- or HCO_3^- come from the roots to the nutrient solution (Krauss, 1951, 1953) (Fig. 12.8). An increasing proportion of the bases are then associated with OH^- or HCO_3^- and, as an example, KOH is much more "basic" than K sulfate.

Eaton (1941b) showed that additions of dilute HNO_3 simultaneously maintained the desired H^+ ion concentration and the original level of NO_3^- in the solution.

Chelators

For pH values of 7 and above, certain elements, such as Cu^{2+}, Zn^{2+}, and Mn^{2+}, tend to become unavailable. If they are provided in the form of chelates, for example, as salts of EDTA, they become more soluble and available as the pH rises into the alkaline range.

Replenishment

Depending on the volume of solution available for the plants in a culture, on the size of the plants, and on how long the nutrient solution has been in use, finally certain salts will have to be added or, better still, the old solution may need to be removed and a fresh one substituted.

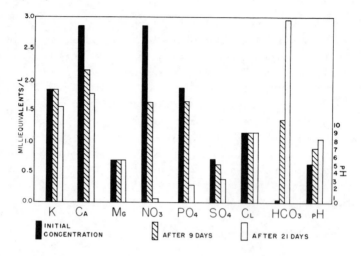

Fig. 12.8

Levels of principal ions in solution after the 1st, 9th, and 21st days of a mass culture of *Scenedesmus obliquus* (Turp.) Kütz. Note particularly the decrease in NO_3 concentration with time and the concomitant rise in HCO_3 in the nutrient solution. (Krauss, 1951.)

Trace Element Toxicity

Particularly in studies of the effects of excesses of salts on plant growth, the nutrient or other salts may contain sufficiently high concentrations of various trace elements (e.g., Zn and B) (Ayers and Hatcher, 1939) to cause toxicity. Anion exchange resins have been used successfully for the removal of Zn^{2+} from phosphate solutions (Wilson and Reisenauer, 1969).

LITERATURE CITED

Albrecht, W. A. 1933. Inoculation of legumes as related to soil acidity. J. Amer. Soc. Agron. 25:512–522.

Albrecht, W. A. 1937. Physiology of root nodule bacteria in relation to fertility levels of the soil. Proc. Soil Sci. Soc. Amer. 2:315–327.

Albrecht, W. A. 1940. Adsorbed ions on the colloidal complex and plant nutrition. Proc. Soil Sci. Soc. Amer. 5:8–16.

Albrecht, W. A. 1946. Colloidal clay cultures – Preparation of the clay and procedures in its use as a plant growth medium. Soil Sci. 62:23–31.

Albrecht, W. A., and R. A. Schroeder. 1939. Colloidal clay culture for refined control of nutritional experiments with vegetables. Proc. Amer. Soc. Hort. Sci. 37:689–692.

Alexander, L. J., V. H. Morris, and H. C. Young. 1939. Growing plants in nutrient solution. Ohio Agr. Exp. Sta. Spec. Circ. 56.

Arnon, D. I. 1937. Ammonium and nitrate nitrogen nutrition of barley at different seasons in relation to hydrogen ion concentration, manganese, copper, and oxygen supply. Soil Sci. 44:91–121.

Arnon, D. I., and D. R. Hoagland. 1939. A comparison of water culture and soil as media for crop production. Science 89:512–514.

Arnon, D. I., and K. A. Grossenbacher. 1947. Nutrient culture of crops with the use of synthetic ion-exchange materials. Soil Sci. 63:159–182.

Arnon, D. I., and D. R. Hoagland. 1940. Crop production in artificial culture solutions and in soils with special reference to factors influencing yields and absorption of inorganic nutrients. Soil Sci. 50:463–584.

Arnon, D. I., and C. M. Johnson. 1942. Influence of hydrogen ion concentration on the growth of higher plants under controlled conditions. Plant Physiol. 17:525–539.

Arnon, D. I., and W. R. Meagher. 1947. Factors influencing availability of plant nutrients from synthetic ion-exchange materials. Soil Sci. 64:213–221.

Ayers, A. D., and J. T. Hatcher. 1939. Quantities of boron and zinc found in salts used in the preparation of culture solutions. Proc. Amer. Soc. Soil Sci. 4:314–315.

Ballentine, J. E. M., and B. J. Forde. 1970. The effect of light intensity and temperature on plant growth and chloroplast ultrastructure in soybean. Amer. J. Bot. 57: 1150–1159.

Bancroft, W. D. 1902. Synthetic analysis of solid phase. J. Phys. Chem. 6:178.

Beckenbach, J., C. H. Wadleigh, and J. W. Shive. 1936. Nutritional studies with corn. I. A statistical interpretation of the nutrient ion effect upon growth in artificial culture. Soil Sci. 41:469–489.

Biekart, H. M., and C. H. Connors. 1935. The greenhouse culture of carnations in sand. New Jersey Agr. Exp. Sta. Bull. 588.

Blanchard, F. A., and V. M. Diller. 1950. Technique for growing plants with roots in a sterile medium. Plant Physiol. 25:767–769.

Brandt, A. E. 1937. Factorial design. J. Amer. Soc. Agron. 29:658–667.

Connors, C. H., and V. A. Tiedjens. 1940. Chemical gardening for the amateur. W. H. Wise, New York.

Converse, C. D., N. Gammon, and J. D. Sayre. 1943. The use of ion exchange materials in studies on corn nutrition. Plant Physiol. 18:114–121.

Crone, G. 1902. Ergebnisse von Untersuchungen über die Wirkung der Phosphorsäure auf die höhre Pflanzen und eine neue Nahrlösung. Sitzungber. Niederrhein. Ges. Natur- und Heilkunde Bonn 1902:167–173.

Culbertson, R. E., T. Nakayama, and M. Hiramatsu. 1957. Hydroponic vegetable production in Japan — An operation of the U.S. Army Quartermaster Corps. Market Growers J. 86:16–19.

Davidson, O. W. 1946. Large-scale soilless culture for plant research. Soil Sci. 62:71–86.

Dep. of War. 1946. Nutriculture. War Dep. Tech. Manual 20-500.

Dunlap, A. A. 1936. Sand culture of seedlings. Connecticut Agr. Exp. Sta. Bull. 380.

Eaton, F. M. 1936. Automatically operated sand-culture equipment. J. Agr. Res. 53: 433–444.

Eaton, F. M. 1941a. Plant culture equipment. Plant Physiol. 16:385–392.

Eaton, F. M. 1941b. Use of nitric acid in control of pH and nitrate levels in nutrient solution. Plant Physiol. 16:834–836.

Eaton, F. M. 1942a. Sand culture methods. Chron. Bot. 7:200–201.

Eaton, F. M. 1942b. Toxicity and accumulation of chloride and sulfate salts in plants. J. Agr. Res. 64:357–399.

Eaton, F. M., and J. E. Bernardin. 1962. Soxhlet-type automatic sand cultures. Plant Physiol. 37:357.

Ellis, C., M. W. Swaney, and T. Eastwood. 1947. Soilless growth of plants. 2nd ed. Van Nostrand Reinhold, New York, N. Y. 277 pp.

Feller, W. 1969. Are life scientists overawed by statistics? Sci. Res. 4:24—29.

Gauch, H. G., and C. H. Wadleigh. 1943. A new type of intermittently-irrigated sana culture equipment. Plant Physiol. 18:543—548.

Gauch, H. G., and R. C. Weigel. 1968. Mineral nutrition research with soybeans in Maryland. Soybean Dig. 28:14—16.

Gericke, W. F. 1942. The complete guide to soilless gardening. Prentice-Hall, Englewood Cliffs, N. J. 285 pp.

Gilbert, S. G., and J. W. Shive. 1945. The importance of oxygen in the nutrient substrate for plants — Relation of the nitrate ion to respiration. Soil Sci. 59:453—461.

Graham, E. R., and W. A. Albrecht. 1943. Nitrate absorption by plants as an anion exchange phenomenon. Amer. J. Bot. 30:195—198.

Hamner, C. L. 1940. Growth responses of Biloxi soybeans to variation in relative concentrations of phosphate and nitrate in the nutrient solution. Bot. Gaz. 101:637—649.

Hamner, K. C., C. B. Lyon, and C. L. Hamner. 1942. Effect of mineral nutrition on the ascorbic-acid content of the tomato. Bot. Gaz. 103:586—616.

Harrisson, J. W. E., R. J. Myers, and D. S. Herr. 1943. Purified mineral-free water for pharmaceutical purposes. J. Amer. Pharm. Ass. 32:121—128.

Hewitt, E. J. 1952. Sand and water culture methods used in the study of plant nutrition. Commonwealth Agr. Bur., Farnham Royal, Bucks, England. 241 pp.

Hewitt, E. J. 1966. Sand and water culture methods used in the study of plant nutrition. Commonwealth Agr. Bur., Farnham Royal, Bucks, England 547 pp.

Hoagland, D. R. 1920. Optimum nutrient solutions for plants. Science 52:562—564.

Hoagland, D. R., and D. I. Arnon. 1938. Growing plants without soil by the water-culture method. C.R.E.A. News Letter 17:12—21.

Hoagland, D. R., and D. I. Arnon. 1939. The water-culture method for growing plants without soil. Smithsonian Inst. Rep. 1938:461—487.

Hoagland, D. R., and D. I. Arnon. 1948. Some problems of plant nutrition. Sci. Monthly 67:201—209.

Hoagland, D. R., and D. I. Arnon. 1950. The water-culture method for growing plants without soil. California Agr. Exp. Sta. Circ. 347(rev.).

Jenny, H. 1946. Adsorbed nitrate ions in relation to plant growth. J. Colloid Sci. 1:33—47.

Jenny, H., and R. Overstreet. 1939. Cation interchange between plant roots and soil colloids. Soil Sci. 47:257—272.

Johnston, E. S. 1920. Nutrient requirement of the potato plant grown in sand cultures treated with "type I" solutions. Soil Sci. 10:389—408.

Johnston, E. S. 1927. An apparatus for controlling the flow of nutrient solutions in plant cultures. Plant Physiol. 2:213—215.

Johnston, E. S. 1932. Growing plants without soil. Smithsonian Inst. Pub. 3156, Smithsonian Rep. 1931:381—387.

Kiplinger, D. C., and A. Laurie. 1948. Gravel culture for growing ornamental greenhouse crops. Ohio Agr. Exp. Sta. Res. Bull. 679.

Klotz, L. J. 1944. A simplified method of growing plants with roots in nutrient vapors. Phytopathology 34:507—508.

Knop, W. 1859. Ein Vegetationversuch. Landwirt. Vers.-Sta. 1:181—202.

Knop, W. 1865. Quantitative Untersuchung über die Ernährungsprocess der Pflanzen. Landwirt. Vers.-Sta. 7:93—107.

Krauss, R. W. 1951. Limiting factors affecting the mass culture of Scenedesmus obliquus (Turp.) Kütz. in an open system. Ph. D. Thesis, Univ. of Maryland, College Park.

Krauss, R. W. 1953. Inorganic nutrition of algae, pp. 85—102. In J. S. Burlew (ed.) Algal culture from laboratory to pilot plant. Carnegie Inst. Washington Pub. 600.

Laurie, A. 1940. Soilless culture simplified. Whittelsey House, New York.

Laurie, A., and D. C. Kiplinger. 1940. Growing ornamental greenhouse crops in gravel culture. Ohio Agr. Exp. Sta. Bull. 616.

Leiser, A. T., A. C. Leopold, and A. L. Shelley. 1960. Evaluation of light sources for plant growth. Plant Physiol. 35:392—395.

Livingston, B. E., and W. E. Tottingham. 1918. A new three-salt nutrient solution for plant cultures. Amer. J. Bot. 5:337—347.

Livingston, B. E., K. F. Kellerman, and W. Crocker. 1919. A plan for cooperative research on the salt requirements of representative agricultural plants, 2nd ed. Prepared for a special committee of the Division of Biology and Agriculture of the National Research Council. 54 pp.

Lyon, C. B., and C. R. Garcia. 1944. Anatomical responses of tomato stems to variations in the macronutrient anion supply. Bot. Gaz. 105:394—405.

McCall, A. G. 1916. The physiological balance of nutrient solutions for plants in sand cultures. Soil Sci. 2:207—255.

Magistad, O. C., A. D. Ayers, C. H. Wadleigh, and H. G. Gauch. 1943. Effect of salt concentration, kind of salt, and climate on plant growth in sand cultures. Plant Physiol. 18:151—166.

Maynard, D. N., A. V. Barker, and H. F. Vernell. 1970. A semiautomatic sand culture system for greenhouse plant nutrition research. Agron. J. 62:304—306.

Miller, E. C. 1938. Plant physiology (with reference to the green plant), 2nd ed. McGraw-Hill, New York. 1201 pp.

Nightingale, G. T. 1942. Nitrate and carbohydrate reserves in relation to nitrogen nutrition of pineapple. Bot. Gaz. 103:409—456.

Offutt, E. G., R. K. Calfee, and J. S. McHargue. 1939. An apparatus made from glass for the continuous watering of pot cultures. J. Amer. Soc. Agron. 31:725—728.

Osmond, C. A., and R. B. Clark. 1970. An improved method for growing mature corn (*Zea mays* L.) by solution culture. Agron. J. 62:432—433.

Pfeffer, W. 1900. The physiology of plants — A treatise upon the metabolism and sources of energy in plants, Vol. I. Transl. and edited by A. J. Ewart. Clarendon, Oxford. 632 pp.

Phillips, A. H. 1940. Gardening without soil. C. Arthur Pearson, London. 139 pp.

Richards, F. J. 1944. Mineral nutrition of plants. Annu. Rev. Biochem. 13:611—630.

Robbins, W. R. 1928. The possibilities of sand culture for research and commercial work in horticulture. Proc. Amer. Soc. Hort. Sci. 25:368—370.

Robbins, W. R. 1946. Growing plants in sand cultures for experimental work. Soil Sci. 62:3—22.

Rohrbaugh, L. M. 1942. Effects of light quality on growth and mineral nutrition of bean. Bot. Gaz. 104:133—151.

Sachs, J. 1859. Bericht über die physiologische Thätigkeit an der Versuchsstation in Tharandt. I. Über den Einfluss der chemischen und physikalischen Beschaffenheit des Bodens auf die Transpiration der Pflanzen, Landwirt. Vers.-Sta. 1:203—240.

Sachs, J. 1860. Bericht ˙ über die physiologische Thätigkeit an der Versuchsstation in Tharandt. IV. Vegetations-versuche mit Ausschluss des Bodens über die Nährstoffe und sonstigen Ernährungsbedingungen von Mais, Bohnen und anderen Pflanzen, Landwirt. Vers.-Sta. 2:219—268.

Schlenker, F. S. 1942. Availability of adsorbed ions to plants growing in quartz sand substrate. Soil Sci. 54:247—251.

Schreinemakers, F. A. H. 1893. Graphische Ableitungen aus den Lösungs-Iostherm eines Doppelsalzes und seiner Komponenten und mögliche Formen der Umwandlungskurve. Z. Phys. Chem. 11:75—109.

Schreiner, O., and J. J. Skinner. 1910. Ratio of phosphate, nitrate, and potassium on absorption and growth. Bot. Gaz. 50:1–30.

Shive, J. W. 1914. A study of physiological balance in nutrient media. Physiol. Res. 1:327–397.

Shive, J. W. 1915. A three-salt nutrient solution for plants. Amer. J. Bot. 2:157–160.

Shive, J. W. 1917. A study of physiological balance for buckwheat grown in three-salt solutions. New Jersey Agr. Exp. Sta. Bull. 319.

Shive, J. W., and W. R. Robbins. 1937. Methods of growing plants in solution and sand cultures. New Jersey Agr. Exp. Sta. Bull. 636.

Shive, J. W., and A. L. Stahl. 1927. Constant rates of continuous solution renewal for plants in water cultures. Bot. Gaz. 84:317–323.

Spencer, E. L., and J. W. Shive. 1933. The growth of *Rhododendron ponticum* in sand cultures. Bull. Torrey Bot. Club 60:423–439.

Stout, P. R., and R. Overstreet. 1950. Soil chemistry in relation to inorganic nutrition of plants. Annu. Rev. Plant Physiol. 1:305–342.

Stuart, N. W. 1948. Growing plants without soil. Sci. Monthly 66:273–282.

Thomas, M., R. H. Hendricks, J. O. Ivie, and G. R. Hill. 1943. An installation of large sand-culture beds surmounted by individual air-conditioned greenhouses. Plant Physiol. 18:334–344.

Thomas, W. H. 1952. Boron contents of floral parts and the effects of boron on pollen germination and tube growth of *Lillium* species. M. S. Thesis, Univ. Maryland, College Park.

Tottingham, W. E. 1914. A quantitative chemical and physiological study of nutrient solutions for plant cultures. Physiol. Res. 1:133–245.

Turner, W. I., and V. A. Henry. 1939. Growing plants in nutrient solutions. Wiley, New York. 154 pp.

Vlamis, J., and D. E. Williams. 1962. Ion competition in manganese uptake by barley plants. Plant Physiol. 37:650–655.

Wadleigh, C. H., and W. H. Tharp. 1940. Factorial design in plant nutrition experiments in the greenhouse. Arkansas Agr. Exp. Sta. Bull. 401.

Went, F. W., and L. Cosper. 1945. Plant growth under controlled conditions. VI. Comparison between field and air-conditioned greenhouse culture of tomatoes. Amer. J. Bot. 32:643–654.

Wilson, D. O., and H. M. Reisenauer. 1969. Removal of zinc from phosphate solutions by anion-exchange. Plant Physiol. 44:1205–1206.

Withrow, R. B., and J. P. Biebel. 1938. Nutrient solution methods of greenhouse crop production. Indiana (Purdue Univ.) Agr. Exp. Sta. Circ. 232 (rev.):1–20.

Withrow, R. B., and A. P. Withrow. 1948. Nutriculture. Indiana (Purdue Univ.) Agr. Exp. Sta. S. C. 328.

Wittwer, S. H., R. A. Schroeder, and W. A. Albrecht. 1947. Interrelationships of calcium, nitrogen, and phosphorus in vegetable crops. Plant Physiol. 22:244–256.

Woodward, J. 1969. Some thoughts and experiments concerning vegetation. Phil. Trans. Roy. Soc. (London) 21:193–227.

CHAPTER 13

Deficiency Symptoms

BACKGROUND

Diagnostic Criteria for Plants and Soils, edited by Chapman (1966a), contains a wealth of information on deficiency symptoms for various essential elements, the concentrations of these elements associated with deficiencies in plants, and methods for overcoming deficiencies. Excellent color and black-and-white pictures of deficiencies are available in *Hunger Signs in Crops—A Symposium* (Hambidge, 1941; Sprague, 1964).

Symptoms Produced and Identified in Artificial Culture

As Arnon (1951) noted,

There seems to be a basis then for reaching the paradoxical conclusion, that the natural growth medium of land plants, the soil, is least adapted for the study of the indispensable nature of plant nutrients. It is with artificial nutrient media, water and sand cultures, that the essential status of the various elements found in plants and derived from the soil was established. The recent history of plant nutrition offers no case of a discovery of a new essential element through soil treatments. For every one of the micronutrients, for example, evidence of their indispensability was available from nutrient solution experiments, well in advance of any responses reported from the field. The artificial culture technique continues to be a powerful and discriminating tool in evaluating the indispensability of inorganic nutrients in plant nutrition.

In order to test the essentiality of a given element for plant growth, it is necessary that the medium be as free as possible of the element under study, and that all other known essential elements be present in adequate amounts. These conditions can be met only by using artificial culture techniques, since it is impossible to treat a soil chemically to remove all traces of an element without drastically altering the original soil. It is often very difficult to provide an environment free of a given element even when an artificial culture technique is employed. Special consideration must be given to the choice of containers (e.g., Pyrex glass contains appreciable B), to the recrystallization of salts, and to the use of the purest water it is possible to produce. Stout and Arnon (1939) described the procedures and precautions for studying the roles of Cu, Mn, and Zn, and this report lists the difficulties of this type of investigation. Various methods have been compared for removing Co and V from salts used in nutrition studies (Hewitt et al., 1965).

Effect

Effects on Enzymes. Brown and Hendricks (1952) postulated that, if a plant is deficient in an element required for one or more enzymes, the deficiency will be manifested by a change in activity of one or more enzymes. Studying corn, wheat, tobacco, lupines, and soybeans, they found that ascorbic acid oxidase and catalase activities were markedly reduced by limited Cu^{2+} supply; activity of peroxidase was essentially unaffected.

Deficiencies of N, P, K^+, Ca^{2+}, Mg^{2+}, S, and Fe^{2+} decreased the activity of NR in *Triticum aestivum* L. seedlings; the first six of these deficiencies had a similar effect on glutamine synthetase. Glutamic acid dehydrogenase activity was decreased by N, P, and S deficiencies, unchanged by K^+ deficiency, and increased by Ca^{2+}, Mg^{2+}, and Fe^{2+} deficiencies (Harper and Paulsen, 1969a). Glutamic-oxaloacetic transaminase was unaffected by macronutrient deficiencies. It was concluded that nutritional deficiencies apparently had depleted the endogenous amino acid pools and therefore the incorporation of exogenous [^{14}C]-leucine was greater than would have been expected in view of the decreased enzymic activities under the nutrient deficiency conditions.

Low concentrations of Mg^{2+} stimulated and high concentrations inhibited the adenine phosphoriboxyltransferase reaction (Gadd and Henderson, 1970).

In a study of the effect of micronutrient deficiencies on enzymes in wheat seedlings, Harper and Paulsen (1969b) noted that the activity of NR was decreased by deficiencies of Mo, Zn^{2+}, and Cl^-; NO_3^- accumulated in Mo-deficient seedlings, declined in Zn-deficient seedlings, and was unaffected by the other (Mn, Cu, B, and Cl) micronutrient deficiencies. Glutamic acid dehydrogenase activity was decreased by Mo deficiency—the only deficiency affecting this enzyme. Glutamine synthetase activity was decreased only by Cu deficiency, whereas glutamic-oxaloacetic transaminase was unaffected by any of the deficiencies.

In maize and radish Fe deficiency caused a depression in catalase, peroxidase, and aldolase and, also, in radish, phosphorylase (Agarwala et al., 1965). Acid phosphatase and ribonuclease in both species, and transaminase in radish, were stimulated by Fe deficiency. Inasmuch as aldolase and phosphorylase were depressed by Fe deficiency, it was concluded that Fe^{2+} is required even for the synthesis of nonheme enzymes.

ATP sulfurylase (ATP: sulfate adenlyl transferase, E.C. 2.7.7.4) catalyzes the production of adenosine-5'-phosphosulfate (ASP) from ATP and SO_4^{2-}. Its discovery and distribution were discussed by Gregory and Robbins (1960), and the enzyme was first reported in a higher plant by Asahi (1964). In a physiological study of this important enzyme, which presumably must occur in all higher plants, Adams and Rinne (1969) found that the level of S nutrition markedly influenced ATP sulfurylase activity despite the absence of visual symptoms of S deficiency in soybean plants. Wilson (1962) reviewed SO_4^{2-} metabolism and reduction in plants.

In a study of nodulated and nonnodulated soybean plants, the capacity of the seedlings to fix N_2 was less than their ability to utilize NO_3^- (Wooding et al., 1970). It was also reported that S deficiency had little effect on the concentration of hemoglobin in nodules, but number of nodules and amount of N_2 fixed were markedly reduced.

Several studies deal with effects of B deficiency on enzymes in roots of *Vicia faba* var. *minor*. Although normal and B-deficient roots showed the same distributions of phosphatases, pyrophosphatases, and ATPases, B-deficient roots had higher concentrations of the last-mentioned two enzymes than did B-sufficient roots (Hinde and Finch, 1966). Amino acid-dependent ATP–pyrophosphate exchange activity in extracts of root sections decreased away from the tip, and the activity of this enzyme was less in B-deficient than in normal sections. This change in enzymic activity could be detected prior to a reduction in growth rate of B-deficient roots (Hinde et al., 1966). As compared with B-sufficient roots, B-deficient roots showed an increased ability to incorporate [^{32}P]-phosphate into nucleic acids (Cory et al., 1966). This effect was most marked 2–5 mm from the tip where it was observable after as little as 4 hr in B-deficient medium.

Effect on Composition of Plants. Deficiencies of K, P, and S result in a disturbance of protein metabolism and accumulation of amino acids and soluble organic N compounds (Steinberg, 1951). It was concluded that a deficiency of any essential inorganic nutrient usually results in reduced protein synthesis.

Effect of Photoperiod. With certain species of plants, Warington (1933) observed that B deficiency symptoms developed more quickly during the long days of summer than the relatively shorter days of winter. In a 16-hr photoperiod, but not in a 9-hr one, *Xanthium,* tomato, sunflower, and buckwheat plants developed B-deficiency symptoms (MacVicar and Struckmeyer, 1946).

However, with regard to the development of B deficiency in tomato root tips, high light intensity, as compared with low, was more causally related to increasing the rate of appearance of B-deficiency symptoms than was a lengthening of the photoperiod (MacInnes and Albert, 1969).

Effect of Light Intensity. Reduced light intensity, as compared with a higher light intensity, reduced the B-deficiency symptoms of *Lemna pausicostata* as determined by dry weight production (Tanaka, 1966).

DEFICIENCY SYMPTOMS

Diagnostic Value

Although it is somewhat rare under field conditions to have a deficiency of only one element rather than two or more, occasionally only one element is sufficiently deficient to result in its clear-cut deficiency symptoms. When this occurs, it is generally fairly easy for someone familiar with deficiency symptoms for the various essential elements to determine which element is deficient. There are numerous descriptions and photographs of deficiency symptoms for many species of plants. One of the most complete references is *Hunger Signs in Crops—A Symposium* (Hambidge, 1941; Sprague, 1964).

In many cases of deficiency under field conditions, two or more essential elements are simultaneously deficient. This situation is likely to exist under conditions of (1) low fertility in general, and (2) soil conditions, such as alkalinity, that render elements such as Zn^{2+}, Cu^{2+}, Mn^{2+}, and B unavailable. When two or more elements are deficient simultaneously and their symptoms are being expressed, the composite picture, or syndrome, may resemble no given deficiency. Under such conditions it is generally impossible visually to determine which elements are responsible for the symptoms.

Symptomatology

Yellowing or chlorosis, which is sometimes followed by necrosis of chlorotic areas of leaves, is a symptom associated with deficiencies of various essential elements. Location and extent of chlorosis on leaves vary, depending on the element that is deficient. Depending on mobility or lack of mobility of an essential element, chlorosis appears on the lower or the upper leaves, respectively.

Deficiencies of certain essential elements result in the formation of anthocyanin pigments (Blank, 1947) in leaves or stems, or both.

Nitrogen. Inasmuch as N is mobile within a plant, a deficiency results in movement of N from older (lower) leaves to younger (upper) leaves with the result that the older, lower leaves turn yellow. In monocots, such as corn, the yellowing starts at the tip of the leaves and proceeds toward the base—primarily with a yellowing of the central portion of the leaf rather than the margins. In extreme cases of N deficiency, all leaves on the plant may become yellow. Jones (1966) lists internal concentrations of N associated with deficiency in various crops.

Phosphorus. In many plant species P deficiency induces the formation of anthocyanin pigments and the leaves acquire a purple color. In corn, purpling occurs primarily along margins of the leaf and on the lower stalk of the plant. In tomatoes purpling occurs chiefly in veins on the underside of the leaf. Sometimes the only symptom in certain plants is an intensely deep-green, almost bluish-green, color of the leaves.

On P-deficient media the growth of duckweed *(Spirodela oligorrhiza)* soon slowed and eventually ceased (Reid and Bieleski, 1970). The decreased growth rate could be detected within 6 hr after transfer to P-deficient media. During P-deficiency the older leaves became chlorotic, but newly formed leaves were dark green and contained a high concentration of anthocyanin. In deficiency the photosynthetic rate fell gradually, roots elongated, and chloroplasts became filled with starch.

Purpling is not indicative solely of P deficiency, since this symptom is associated with Mg^{2+} deficiency in cotton and with B deficiency in clover. Bingham (1966) lists P-deficiency symptoms for many crops, and the P concentrations associated with deficiency, normalcy, and toxicity.

Potassium. The most characteristic general symptom of K-deficiency is that of tip and marginal scorch of the most recently matured leaves (Ulrich and Ohki, 1966). The tip and margin first become yellow, then turn brown and appear scorched, and finally become dried up, brittle, and sometimes ragged in appearance as a result of loss of some of the dead tissue. Tip and marginal scorch also appear on monocots, such as corn, and, since the tip and margins are affected first, this symptom is readily distinguished from N deficiency in which yellowing proceeds up the middle portion of the blade. However, in some plants, particularly legumes, the first sign of K^+ deficiency is a white speckling or freckling of the leaf blades, as reported by Ulrich and Ohki (1966). These investigators describe K-deficiency symptoms for many crops and list the concentrations of K^+ characteristic of deficiency, normal growth, and even toxicity. They list critical K^+ levels (sufficiency values) for various crops.

In field-grown hybrid corn plants in Ohio, K deficiency was associated with an accumulation of Mo in leaves (Jones, 1965). Plants with leaf symptoms of K deficiency had about four times the concentration of Mo found in normal plants (i.e., 4 versus 1 ppm of Mo).

The effects of various deficiencies on chloroplast structure are illustrated in a recent book (Devlin and Barker, 1971), which also includes effects of deficiencies of N, K^+, Mg^{2+}, Zn^{2+}, and P.

Calcium. When it is possible to observe roots, Ca-deficiency symptoms may be observed in roots prior to the appearance of any symptoms on tops. Root tips appear slimy or turn black. The plant may start to form new roots near the base of the stem, but the new roots do not make such growth before showing the same deficiency symptoms (Gauch, 1940).

Ca deficiency occurs in various crops, and Chapman (1966b) lists methods for correcting the deficiency. He also describes Ca-deficiency symptoms, and concentrations of Ca^{2+} associated with deficiency, normal growth, and toxicity.

In tops the first effect of Ca deficiency is usually death of the terminal bud or growing point. The deficiency may then affect axillary buds. In tobacco, tips and margins of the young leaves are cupped downward and the blade portion is very wrinkled and misshapened. In corn the tip of an emerging leaf may stick to the blade of the leaf below it, and margins of the leaves are scalloped, torn, and ragged-looking (Fig. 13.1).

For tomato plants the critical concentration of Ca^{2+} below which cells in the shoot apices became necrotic was 0.16–0.17% on a dry weight basis (Kalra, 1956).

Although Ca-deficiency symptoms on apples have not appeared under field conditions, symptoms can be produced in artificial culture (Fig. 13.2). Symptoms are expressed on fruit in the absence of symptoms on leaves. This situation is reminiscent of the effect of B on root and shoot apexes and on flower buds even though leaves may be free of B-deficiency symptoms.

Ca is relatively immobile in plants, and usually the element does not move out of older, lower leaves to younger, upper leaves. Owing to the immobility, symptoms show at extremities of the plant—terminal bud, youngest leaves, and root tips.

At the cellular level Marinos (1962) noted that the first indisputable signs of structural abnormalities in Ca deficiency appear when the nuclear envelope and the plasma and vacuolar membranes break up and "structureless" areas appear in cells. These changes were followed by disorganization of other structures such as mitochondria and Golgi apparatus; plastids were more persistent although eventually they also disintegrated.

Magnesium. In contrast with Ca^{2+}, Mg^{2+} is mobile. In deficiency Mg^{2+} moves from older to younger leaves with the result that older leaves show interveinal yellowing or interveinal chlorosis. Interveinal areas may also become necrotic if the deficiency is severe enough.

In cotton, Mg^{2+} deficiency induces formation of anthocyanin pigments and a reddish coloration of leaves.

Embleton (1966) describes Mg^{2+}-deficiency symptoms for many crops,

Fig. 13.1
Ca deficiency in corn plant. Note particularly the torn, ragged edges of the leaves.

lists the concentrations associated with deficiency, normal growth, and toxicity, and presents methods for overcoming Mg^{2+} deficiency in various crops.

Sulfur. A general, overall yellowing of leaves occurs during a deficiency of S but, in contrast with N deficiency in which older, lower leaves are affected, the uppermost, youngest leaves are chlorotic. Thus to distinguish N from S deficiency, the portion of the plant from which chlorotic leaves were obtained must be known.

Leaves can readily absorb such S compounds as SO_2. In greenhouses at the University of Maryland, striking S-deficiency symptoms have often been produced and then, when atmospheric conditions cause smoke from a nearby power plant to settle in the greenhouse, symptoms literally disappear in 24 hr or less.

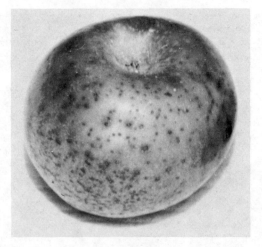

Fig. 13.2
Advanced stage of Ca-deficiency symptoms on a York Imperial apple from a tree grown in controlled nutrient culture. Leaf symptoms of Ca^{2+} deficiency did not develop on the tree. The first symptom on the fruit is an orange-yellow skin color on which the lenticels become very prominent, darken, develop an almost black halo, and become sunken. In extreme cases the fruit cracks, physiological and pathological breakdown occurs, and the fruit drops. Courtesy of C. B. Shear, U.S. Dept. Agriculture, Beltsville, Md.

SO_4^{2-} is a better indicator of the S status of a plant than is organic or total S according to Eaton (1966b). He describes S-deficiency symptoms for various crops and reports concentrations of S in plants from deficiency to excess.

Sodium. When *Atriplex vesicaria* (bladder saltbush) received no Na^+ in the nutrient solution, some plants died in approximately 35 days (Brownell, 1965). In approximately 20 days, Na-deficient leaves of this halophytic plant became chlorotic and developed necrotic patches at their tips and along their margins, after which little further growth was made. Li^+, K^+, or Rb^+ could not substitute for Na^+.

Chlorine. Chlorine is the most recently confirmed essential element for higher plants.

Eaton (1942) obtained clear-cut evidence of growth increases from addition of low concentrations of Cl^- (3 meq/liter), particularly for cotton and tomato. More recently, Corbett and Gausman (1960) obtained a 25% increase in potato tubers with approximately 8 versus 0.7 meq Cl^-/liter. In contrast with Na^+, which has been shown to be required only for a halophytic, higher plant (Brownell, 1965), Stout et al. (1956) demonstrated a Cl^- requirement for tomato, lettuce, and cabbage. In the absence of Cl^-, leaves became chlorotic and then necrotic; growth was exceedingly restricted and plants did not set fruit. When Cl^- was added to Cl-deficient plants, growth was resumed. Cl^--deficiency symptoms normally developed in approximately 1 week, but as little as 1 μmole of Cl^- per plant delayed symptoms for another week. Dried tomato plants showing Cl deficiency contained 200 ppm of Cl^- on a dry weight basis. Thus for each ton of dry tomato plants produced, a minimum of 200 g of Cl^- is required.

Johnson et al. (1957) stressed the extreme care required in experiments with low levels of Cl^-, since plants may receive sufficient Cl^- via the atmosphere. Stout and Johnson (1957) estimated that most crops require approximately 1 lb of Cl^- for each 4000 lb of dry matter—approximately 0.7 meq Cl^-/100 g dry matter. Large crops would presumably require approximately 5 lb or more per acre. Inasmuch as rain delivers 12–35 lb Cl^- per acre per year (Eriksson, 1952), it appears unlikely that field crops in general would benefit from added Cl^-. It is possible of course that some crops may be shown to have high Cl^- requirements.

In tomato, Cl^--deficiency symptoms include: wilting of leaflet blade tips; and, by chlorosis, progressive bronzing and necrosis basipetally in areas proximal to the wilted leaves (Broyer et al., 1954). Severely Cl-deficient tomato plants fail to produce fruit.

In sugar beets Cl-deficiency symptoms appear on young and recently matured leaves, are similar to those of Mn deficiency, and are visible only by transmitted light (Ulrich and Ohki, 1956). The symptoms appear as a netted mosaic pattern with the green veins forming the netting. Later, interveinal areas appear as smooth flat depressions, light green or yellow, and present a striking contrast to the green veins with a "raised" appearance. Secondary roots are characterized by a stubby type of growth.

Wilting is the most frequently observed symptom of Cl deficiency, and Eaton (1966a) reports an instance in which wilting was observed on leaves of sugar beet plants receiving 1.0 meq Cl^-/liter, but not in plants receiving 5 meq Cl^-/liter in the nutrient solution. He describes Cl-deficiency symptoms in many plants and lists the concentrations of Cl^- in Cl-deficient plants. He states that, in an instance of suspected or demonstrated deficiency, an addition of 20 lb Cl^-/acre to low-Cl^- soils should meet all requirements, and that such an application could certainly do no harm.

Iron. Being relatively immobile, Fe deficiency appears first on the youngest leaves of the plant. It may start as interveinal chlorosis but, if severe enough, entire leaf blades become yellow to whitish yellow (Fig. 13.3). Veins are the last portion of the leaf to lose chlorophyll. In some cases chlorotic portions of the leaf or the entire leaf become necrotic. Fe deficiency produces a type of leaf chlorosis that is usually fairly easy to diagnose (Wallihan, 1966).

Wallihan (1966) discusses the visual symptoms of Fe deficiency for various crops, tissue analysis for Fe^{3+}, and control of Fe deficiency by foliar sprays,

Fig. 13.3
Fe deficiency in snap bean plants showing extreme chlorosis of the upper, younger leaves.

stem injection, or application of Fe salts to soil. Fe chelates, as well as $FeSO_4$, can be used to control Fe deficiency in yards and gardens (Locke and Eck, 1965).

Boron. Similar to Ca^{2+}, B is quite immobile in most species, and symptoms appear at extremities of tops and roots. Root tips are the first part of the plant to exhibit symptoms—a translucent, slimy appearance. The terminal growing point of the stem, axillary buds, and flower buds are affected next. They turn yellow, then become necrotic, and may fall from the plant.

One of the few reports indicating mobility for B deals with broccoli. Benson et al. (1961) found that under B-deficiency conditions B moves out of younger leaves to the youngest leaves; concentration of B in the 22nd leaf from the base of the plant changed from 33 to 2 ppm.

The youngest leaves of the plant are often very puffy, quite wrinkled, and misshapened. In tobacco the base of the blade of young leaves becomes necrotic. Inasmuch as some of the other symptoms are quite similar, appearance of the youngest leaves helps to differentiate B from Ca deficiency. Under field conditions similarities in symptoms for these two elements is rarely if ever a problem, since Ca deficiency is only rarely encountered. However, B deficiency in one or more crops has been reported from every one of the 48 continental states in the United States.

There has undoubtedly been more research on B than on any other trace element (Bradford, 1966). Some of the more important B deficiencies are "cracked stem" of celery, top sickness of tobacco, heartrot of beet, white top of alfalfa, and brown rot of cauliflower. Owing to their high B requirements, beets and alfalfa are good indicators of B deficiency in soils. Gopal (1968) has recently described the effects of B deficiency in peanuts. Bradford (1966) discusses in detail the B-deficiency symptoms for various crops, concentrations of B in plants associated with B deficiency, and methods for overcoming B deficiency in various crops.

B has long been implicated in the reproductive phase of growth of higher plants (Gauch and Dugger, 1954). It has also been shown to be required in the reproductive processes of a fern [*Dryopteris dentata* (Forsk.) C. Chr.] (Fig. 13.4) and *Selaginella apoda* (Fig. 13.5) (Bowen and Gauch, 1965).

Excised leaves of B-deficient snap bean leaves, *Phaseolus vulgaris* L. 'Black Valentine,' lose water less rapidly initially than B-sufficient leaves, and the former have a higher concentration of pectins and (possibly) pentosans than the latter, as reported by Baker et al. (1956). They also observed that B-deficient snap bean leaves have abnormal stomata (Fig. 13.6) consisting of one guard cell rather than two.

In snap beans, leaves of B-deficient plants become very thick, wrinkled, and leathery to the touch (Fig. 13.7) as compared with leaves of B-sufficient plants which are smooth and thin. In B-deficient plants branches with flower buds show many aborted buds, prominent white streaks on the stem leading to aborted flower buds, and small, unexpanded, whitish trifoliolate leaves (Fig. 13.8).

It was noted that potentially necrotic spots on petioles of B-deficient celery exhibit a blue-white fluorescence under ultraviolet irradiation (Spurr, 1952). The compounds responsible for this fluorescence were identified as caffeic and chlorogenic acids (Perkins and Aronoff, 1956).

Later, Dear and Aronoff (1965) reported that in sunflower leaves and tips onset of B deficiency and necrosis were characterized by an excessive concentration of caffeic acid. The ratio of caffeic to chlorogenic acid was high, being approximately 10-fold for leaves and 4-fold for tips.

Manganese. Symptoms for Mn deficiency resemble those for Mg deficiency except that in the case of Mn^{2+} interveinal chlorosis appears on the youngest rather than on the oldest leaves. Again, it is important to know whether leaves come from the upper or lower portion of the plant in order to distinguish clearly between Mn^{2+} and Mg^{2+}

In Mn-deficient higher plants, there may occur either a reduction in the

Fig. 13.4
Undersurfaces of leaves of *Dryopteris dentata* sporophytes showing profuse sori with typical indusia on leaves from plus-B cultures (top), and the sparse sori with incomplete or missing indusia on leaves from minus-B cultures (bottom). (Bowen and Gauch, 1965.)

Fig. 13.5
Strobili of *Selaginella apoda* sporophytes receiving no B (left) or 0.5 ppm of B (right) in the nutrient solution. Numerous, apparently normal strobili formed on plants receiving B, whereas very few, abnormal-appearing, apparently aborted strobili formed on plants receiving no B. (Bowen and Gauch, 1965.)

number of chloroplasts or a disorganization of the chloroplast with a resulting low concentration of chlorophyll (Homann, 1967).

Labanauskas (1966) discusses the nature of Mn-deficiency symptoms on various kinds of plants and lists the concentrations of Mn^{2+} associated with deficiency, normal growth, and toxicity.

Zinc. Zn deficiencies have been reported for many crops under field conditions. Using a subsoil from a "little-leaf" orchard and also water cultures, Hoagland et al. (1936) produced and described symptoms of Zn deficiency, including little-leaf, in apricot, tobacco, squash, corn, mustard, tomato, sunflower, and cotton. Alfalfa was much less susceptible to Zn deficiency under these conditions than were other crops. Later, results corroborated early findings with regard to little-leaf and Zn deficiency, as well as the ability of alfalfa to obtain sufficient Zn^{2+} when other plants could not (Hoagland et al., 1937).

Corn plants may show Zn deficiency even though other crops do not show it when grown in the same soil. In corn, Zn deficiency results in white or yellow emerging leaves—a condition called "white bud." In the early seedling stage,

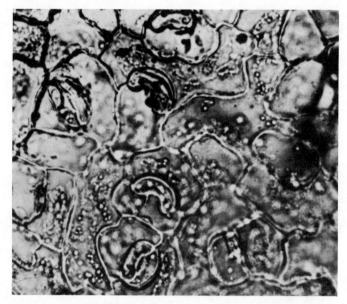

Fig. 13.6
Photomicrograph of the epidermis showing abnormal stomata of a B-deficient bean plant. Note that some stomata consist of only one guard cell. (Baker et al., 1956.)

Fig. 13.7
B-sufficient (left) and B-deficient (right) snap bean leaves. As compared with B-sufficient leaves, those from B-deficient plants were very thick, wrinkled, and leathery to the touch.

Fig. 13.8
Flowering branch on a B-deficient snap bean plant showing small, whitish, unexpanded trifoliolate leaves, aborted flower buds, and whitish streaks on the stem leading to the aborted flower buds.

light yellow streaks appear between viens. Later, small white spots of inactive or dead tissue develop, while some small white areas are present which never develop chlorophyll (Barnette et al., 1936). Margins and sheaths of older leaves may develop a purplish color. If stalks are split down the middle, lower nodes often show a dark purplish discoloration—almost black in severe Zn deficiency (Winters and Parks, 1956).

Rosette is a Zn deficiency of pecan trees characterized by mottled chlorosis and crinkling of newer tipmost leaves and shortening of internodes of new shoot growth (Finch, 1936). Shortening of internodes causes a rosette type of growth. Affected leaflets do not reach normal size and are frequently malformed. Leaves may abscise, and leafless shoots often die back. New growth may arise below the dead area but generally it develops the same symptoms— at least by the subsequent season.

Zn^{2+} deficiency results in a disorder known as little-leaf in peaches (Dickey and Blackmon, 1940). Foliage symptoms include chlorosis, crinkling of leaves, little-leaf, a red-to-purple coloration of some leaves, and premature abscission of some leaves. Twig symptoms include rosette and death of twigs.

Zn^{2+} deficiency has been a problem with citrus in which the deficiency causes little-leaf or rosette. There may not necessarily be a deficiency of Zn^{2+} in the soil, but alkalinity or some other abnormality can result in unavailability of Zn^{2+}. Foliar application of Zn^{2+} is the only practical solution when Zn^{2+} is unavailable and the soil condition cannot readily be changed.

Peaches, pineapple, and nut trees may show Zn^{2+} deficiencies which, as in citrus, can generally be most readily corrected by foliar application of Zn.

Reed (1939) studied the detailed changes in tomato leaf structure caused by Zn^{2+} deficiency. The most conspicuous indications of degeneration were in the spongy parenchyma where more of the cells contained aggregations of Ca oxalate crystals than did those of healthy leaves.

Zn^{2+} deficiency in snap and lima beans is characterized by the appearance of reddish-brown spots about the size of a pinhead on primary leaves; next there is a slight interveinal yellowing of older trifoliolate leaves, as reported by Lingle and Holmberg (1956). They also noted that interveinal yellowing appeared on younger leaves of tomato but on older leaves of sugar beet.

Zn^{2+} deficiency was observed on peaches and nectarines in Central Otago in New Zealand (Bollard, 1953). The symptoms included leaf chlorosis, reduction in leaf size, rosetting, and dieback of leaders. A foliage spray of 3 lb $ZnSO_4$ mixed with 4 lb hydrated lime in 50 gal of water largely rectified the disorder.

Chapman (1966c) discusses Zn^{2+}-deficiency symptoms in numerous crops and lists concentrations of Zn^{2+} in plants associated with deficiency, normalcy, and toxicity. He also discusses the use of Zn^{2+} foliar sprays and Zn^{2+} additions to soil for overcoming Zn deficiencies.

Copper. Lipman and Mackinney (1931) and Sommer (1931), among others, showed Cu^{2+} to be essential for higher plants. In detailed studies of tomato leaf structure as affected by Cu deficiency, Reed (1939) noted that the most characteristic feature was formation of lacunae followed by lysis of palisade cells, resulting in columniated parenchyma totally unlike that observed in plants from cultures lacking other essential elements. Evidence for essentiality of Cu^{2+} and its known functions were reviewed by Arnon (1950).

In citrus a physiological disease, known as exanthema or dieback, is caused by a deficiency of Cu (Reuther and Labanauskas, 1966). In fact, interest in Cu was initially stimulated by reports that exanthema of citrus in Florida could be controlled by soil applications of $CuSO_4$ or by Bordeaux spray (Grossenbacher, 1916; Floyd, 1917). Dieback is a frequent symptom of Cu deficiency, and rosetting may precede dieback, as observed by Reuther and Labanauskas (1966). These investigators discuss specific Cu-deficiency symptoms in many crops, list plant concentrations associated with Cu deficiency, and outline methods of overcoming deficiency. A low concentration of Cu^{2+} in forage causes certain physiological diseases in livestock—particularly ruminants (Reuther and Labanauskas, 1966).

In Cu-deficient corn, upper leaves exhibit a light yellow-green color, and the tip of the leaf becomes necrotic and dies. This area finally breaks and hangs down as a dry, twisted portion. In slight deficiency upper leaves dry at margins near the base of the leaf (Berger, 1962).

Poor tree growth, leaf symptoms, rough bark, shriveling of kernels, and gumming of tree trunks were symptoms of Cu deficiency on almond trees (Kester et al., 1956). Spray applications of 1 lb $CuSO_4$/100 gal of water produced a marked improvement in the trees.

Molybdenum. Considering that a few parts per billion in a nutrient solution or a soil solution is sufficient for normal growth of plants, it is surprising that Mo deficiency has been observed under field conditions. Less than 1 ppm of Mo (dry wt basis) is sufficient in leaves of 8-wk-old tomato plants (Johnson et al., 1952; Johnson, 1966) or 10 ppb (fresh wt basis) (Stout and Meagher, 1948).

As with the essentiality of all required elements, symptoms associated with Mo deficiency were first demonstrated in artificial culture technique. Hewitt and Jones (1947) produced Mo deficiency in savoy cabbage, cauliflower, mustard, and tomatoes. The symptoms on cauliflower (Raleigh, 1950) were identical with whip-tail which is found on certain acid soils in England. For cauliflower and broccoli there is first a puckering and reduction in width of leaf blade, which may become progressively more pronounced until there is merely a fringe of puckered, thickened leaf tissue which is often not more than ½ in. wide along either side of the midrib (Mitchell, 1945). Twenty pounds, or less, of ammonium molybdate per acre as a dressing controls whip-tail.

Mo deficiency causes not only whip-tail of cauliflower but also "yellow spot" of citrus, as reported by Johnson (1966). He noted that when Mo is deficient leguminous plants are in *fact* N deficient and show N-deficiency symptoms; with nonlegumes the plants are in *effect* N deficient even when they have absorbed NO_3^-, since NO_3^- is not reduced and utilized.

Arnon and Stout (1939) were the first investigators to demonstrate a Mo requirement in higher plants. Ten parts of Mo per billion parts of solution permitted normal growth of tomato plants. In the absence of Mo, lower leaves developed a distinct mottling which was different from any other deficiency symptom of tomato. In later stages necrosis at margins and a characteristic involution of the blade accompanied by abscission of blossoms occurred. Piper (1940) confirmed Arnon and Stout's (1939) report on the essentiality of Mo for higher plants.

Using tomato plants and radioactive Mo, Stout and Meagher (1948) noted that Mo accumulated in leaves in those regions in which loss of chlorophyll occurred in the absence of Mo (i.e., interveinal areas).

A few ounces of Mo per acre in a fertilizer overcomes the deficiency unless some soil condition is involved in the unavailability of Mo. The element may also be sprayed onto leaves to correct a deficiency.

The March 1956 issue of *Soil Science* was devoted entirely to Mo in plants, animals, and soil, and it includes excellent color photographs of Mo deficiency in various crops. Mo deficiencies in different crops are also described in detail in other publications (Anderson, 1956; Hewitt, 1956; Reisenauer, 1956; Rubins, 1956; Stout and Johnson, 1956).

LITERATURE CITED

Adams, C. A., and R. W. Rinne. 1969. Influence of age and sulfur metabolism on ATP sulfurylase activity in the soybean and a survey of selected species. Plant Physiol. 44:1241–1246.

Agarwala S. C., C. P. Sharma, and S. Farooq. 1965. Effect of iron supply on growth, chlorophyll, tissue iron and activity of certain enzymes in maize and radish. Plant Physiol. 40:493–499.

Anderson, A. J. 1956. Molybdenum deficiencies in legumes in Australia. Soil Sci. 81:173–182.

Arnon, D. I. 1950. Functional aspects of copper in plants, pp. 89–114. *In* W. D. McElroy and B. Glass (eds.) Copper metabolism: A symposium on animal, plant, and soil relationships. Johns Hopkins Press, Baltimore.

Arnon, D. I. 1951. Growth and function as criteria in determining the essential nature of inorganic nutrients, pp. 313–341. *In* E. Troug (ed.), Mineral nutrition of plants. Univ. Wisconsin Press, Madison.

Arnon, D. I., and P. R. Stout. 1939. Molybdenum as an essential element for higher plants. Plant Physiol. 14:599–602.

Asahi, T. 1964. Sulfur metabolism in higher plants. IV. Mechanisms of sulfate reduction in chloroplasts. Biochim. Biophys. Acta 82:58–66.

Baker, J. E., H. G. Gauch, and W. M. Dugger, Jr. 1956. Effects of boron on the water relations of higher plants. Plant Physiol. 31:89–94.

Barnette, R. M., J. P. Camp, J. D. Warner, and O. E. Gall. 1936. The use of zinc sulphate under corn and other field crops. Florida Agr. Exp. Sta. Bull. 292.

Benson, N. R., E. S. Degman, and I. C. Chmelir. 1961. Translocation and re-use of boron in broccoli. Plant Physiol. 36:296–301.

Berger, K. C. 1962. Micronutrient deficiencies in the United States. Agr. Food Chem. 10:178–181.

Bingham, F. T. 1966. Phosphorus, pp. 324–361. *In* H. D. Chapman (ed.) Diagnostic criteria for plants and soils. Div. Agr. Sci., Univ. California, Berkeley.

Blank, F. 1947. The anthocyanin pigments of plants. Bot. Rev. 13:241–317.

Bollard, E. G. 1953. Zinc deficiency in peaches and nectarines. New Zealand J. Sci. Tech. 35A:15–18.

Bowen, J. E., and H. G. Gauch. 1965. The essentiality of boron for *Dryopteris dentata* and *Selaginella apoda*. Amer. Fern J. 55:67–73.

Bradford, G. R. 1966. Boron, pp. 33–61. *In* H. D. Chapman (ed.) Diagnostic criteria for plants and soils. Div. Agr. Sci., Univ. California, Berkeley.

Brown, J. C., and S. B. Hendricks. 1952. Enzymatic activities as indications of copper and iron deficiencies in plants. Plant Physiol. 27:651–660.

Brownell, P. F. 1965. Sodium as an essential micronutrient element for a higher plant (*Atriplex vesicaria*). Plant Physiol. 40:460–468.

Broyer, T. C., A. B. Carlton, C. M. Johnson, and P. R. Stout. 1954. Chlorine – A micronutrient element for higher plants. Plant Physiol. 29:526–532.

Chapman, H. D. (ed.). 1966a. Diagnostic criteria for plants and soils. Div. Agr. Sci., Univ. California, Berkeley. 793 p.

Chapman, H. D. 1966b. Calcium, pp. 65—92. *In* H. D. Chapman, (ed.) Diagnostic criteria for plants and soils. Div. Agr. Sci., Univ. California, Berkeley.

Chapman, H. D. 1966c. Zinc, pp. 484—499. *In* H. D. Chapman, (ed.) Diagnostic criteria for plants and soils. Div. Agr. Sci., Univ. California, Berkeley.

Corbett, E. G., and H. W. Gausman. 1960. The interaction of chloride with sulfate and phosphate in the nutrition of potato plants. Agron. J. 52:94—96.

Cory, S., L. R. Finch, and R. W. Hinde. 1966. The incorporation of [^{32}P]-phosphate into nucleic acids of normal and boron-deficient bean roots. Phytochemistry 5:625—634.

Dear, J., and S. Aronoff. 1965. Relative kinetics of chlorogenic and caffeic acids during the onset of boron deficiency in sunflower. Plant Physiol. 40:458—459..

Devlin, R. M., and A. V. Barker. 1971. Photosynthesis. Van Nostrand Reinhold, New York. 304 pp.

Dickey, R. D., and G. H. Blackmon. 1940. A preliminary report on little-leaf of the peach in Florida — A zinc deficiency. Florida Agr. Exp. Sta. Bull. 344.

Eaton, F. M. 1942. Toxicity and accumulation of chloride and sulfate salts in plants. J. Agr. Res. 64:357—399.

Eaton, F. M. 1966a. Chlorine, pp. 98—135. *In* H. D. Chapman, (ed.) Diagnostic criteria for plants and soils. Div. Agr. Sci., Univ. California, Berkeley.

Eaton, F. M. 1966b. Sulfur, pp. 444—475. *In* H. D. Chapman, (ed.) Diagnostic criteria for plants and soils. Div. Agr. Sci., Univ. California, Berkeley.

Embleton, T. W. 1966. Magnesium, pp. 225—263. *In* H. D. Chapman (ed.) Diagnostic criteria for plants and soils. Div. Agr. Sci., Univ. California, Berkeley.

Eriksson, E. 1952. Composition of atmospheric precipitation. II. Sulfur, chloride, iodine compounds. Tellus 4:280—303.

Finch. A. H. 1936. Zinc and other mineral constituents in relation to the rosette disease of pecan trees. J. Agr. Res. 52:363—376.

Floyd, B. F. 1917. Dieback, or exanthema, of citrus trees. Florida Agr. Exp. Sta. Bull. 140.

Gadd, R. E. A., and J. F. Henderson. 1970. The role of magnesium ion in the adenine phosphoriboxyltransferase reaction. Can. J. Biochem. 48:302—307.

Gauch, H. G. 1940. Responses of the bean plant to calcium deficiency. Plant Physiol. 15:1—21.

Gauch, H. G. and W. M. Dugger, Jr. 1954. The physiological action of boron in higher plants: A review and interpretation. Maryland Agr. Exp. Sta. Tech. Bull. A-80.

Gopal, N. H. 1968. Boron deficiency in groundnut (*Arachis hypogaea* L.). Indian J. Agr. Sci. 38:832—834.

Gregory, J. D., and P. W. Robbins. 1960. Metabolism of sulfur compounds (sulfate metabolism). Annu. Rev. Biochem. 29:347—364.

Grossenbacher, J. G. 1916. Some bark diseases of citrus trees in Florida. Phytopathology 6:29—50.

Hambidge, G. 1941. Hunger signs in crops — A symposium. Amer. Soc. Agron. and Nat. Fertilizer Ass., Washington, D. C. 327 pp.

Harper, J. E., and G. M. Paulsen. 1969a. Nitrogen assimilation and protein synthesis in wheat seedlings as affected by mineral nutrition. I. Macronutrients. Plant Physiol. 44:69—74.

Harper, J. E., and G. M. Paulsen. 1969b. Nitrogen assimilation and protein synthesis in wheat seedlings as affected by mineral nutrition. II. Micronutrients. Plant Physiol. 44:636—640.

Hewitt, E. J. 1956. Symptoms of molybdenum deficiency in plants. Soil Sci. 81:159—171.

Hewitt, E. J., and E. W. Jones. 1947. The production of molybdenum deficiency in plants in sand culture with special reference to tomato and *Brassica* crops. J. Pomol. Hort. Sci. 23:254—262.

Hewitt, E. J., D. M. Dames, and C. P. Lloyd-Jones. 1965. A comparison of methods for the removal of cobalt and vanadium from salts used in plant nutritional studies. Plant Physiol. 40:326—331.

Hinde, R. W., and L. R. Finch. 1966. The activities of phosphatases, pyrophosphates and adenosine triphosphatases from normal and boron deficient bean roots. Phytochemistry 5:619—623.

Hinde, R. W., L. R. Finch, and S. Cory. 1966. Amino acid-dependent ATP-pyrophosphate exchange in normal and boron deficient bean roots. Phytochemistry 5:609—618.

Hoagland, D. R., W. H. Chandler, and P. L. Hibbard. 1936. Little-leaf or rosette of fruit trees. V. Effect of zinc on the growth of plants of various types in controlled soil and water culture experiments. Proc. Amer. Soc. Hort. Sci. 33:131—141.

Hoagland, D. R., W. H. Chandler, and P. R. Stout. 1937. Little-leaf or rosette of fruit trees. VI. Further experiments bearing on the cause of the disease. Proc. Amer. Soc. Hort. Sci. 34:210—212.

Homann, P. 1967. Studies on the manganese of the chloroplast. Plant Physiol. 42:997—1007.

Johnson, C. M. 1966. Molybdenum, pp. 286—301. *In* H. D. Chapman (ed.). Diagnostic criteria for plants and soils. Div. Agr. Sci., Univ. California, Berkeley.

Johnson, C. M., G. A. Pearson, and P. R. Stout. 1952. Molybdenum nutrition of crop plants. II. Plant and soil factors concerned with molybdenum deficiencies. Plant and Soil 4:178—196.

Johnson, C. M., P. R. Stout, T. C. Broyer, and A. B. Carlton. 1957. Comparative chlorine requirements of different plant species. Plant and Soil 8:337—353.

Jones, J. B., Jr. 1965. Molybdenum content of corn plants exhibiting varying degrees of potassium deficiency. Science 148:94.

Jones, W. W. 1966. Nitrogen, pp. 310—323. *In* H. D. Chapman , (ed.) Diagnostic criteria for plants and soils. Div. Agr. Sci., Univ. California, Berkeley.

Kalra, G. S. 1965. Responses of the tomato plant to calcium deficiency. Bot. Gaz. 118:18—37.

Kester, D. E., J. G. Brown, and T. Aldrich. 1956. Copper deficiency of almonds. California Agr. 10:13, 16.

Labanauskas, C. K. 1966. Manganese, pp. 264—285. *In* H. D. Chapman (ed.) Diagnostic criteria for soils and plants. Div. Agr. Sci., Univ. California, Berkeley.

Lingle, J. C., and D. M. Holmberg. 1956. Zinc-deficient crops (sweet corn, tomatoes, beans, and sugar beets used in tests for zinc deficiency). California Agr. 10:13—14.

Lipman, C. B., and G. Mackinney. 1931. Proof of the essential nature of copper for higher green plants. Plant Physiol. 6:593—599.

Locke, L. F., and H. V. Eck. 1965. Iron deficiency in plants: How to control it in yards and gardens. U. S. Dep. Agr. Home and Garden Bull. 102.

MacInnes, C. B., and L. S. Albert. 1969. Effect of light intensity and plant size on rate of development of early boron deficiency symptoms in tomato root tips. Plant Physiol. 44:965—967.

MacVicar, R., and B. E. Struckmeyer. 1946. The relation of photo-period to boron requirements of plants. Bot. Gaz. 107:454—461.

Marinos, N. G. 1962. Studies on submicroscopic aspects of mineral deficiencies. I. Calcium deficiency in the shoot apex of barley. Amer. J. Bot. 49:834—841.

Mitchell, K. J. 1945. Preliminary note on the use of ammonium molybdate to control whiptail in cauliflower and broccoli crops. New Zealand J. Sci. Tech. 27A:287—293.

Perkins, H. J., and S. Aronoff. 1956. Identification of blue fluorescent compounds in boron deficient plants. Arch. Biochem. Biophys. 64:506—507.

Piper, C. S. 1940. Molybdenum as an essential element for plant growth. J. Australian Inst. Agr. Sci. 6:162—164.

Raleigh, G. J. 1950. Molybdenum deficiency in Dunkirk silty clay loam. Science 112: 433—434.

Reed, H. S. 1939. The relation of copper and zinc salts to leaf structure. Amer. J. Bot. 26:29—33.

Reid, M. S., and R. L. Bieleski. 1970. Response of *Spirodela oligorrhiza* to phosphorus deficiency. Plant Physiol. 46:609—613.

Reisenauer, H. M. 1956. Molybdenum content of alfalfa in relation to deficiency symptoms and response to molybdenum fertilization. Soil Sci. 81:237—242.

Reuther, W., and C. K. Labanauskas. 1966. Copper, p. 157—179. *In* H. D. Chapman (ed.) Diagnostic criteria for plants and soils, Div. Agr. Sci., Univ. California, Berkeley.

Rubins, E. J. 1956. Molybdenum deficiencies in the United States. Soil Sci. 81:191—197.

Sommer, A. L. 1931. Copper as an essential for plant growth. Plant Physiol. 6:339—345.

Sprague, H. B. 1964. Hunger signs in crops — A symposium, 3rd ed. David Mckay, New York. 461 pp.

Spurr, A. 1952. Fluorescence in UV light in the study of boron deficiency in celery. Science 116:421.

Steinberg, R. A. 1951. Correlations between protein-carbohydrate metabolism and mineral deficiencies in plants, pp. 359—386. *In* E. Truog, (ed.) Mineral nutrition of plants. Univ. Wisconsin Press, Madison.

Stout, P. R., and D. I. Arnon. 1939. Experimental methods for the study of the role of copper, manganese, and zinc in the nutrition of higher plants. Amer. J. Bot. 26:144—149.

Stout, P. R., and C. M. Johnson. 1956. Molybdenum deficiency in horticultural and field crops. Soil Sci. 81:183—190.

Stout, P. R., and C. M. Johnson. 1957. Trace elements, pp. 139—150. *In* Soil, U. S. Dep. Agr. Yearbook 1957.

Stout, P. R., and W. R. Meagher. 1948. Studies of the molybdenum nutrition of plants with radioactive molybdenum. Science 108:471—473.

Stout, P. R., C. M. Johnson, and T. C. Broyer. 1956. Chlorine in plant nutrition — Experiments with plants in nutrient solutions establish chlorine as a micronutrient essential to plant growth. California Agr. 10:10.

Tanaka, H. 1966. Response of *Lemna pausicostata* to boron as affected by light intensity. Plant and Soil 25:425—434.

Ulrich, A., and K. Ohki. 1956. Chlorine, bromine and sodium as nutrients for sugar beet plants. Plant Physiol. 31:171—181.

Ulrich, A., and K. Ohki. 1966. Potassium, pp. 362—393. *In* H. D. Chapman (ed.), Diagnostic criteria for plants and soils. Div. Agr. Sci., Univ. California, Berkeley.

Wallihan, E. F. 1966. Iron, pp. 203—212. *In* H. D. Chapman, (ed.) Diagnostic criteria for plants and soils. Div. Agr. Sci., Univ. California, Berkeley.

Warington, K. 1933. The influence of length of day on the response of plants to boron. Ann. Bot. 47:429—457.

Wilson, L. G. 1962. Metabolism of sulfate: Sulfate reduction. Annu. Rev. Plant Physiol. 13:201—224.

Winters, E., and W. L. Parks. 1956. Zinc deficiency of corn in Tennessee. Com. Fertilizers and Plant Food Ind. 92(3):36, 59.

Wooding, F. J., G. M. Paulsen, and L. S. Murphy. 1970. Response of nodulated and non-nodulated soybean seedlings to sulfur nutrition. Agron. J. 62:277—280.

CHAPTER 14

Occurrence, Prevention, and Alleviation of Deficiencies

OCCURRENCE OF DEFICIENCIES UNDER FIELD CONDITIONS

Macronutrients

Nitrogen. As one of the "big three" of the essential elements, along with P and K, N is used in large amounts by crops. Unless large amounts of high-N organic matter are returned to soil, as manure or by plowing down leguminous cover crops, N deficiency may become apparent. In soils across the country, N is probably most frequently the limiting element in plant growth and yield (Jones, 1966).

With higher-yielding strains of hybrid corn, when all other growth factors are optimal, there appears to be almost no limit to the amount of N a crop can use. In fact, there may be only an economic limitation in that the highest increments of N cause ever-diminishing increases in yield to the point where the highest applications cannot be justified based on crop increases and cost of additional fertilizer.

A good stand of alfalfa adds approximately 150 lb of N per acre per year, and rain approximately 5 lb of N (the latter mostly in the form of NO_3^-).

At one time or another, N deficiency has been observed in every type of crop plant grown in the United States. The deficiency most commonly occurs on sandy soils under high rainfall and on soils low in organic matter (Jones, 1966).

Phosphorus. As in the case of N, a typical crop removes large amounts of P from the soil. Under intensive cultivation it is not surprising that P deficiencies have been reported in various crops from every state.

Ordinarily, the P concentration in the soil solution is quite low and of the order of 1 ppm, as reported by Stout and Overstreet (1950). They calculated that the soil solution might need to be completely recharged with phosphate 10 times a day, during the growing season, to support the growth of most crops. Earlier, Pierre and Parker (1927) also reported that P is low and that of the soils they examined the average concentration was 0.35 ppm P—as inorganic pyrophosphate. For the growth of an average crop, they concluded that soil P would have to be renewed about 250 times if complete removal were followed by complete replenishment. However, with flowing cultures and large volumes of culture solution, Tidmore (1930) showed that corn, sorghum, and tomatoes could secure adequate amounts of phosphate from solutions containing only 0.5 ppm of soluble phosphate; 1.0 ppm resulted in luxury consumption without further increase in growth.

Only under heavy P fertilization and with certain soils have there been reports of a build-up of P in soils. Baker (1968) reported that most of the P absorbed by field crops comes from fixed P which is composed of insoluble compounds such as Al and Fe phosphates.

Soils vary greatly with respect to the amount of P combined with the soil, that is, fixed. Some soils have a very high P fixation capacity, so that relatively large amounts of P are required before appreciable amounts of P are available for plant growth.

P deficiency appears most often in: highly weathered soils usually acid in reaction; calcareous soils in which P, even though present in high concentration, is not readily available; and in many peat and muck soils (Bingham, 1966).

Calcium. Rosette disease of lettuce on soils of serpentine origin and high in exchangeable Mg^{2+} was shown to be a Ca deficiency (Vlamis and Jenny, 1948).

A normal soil condition is one in which Ca^{2+} is the dominant ion on the exchange complex—60–85% of the total BEC—as reported by Chapman (1966b). He noted that in soils low in Ca^{2+}, hence high in H^+, various changes occur: absorbability and availability of the remaining Ca^{2+} decreases; solubilities of Mn^{2+}, Al^{3+}, Cu^{2+}, Ni^{2+}, and other elements may reach toxic concentrations; P reverts to less soluble Fe and Al forms; below pH 5.5 rate of nitrification decreases; lack of Ca^{2+} and acidity both deleteriously affect nodulation of legumes; and, particularly when Na^+ replaces much of the Ca^{2+} on the exchange complex, soil structure deteriorates.

Potassium. K is the third member of the "big three" in fertilizers, and it is used in large amounts by nearly all crops. Tobacco is a heavy user of K^+, and a yield of 1500 lb of leaf per acre removes approximately 115 lb of K^+ calculated as K_2O. Celery draws particularly heavily on K^+ in soil, so that a crop of 350 crates/acre removes approximately 235 lb of K^+—calculated as K_2O.

K deficiency occurs in a wide variety of soils from light, sandy soils to organic soils such as peat and muck (Ulrich and Ohki, 1966). As in the case of

P, K^+ may also be fixed by soil, and soils vary widely in the extent of fixation. Part of the fixed K^+ is available to plants, and part of it is essentially unavailable (Truog and Jones, 1938).

Magnesium. Mg deficiency is most common in acid, sandy soils in areas of moderate rainfall but may occur even in alkaline soils and in peats or mucks (Embleton, 1966). Applications of K^+, particularly if excessive, may bring on or aggravate Mg deficiency.

Sulfur. S deficiency has been reported primarily in the northwestern part of the United States (Eaton, 1966b). Extensive acreages of Nebraska soils that are sandy or low in organic matter may be S deficient (Alexander and Knudsen, 1965). Much of the readily available S is held in the organic fraction in soils, and thus sandy soils or those low in organic matter as a result of erosion are most likely to be S deficient. Rainfall supplies 4–12 lb of S per acre each year. Alfalfa is one of the crop plants most likely to develop S deficiency. Fifty pounds of actual S per acre is adequate for 2 years for most S-deficient soils.

In recent years the use of high-analysis fertilizers containing little carrier SO_4^{2-} has probably contributed to the development of S-deficient soils (Eaton, 1966b).

Micronutrients

The Trace Element Committee of the Council on Fertilizer Application made a study of the known micronutrient deficiencies, by crop, as reported by research workers in each of the 50 states in the United States (Berger, 1962). Their findings are listed in Tables 14.1 and 14.2.

Boron. B deficiencies were reported for 41 states, hence this deficiency is more widespread than that of any other micronutrient (Berger, 1962). Alfalfa was reported to show B deficiency in 38 states and, incidentally, alfalfa led all crops in micronutrient deficiencies. In Wisconsin there are approximately 2,000,000 acres of alfalfa that show B deficiency at one time or another (Berger, 1962).

Light-textured sandy soils, acid soils, acid peat and muck soils, and alkaline soils containing free lime are particularly likely to be deficient in B (Bradford, 1966). In the more humid eastern United States, deficiency is related to leaching of B from soils. In the western half of the United States, there is generally sufficient B in the soil, but its availability is reduced in some cases by the relatively high pH (7.5 or higher) of alkaline soils.

Tourmaline is the principal B mineral in soils, but a considerable portion of the B is combined with organic matter. During a drought B is not readily released from organic matter since activity of microorganisms is low (Berger, 1962).

Iron. Fe deficiencies have been observed on many crops throughout the United States—including Hawaii (Wallihan, 1966).

Table 14.1
States Reporting Micronutrient Deficiencies in One or More Crops[a]

Micronutrient deficiency	Number of states
B	41
Zn	30
Fe	25
Mn	25
Mo	21
Cu	13

[a] Berger (1962).

Table 14.2
Extent of Micronutrient Deficiencies in the United States[a]

Crop[b]	States	Crop[b]	States
B		Cu	
Alfalfa	38	Corn	3
Beet	12	Fruit trees	3
Celery	10	Grasses	3
Clover	13	Onion	7
Cruciferae	25	Small grains	4
Fruit trees	21	Mn	
Fe		Bean	13
Bean	5	Corn	5
Corn	3	Fruit trees	9
Fruit trees	11	Small grains	10
Grasses	7	Spinach	8
Shade trees	7	Zn	
Shrubs	11	Bean	7
Mo		Corn	20
Alfalfa	13	Fruit trees	12
Clover	6	Nut trees	10
Cruciferae	9	Onion	4
Soybean	3	Potato	3

[a] Berger (1962).
[b] Deficiencies also observed on many other crops but with less frequency.

Generally, there is sufficient Fe^{3+} in the soil but, for one reason or another, it is unavailable to plants. In soils as alkaline as pH 7.5 or higher, Fe^{3+} becomes quite unavailable. In the reddish-colored, high-Fe soils of Hawaii, Fe^{3+} is often unavailable to pineapple plants because of the high Mn^{2+} concentration in the soil. The latter has been shown to interfere with Fe^{3+} absorption.

In almost all instances the only effective solution to Fe deficiency is foliar application of Fe. Chelated forms of Fe have become very popular because of their effectiveness.

Fe deficiency is a frequent problem in ornamental plants, such as azaleas and rhododendrons. Again, foliar spraying with Fe is usually the most effective treatment.

Copper. Owing to chemical combination with organic matter, Cu^{2+} is most likely to be unavailable in soils containing a high percentage of peat or muck, but deficiency may also occur in alkaline and calcareous soils—even in leached, sandy soils (Reuther and Labanauskas, 1966). Cu deficiencies have been reported for vegetables in the Florida Everglades, and tobacco in Virginia, the Carolinas, Georgia, and Connecticut (Swanback, 1950). In many cases addition of Cu^{2+} to a peat or muck soil is ineffective, since applied Cu^{2+} also becomes unavailable through combination with organic matter. Foliar sprays with Cu^{2+} are generally effective. Cu deficiency is the least widespread of any of the micronutrient deficiencies (Berger, 1962).

The Cu^{2+} concentration in plants may not necessarily indicate whether Cu^{2+} is deficient. Cu-deficient tomato plants may have a higher Cu^{2+} concentration than that of plants not deficient in Cu^{2+} (Bailey and McHargue, 1943). The total Cu^{2+} is less in Cu-deficient than in Cu-sufficient plants, but in Cu-deficient plants Cu^{2+} is not "diluted" by cellulose, and so on, since there is practically no growth. When a very small amount of Cu^{2+} is supplied to Cu-deficient plants, a relatively large amount of growth occurs, and concentration of Cu^{2+} in the plants decreases; total Cu^{2+}, of course, shows an increase. Similar effects on the concentration of essential elements in plants have been reported for other elements. This point must be remembered in making diagnoses based on the chemical composition of plants.

Zinc. Zn deficiencies have been reported for crops growing on sandy soils along the East Coast of the United States and, particularly, in Florida (Barnette et al., 1936). Zn deficiency has been observed most frequently in corn (Barnette et al., 1936) and fruit trees (Berger, 1962) (see also Table 14.2).

There are increasing reports of Zn deficiency in corn as higher and higher yields are obtained. Amounts of Zn^{2+} that were sufficient for the previous lower yields are no longer sufficient. In order to achieve high yields, it is now often necessary to add Zn^{2+} to the soil prior to planting to prevent an irretrievable loss in growth and yield.

In Arizona, Zn deficiency of pecan trees, known as rosette, is common on, but not limited to, alkaline soils (Finch, 1936).

In a California peach orchard showing Zn deficiency, it was calculated that trees and fruit removed approximately 8 oz of Zn^{2+} in 7 years; yet an analysis of the soil showed 3000 lb of Zn per acre within the root zone (Chandler, 1937). Clearly, an analysis of soil for Zn^{2+} would not have indicated that plants growing in it would show Zn deficiency.

In the western United States, Zn^{2+} is sometimes unavailable because of the relatively high pH (7.5 or higher) of alkaline soils. There may be, and usually is, sufficient Zn^{2+} in the soil, but it may be unavailable (Chapman, 1966a). Generally, only foliar sprays are effective in overcoming the deficiency (Chandler, 1937). In acid soils $ZnSO_4$ additions usually correct the deficiency (Barnette et al., 1936).

Manganese. Mn deficiencies have often been reported for the Midwest, particularly for soybeans. Snapbeans are also very susceptible to Mn deficiency.

Overliming may result in Mn deficiency (Labanauskas, 1966). Similar to Fe, Mn^{2+} also becomes unavailable in alkaline soils. Mn deficiency appears in certain crops grown in peat and muck soils (Berger, 1962). Foliar sprays, stem injections, and applications of Mn salts to soils have been used to overcome Mn deficiency (Labanauskas, 1966).

Molybdenum. Mo deficiency was reported several years ago in soils in Australia (Anderson, 1942, 1946) and California (Vanselow and Datta, 1949). Mo deficiency has been reported for cauliflower in western Maryland and in New York (Raleigh, 1950). Mo deficiency may also occur in citrus (Labanauskas, 1966).

It is surprising that Mo deficiency occurs under field conditions. For example, 0.001 ppm of Mo in the nutrient solution was sufficient for growth of alfalfa in water culture (Evans et al., 1950).

Highly significant responses of soybeans were obtained for the first time from Mo treatment of seed at planting time on several Iowa soils during 1968 (deMooy, 1970). Yields were raised 4–5 bu/acre in response to 93g of Mo per hectare at two sites where soil pH ranged from 5.8 to 6.7.

In highly acid soils Mo is sometimes fixed in an unavailable form (Berger, 1962). In contrast with elements such as B, Mn^{2+}, Zn^{2+}, and Cu^{2+}, which may become relatively unavailable at high pH, liming increases the availability of Mo since solubility of molybdates increases with increasing pH. Five-tenths to 2 lb of Na molybdate per acre usually prevents a deficiency if the soil is neither highly acid nor highly alkaline.

Except in those cases in which some soil condition renders Mo unavailable, it may be added to fertilizer or applied as a foliar spray. Several ounces of Mo per acre is all that is generally required for most crops to prevent a deficiency.

Factors Causing Deficiency or Unavailability

Genetic. The yellow stripe maize mutant (ys_1) has an Fe utilization problem in that plants homozygous for yellow stripe are unable to utilize Fe^{3+}. Heterozygous plants are able to use both Fe^{2+} and Fe^{3+} (Bell et al., 1958).

In two varieties of soybeans, the rootstalk was shown to be the controlling factor in utilization of Fe (Brown et al., 1958). The tops of PI-54619-5-1 (PI) and Hawkeye soybeans develop Fe chlorosis when grafted onto PI rootstalk; neither variety develops chlorosis when grafted onto Hawkeye rootstalk. The Hawkeye variety readily takes up Fe^{3+}, whereas the PI variety does not. At the root surface Hawkeye soybean (not susceptible to chlorosis) has a much greater reductive capacity than the PI variety which is susceptible to chlorosis (Brown et al., 1961).

Hawkeye soybeans remain green under high Ca^{2+} and P levels that induce chlorosis in the PI variety (Brown et al., 1959).

Mg chlorosis (i.e., Mg deficiency) in celery, variety Utah 10B, is determined by a single gene. By a simple 3:1 Mendelian ratio, the variety inherits recessively the tendency to become Mg deficient (Pope and Munger, 1953a). For *Helianthus bolanderi* Gray subspecies exilis Heiser, a form endemic on serpentine soils, a high external level of Mg^{2+} is required to bring the internal Mg^{2+} into the sufficiency range (Madhok and Walker, 1969).

Susceptibility to B deficiency in celery is determined by a single gene (Pope and Munger, 1953b).

Studying Fe chlorosis among Manchurian varieties of soybeans, Weiss (1943) showed that Fe utilization is conditioned by a single gene.

The variety of a crop can also determine whether or not Mo deficiency symptoms appear. When grown in the same Dunkirk silty clay loam near Ithaca, New York, Supersnowball cauliflower showed marked whip-tail, whereas there was practically no suggestion of the deficiency with the Improved Holland Erfurt (Snowdrift) variety (Raleigh, 1950).

Experimenting with diallele crosses of perennial ryegrass varieties, Butler and Glenday (1962) showed that the I^- content of varieties was a strongly inherited characteristic.

Interrelationships among Ions. In studies with pineapples, Nightingale (1942) observed that when much NO_3^- was needed extra K^+ was required for NO_3^- absorption—K^+ over and above that necessary for the other functions K^+ performs. NO_3^- level affected P absorption; P level affected NO_3^- absorption. Under field conditions it was noted that it was much more common to find that NO_3^- limited the uptake of P than that P suppressed the absorption of NO_3^-.

DeKock (1955, 1958) regarded Fe deficiency to be attributable to an excess of P; a deficiency of P resulted in Fe toxicity. Further, an excess of Mn^{2+} caused Fe deficiency; a deficiency of Mn^{2+} resulted in a chlorosis characteristic of Fe toxicity (DeKock, 1958). In mustard plants Fe chlorosis occurred when the %P/%Fe ratio was 60 or higher, but not when the ratio was 50 or less (DeKock, 1955).

For soybeans, Somers and Shive (1942) stated that the ratio of soluble Fe^{3+} to soluble Mn^{2+} in the plant should be between 1.5 and 2.5. When the ratio was above 2.5, they obtained symptoms of Mn deficiency ($=$Fe toxicity); when it was below 1.5 the plants showed Fe deficiency ($=$Mn toxicity). They proposed the accompanying scheme for illustrating a possible theoretical equilibrium system in which Fe^{2+} and Mn^{2+} are functionally related.

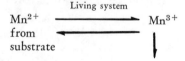

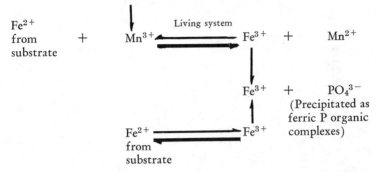

However, Morris and Pierre (1947) and Berger and Gerloff (1947) were unable to confirm the similarity of Fe deficiency and Mn toxicity. Similarly, Mn-toxicity symptoms of barley plants were quite distinct from those of Fe-deficiency symptoms (Agarwala et al., 1964). In *Lespedeza* (Morris and Pierre, 1947), alleviation of Mn toxicity by higher levels of Fe^{3+} reduced the Mn^{2+} concentration in plants with no particular increase in Fe. An excess of Mn^{2+} for potatoes produced symptoms identical with those for "stem streak necrosis" occurring on acid soils (Berger and Gerloff, 1947).

Studying green and albino sunflower leaf tissue, Weinstein and Robbins (1955) reported that low nutrient levels of Fe^{3+} or high nutrient levels of Mn^{2+} reduced catalase and cytochrome oxidase activities. They concluded that induction of Fe deficiency by high Mn^{2+} is a result of direct competition between Mn^{2+} and Fe^{3+} for a position in the heme nucleus of Fe-containing enzymes.

It has frequently been reported that Mg deficiency is accentuated by a high level of K^{+} supply (Harley, 1947; Boynton, 1947; Cain, 1948). Heavy applications of K^{+} fertilizers may reduce absorption of Mg^{2+} by tobacco and give rise to "sand drown"—a Mg deficiency (Garner et al., 1923).

Addition of K^{+} fertilizer to grain crops on acid muck soils may increase Fe^{3+} accumulation to the extent of toxicity, or reduce Ca^{2+} and Mg^{2+} concentrations in plants to the point of starvation (Loehwing, 1928).

In field-grown hybrid corn in Ohio, K deficiency was associated with an accumulation of Mo in leaves. Plants with K-deficiency symptoms had about four times the concentration of Mo as normal plants (i.e., 4 versus 1 ppm Mo) (Jones, 1965).

Much more Mo accumulated in leaves and stems of tomato plants when phosphate was present in the external solution (Stout and Meagher, 1948). The concentration of Mo was lowest in leaves and stems when SO_4^{2-} (in the absence of phosphate) was present in the external solution.

High pH. Fe, Mn^{2+}, Cu^{2+}, and Zn^{2+} are less available as the pH approaches 7.5 and higher (Fig. 14.1). B becomes progressively less available

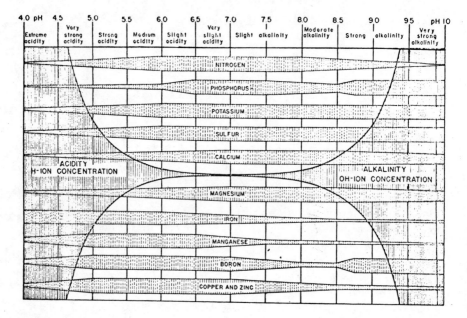

Fig. 14.1

The relation of soil pH to the availability of plant nutrients. Each element is represented by a band whose width at any particular pH value indicates the relative favorableness of this pH value, and associated factors, to the presence of the element in readily available forms. The wider the band, the more favorable the influence of pH on availability. From E. Truog (1951) Soil as a medium for plant growth, pp. 23–55. *In* E. Truog, (ed.) Mineral Nutrition of Plants, Univ. Wisconsin Press, Madison. © 1951 by the Regents of the University of Wisconsin—by permission.

from pH 7.0 to 8.5; from pH 8.5 on, B becomes available again (Truog, 1951). The availability of P decreases from pH 7.5 to 8.5, and from then on P again becomes available.

The addition of $CaCO_3$, or limestone, to soil can have a profound effect on Mn^{2+} availability. Albrecht and Smith (1941) found that when they mixed limestone throughout the soil, so as to modify its reaction, the concentration and total amount of Mn^{2+} in crops was decreased. However, when the limestone was added to the surface soil, the concentration and total amount of Mn^{2+} in crops was increased.

Low pH. Nitrogen, K^+, Ca^{2+}, Mg^{2+}, P, and S become less and less available from a pH of approximately 6.5 down to 4.5. Availabilities of Mn^{2+}, B, Cu^{2+}, and Zn^{2+} are drastically reduced between pH 5 and 4, and at pH 7.5 and higher (Truog, 1951) (Fig. 14.1). Maximal availability of Fe^{3+} occurs between pH 4 and 6; it decreases progressively as the pH rises above 6.0.

In a study of PO_4^{3-} fixation in soil by Fe^{3+} and Al^{3+}, Swenson et al.

(1949) found that precipitation of $H_2PO_4^-$ by Fe^{3+} and Al^{3+} in the presence of OH^- took place in the form of basic Fe or Al phosphate, $Fe(H_2O)_3(OH)_2H_2PO_4$. This was the case even when the solution contained sufficient quantities of PO_4^{3-} to occupy three of the coordination positions of Fe^{3+} and Al^{3+}. Maximal fixation of PO_4^{3-} occurred at pH 2.5–3.5 (Fe) and at pH 3.5–4 (Al). Citrate prevented fixation of PO_4^{3-} by Fe^{3+} and Al^{3+}, and it was reasoned that the beneficial effect of organic matter on the availability of PO_4^{3-} is attributable to the presence of organic acids. McAuliffe et al. (1947), using ^{32}P, demonstrated that there are two reactions in PO_4^{3-} fixation by soils—one proceeding rapidly and one more slowly. The rapid one is an adsorption of PO_4^{3-} on the surface of soil colloids; this surface-held PO_4^{3-} was considered available for plants.

Fixation. K^+ occurs in three forms in soil: readily available K^+ (released by dilute salt solutions), moderately available K^+ (released by 0.5–1 N HCl), and difficulty available or nonavailable K^+ forming part of primary or secondary mineral compounds. Attoe and Truog (1945) described the three availability categories of soil K^+ and their dynamic relationships and transformations from one category to another:

Readily available:
exchangeable K, K in organic
matter, water-soluble K salts.
Often constitutes about 1% of
total in a fertile soil

Moderately available:
Fixed K, biotite K. Constitutes
about 1 or 2% of total

Difficulty available:
Feldspar K, muscovite K. Usually
constitutes 97–98% of total

The first two forms are rather easily available to plants. Truog and Jones (1938) reported that nonavailable K^+ enters the mineral particles (perhaps muscovite) in empty spaces in the silicate lattices.

Synonyms for the several forms, particularly as they apply to K^+, are as follows. *Category A:* Readily available; soluble; exchangeable. *Category B:* Moderately available; fixed. *Category C:* difficultly available; slowly available; nonexchangeable available to plants; nonavailable.

The readily available, exchangeable K^+ is held by montmorillonite; moderately available K^+ is associated with hydrous mica (illite) and biotite; and the slowly or difficultly available K^+ is held by feldspars and muscovite (Truog, 1951).

"Nonexchangeable K available to plants" can be released by laboratory methods such as continued moist contact, alternate freezing and thawing, digestion with 1.0 N HNO$_3$, and electrodialysis, as reported by Reitemeier et al. (1947). They found that in some of the soils tested more than half of the K$^+$ absorbed by ladino clover came from nonexchangeable forms. Conceivably, the H$^+$ ion swarm surrounding roots could act as an efficient releaser of nonexchangeable K$^+$ (Jenny, 1951).

Phosphate may be attached to Al and Fe hydroxides and silicates as an exchangeable anion, and some of the P may be considered readily available. Most of this P is only moderately available. P attached to Fe hydroxide, and that in apatite, is slowly or difficultly availabile (Truog, 1951).

The mechanism of P fixation is not satisfactorily understood, and there may be no single mechanism applicable to all soil conditions (Overstreet and Dean, 1951).

Excess Salts. When neutral salts accumulate to a pronounced degree, resulting in soil salinity, uptake of ions other than the "salinity" ions may be reduced. For example, a pronounced accumulation of Ca^{2+} may interfere with absorption of Mg^{2+}. Thus an interrelationship of certain ions is more clear-cut than those of other pairs of ions.

Earlier it was mentioned that excess Mn^{2+} in certain Fe-rich soils in Hawaii interfered with absorption of Fe^{3+} to the extent that pineapple plants became Fe deficient.

Zn deficiency has been observed in Tennessee most frequently on soils naturally high in PO$_4{}^{3-}$—such as the Maury and Armour of the Central Basin area (Winters and Parks, 1956).

Drought. Drought may cause a deficiency or unavailability of an element in various ways. One of the main effects of drought is on the growth of roots and on their capacity for salt absorption. Nightingale (1942) regarded soil moisture as an important index in considering the efficacy of salt absorption by pineapple. When the soil was dry and the roots were not growing, he showed that there was very little uptake of applied nutrients.

In addition, it has been reported that drought causes a decrease in absorption of B (Berger, 1962).

Excessive Rainfall. An excess of rain, sufficient to carry water to the water table, leaches certain ions from the soil, particularly bases such as Ca^{2+}, Mg^{2+}, K$^+$, Zn^{2+}, Cu^{2+}, and Mn^{2+}, and anions such as Cl$^-$ and NO$_3{}^-$. Ca^{2+}, Mg^{2+}, and NO$_3{}^-$ are the main nutrient ions lost from soils by leachering (Wadleigh and Richards, 1951).

Surface erosion of soils is of greater concern in the loss of nutrients than leaching (Kohnke, 1942). Fippin (1945) estimated that the Mississippi River carries 475,000,000 tons of silt into the Gulf of Mexico during the average year, and that this silt contains 4,500,000 tons of exchangeable bases and 1,500,000 tons of phosphoric acid and N.

Temporary Combination into Microorganismal Protein. Although various nutrients essential for plants could be bound up in large amounts by bacteria and other microorganisms under certain conditions, the outstanding example is N. When high-carbohydrate, low-N plant residues, such as straw, are returned to soil, available NO_3^- is largely consumed by microorganisms involved in degradation of organic matter. Unless sufficient N is applied to fulfill both the needs of microorganisms and the crop, plants may show mild to severe symptoms of N deficiency. When the incorporated organic matter is finally decomposed, N once more becomes available when the microorganisms die.

For a time, then, microbial flora may tie up available forms of nutrients (Leeper, 1952), as has been shown for P (Kaila, 1949) and Zn^{2+} (Millikan, 1942).

Organic Matter. High concentrations of organic matter, as in peat or muck soils, can combine with such a high proportion of the Cu^{2+} that plant growth and yield are reduced (Felix, 1927; Allison et al., 1927; Swanback, 1950).

Loss of Topsoil. Removal of topsoil to aid gravity irrigation may expose subsoils low in available Zn^{2+} (Grunes et al., 1961).

CHLOROSIS

Although a deficiency of any one of various essential elements may cause chlorosis, a deficiency of Fe is the most frequent cause. In some cases Fe chlorotic leaves have been shown to contain higher concentrations of total Fe than nonchlorotic leaves on the same plant. In particular, high external concentrations of Ca^{2+}, lime, and P have been implicated as affecting the Fe nutrition of plants. Also, there has been a great amount of research dealing with the possible existence of natural Fe-chelating compounds in plants and with the fate of externally applied Fe chelates.

In leaves of pea plants, incoming Fe was at first associated with phloem and border parenchyma (Branton and Jacobson, 1962). Although Fe accumulated in vascular tissues, there was no evidence (in chlorotic plants) that iron is inactivated or precipitated in the xylem.

Starting with Fe chlorotic soybean leaves, increase in chlorophyll is the most sensitive indicator of an improvement in Fe nutrition of plants (Simons et al., 1963).

Factors Affecting Chlorosis

Bicarbonate. In soybean, bicarbonate did not appear to be a direct cause of Fe chlorosis and the anion did not directly inactivate cytochrome oxidase (Miller et al., 1960).

Nitrogen Supply. In contrast with corn plants receiving only NO_3^-, Wadleigh and Shive (1939) observed that chlorosis was almost entirely eliminated when plants received equivalent concentrations of NH_4^+ and NO_3 nitrogen.

Variety. In a study of two varieties of soybeans, Brown et al. (1958) noted that rootstock was the controlling factor in utilization of Fe^{3+} from the culture solution. Tops of the susceptible and nonsusceptible varieties developed Fe chlorosis when grafted onto rootstocks of the susceptible variety. Neither variety developed chlorosis when grafted onto nonsusceptible rootstocks.

In another study, Brown et al. (1959) observed that a nonsusceptible (to Fe deficiency) variety of soybeans remained green under high-Ca^{2+} and high-P conditions that induced chlorosis in a susceptible variety.

Soybean plants not susceptible to Fe deficiency had a greater Fe-reducing capacity at the root surface than did roots from a susceptible variety of soybeans (Brown et al., 1961; Ambler et al., 1971). Zn may interfere with absorption of Fe (Ambler, 1969; Ambler et al., 1970).

Elmstrom and Howard (1969) studied Fe-efficient genotypes of soybeans, Hawkeye and A62-9(E-9), and Fe-inefficient ones, PI and A62-10(I-10). A greater amount of Fe was associated with roots of inefficient plants than with roots of efficient plants. This was interpreted as indicating a slower rate of Fe translocation in the former type.

Chelates. Wallace (1956) and Wallace et al. (1955) reported that the entire Fe chelate molecule is absorbed by plants, whereas later investigators (Tiffin and Brown, 1959, 1961; Tiffin et al., 1960; Brown et al., 1960) reported that Fe is differentially absorbed from the chelate at the root surface. Simons et al. (1962) concluded that Fe is separated from chelate at the root surface and that chelate does not enter the plant unless the chelate concentration exceeds that of Fe in the external solution. Then, if the chelate is a strong one, it chelates Fe within the plant, causes internal Fe to be unavailable to the plant, and chlorosis ensues.

A natural chelate was found in exudate from tobacco plants (Schmid and Gerloff, 1961); some of its properties were established, but it was not identified. Tests indicated that it was not an amino acid, an organic acid, or ascorbic acid. Later, natural chelating agents, identified as malic and malonic acids, were found in soybean exudates (Tiffin and Brown, 1962). The principal natural chelate was Fe malate, and it was concluded that Fe is translocated in plants in the chelated form.

Natural chelates in soybean plants can compete favorably for Fe with chelates having stability constants of 10^{25} or less, but not with strong chelates having stability constants of 10^{28} or higher. Simons et al. (1962) listed as weak chelators of iron: ferric nitrilotriacetic acid (FeNTA), ferric N-hydroxyethylenediaminetriacetic acid (FeHEDTA), and ferric ethylenediaminetetraacetic acid (FeEDTA); and as strong ones: ferric diethylenetriaminepentaacetic acid (FeDTPA), ferric cyclohexanediaminetetraacetic acid (FeDHDTA), and ferric ethylenebishydroxyphenylglycine (FeEHPG).

Phytotoxin. Certain strains of *Rhizobium japonicum* produce a phytotoxin which causes chlorosis in soybeans (Johnson et al., 1959; Owens and Wright, 1965a, b). The toxin appears to be an amino compound, but it has not been identified (Owens and Wright, 1965a).

Phosphorus. When leaves have a higher concentration of P than that required for growth and extension, excess P reacts with ions passing through such P-rich tissues, particularly Fe (Biddulph and Woodbridge, 1952). They believed that the principal effect of high P concentrations on chlorosis could be explained on this basis.

Lime-Induced Chlorosis

One of the earliest definitive studies on lime-induced chlorosis was made by Lindner and Harley in 1944. They noted that lime-induced chlorosis was originally thought to be caused by a high $CaCO_3$ concentration in soil which raised pH and made Fe unavailable. They observed, however, that many workers found that the Fe concentrations of normal and lime-induced, chlorotic leaves are similar (Milad, 1924, 1939; Allyn, 1927; Wallace, 1928; Oserkowsky, 1933). A high concentration of leaf K^+ is associated with lime-induced chlorosis, and it was concluded that K^+ replaces some of the Fe of the Fe-containing enzyme responsible for chlorophyll formation (Lindner and Harley, 1944). They stressed that lime-induced chlorosis is different from true Fe deficiency, upset Mn/Fe balance, and upset P/Fe balance.

Thorne et al. (1950) characterized the various features generally associated with lime-induced chlorosis as: (1) occurs under alkaline reaction and high lime, but pH and lime may be about the same around green and chlorotic plants; (2) fine soil texture, high soil moisture, poor soil aeration, and cool soil temperature favor the chlorosis; (3) high soluble phosphate in the soil or plants may inactivate the Fe; (4) leaves of affected plants generally have a higher K/Ca ratio, a higher concentration of total and soluble N, and a higher concentration of some organic acids than green leaves; (5) chlorotic leaves of deciduous trees are usually lower in acid-soluble and ferrous iron than normal leaves, but the differences are not always in favor of the green leaves; and, (6) affected plants respond to treatment with Fe salts.

Lime-induced chlorosis has been ascribed to a variety of causes, as indicated by the points just enumerated. Thorne et al. (1950) concluded that the "oxidation of iron to the ferric state and possible inactivation in protein combinations followed by a disturbance in the productive protein-chlorophyll combination appears as a plausible concept to explain part of the relationships involved in chlorosis."

Weiss (1943) found marked differences in chlorosis among Manchurian varieties of soybeans tested on a calcereous soil. When these varieties were grown in nutrient solution, it was shown that the differential performance was induced only when available Fe was low. Efficient varieties made normal growth and inefficient varieties showed severe chlorosis. On the basis of F_2, F_3, and back-cross populations, it was determined that Fe utilization is conditioned by a single gene.

In soybeans catalase activity was lower in lime-induced chlorotic plants than in normal plants. It was suggested that an Fe-requiring metabolic system appears to predominate in plants susceptible to lime-induced chlorosis, and that these plants are not responsive to Cu (Brown, 1953). Ascorbic acid oxidase activity was a good index of available Cu^{2+} supply regardless of whether or not the plant showed visual Cu-deficiency symptoms. Roots of soybeans resistant to Fe chlorosis have a much greater reductive capacity (for Fe) than those of plants susceptible to chlorosis (Brown et al., 1961).

When Fe is deficient, as in lime-induced chlorosis, citric acid accumulates in leaves even prior to the appearance of symptoms (Iljin, 1951), presumably because aconitase is not fully functional (DeKock, 1958). Fe^{2+} is required for the activity of aconitase (Dickman and Cloutier, 1951). Unless there is sufficient Fe so that some of it is in the Fe^{2+} state (DeKock, 1958), a high percentage of Fe^{3+} may be attached to phosphoproteins (Liebich, 1941).

In *Glycine max* PI-54619-5-1 (PI), citric acid in leaves was increased over that of normal plants by the four types of Fe chlorosis studied: (1) Fe-deficiency-induced chlorosis, (2) HCO_3-induced chlorosis, (3) high-P-induced chlorosis, and (4) high-Mn^{2+}-induced chlorosis (Su and Miller, 1961). In addition, HCO_3-induced and high-Mn-induced chloroses were also characterized by an unusually high malonic acid concentration in the leaves.

Inasmuch as chlorotic leaves, especially in lime-induced chlorosis, may contain as much or more Fe than green leaves, DeKock (1955, 1958) assessed the Fe status of leaves by considering P and Fe analyses together as a ratio. An excess of P indicated a deficiency of Fe; a deficiency of P indicated Fe toxicity (see Table 14.3). Incidentally, DeKock (1958) also noted that an excess of Mn^{2+} results in a chlorosis which is characteristically of the Fe-deficient type, and a deficiency of Mn^{2+} in a chlorosis of the Fe-toxicity type.

With regard to the chlorotic leaves observed by DeKock (1958) from plants growing in highly calcereous soils and in Mn-deficient soils, Thorne et al. (1950) concluded that chlorosis of the former was caused by Fe deficiency, and of the latter by Fe toxicity.

From the appearance of Fe deficiency symptoms on the youngest leaves, it appears that Fe does not move from older to younger leaves during a deficiency of Fe. However, in sorghum, cotton, and bean plants, it was observed that Fe was fairly mobile, that it was translocated by the phloem and xylem, and that it moved particularly to the actively growing, young, developing tissues (Brown et al., 1965).

Table 14.3
The P/Fe Ratio of Mustard Plants in Nutrient Given Varying Amounts of Fe as the Fe—EDTA Chelate[a]

Fe level in nutrient solution (ppm)	Leaf symptoms	P/Fe ratio	Explanation
0	Severe chlorosis	37.7	(Fe deficiency; P toxicity)
2	Healthy green	14.0	
4	Chlorotic	5.3	(Fe toxicity; P deficiency)
8	Severe chlorosis	4.4	(Fe toxicity; P deficiency)

[a] DeKock (1958).

PREVENTION OR ALLEVIATION OF DEFICIENCIES

Inasmuch as deficiencies of essential nutrients reduce growth and yield of crops, it is obviously desirable that deficiencies be prevented if at all possible, or alleviated promptly when and if they appear.

Fertilizers Applied to Soil

If a deficiency can be diagnosed by soil analysis prior to planting, it is easier to apply the necessary fertilizer at that time rather than later, and deficiency can be prevented at the start. This is highly preferable to correcting a deficiency that appears after the crop has begun to grow. Plant growth and yield can never be as good as if the deficiency had never been permitted to occur.

However, it is not always possible to determine in advance that some essential nutrient will prove to be deficient. In such a case the next best thing is a remedial application of the nutrient as early as possible; sometimes the deficient element can be applied as a side dressing. It is also possible to correct certain deficiencies, particularly those of the trace elements, by foliar application to leaves, as discussed in the next section.

Foliar Nutrient Sprays

History. Aquatic vascular plants, for example *Elodea, Lemna,* and *Vallisneria* species, absorb most of their nutrients through the leaves. Although nutrients are not generally administered to leaves of terrestrial plants, it has long been known that nutrients are readily absorbed by leaves of most species. We are concerned, then, with application of nutrients to the leaves of land plants and the significance of such applications.

There has even been an interest in the absorption and translocation of ions by tropical plants growing in soil under natural conditions. Thomasson et al. (1969) reported that ^{134}Cs was highly selected over ^{59}Fe, ^{85}Sr, and ^{185}W by banana and coconut plants. There was a slight but measurable amount of ^{185}W in the banana peel but not in the fruit. There was less translocation of ^{134}Cs to the fruit of coconut than to that of the banana.

Penetration. The penetration of leaf cuticles by various substances depends on many factors including, among others, the nature of the substance, lipophilic substituents, temperature, concentration, relative solubility in organic solvents, and species of plant. Overbeek (1956) concluded that penetration of leaf cuticle by chemicals is unlikely. In contrast, Mitchell et al. (1960) concluded that hydrated cuticle is permeable to a degree to water and to dissolved compounds associated with this water, and highly permeable to nonpolar compounds.

Darlington and Cirulis (1963) found large variations in the permeability of apricot leaf cuticle to different substances. The slow penetration of sugars (2–3% in 48 hr) contrasted with the almost complete penetration of the N-*n*-hexyl- and N-isopropyl-α-chloroacetamides. Penetration could be increased markedly by lipophilic substituents, but secondary effects occurred which did not correlate with chloroform/water partitioning. Although they regarded the cuticle as a formidable barrier to penetration, their comparison with permeability coefficients for sucrose compiled by Davson and Danielli (1952) showed that apricot leaf cuticle had a 10^4 to 10^6 higher permeability than that of certain algal cells.

According to Roberts et al. (1948), the cuticle of McIntosh apple leaves is not continuous but is interspersed with hydrated pectinaceous layers running at right angles to the cuticle. These layers extend in depth to the vein extensions within the leaf and thus form a continuous pathway for water, and substances dissolved in water, from the surface of the leaf to the vascular tissue. Greater absorption of certain pesticides and other model compounds occurred through the lower than through the upper cuticles of apple leaves (Goodman and Addy, 1962).

Apple rootstocks were grown with either 0.02 or 5.0 ppm of Fe to produce very chlorotic or dark green plants, respectively (Delap, 1970). Then, by dipping the shoots of plants from both treatments in solutions containing Fe, an attempt was made to introduce Fe through the leaves rather than the roots. Dipping prevented chlorosis of low-Fe plants and increased growth, but growth did not equal that of plants receiving a higher level of Fe through the roots. Following dipping, Fe was not translocated from the leaves to the roots even though the concentration of Fe in the leaves had been increased by the dipping. However, dipping reduced the concentration of Mn^{2+} in low-Fe roots to one-fourth that in roots of undipped low-Fe plants.

Ectodesmata (plasmodesmata) may provide the pathway for transport of substances from the outside to the interior to tissues, and vice versa, as reported by Franke (1961). He observed that leaf structures such as guard cells, anticlinal walls, and epidermal cells adjacent to leaf veins contain large numbers of ectodesmata, while in neighboring cells these may be low in number or lacking. Later, Franke (1964) stressed that stomatal pores do not function as "portholes" for the entry of solution, but that guard cells do so by virtue of the ectodesmata so characteristic of these cells. His conclusions were based on a study of the entry of labeled sucrose into leaves of *Spinacia oleracea* and *Viola tricolor*. In a study of absorption of $^{59}Fe^{3+}$ by leaves of various species, stomata were shown to play a major role in absorption (Eddings and Brown, 1967). Surfactant-free solutions apparently do not enter stomata (Dybing and Currier, 1961).

Electron micrographs revealed the cuticle and wax (when present) of leaves of various species of plants (Schieferstein and Loomis, 1956). Of 48 plant species, only 28 showed surface wax deposits. Waxes reduce the retention of sprays and could presumably affect the action of herbicidal and nutrient sprays. Growing leaves have a permeable, immature zone in the cuticle through which herbicides, for example, can penetrate rapidly.

With enzymically isolated cells from tobacco leaves, it appeared that the coupling between absorption and metabolism occurred at the cellular level during foliar absorption of mineral nutrients (Jyung et al., 1965).

Substances can not only enter leaves, but they can also be lost by leaves through leaching (Mecklenburg and Tukey, 1964). When the tops of bean plants were subjected to an atomized distilled water mist for 4 days, 30–40% of the recently root-absorbed ^{45}Ca was leached from the foliage. In a study of seven plants species, 21 amino acids and amides and 14 organic acids, in addition to 4 free sugars, polysaccharides and other carbohydrate materials, were found in leachates from leaves (Morgan and Tukey, 1964). Thus inorganic and organic metabolites may be leached from above-ground parts of plants by rain, dew, and mist (Arens, 1934; Tukey et al., 1958; Tukey and Tukey, 1962; Morgan and Tukey, 1964; Tukey, 1970). Ca is leached from bean leaves *(Phaseolus vulgaris)* by a process of ion exchange and diffusion that involves exchange sites within the leaf and on the surface (Mecklenburg et al., 1966).

High humidity increased the absorption of foliar-applied 2,4-D (Pallas, 1958), urea (Volk and McAuliff, 1954), maleic hydrazide (Zukel et al., 1956), and 2-4-D, urea, and 3-amino-1,2,4-triazole (Clor et al., 1963). In addition, Clor et al. (1962) observed that high humidity increased the translocation of 2,4-D and urea in cotton plants.

The kinetics of urea penetration are markedly different from those of other substances. Franke (1967) reported that the rate of penetration of urea exceeds that of ions by 10- to 20-fold, and that rate of entry of urea is independent of concentration. This high permeability of leaves to urea also favors foliar absorption of ions applied together with urea; urea is a promoter of permeability (Franke, 1967).

Binding Sites on Leaves. Using $^{45}CaCl_2$ and $K_2^{35}SO_4$, Yamada et al. (1964a) studied the entry, binding, and retention of Ca^{2+} and S by stomatous, mature, green onion leaves and astomatous surfaces of ripe tomato fruits (Table 14.4). Ca^{2+} was absorbed and retained to a greater degree than was S.

Using the same cuticular membranes, Yamada, et al. (1964b) reported that there was greater permeability from the outer to the inner surfaces of cuticles than in the opposite direction (for ^{45}Ca, ^{86}Rb, ^{35}S, and ^{36}Cl). A dialyzing membrane showed no such differences. Rate of penetration through the cuticular surfaces was directly related to the extent of ion binding on the surface opposite the site of initial entry. It was concluded that it was possible that the greater ion binding on the inside, compared to the outside, of isolated cuticular membranes facilitates foliar absorption.

When Yamada et al. (1965) studied entry and binding of urea by onion leaf and tomato fruit cuticles, they observed that retention of urea was essentially the same on the inner and outer surfaces. The amounts of urea that penetrated tomato fruit cuticle exceeded by 10 to 20 times that of the inorganic ions $(Rb^{2+}, Ca^{2+}, Cl^-, SO_4^{2-})$.

Using microautoradiography, Yamada et al. (1966) studied binding of labeled Ca^{2+}, Cl^-, and urea on cuticular membranes from tomato fruits and onion leaves. In tomato fruit cuticle there was a greater binding of Ca^{2+} and other ions on the inner as compared with the outer surface; urea was bound to a much lesser degree and about equally on the two surfaces. There was a more intense and localized binding of Ca^{2+} and Cl^- on onion leaf than on tomato fruit cuticle. In onion, Ca^{2+}-binding sites were on the outer surface and Cl^- sites on the inner surface above the periclinal cell walls.

The cuticle is a heterogenous membrane in which wax, pectin, and cellulose in varying proportions are contained in a cutin framework. Holloway and Baker (1968) recommended a $ZnCl_2$–HCl solution as a simplified method for obtaining cuticles.

Cuticular membranes were enzymically isolated from tomato fruits and from the dorsal and ventral surfaces of *Euonymus japonicus* to study movement of Fe through cuticles (Kannan, 1969). It was observed that Fe from $FeSO_4$ penetrated more rapidly than Fe from FeEDDHA [ferric ethylenediamine di (*o*-hydroxyphenylacetate)]; urea reduced the penetration of Fe from both sources.

Current Interest and Use. In general, most horticulturists and agronomists do not consider the foliar route of application of macronutrients, such as N, P, and K, to be as economical a route as that of soil application. There is, how-

Table 14.4
Retention of Ions by Outer and Inner Surfaces of Ripe Tomato Fruit and Green Onion Leaf Cuticles[a]

Ion	Surface	Blotting (nmole/cm^2)	Washing (nmole/cm^2)	Exchange (nmole/cm^2)
Ripe tomato fruit cuticle				
Ca^{2+}	Outer	1.6	0.9	0.4
	Inner	18.0	13.0	4.0
SO$_4$$^{2-}$	Outer	0.4	0.01	0
	Inner	1.1	0.02	0
Green onion leaf cuticle				
Ca^{2+}	Outer	58	48	15
	Inner	149	126	33
SO$_4$$^{2-}$	Outer	0.11	0.05	0
	Inner	3.40	0.32	0

[a] Yamada et al., (1964a).

ever, an apparent exception in the case of apple trees. In recent years there has been considerable interest in applying N, as urea, directly to the leaves of apple trees. Foliar applications of N have not proved beneficial for peach trees, apparently because the applied nutrients do not readily penetrate the surfaces of peach leaves.

Early in the spring, when nitrification is still proceeding slowly because of low soil temperature, direct application of N to turf grasses on golf courses has proved beneficial in overcoming chlorosis resulting from N deficiency.

Cracking of Bing and Royal Ann cherries following rains is a very serious problem in many areas. CuSO$_4$, as a spray or dust, appears to check the cracking (Powers and Bollen 1947), apparently by a toughening effect (McLean et al., 1944) on skin.

Advantages. There are two main advantages in applying nutrients directly to leaves. First, this route enables the plant to obtain needed nutrients when some soil condition (low temperature, waterlogging, and so on) prevents nutrient absorption from the soil. Second, foliar application delivers the nutrients directly to the plant and bypasses any fixation processes (for that nutrient) in the soil.

Disadvantages. Whenever possible, it is more economical to apply nutrients to the soil, especially prior to planting, than it is to make a special foliar application.

Although a given formulation for foliar application has worked well in the past, there is always the possibility that under different conditions the nutrient spray may cause damage to leaves.

There have been a few instances in which urea applied as a foliar spray contained biuret as a contaminant; the latter caused pronounced damage to foliage. A contaminant such as biuret is probably much more serious in the case of foliar application than in that of a soil application, since soil has a mitigating influence in the case of many toxicants.

Liquid Fertilizers

There is some interest in liquid fertilizers in the form of concentrates which are diluted prior to application. For plants around the house or for a small greenhouse operation, such a product may possibly be suitable. In such cases cost need not be a deterrent and the advantages may outweigh any consideration of cost.

Although it may be economically sound to use concentrated liquid fertilizers as a source of trace elements, the question of economics arises in connection with macronutrients such as N, P, and K which are required in large amounts by most crops. Calculations of the cost per pound of actual nutrient in liquid fertilizers versus that in conventional dry fertilizers indicate that the cost of liquid fertilizers can be prohibitive.

Although a relatively small amount of nutrients may be quite effective when applied directly to leaves, it is easy to calculate that it would generally be difficult to supply the entire requirements for N, P, and K by foliar applications.

Starter Solutions

It could perhaps be said that the original starter solution was not a solution at all. The case in point is the legendary dead fish that an Indian was supposed to have placed alongside the corn grain he planted. Admittedly, the dead fish probably functioned not only as a "starter," but also supplied nutrients for the continued growth of the plant.

The sole purpose of starter solutions is to provide readily available nutrients for early growth of a transplant or a seedling as it emerges from the seed. Some growers and researchers regard starter solutions as a valuable aid in early growth of seedlings prior to the time that roots come in contact with the main body of fertilizer placed in the soil.

The recommendation for and the use of starter solutions also takes into consideration the fact that the nutrient requirements of the seedling stage may be somewhat different from those for later growth. Therefore most starter solutions are high in P since P is generally regarded to be especially important in promoting root growth. Rapid root growth is believed to be important in obtaining nutrients available throughout the root zone in the soil.

The compositions of two starter solutions used for tomatoes are shown in Table 14.5.

In New York, a starter solution for tomatoes increased early and total yields significantly over those of the control (water only) (Sayre, 1938). Of the various treatments any one that contained 20 oz of ammonium phosphate per 50

Table 14.5
Examples of Starter Solutions Giving Significant Yield Increases for Tomatoes

Treatment	Crop	Location	Investigator
20 oz Ammo Phos A, plus 10 oz nitrate of potash (per 50 gal water; 1 pt per plant)	Tomato	New York	Sayre, 1938
2 lb, 10½ oz Ammo Phos A, 1 lb, 5½ oz KNO_3 (per 50 gal water; ¼ pt per plant)	Tomato	New York	Sayre, 1940

gal of water significantly increased yield. One pint of solution was applied to each transplant. For the best of the treatments the chemicals were estimated to cost 48¢/acre, but this amount increased the early yield of tomatoes by 1.44 tons and the total yield by 1.85 tons. Later (Sayre, 1940), monoammonium phosphate, alone or in combination with KNO_3, increased the earliness and total yield of tomato transplants, and also of field-seeded tomatoes. The cost was approximately 80¢/acre.

In attempting to explain why such a small quantity of nutrients should make such a large difference in yield in a well-fertilized field, Sayre (1938) suggested that the most probable explanation was that the nutrient solution was applied in a particularly available form just at a critical time in the growth of the plants when they were low in minerals. The application enabled the plants to withstand better the shock of transplanting, to become established quickly, and to grow vigorously from the time they were transplanted.

LITERATURE CITED

Agarwala, S. C., C. P. Sharma, and A. Kumar. 1964. Interrelationship of iron and manganese supply in growth, chlorophyll, and iron porphyrin enzymes in barley plants. Plant Physiol. 39:603—609.

Albrecht, W. A., and N. C. Smith. 1941. Calcium and phosphorus as they influence manganese in forage crops. Bull. Torrey Bot. Club 68:372—380.

Alexander, U. U., and D. Knudsen. 1965. Sulfur deficiency of field crops in Nebraska. Nebraska Agr. Exp. Sta. Ext. Circ. 65—164. 7 pp.

Allison, R. V., O. C. Bryan, and J. H. Hunter. 1927. The stimulation of plant response on the raw peat soils of the Florida Everglades through the use of copper sulphate and other chemicals. Florida Agr. Exp. Sta. Bull. 190.

Allyn, W. P. 1927. The relation of lime to the absorption of iron by plants. Proc. Indiana Acad. Sci. 43:405—409.

Ambler, J. E. 1959. Effect of zinc on the translocation of iron in soybean plants. Ph.D. Thesis, Univ. Maryland, College Park.

Ambler, J. E., J. C. Brown, and H. G. Gauch. 1970. Effect of zinc on translocation of iron in soybean plants. Plant Physiol. 46:320—323.

Ambler, J. E., J. C. Brown, and H. G. Gauch. 1971. Sites of iron reduction in soybean plants. Agron. J. 63:95—97.

Anderson, A. J. 1942. Molybdenum deficiency on a South Australian ironstone soil. J. Australian Inst. Agr. Sci. 8:73—75.

Anderson, A. J. 1946. Molybdenum in relation to pasture improvement in South Australia. Council Sci. Ind. Res. (Australian) 19:1—15.

Arens, K. 1934. Die kutikuläre Exkretion der Laubblätter. Jahrb. Wiss. Bot. 80:248—300.

Attoe, O. J., and E. Truog. 1945. Exchangeable and acid-soluble potassium as regards availability and reciprocal relationships. Proc. Soil Sci. Soc. Amer. 10:81—86.

Bailey, L. F., and J. S. McHargue. 1943. Copper deficiency in tomatoes. Amer. J. Bot. 30:558—563.

Baker, D. E. 1968. Crop production related to the capacity of soil to hold plant nutrients. Vegetable Grower's Messenger 20:8.

Barnette, R. M., J. P. Camp, J. D. Warner, and O. E. Gall. 1936. The use of zinc sulphate under corn and other field crops. Florida Agr. Exp. Sta. Bull. 292.

Bell, W. D., L. Bogorad, and W. J. McIlrath. 1958. Response of the yellow-stripe maize mutant (ys_1) to ferrous and ferric iron. Bot. Gaz. 120:36—39.

Berger, K. C. 1962. Micronutrient deficiencies in the United States. Agr. Food Chem. 10:178—181.

Berger, K. C., and G. C. Gerloff. 1947. Manganese toxicity of potatoes in relation to strong soil acidity. Proc. Soil Sci. Soc. Amer. 12:310—314.

Biddulph, O., and C. G. Woodbridge. 1952. The uptake of phosphorus by bean plants with particular reference to the effects of iron. Plant Physiol. 27:431—444.

Bingham, F. T. 1966. Phosphorus, pp. 324—361. In H. D. Chapman (ed.) Diagnostic criteria for plants and soils. Div. Agr. Sci., Univ. California, Berkeley.

Boynton, D. 1947. Magnesium nutrition of apple trees. Soil Sci. 63:53—58.

Bradford, G. R. 1966. Boron, pp. 33—61. In H. D. Chapman (ed.) Diagnostic criteria for plants and soils. Div. Agr. Sci., Univ. California, Berkeley.

Branton, D., and L. Jacobson. 1962. Iron localization in pea plants. Plant Physiol. 37:546—551.

Brown, A. L., S. Yamaguchi, and J. Leal-Diaz. 1965. Evidence for translocation of iron in plants. Plant Physiol. 40:35—38.

Brown, J. C. 1953. The effect of the dominance of a metabolic system requiring iron or copper on the development of lime-induced chlorosis. Plant Physiol. 28:495—502.

Brown, J. C., R. S. Holmes, and L. O. Tiffin. 1958. Iron Chlorosis in soybeans as related to genotype of the rootstalk. Soil Sci. 86:75—82.

Brown, J. C., L. O. Tiffin, R. S. Holmes, A. W. Specht, and J. W. Resnicky. 1959. Internal inactivation of iron in soybeans as affected by root growth medium. Soil Sci. 87:89—94.

Brown, J. C., L. O. Tiffin, and R. S. Holmes. 1960. Competition between chelating agents and roots as factors affecting absorption of iron and other ions by plant species. Plant Physiol. 35:878—886.

Brown, J. C., R. S. Holmes, and L. O. Tiffin. 1961. Iron chlorosis in soybeans as related to the genotype of rootstalk: 3. Chlorosis susceptibility and reductive capacity at the root. Soil Sci. 91:127—132.

Butler, G. W., and A. C. Glenday. 1962. Iodine content of pasture plants. II. Inheritance of leaf iodine content of perennial ryegrass (*Lolium perenne* L.). Australian J. Biol. Sci. 15:183—187.

Cain, J. C. 1948. Some interrelationships between calcium, magnesium and potassium in one-year-old McIntosh apple trees grown in sand culture. Proc. Amer. Soc. Hort. Sci. 51:1—12.

Chandler, W. H. 1937. Zinc as a nutrient for plants. Bot. Gaz. 98:625—646.

Chapman, H. D. (ed). 1966a. Diagnostic criteria for plants and soils. Div. Agr. Sci., Univ. California, Berkeley. 793 pp.

Chapman, H. D. 1966b. Calcium, pp. 65—92. In H. D. Chapman (ed.) Diagnostic criteria for plants and soils. Div. Agr. Sci., Univ. California, Berkeley.

Clor, M. A., A. S. Crafts, and S. Yamaguchi. 1962. Effects of high humidity on translocation of foliar-applied labeled compounds in plants. Plant Physiol. 37:609–617.

Clor, M. A., A. S. Crafts, and S. Yamaguchi. 1963. Effects of high humidity on translocation of foliar-applied labeled compounds in plants. II. Translocation from starved leaves. Plant Physiol. 38:501–507.

Darlington, W. A., and N. Cirulis. 1963. Permeability of apricot leaf cuticle. Plant Physiol. 38:462–467.

Davson, H., and J. F. Danielli. 1952. Permeability of natural membranes. Cambridge Univ. Press, Cambridge.

DeKock, P. C. 1955. Iron nutrition of plants at high pH. Soil Sci. 79:167–175.

DeKock, P. C. 1958. The nutrient balance in plant leaves. Agr. Progress 33:88–95.

Delap, A. V. 1970. Studies in the nutrition of apple rootstocks. The effect on growth and mineral composition of supplying iron through the leaves. Ann. Bot. 34:911–919.

deMooy, C. J. 1970. Molybdenum response of soybeans [*Glycine max* (L.) Merill] in Iowa. Agron. J. 62:195–197.

Dickman, S. R., and A. A. Cloutier. 1951. Factors affecting the activity of aconitase. J. Biol. Chem. 188:379–388.

Dybing, C. D., and H. B. Currier. 1961. Foliar penetration by chemicals. Plant Physiol. 36:169–174.

Eaton, F. M. 1966. Sulfur, pp. 444–475. *In* H. D. Chapman (ed.) Diagnostic criteria for plants and soils. Div. Agr. Sci., Univ. California, Berkeley.

Eddings, J. L., and A. L. Brown. 1967. Absorption and translocation of foliar-applied iron. Plant Physiol. 42:15–19.

Elmstrom, G. W., and F. D. Howard. 1969. Iron accumulation, root peroxidase activity, and varietal interaction in soybean genotypes that differ in iron nutrition. Plant Physiol. 44:1108–1114.

Embleton, T. W. 1966. Magnesium, pp. 225–263. *In* H. D. Chapman (ed.) Diagnostic criteria for plants and soils. Div. Agr. Sci., Univ. California, Berkeley.

Evans, H. J., E. R. Purvis, and F. E. Bear. 1950. Molybdenum nutrition of alfalfa. Plant Physiol. 25:555–566.

Felix, E. L. 1927. Correction of unproductive muck by the addition of copper. Phytopathology 17:49:50.

Finch, A. H. 1936. Zinc and other mineral constituents in relation to the rosette disease of pecan trees. J. Agr. Res. 52:363–376.

Fippin, E. O. 1945. Plant nutrient losses in silt and water in the Tennessee River system. Soil Sci. 60:223–239.

Franke, W. 1961. Ectodesmata and foliar absorption. Amer. J. Bot. 48:683–691.

Franke, W. 1964. Role of guard cells in foliar absorption. Nature 202:1236–1237.

Franke, W. 1967. Mechanisms of foliar penetration of solutions. Annu. Rev. Plant Physiol. 18:281–300.

Garner, W. W., J. E. McMurtrey, C. W. Bacon, and E. G. Moss. 1923. Sand drown, a chlorosis of tobacco due to magnesium deficiency, and the relation of sulphates and chlorides of potassium to the disease. J. Agr. Res. 23:27–40.

Goodman, R. N., and S. K. Addy. 1962. Penetration of excised apple cuticular membranes by radioactive pesticides and other model compounds. Phytopath. Z. 46:1–10.

Grunes, D. L., L. C. Brown, C. W. Carlson, and F. G. Viets, Jr. 1961. Land leveling – It may cause zinc deficiency. North Dakota Farm Res. 21:4–7.

Harley, C. P. 1947. Magnesium deficiency in Kieffer pear trees. Proc. Amer. Soc. Hort. Sci. 50:21–22.

Holloway, P. J., and E. A. Baker. 1968. Isolation of plant cuticles with zinc chloride-hydrochloric acid solution. Plant Physiol. 43:1878—1879.

Iljin, W. S. 1951. Metabolism of plants affected with lime-induced chlorosis (calciose). II. Organic acids and carbohydrates. Plant and Soil 3:339—351.

Jenny, H. 1951. Contact phenomena between adsorbents and their significance in plant nutrition, pp. 107—132. *In* E. Truog (ed.) Mineral nutrition of plants. Univ. Wisconsin Press, Madison.

Johnson, H. W., U. M. Means, and F. E. Clark. 1959. Responses of seedlings to extracts of soybean nodules bearing selected strains of *Rhizobium japonicum*. Nature 183: 308—309.

Jones, J. B., Jr. 1965. Molybdenum content of corn plants exhibiting varying degrees of potassium deficiency. Science 148:94.

Jones, W. W. 1966. Nitrogen, pp. 310—323. *In* H. D. Chapman (ed.). Diagnostic criteria for plants and soils. Div. Agr. Sci., Univ. California, Berkeley.

Jyung, W. H., S. H. Wittwer, and M. J. Bukovac. 1965. Ion uptake by cells enzymically isolated from green tobacco leaves. Plant Physiol. 40:410—414.

Kaila, A. 1949. Biological absorption of phosphorus. Soil Sci. 68:279—289.

Kannan, S. 1969. Penetration of iron and some organic substances through isolated cuticular membranes. Plant Physiol. 44:517—521.

Kohnke, H. 1942. Runoff chemistry: An undeveloped branch of soil science. Proc. Soil Sci. Soc. Amer. 6:492—500.

Labanauskas, C. K. 1966. Manganese, pp. 264—285. *In* H. D. Chapman (ed.). Diagnostic criteria for plants and soils. Div. Agr. Sci., Univ. California, Berkeley.

Leeper, G. W. 1952. Factors affecting availability of inorganic nutrients in soils with special reference to micronutrient metals. Annu. Rev. Plant Physiol. 3:1—16.

Liebich, H. 1941. Quantitativ-chemische Untersuchungen über das Eisen in den Chloroplasten und übrigen Zellbestandteilen von *Spinacia oleracea*. Z. Bot. 37: 129—156.

Lindner, R. C., and C. P. Harley. 1944. Nutrient interrelations in lime-induced chlorosis. Plant Physiol. 19:420—439.

Loehwing, W. F. 1928. Calcium, potassium, and iron balance in certain crop plants in relation to their metabolism. Plant Physiol. 3:261—275.

McAuliffe, C. D., N. S. Hall, L. A. Dean, and S. B. Hendricks. 1947. Exchange reactions between phosphates and soils: Hydroxylic surfaces of soil minerals. Proc. Amer. Soc. Soil Sci. 12:119—123.

McLean, J. G., W. C. Sparks, and A. M. Binkley. 1944. The effect of certain minor elements on yield, size, and skin thickness of potato tubers. Proc. Amer. Soc. Hort. Sci. 44:362—368.

Madhok, O. P., and R. B. Walker. 1969. Magnesium nutrition of two species of sunflower. Plant Physiol. 44:1016—1022.

Mecklenburg, R. A., and H. B. Tukey, Jr. 1964. Influence of foliar leaching on root uptake and translocation of calcium-45 to the stems and foliage of *Phaseolus vulgaris*. Plant Physiol. 39:533—536.

Mecklenburg, R. A., H. B. Tukey, Jr., an d J. V. Morgan. 1966. A mechanism for the leaching of calcium from foliage. Plant Physiol. 41:610—613.

Milad, Y. 1924. The distribution of iron in chlorotic pear trees. Proc. Amer. Soc. Hort. Sci. 21:93—98.

Milad, Y. 1939. Physiological studies in lime-induced chlorisis. Ministry Agr. Egypt Tech. Sci. Service Bull. 211.

Miller, G. W., J. C. Brown, and R. S. Holmes. 1960. Chlorosis in soybean as related to iron, phosphorus, bicarbonate, and cytochrome oxidase activity. Plant Physiol. 35:619—625.

Millikan, C. R. 1942. Studies on soil conditions in relation to root-rot of cereals. Proc. Roy. Soc. Victoria 54:145–195.

Mitchell, J. W., B. C. Smale, and R. L. Metcalf. 1960. Absorption and translocation of regulators and compounds used to control plant diseases and insects. Advance. Pest Control Res. 3:359–436.

Morgan, J. V., and H. B. Tukey, Jr. 1964. Characterization of leachate from plant foliage. Plant Physiol. 39:590–593.

Morris, H. D., and W. H. Pierre. 1947. The effect of calcium, phosphorus, and iron on the tolerance of *Lespedeza* to manganese toxicity in culture solutions. Proc. Soil Sci. Soc. Amer. 12:382–386.

Nightingale, G. T. 1942. Potassium and phosphate nutrition of pineapple in relation to nitrate and carbohydrate reserves. Bot. Gaz. 104:191–223.

Oserkowsky, J. 1933. Quantitative relation between chlorophyll and iron in green and chlorotic pear leaves. Plant Physiol. 9:440–468.

Overbeek, J. van. 1956. Absorption and translocation of plant growth regulators. Annu. Rev. Plant Physiol. 7:355–372.

Overstreet, R., and L. A. Dean. 1951. The availability of soil anions, pp. 79–105. *In* E. Truog (ed.) Mineral nutrition of plants. Univ. Wisconsin Press, Madison.

Owens, L. D., and D. A. Wright. 1965a. Rhizobial-induced chlorosis in soybeans: Isolation, production in nodules, and varietal specificity of the toxin. Plant Physiol. 40:927–930.

Owens, L. D., and D. A. Wright. 1965b. Production of the soybean-chlorosis toxin by *Rhizobium japonicum* in pure culture. Plant Physiol. 40:931–933.

Pallas, J. 1958. Effects of temperature and humidity on the absorption and translocation of 2,4-dichlorophenoxyacetic acid and benzoic acid. Ph.D. Thesis, Univ. California, Davis.

Pierre, W. H., and F. W. Parker. 1927. Soil phosphorus studies: II. The concentration of organic and inorganic phosphorus in the soil solution and soil extracts and the availability of the organic phosphorus to plants. Soil Sci. 24:119–128.

Pope, D. T., and H. M. Munger. 1953a. Heredity and nutrition in relation to magnesium deficiency chlorosis in celery. Proc. Amer. Soc. Hort. Sci. 61:472–480.

Pope, D. T., and H. M. Munger. 1953b. The inheritance of susceptibility to boron deficiency in celery. Proc. Amer. Soc. Hort. Sci. 61:481–486.

Powers, W. L., and W. B. Bollen. 1947. Control of cracking of fruit by rain. Science 105:334–335.

Raleigh, G. J. 1950. Molybdenum deficiency in Dunkirk silty clay loam. Science 112:433–434.

Reitemeier, R. F., R. S. Holmes, J. C. Brown, L. Klipp, and R. Q. Parks, 1947. Release of nonexchangeable potassium by greenhouse, Neubauer, and laboratory methods. Proc. Soil Sci. Soc. Amer. 12:158–162.

Reuther, W., and C. K. Labanauskas. 1966. Copper, p. 157–179. *In* H. D. Chapman (ed.). Diagnostic criteria for plants and soils. Div. Agr. Sci., Univ. California, Berkeley.

Roberts, E. A., M. D. Southwick, and D. H. Palmiter. 1948. A microchemical examination of McIntosh apple leaves showing relationship of cell wall constituents to penetration of spray solutions. Plant Physiol. 23:557–559.

Sayre, C. B. 1938. Use of nutrient solutions and hormones in the water for transplanting tomatoes and their effect on earliness and total yields. Proc. Amer. Soc. Hort. Sci. 36:732–736.

Sayre, C. B. 1940. Nutrient or starter solutions and vitamin B for transplanting tomatoes. Proc. Amer. Soc. Hort. Sci. 38:489–495.

Schieferstein, R. H., and W. E. Loomis. 1956. Wax deposits on leaf surfaces. Plant Physiol. 31:240–247.

Schmid, W. E., and G. C. Gerloff. 1961. A naturally occurring chelate of iron in xylem exudate. Plant Physiol. 36:226–231.

Simons, J. N., R. Swidler, and H. M. Benedict. 1962. Absorption of chelated iron by soybean roots in nutrient solutions. Plant Physiol. 37:460–466.

Simons, J. N., R. Swidler, and H. M. Benedict. 1963. Kinetic studies of regreening of iron-deficient soybeans. Plant Physiol. 38:667–674.

Somers, I. I., and J. W. Shive. 1942. The iron-manganese reaction in plant metabolism. Plant Physiol. 17:582–602.

Stout, P. R., and W. R. Meagher. 1948. Studies on the molybdenum nutrition of plants with radioactive molybdenum. Science 108:471–473.

Stout, P. R., and R. Overstreet. 1950. Soil chemistry in relation to inorganic nutrition of plants. Annu. Rev. Plant Physiol. 1.305–342.

Su, L., and G. W. Miller. 1961. Chlorosis in higher plants as related to organic acid content. Plant Physiol. 36:415–420.

Swanback, T. R. 1950. Copper in tobacco production. Connecticut Agr. Exp. Sta. Bull. 535.

Swenson, R. M., C. V. Cole, and D. H. Sieling. 1949. Fixation of phosphate by iron and aluminum and replacement by organic and inorganic ions. Soil Sci. 67:3–22.

Thomasson, W. N., W. E. Bolch, and J. F. Gamble. 1969. Uptake and translocation of ^{134}Cs, ^{59}Fe, ^{85}Sr, and ^{185}W by banana plants and a coconut plant following foliar application. Bioscience 19(7):613–615.

Thorne, D. W., F. B. Wann, and W. Robinson. 1950. Hypotheses concerning lime-induced chlorosis. Proc. Soil Sci. Soc. Amer. 15:254–258.

Tidmore, J. W. 1930. Phosphate studies in solution cultures. Soil Sci. 30:13–31.

Tiffin, L. O., and J. C. Brown. 1959. Absorption of iron from iron chelate by sunflower roots. Science 130:274–275.

Tiffin, L. O., and J. C. Brown. 1961. Selective absorption of iron from iron chelates by soybean plants. Plant Physiol. 36:710–714.

Tiffin, L. O., and J. C. Brown. 1962. Iron chelates in soybean exudate. Science 135:311–313.

Tiffin, L. O., J. C. Brown, and R. W. Krauss. 1960. Differential absorption of metal chelate components by plant roots. Plant Physiol. 35:362–367.

Truog, E. 1951. Soil as a medium for plant growth, pp. 23–55. In E. Truog (ed.), Mineral nutrition of plants. Univ. Wisconsin Press, Madison.

Truog, E., and J. R. Jones. 1938. Fate of soluble potash applied. Ind. Eng. Chem. 30:882–885.

Tukey, H. B., Jr. 1970. The leaching of substances from plants. Annu. Rev. Plant Physiol. 21:305–324.

Tukey, H. B., Jr., and H. B. Tukey. 1962. The loss of organic and inorganic materials by leaching from leaves and other above-ground plant parts, pp. 289–302. In Radioisotopes in soil-plant nutrition studies. Int. At. Energy Agency, Vienna.

Tukey, H. R., Jr., H. B. Tukey, and S. H. Wittwer. 1958. Loss of nutrients by foliar leaching as determined by radioisotopes. Proc. Amer. Soc. Hort. Sci. 71:496–505.

Ulrich, A., and K. Ohki. 1966. Potassium, pp. 362–393. In H. D. Chapman (ed.) Diagnostic criteria for plants and soils. Div. Agr. Sci., Univ. California, Berkeley.

Vanselow, A. P., and N. P. Datta. 1949. Molybdenum deficiency of the citrus plant. Soil Sci. 67:363–375.

Vlamis, J., and H. Jenny. 1948. Calcium deficiency in serpentine soils as revealed by adsorbent technique. Science 107:549.

Volk, B., and C. McAuliffe. 1954. Factors affecting the foliar absorption of N^{15}-labeled urea by tobacco. Proc. Soil Sci. Soc. Amer. 18:308–312.

Wadleigh, C. H., and L. A. Richards. 1951. Soil moisture and the mineral nutrition of plants, pp. 411–450. *In* E. Truog (ed.), Mineral nutrition of plants. Univ. Wisconsin Press, Madison.

Wadleigh, C. H., and J. W. Shive. 1939. Organic acid content of corn plants as influenced by pH of substrate and form of nitrogen supplied. Amer. J. Bot. 26:244–248.

Wallace, A. 1956. Symposium on the use of metal chelates in plant nutrition. Introduction: Metal chelates in agriculture. National Press, Palo Alto, Calif.

Wallace, A., C. P. North, R. T. Mueller, L. M. Shannon, and N. Hemaidan. 1955. Behavior of chelating agents in plants. Proc. Amer. Soc. Hort. Sci. 65:9–16.

Wallace, T. 1928. Investigations on chlorosis of fruit trees. II. The composition of leaves, bark, and wood of current season's shoots in cases of lime-induced chlorosis. J. Pomol. Hort. Sci. 7:172–183.

Wallihan, E. F. 1966. Iron, pp. 203–212. *In* H. D. Chapman (ed.). Diagnostic criteria for plants and soils. Div. Agr. Sci., Univ. California, Berkeley.

Weinstein, L. H., and W. R. Robbins. 1955. The effect of different iron and manganese nutrient levels on the catalase and cytochrome oxidase activities of green and albino sunflower leaf tissue. Plant Physiol. 30:27–32.

Weiss, M. G. 1943. Inheritance and physiology of efficiency in iron utilization in soybeans. Genetics 28:253–268.

Winters, E., and W. L. Parks. 1956. Zinc deficiency of corn in Tennessee. Commercial Fertilizers and Plant Food Ind. 92(3):36, 59.

Yamada, Y., M. J. Bukovac, and S. H. Wittwer. 1964a. Ion binding by surfaces of isolated cuticular membranes. Plant Physiol. 39:978–982.

Yamada, Y., S. H. Wittwer, and M. J. Bukovac. 1964b. Penetration of ions through isolated cuticles. Plant Physiol. 39:28–32.

Yamada, Y., S. H. Wittwer, and M. J. Bukovac. 1965. Penetration of organic compounds through isolated cuticular membranes with special reference to C^{14} urea. Plant Physiol. 40:170–175.

Yamada, Y., H. P. Rasmussen, M. J. Bukovac, and S. H. Wittwer. 1966. Binding sites for inorganic ions and urea on isolated cuticular membrane surfaces. Amer. J. Bot. 53:170–172.

Zukel, J. W., A. E. Smith, G. M. Stone, and M. E. Davies. 1956. Effects of some factors on rate of absorption of maleic hydrazide. Plant Physiol. 31(Suppl.):xxi.

CHAPTER 15

Excesses of Salts

SOIL SALINITY

In the late 1930s the United States Regional Salinity Laboratory was organized at Riverside, California, to study responses of plants and soils to salinity, and to develop better methods for combating the effects of salinity. Much of the research in salinity, particularly from this laboratory, has been discussed in review articles (Magistad, 1945; Hayward and Magistad, 1946; Hayward and Wadleigh, 1949; Hayward and Bernstein, 1958) and in a book entitled, *Diagnosis and Improvement of Saline and Alkali Soils* (Richards, 1954). Other reviews of salinity have been provided by Strogonov (1964) and Boyko (1966, 1968).

Diagnostic Criteria for Plants and Soils, edited by Chapman (1966a), contains a wealth of information on the nature, diagnosis, and reclamation of alkali and saline soils. Readers interested in the various aspects of salinity should consult this work, and Chapman's (1966d) chapter on alkali and saline soils.

Characterization

In regions of limited rainfall in which irrigation is practiced, salts tend to accumulate—sometimes to such a concentration that the soil is barren. Lesser degrees of salt accumulation impair plant growth slightly to severely, depending on the salt concentration. Although it is difficult to draw a line between soils that are saline and those that are not, nevertheless some general guidelines help to characterize soils with and without appreciable quantities of soluble salts.

Definitions

Saline Soil. All soils, even in the humid eastern portion of the United States, contain soluble salts, but a saline soil is one that contains an excess of salts (Chapman, 1966d).

A saline soil is "a nonalkali soil containing soluble salts in such quantities that they interfere with the growth of most crop plants. The electrical conductivity of the saturation extract is greater than 4 millimhos per cm. at 25°C, and the exchangeable sodium percentage is less than 15. The pH reading of the saturated soil is usually less than 8.5. Most of the soils formerly known as white alkali would fall under this category" (Richards, 1954).

In general, it is agreed that a soil is saline if it contains in excess of 0.1% soluble salt (0.1% equals 2000 lb of salt in the 0- to 6-in. layer of soil) (Magistad, 1945; Chapman, 1966d). This concentration of salt is sufficient to reduce appreciably the growth and yield of most crops. However, the growth inhibitory effect of this concentration can be altered by various factors. If a fairly high moisture level is maintained in the soil most of the time, the concentration of salts will be reduced and growth not as seriously affected as if the soil were permitted to become quite dry (Ayers et al., 1943). Distribution of salts in the soil profile can also affect the extent to which crop growth is inhibited. If there is a relatively salt-free portion of soil in the root zone, plant roots grow primarily in the low-salt portion and are not much affected by salts at greater depth. The nature of the salts that contribute to the salinity can also affect the response of a crop, since certain salts, such as Na^+ or Mg^{2+} salts, are more toxic than, for example, Ca^{2+} salts. Last, some crops tolerate salinity better than others.

Alkali Soil. An alkali soil is one "that contains sufficient exchangeable sodium to interfere with the growth of most plants, either with or without appreciable quantities of soluble salts" (Richards, 1954).

Whereas a Ca-saturated soil, for example, tends to be well aggregated, have lots of air, and be readily permeable to water, a Na-saturated soil has the opposite characteristics and is of very poor physical structure.

Saline-Alkali Soil. A saline-alkali soil is one "containing sufficient exchangeable sodium to interfere with the growth of most crop plants and containing appreciable quantities of soluble salts. The exchangeable sodium percentage is greater than 15, and the electrical conductivity of the saturation extract is greater than 4 millimhos per cm. at 25°C. The pH reading of the saturated soil is usually less than 8.5. Soils of the so-called black alkali type would come under this classification" (Richards, 1954).

Nonsaline Alkali Soil. A nonsaline-alkali soil is one "that contains sufficient exchangeable sodium to interfere with the growth of most crop plants and does not contain appreciable quantities of soluble salts. The exchangeable sodium percentage is greater than 15, and the electrical conductivity of the saturation extract is less than 4 millimhos per cm. at 25 °C. The pH reading of the saturated soil paste is usually greater than 8.5" (Richards, 1954).

Location and Cause of Salinity

Salinity, alkalinity, or a combination of the two, occur primarily in the semiarid western United States where there is insufficient rainfall and, in many instances, irrigation water is applied. In the 19 western states (Magistad and Christiansen, 1944), there are approximately 20,000,000 acres of irrigated land. In much of this land there is some degree of salinity ranging from slight to severe with regard to reduction of crop growth and yield.

The soils contain some soluble salts and the river or well waters also contain salt. When applied irrigation water is used by plants, and also when it evaporates, the salts it carried are left behind. When there is limited rainfall and an insufficient application of irrigation water to leach accumulated salts away, the soil becomes more saline with time. In some cases the water table is too high for applied irrigation water to do an effective job of leaching away the salts no matter how much water is applied. In still other cases rate of percolation of water through the soil is too slow to achieve an adequate leaching. Good soil drainage is generally the key to alleviating soil salinity. Unless good drainage can be provided, the salinity condition can only become increasingly worse.

Chapman (1966d) reports that saline soils owe their origin to one or a combination of the following: (1) capillary rise of water (carrying dissolved salt) from water tables near the soil surface; (2) salt accumulation from irrigation water, particularly when subsoil leaching is insufficient to remove the salt; (3) prevailing winds from the ocean which carry fine spray short distances inland; (4) evaporation of inland seas and lakes; (5) inundation of land by seawater; and (6) inland basins lacking a drainage outlet and subject to periodic flooding and evaporation.

Insidious Nature of Salinity

Unless the degree of salinity is such as to result in extremely poor crop growth and production, the significance of salinity is often overlooked. There are at least two reasons for the lack of appreciation of the significance of salinity. First, there is no available basis of comparison, since crop growth on the same soil type, without accumulated salts, is not available for comparison. Second, the build-up of salinity is an insidious development which is of comparatively little significance initially but which slowly and continuously takes on ever-increasing significance. Ultimately, salinity can build to a point where crops can no longer be grown successfully on the land.

Ions Causing Salinity

Sodium. Na^+ is generally the prominent cationic component of the soil solution in saline soils (Lunt, 1966). One of the major effects of Na^+ is on soil structure—the effect being primarily a dispersion of soil colloids. Associated with this change in the aggregation of soil particles is a decrease in soil aeration. Incidentally, poor aeration appears to be associated with increased translocation of Na^+ to the tops of plants, since Na^+ exclusion (to the tops) is dependent on adequate aeration around the roots (Lunt, 1966).

Calcium. When Ca^{2+} is associated with SO_4^{2-} in a salinity situation, the concentration of Ca^{2+} may not be very high owing to the relatively low solubility of $CaSO_4$, that is, approximately 25–30 meq/liter. However, when Ca^{2+} is associated with Cl^-, its concentration can be very high. There are few if any specific symptoms associated with excesses of Ca^{2+} (Chapman, 1966b). Symptoms are generally caused by the associated anion, for example, Cl^- or SO_4^{2-}. High levels of Ca^{2+} in a nutrient solution were lethal to orchard grass when the associated anion was either Cl^- or NO_3^- (Wadleigh et al., 1951).

Ca^{2+} excess in soils is usually associated with excesses of soluble salts (e.g., $CaCl_2$) or $CaCO_3$, as observed by Chapman (1966b). He noted that excess lime can be eliminated by $(NH_4)_2SO_4$ or other acidifying agents only when it is present in relatively low concentrations in the soil. When excess soluble salts of Ca^{2+} are present, correction consists of leaching the salt out of the soil.

Magnesium. Inasmuch as $MgSO_4$ and $MgCl_2$ are highly soluble, Mg^{2+} may exist in a high concentration whether in association with SO_4^{2-} or Cl^- ions. Mg^{2+} occurs in toxic concentrations in certain soils and certain locations, particularly in the semiarid regions of the United States. Mg^{2+} applications may also result in Mg^{2+} toxicity in soils low in Ca^{2+} unless supplemental Ca^{2+} is applied with the Mg^{2+}, as reported by Embleton (1966). He reported one soil in which more than 90% of the cation exchange capacity was satisfied by Mg^{2+}. The soil was almost completely unproductive.

There has been far more agricultural interest in Mg^{2+} deficiency than in Mg^{2+} toxicity. With rare exceptions specific symptoms indicative of Mg^{2+} toxicity have not been observed (Embleton, 1966).

Potassium. Although K^+ rarely contributes significantly to a salinity condition, it may do so if the parent rock was unusually high in K^+ or underground water moved through K-bearing salts from a higher elevation on its way to a given soil.

Almost without exception, K^+ excess occurs as the result of excessive K^+ applications, according to Ulrich and Ohki (1966). They note that one explanation for the paucity of cases of K^+ toxicity is the fact that K^+ is fixed in exchangeable and nonexchangeable forms in soils. Thus K^+ is not often excessively absorbed by plants.

Chloride. Cl^- salts are frequently involved, either partly or almost wholly, in salinity conditions. Cl^- salts may be, and often are, associated with accumulations of SO_4^{2-}, HCO_3^-, and CO_3^{2-} ions (Eaton, 1966a).

According to Eaton (1966a), symptoms of Cl^- excess include burning and firing of leaf tips and margins, bronzing, premature yellowing and abscission of leaves and, less frequently, chlorosis. He describes Cl^- toxicity symptoms for various crops and indicates the concentrations of Cl^- in plants associated with toxicity. For most plants the internal concentration of Cl^- closely reflects the external concentration. Beets, barley, flax, cotton, and tomatoes are in the high-tolerance group with regard to Cl^- (Eaton, 1966a).

Sulfate. As indicated earlier, the SO_4^{2-} concentration may be relatively low if the anion is associated with a high concentration of Ca^{2+}, since Ca^{2+} and SO_4^{2-} tend to precipitate as $CaSO_4$ with only 25–30 meq/liter of SO_4^{2-} in solution (Eaton, 1966b). In association with Na^+ or Mg^{2+}, the SO_4^{2-} concentration in the soil solution may be quite high.

There are few instances of specific symptoms indicative of SO_4^{2-} excess, and in fact the symptoms sometimes resemble those resulting from an excess of B, reported by Eaton (1966b). He noted that if drainage of the soil is possible soluble SO_4^{2-} salts can usually be leached out.

Although plants ordinarily absorb S as SO_4^{2-}, SO_2 readily enters leaves and, in areas of high SO_2 concentration, rather specific leaf symptoms develop during SO_2 toxicity (Eaton, 1966b).

Carbonate and Bicarbonate. Depending on pH, only HCO_3^- may be present, only CO_3^{2-} may be present, or there may be various proportions of these anions (Pratt, 1966). When CO_3^{2-} alone is present, pH is high, organic matter is brought into solution to seep to and accumulate in the surface of the soil, and a condition known as "black alklai" results.

Absence of CO_3^{2-} or HCO_3^- in the soil has no adverse effect on plants, but phytotoxicity results when either of these ions is present in high concentration (Pratt, 1966). Except for highly acid soils, HCO_3^- is present in soil, but CO_3^{2-} is present in measurable concentrations only in soils with a pH approximately 8.5 or higher.

HCO_3^- has been associated with Fe chlorosis in many plants. The actual amount of insoluble carbonates in calcareous soils is not as important as its presence (Pratt, 1966). High-lime Fe chlorosis is associated with calcareous soils, but some of these soils do not produce high-lime Fe chlorosis. Thus, lime concentration per se is not a clear-cut diagnostic index, according to Pratt (1966). He stated that Na soils containing high lime, that is, containing $CaCO_3$, can be improved only by acidification to dissolve the $CaCO_3$, so that Ca^{2+} can replace Na^+ on the exchange complex. He added further that the

HCO_3^- ion may not readily enter root cells, but that it would not need to enter in order to produce a high HCO_3^- concentration inside cells. Inasmuch as HCO_3^- is produced by respiration, high external concentrations of HCO_3^- could cause an accumulation of metabolically produced HCO_3^- inside cells.

Growth of beets was reduced less by HCO_3^- than was bean growth (Brown and Wadleigh, 1955). Comparison of cation accumulation in bean and beet leaves showed that treatment and chlorosis were *not* correlated with any particular cation or with the K/Ca ratio in *both* species, but rather with monovalent cations or the (Na + K)/(Ca + Mg) ratio. When given to bean plants, $NaHCO_3$ resulted in lowered Fe activity and Ca^{2+} concentration in leaves and enhanced K^+ concentration (Wadleigh and Brown, 1952). Along with accumulation of K^+, citric acid accumulates in leaves showing HCO_3^--induced chlorosis. It was concluded that the primary effect of the HCO_3^- ion is brought about through its effect on protoplasmic consistency of the absorbing cells of roots, so that bean plants accumulate relatively more monovalent cations and relatively less divalent cations.

Nitrate. Toxicity from excess NO_3^- usually occurs as a result of overfertilization, and this is a rather rare event (Jones, 1966).

There is little evidence of ground water pollution by NO_3^-, even in aquifers directly beneath large cattle feedlots (Anonymous, 1970). The manure pack seems to provide an effective barrier to water movement. Only twice during a 2-yr sampling period did the NO_3^- concentration rise as high as 10 ppm, far below the safety level of 45 ppm set by the United States Public Health Service. It was noted that even though scientists may find 3–17 ppm of N in tile drain water, sampling was not systematically conducted before farmers started to use the present higher applications of fertilizer. Therefore there is no way of knowing whether or not there has actually been an increase in NO_3^- in ground waters. Fertilizers probably contribute to the NO_3^- concentration of groundwater, but it appears that it will be some years before the concentration will be high enough to be of real concern.

In a comprehensive review of agricultural pollution, Wadleigh (1968) stated that conclusive evidence is lacking that chemical fertilization of fields results in high NO_3^- levels in well waters. He added, however, that natural nitrification processes in soils and nitrification of sewage effluent and animal wastes seem to be major contributors when NO_3^- is found in groundwater. He concluded that much more definitive information is needed.

Other Ions Sometimes in Excess but Not Usually Contributing to Salinity

Ammonium. In tomato plants accumulation of toxic concentrations of NH_4^+ ion resulted in morphological modification of leaf chloroplasts (Puritch and Barker, 1967). In later stages the lamellae of the grana swelled and some disappeared.

When cucumber plants were cultured with 20 and 200 mg NH_4 nitrogen per liter for 5 days, incorporation of photosynthesized ^{14}C (in the leaves) not only into starch but also into other higher polymers was suppressed by NH_4^+ toxicity (Matsumoto et al., 1969). Incorporation into the 80% ethanol fraction was higher in the 200-mg than in the 20-mg NH_4 nitrogen per liter treatment. NH_4^+ toxicity resulted in an increase in uridine diphosphate glucose (UDPG) and a decrease in UDPX (uridine compound containing sugar or sugar derivative). Glucose-6-phosphate was lower in NH_4^+-injured than in normal plants, whereas glucose-1-phosphate and fructose-6-phosphate were higher in injured plants.

Arsenic. As is nonessential for plants and usually the concentration of As in plants is exceedingly low (Liebig, 1966). It was noted, however, that As may be present in phytotoxic concentrations in soils following prolonged use of arsenical sprays. Phytotoxically high concentrations of As in the soil may be amerliorated or negated by the application of various compounds, for example, $ZnSO_4$ (Thompson and Batjer, 1950).

Boron. Inasmuch as B can be toxic to many crops when present as a few parts per million, B does not exist in high enough concentrations to cause salinity or to contribute to a salinity condition (Bradford, 1966a).

A strain of *Penicillium notatum* was isolated from a saturated Ca acetate solution, which showed a high tolerance to B in several states of oxidation (Roberts and Siegel, 1967). It was concluded that B tolerance is generally related to high performance under salt stress.

Toxic levels of B for peanut *(Arachis hypogea* L. var. TMV-2) resulted in reduced activity levels of cytochrome oxidase and catalase with less effect on peroxidase (Gopal, 1969). Along with these effects on enzymes, high B interfered with the ability of Fe to complex with proteins. That is, B complexed with proteins and prevented Fe from doing so. As a result, concentrations of the Fe–porphyrin enzymes cytochrome oxidase and catalase were lowered.

Peanut seeds treated with 25-ppm B solution germinated more quickly and grew more than those placed in deionized water for the first 12 days (Gopal and Rao, 1969). Between 12 and 21 days, the B-treated seedlings fell behind those growing in deionized water as B toxicity finally gained ascendency.

The early stages of B toxicity generally involve leaf-tip yellowing, but other injuries can induce this symptom, hence the need for plant analyses (Bradford, 1966a). Also, beneficial and injurious effects of B in plants may overlap (Eaton, 1944), so that, again, it is impossible to place much reliance on symptoms in early stages of B toxicity. Moderate-to-acute stages of B toxicity are reasonably specific with regard to symptoms, at least for experts in diagnosis.

Bradford (1966a) describes specific B toxicity symptoms for many crops and also lists the concentrations of B associated with toxicity. In addition, he lists plants that are B-sensitive, B-semitolerant, and B-tolerant. He notes that if excess B has come from irrigation water the best solution is to find another source of water if possible. Leaching and moderate liming may also prove to be helpful measures.

Copper. Instances of toxic concentrations of Cu^{2+} have been reported in a few isolated instances. Usually Cu^{2+} excess occurs in highly leached, light, sandy soils, in very sandy acid soils, or in soils that have accumulated excess Cu^{2+} as a result of Cu^{2+} sprays on plants (Reuther and Labanauskas, 1966).

In *Chlorella pyrenoidosa*, Cu^{2+} binds to cytoplasmic membranes, and cells are then unable to divide (Nielsen et al., 1969). Cells also become saturated with assimilation products which have a depressant effect on the rate of photosynthesis. There was a significant deleterious effect of Cu^{2+} when the concentration was as low as 1.0 μg/liter.

Another study indicated that excess Cu^{2+} may alter membrane integrity in *Chlorella,* and that this alteration could be the primary focal point of Cu^{2+} action. Excess Cu^{2+} supplied to nongrowing cells of a normal, green *Chlorella* caused a reduction in total pigments and a blue shift of chlorophyll absorption—concurrent with inhibition of photosynthesis (Gross et al., 1970). Chlorophyll-less yellow and white mutant strains of the same alga showed a rise in nonspecific absorption (i.e., change in light scatter) 5–10 min after the addition of $CuSO_4$; concomitantly there was a lowering of packed cell volume and a rise in respiration. Glutathione prevented all Cu^{2+}-induced changes, whereas $MnCl_2$ protected only partially.

Fluoride. Although widely distributed in nature and essential for animals, F^- is nonessential for plants, as reported by Brewer (1966a). He noted that toxicity from F_2 may occur as the result of industrially polluted air or of acid soils of moderate F^- concentration but low in available Ca^{2+}. Marginal necrosis was reported as the most common symptom of F^- injury. With low concentrations of F^-, some plants show interveinal chlorosis. Inasmuch as interveinal chlorosis and marginal necrosis can result from other causes, visual diagnosis must usually be confirmed by chemical analyses of leaves or other plant parts (Brewer, 1966a).

F^- accumulates primarily in chloroplasts (Chang and Thompson, 1966). The effect of F^- injury on glucose breakdown was studied by means of C_6/C_1 ratios in normal and F^--injured leaves of *Polygonum orientale* L. and *Chenopodium murale* L. (Ross et al., 1962). Decreased ratios were observed in damaged leaves, indicating a relative increase in importance of the pentose phosphate pathway. The decreased ratios were believed to result from an inhibition of glycolysis, probably principally affecting enolase. In fact, enolase has been reported to be inhibited by very low F^- concentrations, especially in the presence of phosphate and Mg^{2+} ions (Miller, 1958).

Iron. Depending primarily on the parent material from which soil was formed, Fe^{3+} may be present in relatively high concentration in soils (Wallihan, 1966). Fe toxicity has not been much in evidence under natural conditions, but when it occurs the symptoms usually first appear as necrotic spots on leaves (Wallihan, 1966).

The concentration of Fe^{3+} in a soil solution is markedly affected by pH, since pH values of 7 or higher drastically reduce the availability of Fe to plants because of the precipitation of Fe in the soil.

Lead. Pb is nonessential for plants, and the element ordinarily occurs in very low concentrations in plants and soils. Brewer (1966b) noted that, as in the case of As, Pb^{2+} may occur in soils in high concentrations as a result of repeated use of Pb-containing sprays. Pb has been used over the years primarily as Pb arsenate and, in general, the phytotoxicity is believed to come from the arsenic rather than the Pb.

Root growth of sheep fescue *(Festuca ovina)* in solution culture was measurably retarded by 10 ppm of Pb^{2+}, markedly reduced by 30 ppm, and stopped by 100 ppm (Wilkins, 1957).

Lithium. Li is widely distributed in nature, is nonessential for plants, and generally occurs in plants at concentrations of 1.0 ppm or lower (Bradford, 1966b).

Manganese. Depending on the parent material from which a soil was formed, Mn^{2+} may occur in toxic concentrations in the soil (Labanauskas, 1966). Mn toxicity sometimes occurs in acid (below pH 5) soils (Hale and Heintze, 1946), and the Mn^{2+} concentration of vegetation may be 10 to 100 times higher than normal. Mn excess and toxicity can often be overcome by liming (Labanauskas, 1966). In certain soils in Hawaii, the concentration of Mn^{2+} is sufficiently high to interfere with the absorption of Fe^{3+} even in the high-Fe "red" soils.

Excessive internal concentrations of Mn^{2+} are reported to cause a physiological disease of apples called internal bark necrosis (Berg et al., 1958).

Mn toxicity symptoms may be an expression of auxin deficiency caused in turn by an increased activity of IAA oxidase along with a subdued activity of IAA oxidase inhibitor (Morgan and Hall, 1963; Morgan et al., 1966). In the absence of or during the inactivation of IAA oxidase inhibitor by high Mn^{2+}, oxidation of IAA by endogenous peroxidases is accelerated (Stonier et al., 1968).

Barley plants were considerably less susceptible to Mn^{2+} toxicity when Si was added to the synthetic culture medium (Williams and Vlamis, 1957). In later work Ca^{2+}, Mg^{2+}, and NH_4^+ were shown to be more effective than

K^+ and Na^+ in repressing Mn^{2+} uptake by barley (Vlamis and Williams, 1962).

Molybdenum. Mo excess or toxicity is rarely observed under field conditions. In those instances in which Mo excess has occurred, it has resulted in molybdenosis, teart disease, or peat scours (Johnson, 1966). Excesses of Mo have caused poisoning of cattle when pasture grasses contained 20 ppm or more of Mo (Ferguson et al., 1943; Barshad, 1948).

Nickel. A few parts per million of Ni^{2+} in the culture solution are toxic for many kinds of plants, as reported by Vanselow (1966). He noted that there are soils in scattered regions of the United States that have sufficient concentrations of Ni^{2+} to be phytotoxic. He reported that soils normally contain approximately 100 ppm of Ni^{2+}; soils derived from ultrabasic igneous rocks (serpentine) may contain as much as 5000 ppm of Ni^{2+}, definitely a phytotoxic concentration.

Phosphorus. P toxicity results from overfertilization with P-containing fertilizers, as observed by Bingham (1966). He noted that excessive concentrations of P in the soil may upset Cu^{2+}, Zn^{2+}, and Fe^{3+} metabolism in plants. Further, citrus species are especially sensitive to P excess and show both Cu^{2+}- and Zn^{2+}-deficiency symptoms. The possibility of Cu^{2+}, Zn^{2+}, or Fe^{3+} deficiencies must be considered when high rates of P are applied, for example, following the high rates of P fertilization required to overcome P deficiency in trees.

Selenium. For many years the chief interest in Se centered on its toxicity to animals. With most elements present in excessive concentrations in plants, visual symptoms appear, but this is not the case with Se (or, incidentally, with Mo or NO_3 at concentrations toxic to animals) (Ganje, 1966); 5 ppm of Se is potentially dangerous to animals. The quantity of Se in plants varies from traces to as high as 15,000 ppm on a dry weight basis (Ganje, 1966).

When soils are formed from seleniferous parent material, appreciable concentrations of Se may be present (Ganje, 1966). In *Astragalus* roots absorption of selenate and selenite is favored by $CaCl_2$ (Ulrich and Shrift, 1968). There is little that can be done to seleniferous soils or about Se-containing plants. Prevention of overgrazing may be helpful but, in general, it is best to keep livestock away from Se-containing plants (Ganje, 1966).

Most *Astragalus* species are nonselenium accumulators (Shrift, 1969). The preponderant soluble Se compound in accumulators is Se-methylselenocysteine ($CH_3SeCH_2CHNH_2COOH$); in nonaccumulators it is Se-methylselenomethionine [$(CH_3)_2SeCH_2CH_2CHNH_2COOH$]. Although for some time it was widely believed that Se compounds interfered with S metabolism by a typical antimetabolite action, this belief is no longer tenable. S and Se are not always metabolically equivalent.

Possibly, accumulator species have evolved a detoxification mechanism whereby Se that is absorbed is metabolized into innocuous compounds such as Se-methylselenocysteine and selenocystathionine—nonprotein amino acids known to occur in large amounts in these plants (Peterson and Butler, 1967). The detoxification mechanism is considered absent in nonaccumulator species which presumably metabolize Se into analogs of the protein S amino acids methionine and cysteine, so that proteins are produced with altered structures and impaired functions.

Callus cultures from five Se accumulator and three nonaccumulator species of *Astragalus* have been obtained (Ziebur and Shrift, 1971). Those from accumulator species characteristically retained their tolerance to high concentrations of selenate and selenite, whereas calluses derived from nonaccumulator species were markedly inhibited by these two forms of Se. Cultures of *Astragalus* callus showed that some physiological characteristics of the intact plant were retained by the callus, whereas others may have been lost. It was concluded that neither the detoxification hypothesis (Peterson and Butler, 1967) nor callus culture experiments suggest any reason for the restriction of accumulator species to seleniferous soils.

Zinc. Certain acid peats, and also soils derived from rocks high in Zn^{2+}, may contain toxic concentrations of Zn^{2+} (Chapman, 1966c). It was reported that certain sandy soils in Florida have accumulated toxic concentrations of Zn^{2+} from the widespread use of Zn^{2+} in foliar sprays and fertilizers. Numerous workers have shown that an excess of Zn^{2+} may cause Fe deficiency and Fe chlorosis. In many cases liming of an acid soil to near neutrality has overcome Zn^{2+} excess in soils (Chapman, 1966c).

Saginaw navy beans *(Phaseolus vulgaris* L.) grew normally with 5 ppm Zn^{2+}, while growth of the Sanilac variety was extremely reduced by this high concentration (Polson and Adams, 1970).

Irrigation Waters

Total salt and water quality appraisal is authoritatively discussed in detail by Eaton (1966c). He clearly indicates some of the complexities involved in evaluating total salt and water quality, and thus his contribution to the subject is vital to all concerned with this aspect of the salinity problem. Over the years the salts in irrigation water contribute far more salt to a salinity condition than do the initial soluble salts of the soil and the salts formed by breakdown of the soil (Eaton, 1966c). In Fig. 15.1 the compositions of certain river waters are shown. It can be seen that some of these waters contribute large amounts of salt.

Inasmuch as water for a given irrigation district is withdrawn upstream from the area, and the drainage waters are returned downstream, the salt concentration of river water increases from its source to the mouth of the river. In Fig.

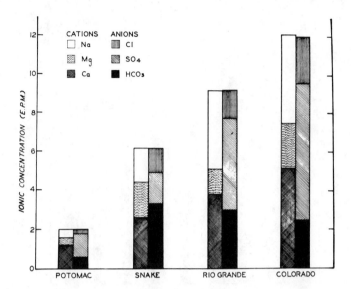

Fig. 15.1
Weighted mean salt content of representative rivers in eastern and western United States. (E.P.M., equivalents per million.) From O. C. Magistad (1941) Ion and plant relationships in western arid soils, Soil Sci., 51:461–471. © 1941, The Williams and Wilkins Co., Baltimore, Md.

15.2 note how the concentration of salts increases in Rio Grande River water at the various locations along the course of the river.

Irrigation waters may be roughly divided into three categories depending on electrical conductivity $(K \times 10^5$ at 25 °C), parts per million of B, percentage of Na^+, and milliequivalents of Cl per liter (See Table 15.1).

Methods of Measuring Salinity

Hydrometer. For irrigation waters or drainage waters, a hydrometer may be used to measure the overall salt concentration. Unless one salt, such as NaCl, is known to predominate, the measurement cannot readily be converted into specific units of concentration. The hydrometric measurement is fast, however, and can be used to monitor the concentration and changes in concentration of irrigation or drainage waters.

Conductivity. This measurement is based on the simple fact that capacity of a solution to conduct an electric current is a function of salt concentration. It is a rapid and highly reproducible, accurate measurement.

By use of specially designed electrodes, it is also possible to determine the conductivity of soil—usually as a soil paste in a specially designed "conductivity cup."

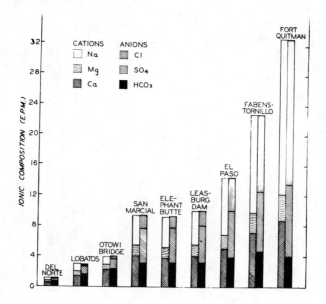

Fig. 15.2

Weighted mean salt content of Rio Grande River at different gauging stations. From O. C. Magistad (1941) Ion and plant relationships in western arid soils, Soil Sci., 51:461–471. © 1941, The Williams and Wilkins Co., Baltimore, Md.

Analytical. The hydrometric and conductometric measurements, while rapid, provide no information on the qualitative nature of the salts in a solution. They measure only the quantitative aspect of salts in solution.

Although they are inevitably much slower, appropriate methods can be used in the analysis of solutions or water extracts of soils. Considerable time is re-

Table 15.1

Three Broad Classes of Irrigation Water, Based on Electrical Conductivity, Na Percentage, and Concentrations of B and Cl^- [a]

Measurement	Class I Excellent to good	Class II Good to injurious	Class III Injurious to unsatisfactory
Electrical conductivity ($K \times 10^5$ at 25° C)	Less than 100	100–300	Over 300
B (ppm)	0.5	0.5–2.0	Over 2
Na (%)	60[b]	60–75	Over 75
Cl (meq/liter)	5	5–10	Over 10

[a] Richards (1954).

[b] Not much Na^+ goes onto the exchange complex until the Na^+ percentage is over 60%.

quired for a more-or-less complete analysis including Ca^{2+}, Mg^{2+}, Na^+, K^+, Cl^-, SO_4^{2-}, PO_4^{3-}, NO_3^-, HCO_3^-, and CO_3^{2-}. Sometimes only certain of these constituents are of interest, for example, the concentration of Na^+ or of Cl^-.

More rapid methods have been developed over the years for determining some of the constituents just mentioned. The advent of flame emission and absorption spectrophotometry greatly facilitated the rapidity with which Ca^{2+}, Mg^{2+}, K^+, and Na^+ could be determined. With flame spectrophotometry it is possible to analyze 25 to 50 samples per hour for any one of the cations just mentioned.

If, when a complete analysis of a solution is made, the results are expressed on a milliequivalent per liter basis, the sum of the cations should equal the sum of the anions. Such a comparison provides a quick check of the overall accuracy of the analyses.

LITERATURE CITED

Anonymous. 1970. Nitrates in ground water? Farm J. March, p. 51.

Ayers, A. D., C. H. Wadleigh, and O. C. Magistad. 1943. The interrelationships of salt concentration and soil moisture content with the growth of beans. J. Amer. Soc. Agron. 35:796—810.

Barshad, I. 1948. Molybdenum content of pasture plants in relation to toxicity to cattle. Soil Sci. 66:187—195.

Berg, A., G. Clulo, and C. R. Orton. 1958. Internal bark necrosis of apple resulting from manganese toxicity. West Virginia Agr. Exp. Sta. Bull. 414T.

Bingham, F. T. 1966. Phosphorus, pp. 324—361. In H. D. Chapman (ed.) Diagnostic criteria for plants and soils. Div. Agr. Sci., Univ. California, Berkeley.

Boyko, H. (ed.). 1966. Salinity and aridity — New approaches to old problems. Monogr. Biol. Vol. XVI. W. Junk, The Hague. 408 p.

Boyko, H. (ed.). 1968. Saline irrigation for agriculture and forestry (Proceedings of the international symposium on plant growing with highly saline or sea-water, with and without desalination, held in Rome, 5—9 Sept., 1965). World Acad. Art and Sci. Vol. IV. W. Junk, The Hague. 323 p.

Bradford, G. R. 1966a. Boron, pp. 33—61. In H. D. Chapman (ed.). Diagnostic criteria for plants and soils. Div. Agr. Sci., Univ. California, Berkeley.

Bradford, G. R. 1966b. Lithium, pp. 218—224. In H. D. Chapman (ed.) Diagnostic criteria for plants and soils. Div. Agr. Sci., Univ. California, Berkeley.

Brewer, R. F. 1966a. Fluorine, pp. 180—196. In H. D. Chapman (ed.) Diagnostic criteria for plants and soils. Div. Agr. Sci., Univ. California, Berkeley.

Brewer, R. F. 1966b. Lead, pp. 213—217. In H. D. Chapman (ed.) Diagnostic criteria for plants and soils. Div. Agr. Sci., Univ. California, Berkeley.

Brown, J. W., and C. H. Wadleigh. 1955. Influence of sodium bicarbonate on the growth and chlorosis of garden beet. Bot. Gaz. 116:201—209.

Chang, C. W., and C. R. Thompson. 1966. Site of fluoride accumulation in navel orange leaves. Plant Physiol. 41:211—213.

Chapman, H. D. (ed.). 1966a. Diagnostic criteria for plants and soils. Div. Agr. Sci. Univ. California, Berkeley, Calif. 793 pp.

Chapman, H. D. 1966b. Calcium, pp. 65–92. *In* H. D. Chapman (ed.) Diagnostic criteria for plants and soils. Div. Agr. Sci., Univ. California, Berkeley.

Chapman, H. D. 1966c. Zinc, pp. 484–499. *In* H. D. Chapman (ed.) Diagnostic criteria for plants and soils. Div. Agr. Sci., Univ. California, Berkeley.

Chapman, H. D. 1966d. Alkali and saline soils, pp. 510–532. *In* H. D. Chapman (ed.) Diagnostic criteria for plants and soils. Div. Agr. Sci., Univ. California, Berkeley.

Eaton, F. M. 1944. Deficiency, toxicity, and accumulation of boron in plants. J. Agr. Res. 69:237–277.

Eaton, F. M. 1966a. Chlorine, pp. 98–135. *In* H. D. Chapman (ed.) Diagnostic criteria for plants and soils. Div. Agr. Sci., Univ. California, Berkeley.

Eaton, F. M. 1966b. Sulfur, pp. 444–475. *In* H. D. Chapman (ed.) Diagnostic criteria for plants and soils. Div. Agr. Sci., Univ. California, Berkeley.

Eaton, F. M. 1966c. Total salt and water quality appraisal, pp. 501–509. *In* H. D. Chapman (ed.). Diagnostic criteria for plants and soils. Div. Agr. Sci., Univ. California, Berkeley.

Embleton, T. W. 1966. Magnesium, pp. 225–263. *In* H. D. Chapman (ed.) Diagnostic criteria for plants and soil. Div. Agr. Sci., Univ. California, Berkeley.

Ferguson, W. S., A. H. Lewis, and S. J. Watson. 1943. The teart pastures of Somerset. I. The cause and cure of teartness. J. Agr. Sci. 33:44–51.

Ganje, T. J. 1966. Selenium, pp. 394–404. *In* H. D. Chapman (ed.). Diagnostic criteria for plants and soils. Div. Agr. Sci., Univ. California, Berkeley.

Gopal, N. H. 1969. Effect of boron toxicity on iron, heme enzymes and boron-protein complexes in groundnut. Indian J. Exp. Biol. 7:187–189.

Gopal, N. H., and I. M. Rao. 1969. Effect of excess boron supply on germination and seedling growth of groundnut (*Arachis hypogaea* Linn.). Plant and Soil 31:188–192.

Gross, R. E., P. Pugno, and W. M. Dugger. 1970. Observations on the mechanism of copper damage in *Chlorella*. Plant Physiol. 46:183–185.

Hale, J. B., and S. G. Heintze. 1946. Manganese toxicity affecting crops on acid soils. Nature 157:554.

Hayward, H. E., and L. Bernstein. 1958. Plant-growth relationships on salt-affected soils. Bot. Rev. 24:584–635.

Hayward, H. E., and O. C. Magistad. 1946. The salt problem in irrigation agriculture (Research at the United States Regional Salinity Laboratory). U. S. Dep. Agr. Misc. Pub. 607.

Hayward, H. E., and C. H. Wadleigh. 1949. Plant growth on saline and alkali soils. Advance. Agron. 1:1.

Johnson, C. M. 1966. Molybdenum, pp. 286–301. *In* H. D. Chapman (ed.) Diagnostic criteria for plants and soils. Div. Agr. Sci., Univ. California, Berkeley.

Jones, W. W. 1966. Nitrogen, pp. 310–323. *In* H. D. Chapman (ed.) Diagnostic criteria for plants and soils. Div. Agr. Sci., Univ. California, Berkeley.

Labanauskas, C. K. 1966. Manganese, pp. 264–285. *In* H. D. Chapman (ed.) Diagnostic criteria for plants and soils. Div. Agr. Sci., Univ. California, Berkeley.

Liebig, G. F., Jr. 1966. Arsenic, pp. 13–23. *In* H. D. Chapman (ed.) Diagnostic criteria for plants and soils. Div. Agr. Sci., Univ. California, Berkeley.

Lunt, O. R. 1966. Sodium, pp. 409–432. *In* H. D. Chapman (ed.) Diagnostic criteria for plants and soils. Div. Agr. Sci., Univ. California, Berkeley.

Magistad, O. C. 1941. Ion and plant relationships in western arid soils. Soil Sci. 51:461–471.

Magistad, O. C. 1945. Plant growth relations on saline and alkali soils. Bot. Rev. 11:181–230.

Magistad, O. C., and J. E. Christiansen. 1944. Saline soils — Their nature and management. U. S. Dep. Agr. Circ. 707.

Matsumoto, H., N. Wakiuchi, and E. Takahashi. 1969. The suppression of starch synthesis and the accumulation of uridine diphosphoglucose in cucumber leaves due to ammonium toxicity. Physiol. Plantarum 22:537–545.

Miller, G. W. 1958. Properties of enolase in extracts from pea seeds. Plant Physiol. 33: 199–206.

Morgan, P. W., and W. C. Hall. 1963. Indoleacetic acid oxidizing enzyme and inhibitors from light-grown cotton. Plant Physiol. 38:365–370.

Morgan, P. W., H. E. Joham, and J. V. Amin. 1966. Effect of manganese toxicity on the indoleacetic acid oxidase system of cotton. Plant Physiol. 41:718–724.

Nielsen, E. S., L. Kamp-Nielsen, and S. Wium-Andersen. 1969. The effect of deleterious concentrations of copper on the photosynthesis of *Chlorella pyrenoidosa*. Physiol. Plantarum 22:1121–1133.

Peterson, P. J., and G. W. Butler. 1967. Significance of selenocystathionine in an Australian selenium-accumulating plant, *Neptunia amplexicaulis*. Nature 312: 599–600.

Polson, D. E., and M. W. Adams. 1970. Differential response of navy beans (*Phaseolus vulgaris* L.) to zinc. I. Differential growth and elemental composition at excessive Zn levels. Agron. J. 62:557–560.

Pratt, P. F. 1966. Carbonate and bicarbonate, pp. 93–97. *In* H. D. Chapman (ed.) Diagnostic criteria for plants and soils. Div. Agr. Sci., Univ. California, Berkeley.

Puritch, G. S., and A. V. Barker. 1967. Structure and function of tomato leaf chloroplasts during ammonium toxicity. Plant Physiol. 42:1229–1238.

Reuther, W., and C. K. Labanauskas. 1966. Copper, pp. 157–179. *In* H. D. Chapman (ed.) Diagnostic criteria for plants and soils. Div. Agr. Sci., Univ. California, Berkeley.

Richards, L. A. (ed.). 1954. Diagnosis and improvement of saline and alkali soils. U. S. Dep. Agr. Handbook 60, Washington, D. C.

Roberts, K., and S. M. Siegel. 1967. Experimental microbiology of saturated salt solutions and other harsh environments. III. Growth of salt-tolerant *Penicillium notatum* in boron-rich media. Plant Physiol. 42:1215–1218.

Ross, C. W., H. H. Wiebe, and G. W. Miller. 1962. Effect of fluoride on glucose catabolism in plant leaves. Plant Physiol. 37:305–309.

Shrift, A. 1969. Aspects of selenium metabolism in higher plants. Annu. Rev. Plant Physiol. 20:475–494.

Stonier. T., F. Rodriguez-Tormes, and Y. Yoneda. 1968. Studies on auxin protectors. IV. The effect of manganese on auxin protector-I of the Japanese morning glory. Plant Physiol. 43:69–72.

Strogonov, B. P. 1964. Physiological basis of salt tolerance of plants (Transl. from Russian). Davey, New York.

Thompson, A. H., and L. P. Batjer. 1950. Effect of various soil treatments for correcting arsenic injury of peach trees. Soil Sci. 69:281–290.

Ulrich, A., and K. Ohki. 1966. Potassium, pp. 362–393. *In* H. D. Chapman (ed.) Diagnostic criteria for plants and soils. Div. Agr. Sci., Univ. California, Berkeley.

Ulrich, J. M., and A. Shrift. 1968. Selenium absorption by excised *Astragalus* roots. Plant Physiol. 43:14–20.

Vanselow, A. P. 1966. Nickel, pp. 302–309. *In* H. D. Chapman (ed.) Diagnostic criteria for plants and soils. Div. Agr. Sci., Univ. California, Berkeley.

Vlamis, J., and D. E. Williams. 1962. Ion competition in manganese uptake by barley plants. Plant Physiol. 37:650–655.

Wadleigh, C. H. 1968. Wastes in relation to agriculture and forestry. U. S. Dep. Agr. Misc. Pub. 1065.

Wadleigh, C. H., and J. W. Brown. 1952. The chemical status of bean plants afflicted with bicarbonate-induced chlorosis. Bot. Gaz. 113:373–392.

Wadleigh, C. H., H. G. Gauch, and M. Kolisch. 1951. Mineral composition of orchard grass grown on Pachappa loam salinized with various salts. Soil Sci. 72:275–282.

Wallihan, E. F. 1966. Iron, pp. 203–212. *In* H. D. Chapman (ed.) Diagnostic criteria for plants and soils. Div. Agr. Sci., Univ. California, Berkeley.

Wilkins, D. A. 1957. A technique for the measurement of lead tolerance in plants. Nature 180:37–38.

Williams, D. E., and J. Vlamis. 1957. The effect of silicon on yield and manganese-54 uptake and distribution in the leaves of barley plants grown in culture solutions. Plant Physiol. 32:404–408.

Ziebur, N. K., and A. Shrift. 1971. Response to selenium by callus cultures derived from *Astragalus* species. Plant Physiol. 47:545–550.

CHAPTER 16

Applied Aspects of Salinity

EFFECTS OF SALINITY

Effects on Plants

Appearance and Symptoms. The principal effect of salinity on plant growth is one of stunting, although plant growth may be completely inhibited (Fig. 16.1) if the salt concentration is high enough (Hayward and Magistad, 1946). Unless the salt concentration is high enough to result in a burning or firing of leaves, there may be no symptom other than stunting.

Toxic concentrations of B in soil generally produce quite specific leaf symptoms which are characteristic of an excess accumulation of B. When the effects of salinity on plants first began to receive special and major consideration at the United States Regional Salinity Laboratory, Riverside, California, it was hoped that research would establish definitive symptoms indicative of excesses of Ca^{2+}, of Na^+ of Cl^-, etc. For the most part, specific symptoms were not obtained.

The frequency and length of root hairs of citrus were reduced by high concentrations of Cl^- salts, and there were numerous anatomical alterations (Hayward and Blair, 1942). Growth of guayule was much more reduced by excess concentrations of Mg^{2+} than by excesses of Na^+ or Ca^{2+} (Wadleigh and Gauch, 1944). On a weight basis, Cl-salts were more toxic than SO_4-salts for peaches (Hayward and Long, 1942; Hayward, *et al.*, 1946) and red kidney beans (Gauch and Wadleigh, 1944), whereas flax (Hayward and Spurr, 1944b), guayule (Wadleigh and Gauch, 1944), and some grasses were more severely

Fig. 16.1
The effect of various concentrations of salt in a salinity area on the growth of barley. Where the salt concentration was too high, there was no germination or the plants failed to survive. (Hayward and Magistad, 1946.)

affected by SO_4- than by Cl-salts. Irish potatoes are especially sensitive to Cl^- (Masaewa, 1936). He noted that plants which are Cl-sensitive are also lime-sensitive—hence the parallelism between "chlorophobia" and "calciphobia." Application of Cl^- narrowed the K/Ca ratio, and this was regarded as the main detrimental effect on the plant. The growth of Dallis grass was very drastically reduced by 12 meq of $NaHCO_3$/liter in a nutrient solution, whereas the growth of Rhodes grass was as good in the presence of HCO_3 as in its absence (Fig. 16.2) (Gauch and Wadleigh, 1951).

Growth. When plants are grown in artificial culture and irrigated with nutrient solutions containing various increments of added salt, growth is closely related to the OP of the solutions (Gauch and Wadleigh, 1944, 1945). Results obtained with Dwarf Red Kidney bean plants serve to illustrate this relationship (Figs. 16.3 and 16.4).

In a study of the soil solution from saline soils, Magistad and Reitemeier (1943) noted a relationship between plant growth and OP of the soil solution, which was similar to and of the same order of magnitude as that observed in sand and solution culture experiments. Above 40 atm concentration soils were barren. Normal, fertile, irrigated soils had a soil solution concentration, at the

Fig. 16.2

Growth and appearance of Rhodes and Dallis grasses receiving pH 8.0 base nutrient solution or pH 8.0 base nutrient solution containing 12 meq $NaHCO_3$/liter. Growth of Rhodes grass was essentially unaffected by $NaHCO_3$, whereas that of Dallis grass was drastically reduced. From H. G. Gauch and C. H. Wadleigh (1951) Bot. Gaz., 112:259–271. The Univ. of Chicago Press and copyright by the University of Chicago—by permission.

WP, of 1.3–1.8 atm, conductance values $(K \times 10^5$ at 25°C) of 200–350, from 2000–4000 ppm and 30–50 meq of salts per liter. Incidentally, the OP of a soil solution may be closely approximated by the relationship: OP = 0.321 (EC $\times 10^3$) (Wadleigh et al., 1951). Incidentally, it is very helpful to remember the following relationships in order to convert one expression of concentration to another.

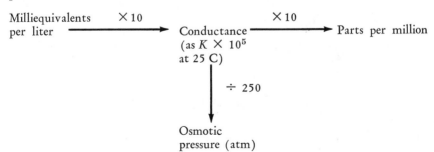

The above relationship is a generalized one which holds only for a mixture of salts having an average equivalent weight of 100. This is roughly true for most soil solutions and nutrient solutions.

Fig. 16.3
Dwarf Red Kidney bean plants, grown in solution culture, showing the influence of various increments of salt added to a base nutrient (B.N.) solution. Osmotic pressure of the B.N. solution was 0.5 atm; 1, 2, 3, and 4 indicate the atmospheres of osmotic pressure attributable to each added salt, that is, Na_2SO_4, NaCl, $CaCl_2$, $MgCl_2$, or $MgSO_4$. From H. G. Gauch and C. H. Wadleigh (1944) Bot. Gaz., 105:379–387. The Univ. of Chicago Press and copyright by the University of Chicago—by permission.

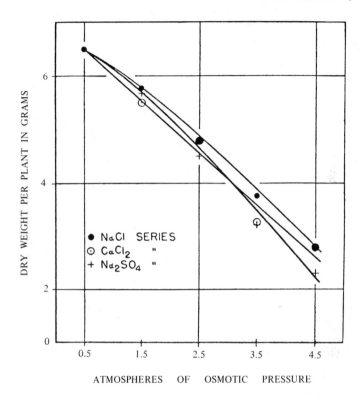

Fig. 16.4

Average dry weights of Dwarf Red Kidney bean plants as a function of the osmotic pressures of solutions to which NaCl, CaCl₂, or Na₂SO₄ was added. From H. G. Gauch and C. H. Wadleigh (1945) Soil Sci., 59:139–153. © 1945, The Williams and Wilkins Co., Baltimore, Md.

Wadleigh and Ayers (1945) studied the effect of moisture stress and of salinity on the growth of Dwarf Red Kidney bean plants. They concluded that growth reduction was the result of reduced hydration of protoplasmic proteins regardless of whether the stress was attributable to osmotic or physical retention of soil water.

In soils water is withheld from plants both physically by soil colloids and osmotically by dissolved salts in the soil solution. Plant growth is a function of the summation of these two water-withholding forces. If each component is expressed in atmospheres, they can be added together, and the two components constitute total soil moisture stress (TSMS) (Wadleigh et al. 1946; Wadleigh and Gauch, 1948)—a term synonymous with integrated moisture stress introduced by Wadleigh and Ayers (1945) and Wadleigh (1946). Plant growth is a function of TSMS, and it makes no difference what proportion of the stress is physical and what proportion is osmotic. This can be seen in re-

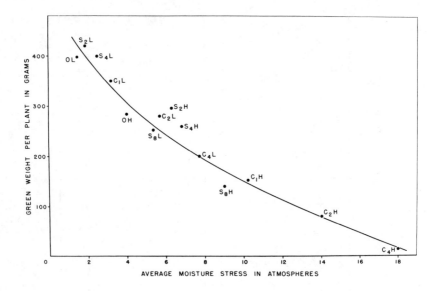

Fig. 16.5

Relationship between average soil moisture stress and growth of guayule plants. C, Added NaCl; S, added Na_2SO_4; subscripts 0, 2, 4, and 8 represent 0, 0.2, 0.4, and 0.8% added salt, respectively; L, low soil moisture tension; H, high soil moisture tension. (Wadleigh, et al., 1946.)

sults obtained with guayule (Fig. 16.5) (Wadleigh et al., 1946). Growth and yield of Dwarf Red Kidney bean plants was reduced by salinity or by high soil moisture stress (Ayers et al., 1943). For example, the yield of seed from the plants (Fig. 16.6) was about the same with medium soil moisture tension and 1000 ppm of added NaCl as it was with low soil moisture tension and 2000 ppm of added NaCl. This is another example of the response of plants to salinity, moisture tension, or both.

Osmotic and specific ion effects have been separated by growing Dwarf Red Kidney bean plants in base nutrient solution to which was added, at isosmotic concentrations, NaCl, $CaCl_2$, $MgCl_2$, or Carbowax polyethylene glycol having a molecular weight of approximately 20,000 (C20M) (Lagerwerff and Eagle, 1961). The Carbowax-treated plants yielded better than those receiving isosmotic salt solutions, and it was concluded that the lower yields in the presence of added salts were attributable to specific effects of the salts.

The organic acid anion concentration of plants has been related to their growth; any treatment that reduced organic acids also reduced plant growth, as reported by Noggle (1966). He noted that Cl^- salts reduced organic acids more than SO_4^{2-} salts did, and that Cl^- salts reduced growth more than SO_4^{2-} salts did. He concluded that increasing the Cl^- content of the soil may reduce yields before the salt concentration is sufficiently high enough to cause any osmotic stress.

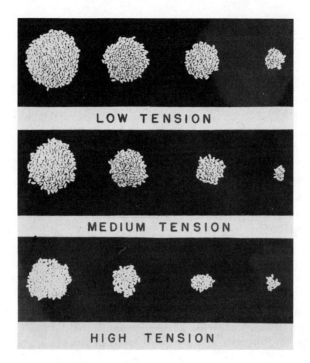

Fig. 16.6

Yields of Dwarf Red Kidney bean plants, grown in nonsalinized soil and in soil to which three levels of NaCl were added, as affected by three soil moisture tensions. From A. D. Ayers et al. (1943) J. Amer. Soc. Agron. 35:796–810.

Reduction of growth by salinity cannot be attributed solely to water stress, since Bernstein (1961) noted that as the OP of solutions around roots increased the OP of cells in roots and above-ground parts of cotton and pepper plants also increased. An osmotic adjustment process was occurring, and Bernstein (1961) concluded that this adjustment process may be a likely limiting factor for growth under saline conditions. The osmotic adjustment depended in part on increased accumulation of ions and also on substitution of monovalent for polyvalent ions. In this connection, osmotic potential has been shown to be inversely proportional to the relative water content of leaves, as reported by Gardner and Ehlig (1965). They also noted that the major change in leaf dimensions accompanying changing water content occurred in leaf thickness; only a slight change occurred in leaf areas.

KCl and $CaCl_2$ suppressed the growth of a mangrove, *Avicennia marina*, but NaCl increased the growth of plants over that of plants in nutrient solution without added NaCl (Connor, 1969). The optimal level of NaCl was approximately $1\frac{1}{2}\%$.

The osmotic potential in plants decreases as the osmotic potential of the solution around roots increases (Bernstein, 1961, 1963). It was also shown that a change in osmotic potential in plants occurs within a day after the roots are exposed to a new external osmotic concentration, and osmotic potential differences between the plant and the root media are thereby maintained. Turgor pressure does not decrease, hence salinity-induced growth reduction cannot be attributed to reduction of turgor (Bernstein, 1961, 1963; Ehlig et al., 1968).

A close relationship was demonstrated among soil water potential, leaf water potential, and plant growth (Gardner and Nieman, 1964).

Shalhevet and Bernstein (1968) divided the root system of alfalfa by a horizontal wax layer into two equal-depth sections which were separately salinized and irrigated. Mean salinity of the root zone correlated with growth reduction and was a good measure of effective salinity.

Working with bean *(Phaseolus vulgaris* var. Blue Lake) and barley *(Hordeum vulgare* var. Liberty) plants with roots split equally between two differentially salinized nutrient solutions, Kirkham et al. (1969) observed that the degree of osmotic adjustment and rate of growth were functions of the proportion of the root system exposed to saline conditions. That is, osmotic potentials of plants with half their roots in saline solution and half in nonsaline solution were half-way between those of plants in saline or in nonsaline solution; the same relationship held for dry weight of plants. Reduction in growth appeared to be caused mainly by excess concentration of salts and not by reduced water availability or reduced turgor. There was a higher resistance to water flow under high-salt conditions, and it was suggested that perhaps the high concentration of salts reduced the permeability of root membranes.

Salinity-induced growth reductions therefore appear to be related to lowered osmotic potentials in plants per se and not to reduced turgor or a change in the *differential* between the osmotic potential of the plant and that of the external medium.

Red mangrove *(Rhizophora mangle)* was grown with different salt concentrations ranging from a dilute soil solution to salt concentrations approximating those found in soils under the influence of seawater (Stern and Voight, 1959). The application of salts increased the dry matter production, particularly at a salt concentration approximately that of seawater. It was concluded that the early development of mangrove seedlings is favored by a high salt concentration.

Growth of individual intact cotton leaves was also observed to be a function of the TSMS to which plants were subjected (Figs. 16.7, 16.8, and 16.9) (Wadleigh and Gauch, 1948). Expansion of leaves ceased when TSMS was approximately 15 atm (Fig. 16.10), and this observation constituted a biological confirmation of the 15-atm value of soils at the PWP as determined by Richards and Weaver (1944).

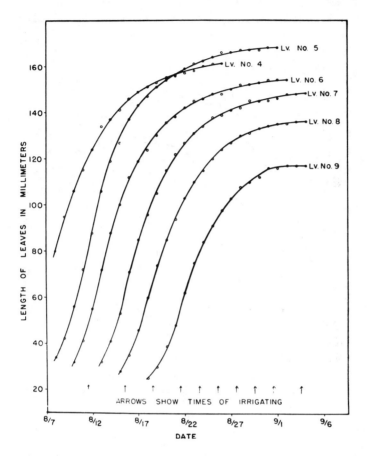

Fig. 16.7

Growth of individual cotton leaves of plants growing in a nonsaline soil. (Wadleigh and Gauch, 1948.)

Using a thermocouple psychrometer, Boyer (1968) showed that a minimum turgor of 6.5 bars was required for the enlargement of intact sunflower leaves, and that growth occurred only when the leaf water potential was above −3.5 bars.

Under certain conditions the thermocouple psychrometer may prove unreliable. Owing to salt secretion by cotton leaves, which could not be eliminated by washing the leaves since salts were secreted during equilibration periods, estimates of the water potential with a thermocouple psychrometer may be in error by 400–500% (Klepper and Barrs, 1968). In view of this phenomenon, it was concluded that a thermocouple psychrometer is not suitable for measuring the water potential of cotton leaves when the value is above approximately −10 bars. A pressure chamber (Scholander et al., 1964, 1965; Boyer, 1967) obviates the error from salt secretion by leaves.

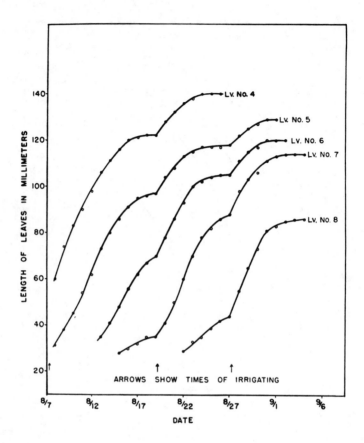

Fig. 16.8

Growth of individual cotton leaves of plants growing in a saline soil. (Wadleigh and Gauch, 1948.)

Bean plants were exposed to salinity either during the light or the dark period of the day (Meiri et al., 1970). With relatively mild evapotranspiration, transpiration during salt absorption did not markedly affect salt uptake or plant growth. In all cases duration of exposure to salinity was the main factor; salinity reduced the transpiration rate.

For *Glycine wightii* and *G. tomentella,* the general picture of salt stress appears to be one of immediate growth reduction through an initial water stress approximately proportional to the concentration of salt applied (Wilson et al., 1970). At 40 meq NaCl/liter, osmotic adjustment occurred without significant tissue injury. At 80 meq NaCl/liter some osmotic adjustment occurred, but at 160 meq NaCl/liter a rapid and excessive Cl^- accumulation occurred which injured the leaves and the growth rate was progressively reduced.

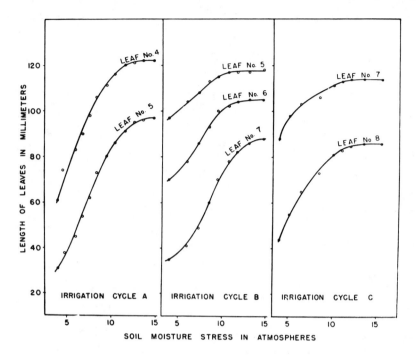

Fig. 16.9

Growth of individual cotton leaves of plants growing in a saline soil as a function of total soil moisture stress. (Wadleigh and Gauch, 1948.)

A *Pencillium* mutant grew in KCl- and H_3BO_3-saturated media in the presence of monosaccharides, but growth rates were greatly enhanced by inosine-5'-phosphate (IMP) (Siegel, 1969). Boric acid blocked sporulation under all conditions. KCl permitted spores of abnormal color to form without IMP, whereas spores of normal color appeared in its presence. With KCl and sucrose there was some growth without IMP, but sporulation occurred only when IMP was added.

Saltbush (*Atriplex halimus* L.) was grown in culture solutions to which NaCl or Na_2SO_4 had been added (Gale and Poljakoff-Mayber, 1970). OP of sap, growth, morphology, and leaf gaseous exchange (transpiration and photosynthesis) were studied. Salinity (1) increased leaf area and succulence at a relatively low concentration of salt, (2) increased stomatal resistance to water vapor loss and CO_2 uptake, and (3) caused changes in mesophyll resistance to CO_2 uptake. It was concluded that these responses to salinity tended to counteract one another in their overall effects on growth. Which factor dominates is determined by the concentration and type of salinity and also possibly by climatic conditions.

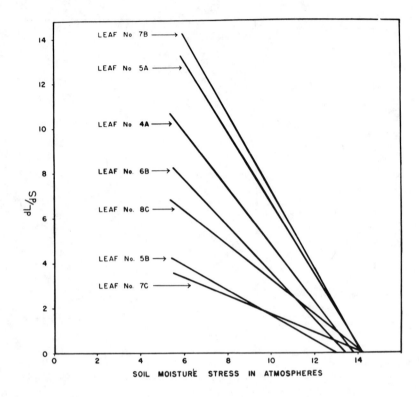

Fig. 16.10

Rate of change in length of cotton leaves as related to total soil moisture stress, showing that leaf enlargement ceased at approximately 14 atm of total stress. (Wadleigh and Gauch, 1948.)

In another study with saltbush, the plants were grown in saline nutrient solution in growth chambers with 27 or 65% relative humidity (Gale et al., 1970). Salinity ranged from 0 to −20 atm of NaCl. Under low humidity typical optimal growth curves were obtained with maximal growth at about −5 atm. Under humid conditions no such optimal curves were obtained; growth was greatest in nonsaline controls and decreased with each increment in salinity. It was concluded that low concentrations of NaCl are apparently required by *Atriplex halimus* in order to overcome the stress caused by low relative humidity. Under high relative humidity salinity is harmful to growth at all concentrations.

Wilson (1970) studied the growth of well-nodulated plants of *G. wightii* subjected to 14 days of salinity ranging from 0 to 148 meq NaCl/liter of nutrient solution, as compared with growth of similarly treated nitrogen-fertilized plants. The latter showed less tissue injury and less reduction in growth rate at high salinity than did the nodulated plants. During salinity treatment the development of new nodules, and fixation by the existing nodules, were greatly reduced.

Incidentally, with respect to leaves, Milthorpe's *The Growth of Leaves* (1956) is unique in covering this important topic.

Composition. The presence of excessive concentrations of Cl^- or other ions sometimes results in very high concentrations of these ions in plants. In general, fruit trees have a relatively low tolerance for salinity. In a study of the responses of peach trees to excesses of various salts, Bernstein et al. (1956) observed that 50–60 meq $Cl^-/100$ g dry matter in peach leaves constituted a lethal concentration of Cl^-. They attributed about half the growth reduction in peach trees, under the conditions of their experiment, to Cl^- and the other half to OP effects.

In Dwarf Red Kidney bean plants receiving NaCl or $CaCl_2$ increments of added salt, the percentage of protein diminished with increments of added salt (Wadleigh and Gauch, 1942).

Most ions accumulate primarily in leaves but, except for halophytes in which distribution is quite uniform throughout the plant, Na^+ accumulates primarily in roots. In Dwarf Red Kidney bean plants, Ca^{2+} greatly predominated over Na^+ in leaves and stems, while in the NaCl series the roots were relatively high in Na^+; in the $CaCl_2$ series they were relatively high in Ca^{2+} (Gauch and Wadleigh, 1942).

The outstanding feature of mangroves is their adaptation to tidal zones of tropical seas. Some genera, such as *Aegialitis* and *Avicennia,* eliminate considerable quantities of salt through special glands on the leaves—a property other species such as *Rhizophora* and *Sonneratia* do not share (Scholander et al., 1962). Xylem sap of salt-secreting species contains approximately 0.2–0.5% NaCl, a concentration exceeding that of nonsecreting species by approximately 10-fold. Salt may be kept out of nonsecreting species by exclusion, that is, by active transport outward.

Some mangroves maintain very low salt concentrations in the xylem sap by excluding salts almost completely; others absorb considerable quantities of NaCl from seawater and excrete it by special glands in the leaves (Scholander et al., 1966). It was also reported that xylem sap of halophytic plants growing in seawater may have 30–60 atm of pressure.

In citrus, Na^+ fluxes in the free space of roots of the salt-sensitive Sweet Lime variety and the salt-resistant Cleopatra Mandarin variety were nearly equal (Greenshpan and Kessler, 1970). However, Na^+ fluxes from intracellular spaces were considerably lower in roots of Cleopatra Mandarin than in those of Sweet Lime. Rates of Na^+ influx and efflux were much slower in Cleopatra Mandarin than in the Sweet Lime variety, and these findings are in good agreement with the relative sensitivity of these citrus varieties to salinity.

Nutrient Absorption. In this section we are primarily concerned with the effects of salinity and alkalinity on the absorption of essential nutrients. A high external concentration of Ca^{2+} may drastically reduce the absorption of Mg^{2+}, and vice versa. A high external concentration of Na^+ may adversely affect absorption of K^+. With regard to anions external Cl^- may affect particularly the absorption of NO_3^-. External NO_3^- concentration may affect internal concentrations of Cl^- in plants. In snapbeans, there was an inverse relationship between concentrations of external NO_3^- and internal Cl^- (Weigel, 1968) (Table 16.1). In the low-N, medium-N, and high-N groups, there were instances in which essentially the same external Cl^- concentration (e.g., approximately 9 meq/liter) existed and yet the highest, medium, and lowest internal concentrations correlated with lowest, medium, and highest external NO_3^- levels (Weigel, 1968). These data have been corroborated by experiments with snap beans in which the external concentration of Cl^- was constant at 10 meq/liter (Hsiao and Gauch, 1971). Characteristically, soybeans do not readily absorb Cl^-, and the effect of NO_3^- in reducing internal Cl^- concentrations was much less marked in soybeans (Weigel, 1970) than in snap beans (Weigel, 1968).

Alkalinity of the soil markedly reduces the availability, hence absorption, of Zn^{2+}, Cu^{2+}, Mn^{2+}, and B.

Water Absorption. Water absorption has been shown to be a function of the OP of solutions or of the DPD of soils. In soils DPD has physical and osmotic origins. The osmotic component is variable depending on the concentration of soluble salts in the soil. The physical component varies according to type of soil and percent of water. Air-dry soil of Oswego silt loam holds water with a tension or negative pressure of approximately 1000 atm (Shull, 1916).

Hayward et al. (1942) devised a potometer for measuring rate of water entry into roots at any location on the root (Figs. 16.11 and 16.12). Using this apparatus, Hayward and Spurr (1944a) studied the effects of various concentrations of NaCl, Na_2SO_4, $CaCl_2$, sucrose, and mannitol on the rates of entry of water into corn roots (Figs. 16.13 and 16.14). Rates of water intake were very closely correlated with OP values of the solutions—the greater the OP of the solution, the lower the rate of water intake by corn roots. Water intake was reduced as much as 80% when the OP of the solution was raised from 0.8 to 4.8 atm and intake was completely inhibited at 6.8 atm. It was concluded that water may *not* be equally available from FC to the WP.

Growth and water absorption are generally closely related. It is not as easy to study water absorption by roots in soils as by those grown in solution culture. It is possible, however, to calculate the TSMS in soils (Wadleigh, 1946) and to relate TSMS to growth (Wadleigh and Gauch, 1948). When this relationship is studied, as was the case with guayule (Wadleigh et al., 1946), it is clear that the greater the TSMS, the greater the reduction in plant growth (Fig. 16.5).

Table 16.1

The Effect of NO_3^- Level in the Nutrient Solution on the Concentration of Cl^- in Leaves of Snapbean Plants[a]

Level of variable[b]			Cl^- concentration	
N	P	S	Nutrient solution (meq/liter)	Leaves, (meq/ 100 G dry matter)
L	L	L[b]	18.8	218
L	L	M	17.9	218
L	L	H	13.9	180
L	M	L	17.9	262
L	H	L	13.9	256
L	M	M	17.0	254
L	M	H	13.0	198
L	H	M	13.0	220
L	H	H	9.0	198
			Average	223
M	L	L	14.8	75
M	L	M	13.9	73
M	L	H	9.9	50
M	M	L	13.9	73
M	H	L	9.9	83
M	M	M	13.0	60
M	M	H	9.0	62
M	H	M	9.0	47
M	H	H	5.0	25
			Average	61
H	L	L	9.8	11
H	L	M	8.9	11
H	L	H	4.9	5
H	M	L	8.9	27
H	H	L	4.9	8
H	M	M	8.0	25
H	M	H	4.0	9
H	H	M	4.0	4
H	H	H	0.0	3
			Average	11

[a] Weigel (1968).
[b] L, low (14 ppm N); M, medium (70 ppm N); H, high (140 ppm N).

Haise et al. (1955) reported that wheat roots are capable of absorbing soil moisture at tensions exceeding 26 atm.

Water Relations.   A thermocouple psychrometer can be used to determine DPD and OP on the same sample from a single leaf of most plants (Ehlig, 1962). As soil moisture was depleted, values for DPD and OP increased whereas those for TP decreased.

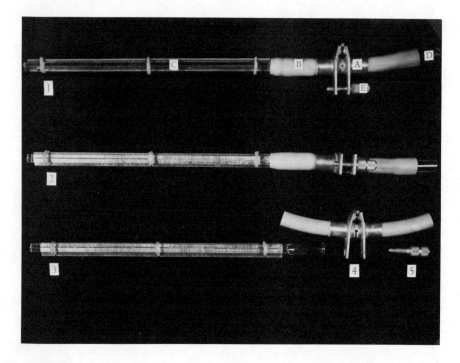

Fig. 16.11
Potometer for measuring the rate of water uptake by various sections of a root. A, Brass holder; B, rubber chamber; C, calibrated capillary tube and attached scale; D, glass plug; E, set screw. From H. E. Hayward et al. (1942) Bot. Gaz., 104:152–160. The Univ. of Chicago Press and copyright by the University of Chicago—by permission.

Addition of polyethylene glycol 6000 to the nutrient solution resulted in water relations in pepper plants similar to those expected in soil at the same water potentials (Kaufmann and Eckard, 1971). Xylem pressure potential in the root and leaf became more negative during a 24-hr treatment, whereas osmotic potential of the root xylem sap remained constant. Decrease in pressure potential was closely correlated with decrease in osmotic potential of the nutrient solution. In contrast, addition of polyethylene glycol 400 to the nutrient solution resulted in a reduction of osmotic potential in the root xylem sap; this osmotic adjustment in the xylem was large enough to establish an osmotic gradient for entry of water and cause guttation at a nutrient solution osmotic potential of −4.8 bars. These studies indicate that larger polyethylene glycol molecules, such as polyethylene glycol 6000, are more useful in simulating soil water stress than smaller molecules such as polyethylene glycol 400. It was also found that polyethylene glycols having molecular weights of 1000 or higher are not absorbed in significant quantities by healthy roots, whereas PEG-200 and mannitol were absorbed (Lawler, 1970).

Fig. 16.12
Potometers attached at three positions on a corn root, and one potometer (at bottom) as a
check. From H. E. Hayward et al., (1942) Bot. Gaz., 104:152–160. The Univ. of Chicago Press and
copyright by the University of Chicago—by permission.

The OP of sap within live cells can be measured by a freezing-slide apparatus
for visual observation of frozen water and melting ice, particularly in guard
and epidermal cells (Bearce and Kohl, 1970). With this apparatus tissue strips
are quick-frozen (at a cooling rate of $33°C/0.5$ min) and then warmed very
slowly (at a rate of $2°C/min$) for observation of melting points.

Respiration. HCO_3^- inhibited respiration of excised bean roots, and the
effect could not be fully explained as a direct inhibition of the cytochrome system (Miller and Thorne, 1956).

Flowering. In sand cultures it was noted that salinity delayed flowering of
flax (Hayward and Spurr, 1944b) but hastened that of Elberta peach trees
(Hayward et al., 1946).

Stem Anatomy. Working with flax, Hayward and Spurr (1944b) observed
reduced cambial activity and smaller xylem vessels and phloem fibers in stems

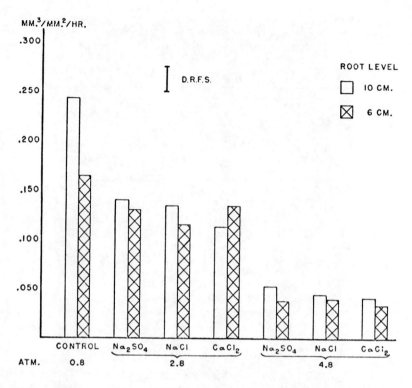

Fig. 16.13

Effect of inorganic substrates on rate of entry of water into adventitious roots of corn at 6- and 10-cm levels. Na₂SO₄, NaCl, and CaCl₂ were added individually to base nutrient solution (0.8 atm) to produce isosmotic pressures of 2.8 and 4.8 atm. Difference required for significance (D.R.F.S.) as derived from pooled error variance was 0.0263 at the 5% level of significance. From H. E. Hayward and W. B. Spurr (1944a) Bot. Gaz., 106:131–139. The Univ. of Chicago Press and copyright by the University of Chicago—by permission.

of plants subjected to an additional 4 atm of NaCl added to a base nutrient so-lution as compared with the no-salt-added control (Fig. 16.15).

Metabolic. Salinity decreased the incorporation of amino acids into pro-teins and altered the normal metabolic pathways (Kahane and Poljakoff-Mayber, 1968). Pronounced differences were observed between the effects of NaCl and Na₂SO₄, inasmuch as kinetin inhibited the incorporation of amino acids in Na₂SO₄-stressed pea roots but promoted their uptake and incorporation in NaCl-stressed roots.

In relation to their size, the leaves of 12 species of plants subjected to salin-ity had about the same photosynthetic and respiratory capabilities as normal control leaves from plants not subjected to salinity (Nieman, 1962). In a later study, Nieman (1965) reported that salinity reduced the rates of RNA and

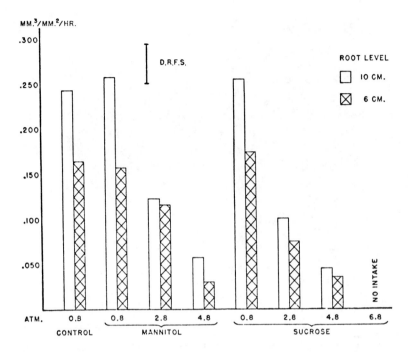

Fig. 16.14

Effect of organic substrates on rate of entry of water into adventitious roots of corn at 6- and 10-cm root levels. Mannitol and sucrose were added to tap water to produce the desired OP values. Difference required for significance (D.R.F.S.) as derived from pooled error variance was 0.0452 at the 5% level of significance. From H. E. Hayward and W. B. Spurr (1944a) Bot. Gaz., 106:131–139. The Univ. of Chicago Press and copyright by the University of Chicago—by permission.

protein synthesis and cell enlargement of bean leaves. He found that these three processes were merely delayed by salinity, since the maximal values finally approached those attained earlier by controls. He concluded that salinity affects growth by imposing a water stress, and that reduced rates of cell enlargement and of RNA and protein syntheses are the primary effects of salinity.

The presence of NaCl has been reported to favor the synthesis of amino acids over that of organic acids during incorporation of $^{14}CO_2$ in the dark (Joshi et al., 1962).

With regard to the activity of various enzymes, the effect of growing plants in saline media is different from the direct effect of salt in an assay mixture (Hasson-Porath and Poljakoff-Mayber, 1969). Earlier, Porath and Poljakoff-Mayber (1964) reported that NaCl or Na_2SO_4 decreased the specific activity of malate dehydrogenase in pea roots. Later work indicated that Na_2SO_4 re-

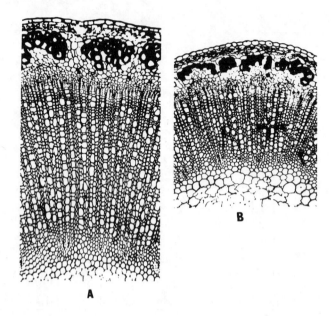

Fig. 16.15
(A) Transection of a median internode of the main stem of flax from a control plant (0.5 atm OP) showing development of phloem fibers and amount of secondary thickening of the axis. (B) Transection of a corresponding internode from a plant grown on NaCl substrate at 4.5 atm OP. Formation of wood fibers in the xylem region is shown. Comparison with (A) indicates the inhibitive effect of high concentrations of salt on cambial activity and size of cells. From H. E. Hayward and W. B. Spurr (1944b.) J. Amer. Soc. Agron. 36:287–300—by permission.

duced the activity of malic acid dehydrogenase more drastically than NaCl did (Hasson-Porath and Poljakoff-Mayber, 1969).

Concentrations of 18 enzymes were determined in leaves, stems, and roots of 11-day-old pea seedlings grown in a liquid medium or in the same medium containing, in addition, 5 atm of either NaCl, KCl, Na_2SO_4, or K_2SO_4 (Weimberg, 1970). Although plants grown in saline media were stunted, specific activities of enzymes were the same in a given tissue of all plants. The electrophoretic pattern of isozymes of malate dehydrogenase was unaltered by growth of plants in a saline medium. The isozyme pattern of peroxidase from roots of salt-grown plants was altered in that two of the five detectable isozymes migrated a little more slowly than those in extracts from nonsaline plant tissues.

Pea (*Pisum sativum*) plants were grown in media salinized with either NaCl or Na_2SO_4 (Hasson-Porath and Poljakoff-Mayber, 1971). With salinization ATP decreased, ADP increased, and AMP was unaffected, as compared with unsalinized plant roots.

In seeds of crested wheatgrass, some phosphorylation occurred with a water potential of -880 atm, but seeds did not incorporate ^{32}P into NAD, ATP, and uridine diphosphate until the water potential reached -130 atm (Wilson and Harris, 1968). Metabolic patterns may be altered when water availability decreases, but there is little evidence indicating whether the free energy of water per se is the important factor, or whether other changes associated with water availability are involved (Boyer, 1969).

Root Growth. Growth of roots into saline soil was studied by Wadleigh et al. (1947). They grew Red Kidney bean, Mexican June corn, California common alfalfa, and Shafter Acala cotton plants in Fallbrook loam in boxes 1 ft square and 3 ft deep. The surface 6 in. contained no added salt; the next five 6-in. layers, in descending order, contained 0.05, 0.1, 0.15, 0.2, and 0.25% NaCl (dry wt basis). Few bean roots penetrated soil with 0.1% salt; only a few corn roots entered the 0.2% salt; a few alfalfa roots entered the 0.25% layer; and cotton roots were abundant in all layers. Water was removed from each layer to such a degree that final OP values of all layers in each soil column were nearly uniform. The critical OP values of the soil solution were: 7–8 atm for beans; 10.5–11.5 for corn; 12–13 for alfalfa; and 16–17 for cotton.

Photosynthesis. Cl^- per se may reduce photosynthetic activity, as has been reported for potatoes (Rémy and Liesegang, 1926) and, possibly, Dwarf Red Kidney bean plants (Wadleigh and Ayers, 1945). Moisture stress has also been related to reduced photosynthetic activity (Schneider and Childers, 1941). Cl^- reduced chlorophyll in potatoes (Basslavskaya and Syroeshkina, 1936), which might presumably reduce photosynthesis as well.

Starch and Sugar. Moisture stress has been reported to lower the percentage of starch in plants (Magness et al., 1932; Spoehr and Milner, 1939; Wadleigh et al., 1943; Wadleigh and Ayers, 1945) and to increase the concentration of sugars (Magness et al., 1932).

Effects on Soils

Structure, Aeration, and Permeability to Water. When the BEC of a soil has a high Na^+ percentage, soil colloids tend to be highly dispersed, there is little air space, and penetration of water into and percolation of water through such a soil is extremely limited. By contrast, a high percentage of Ca^{2+} on the exchange complex results in a well-aggregated soil having good physical characteristics and good water permeability.

Nutrient Availability. Inasmuch as the exchange complex of the soil can hold only a given amount of ions, the higher the proportion of Na^+ the lower the proportions of Ca^{2+}, Mg^{2+}, and K^+. Therefore a high percentage saturation of the base exchange complex with Na^+ not only means that the soil will be highly dispersed and of poor physical structure, but that plants will have reduced amounts of Ca^{2+}, Mg^{2+}, and K^+ on the exchange complex.

The lowering of the amounts of these last three ions means that their uptake by plants will probably be reduced. In addition, Na-saturated soils are often quite low in O_2, and this in turn deleteriously affects nutrient absorption in general.

METHODS FOR COMBATING SALINITY

The most immediately practical solution for overcoming salinity involves the selection and use of salt-tolerant species or varieties of plants. Field observations reveal that some crops can tolerate salinity conditions better than others. For that reason barley, alfalfa, sugar beets, table beets, asparagus, and certain other crops are selected whenever possible for more highly saline areas.

Actually, alfalfa is not very salt tolerant in the seedling stage, but becomes quite salt tolerant later on. In general, in saline areas salts are removed from the surface layer of soil by irrigation and leaching, alfalfa is planted, and the crop passes through the salt-sensitive seedling stage of growth before salts again move up into the upper layer of soil.

With regard to soil there are several approaches that can be used to combat salinity, and sometimes a combination of approaches proves especially helpful. In situations in which the water table is sufficiently low so that salts may be leached out of a good portion of the soil and the soil is freely permeable to water, one of the simplest remedial approaches is leaching the salts with applied water. This remedial approach is undoubtedly more widely practiced than any other.

In some cases of salinity, for example, soil highly saturated with Na^+, it is necessary to apply a soil amendment. When there is an excess of Na^+, the best remedy involves addition of Ca^{2+}; the latter drives off and replaces Na^+ on the exchange complex and thus improves the aggregation and physical structure of the soil. Aggregation of soil colloids greatly facilitates the entry and percolation of water through soil. The more Ca^{2+} that is added, the greater the aggregation and the greater the rate of water entry into soil. Increasing rates of aggregation and water percolation result in an ever-increasing rate of improvement in the salinity condition. $CaSO_4$ is the usual Ca^{2+} amendment.

Large applications of manure or plowed-down cover crops also improve the physical structure of soil, improve the rate of water entry and, finally, result in a leaching-out of accumulated salts.

Recently, it was reported that bean plants *(P. vulgaris* L.) of a very salt-sensitive species show no damage from a NaCl concentration $1/10$ that of seawater if the Ca^{2+} concentration of the nutrient solution is 1.0 mM or higher, but that damage is severe at lower concentrations of Ca^{2+} (LaHayte and Epstein, 1969). In leaves Na^+ concentration dropped from 3.2 mg $Na^+/100$ mg (dry wt) at 0.1 mM $CaSO_4$ to 0.2 mg $Na^+/100$ mg at 3.0 mM. At less than 3.0 mM $CaSO_4$, a massive breakthrough of Na^+ into the leaves occurred.

Combination of Approaches

Inasmuch as the growing of a salt-tolerant crop, leaching, and application of a soil amendment may individually be helpful in combating salinity, a combination of two or more of these approaches is often strikingly effective.

SPECIES AND VARIETAL DIFFERENCES IN SALT TOLERANCE

In an excellent review article, Hayward and Wadleigh (1949) discuss salt tolerances of crops. According to Hayward and Bernstein (1958), the mechanisms whereby Cl^- or Na^+ ions are specifically toxic to sensitive species remain unknown. They added that the mechanism of salt toxicity, and the distinguishing features of salt tolerance, appear to be the major tasks for research on salt tolerance of plants.

There was considerable variation in salt tolerances of varieties of strawberry clover (Gauch and Magistad, 1943), an indication that breeding for salt tolerance may be a fruitful approach to obtaining tolerance. Similar variations in salt tolerances have been reported for varieties of other crops.

Desert saltbush [*Atriplex polycarpa* (Torr.) S. Wats.] plants were grown for 6 wk in water culture in four solutions containing NaCl to produce osmotic potentials of -0.8, -10.0, -20.0, and -30.0 atm (Chatterton et al., 1970); 10 and 13% of the Na^+ and Cl^-, respectively, entered the plants. Although these plants absorb large quantities of salt, it appeared that their adaptation to salinity may involve very little salt in photosynthetically active tissues, the absorbed salt being localized in trichomes.

The concentration of salts in vesiculated hairs of *A. halimus* L. was measured, and it was remarkably higher than that of the leaf sap and xylem exudate (Mozafar and Goodin, 1970). Despite their unusually high salt concentration, these hairs, when immersed in water, were unable to absorb water—in apparent contradiction to the previously held hypothesis that these hairs make it possible for a plant to absorb water from the atmosphere. When plants were grown under saline conditions, the salt concentration of hairs increased from 2.3 to 11.6 M Na + K, but salt concentration of expressed leaf sap from young leaves did not change significantly. This observation indicates that in *A. halimus* the vesiculated hairs play a significant role in removing salt from the remainder of the leaf and in preventing accumulation of toxic salts in parenchyma and vascular tissues. Thus a nearly constant salt concentration is maintained in leaf cells other than hairs.

A high ionic concentration in the medium stimulated salt secretion in *Aeluropus litoralis* (Willd.) Parl. (Pollak and Waisel, 1970). A positive correlation was found between the outer NaCl concentration and the amount of Na^+ secreted and/or leaf concentrations. Na^+ secretion showed high efficiency in excluding excess Na^+ from leaf tissues.

PLANT INDICATORS (OF SALTS OR IONS) OR "ACCUMULATOR" PLANTS

Indicators or Accumulators

Salinity. It would be of diagnostic value in the field if it were found that certain species of plants indicated or delineated saline areas. It would also be helpful if the presence of certain species or symptoms indicated the nature of the salts contributing to a salinity condition. These hopes have not been supported by a search for plant indicators.

It is true that certain species are often observed growing in saline environments. The most frequent ones are:

Allenrolfea occidentalis, Atriplex spp. (saltbush), *Kochia vestita* (white sage), *Salicornia rubra* (red samphire), *Sarcobatus vermiculatus* (greasewood), *Suaeda* spp., six genera of Chenopodiaceae, *Distichlis spicata* (salt grass), and *Sporobolus airoides* (alkali sacaton or Tussock grass). However, all of these species grow even better in a nonsaline than in a saline environment, and thus the presence of one or more of them is not necessarily an indication of salinity. The important clue for salinity is not the presence of any one or more of the "indicator" species, but the *absence* of salt-sensitive species of plants.

In a study of types of vegetation in Escalante Valley, Utah, as indicators of soil conditions, Shantz and Piemeisel (1940) indicated the plant communities on nonsaline and saline soils as follows.

Plant communities on nonsaline soils	
Sagebrush association	*Artemisia tridentata*
Galleta association	*Hilaria jamesii*
Little rabbitbrush association	*Chrysothamnus* spp.
Winterfat association	*Eurotia lanata*
Fourwing saltbush community	*Atriplex canescens*
Juniper association	*Juniperus utahensis*

Plant communities on saline soils	
Shadscale association	*Atriplex confertifolia*
Greasewood association	*Sarcobatus vermiculatus*
Saltgrass association	*Distichlis spicata*
Pickleweed association	*Allenrolfea occidentalis*
Red samphire community	*Salicornia rubra*
Saltsage community	*Atriplex falcata*
Mixed vegetation on sand dunes or hummocks	*Chrysothamnus* spp. and others

Second, specific leaf symptoms indicative of the kind of salt present have not generally been observed. The symptoms are similar whether the salinity is caused by NaCl or Na_2SO_4. An excess of B produces leaf symptoms which can usually be recognized and diagnosed when B alone is present in a toxic concentration.

Zinc. Robinson et al. (1947) determined the concentration of Zn^{2+} in 30 plants species growing on slime ponds containing approximately 12.5% zinc. The Zn^{2+} concentration ranged from 39 ppm in fruit of false Solomon's seal *(Smilacina racemosa)* to 5400 ppm in horsetail *(Equisetum arvense)*. Poplar *(Populus grandidentata)* and ragweed *(Ambrosia artemisiifolia)* were considered worthy of further study as indicators of Zn ore deposits, since the leaves of these species had high concentrations of Zn^{2+} (1800 ppm for each) when growing in a slime pond high in Zn. A ragweed plant on a Zn mineral outcrop near Zinc, Arkansas, contained 3800 ppm of Zn^{2+} (Robinson and Edgington, 1945).

Thallium. "Frenching" of tobacco was first reported to be a physiological disease by Clinton (1915). There have been reports that Tl toxicity is the cause of frenching. Of 33 elements studied, Spencer (1937) found that only Tl caused chlorosis and strap-shaped leaves characteristic of this disorder. In water culture 0.067 ppm of Tl produced typical chlorosis in tobacco; 0.38 ppm were required for the effect in a light, sandy loam. In soil the disorder can be controlled by the addition of N salts, a dilute solution of Al sulfate, or by the addition of K iodide. Species susceptible to frenching are also ones that are sensitive to Tl. *Nicotiana langsdorffii, N. rustica,* and *Lycopersicon esculentum,* which show only faint symptoms, are less sensitive to Tl than the extremely sensitive Turkish tobacco. *Nicotiana glutinosa* and *N. glauca,* which are not susceptible to frenching, are little affected by doses of Tl that are very toxic for Turkish tobacco.

Even with spectographic analysis, however, it has not been possible to relate frenching to Tl toxicity (Spencer and Lavin, 1939). Tl could not be detected in frenched or healthy plants growing in the field. When 0.1 or 0.2 ppm of Tl was added to a nutrient solution for tobacco plants, Tl could be detected and the plants showed, respectively, faint mottling and strap leaf formation.

Fluorine. In general, very little F^- appears to be absorbed by plants (Robinson and Edgington, 1945). Leaves of citrus trees located close to gaseous sources of F_2 may, however, contain high concentrations of F^- (Haas and Brusca, 1955). Fluorides occur in certain crude phosphatic materials from which fertilizers are prepared. Haas and Brusca (1955) produced and described F-toxicity symptoms and cautioned that the tip burn could easily be confused with Cl^- tip burn.

Leaves of gladiolus, tomato, rose, and pine contained from 230 to 949 ppm of F^- on a dry weight basis (Gwirtsman et al., 1957), whereas tissues of most plants contain approximately 0.1–10 ppm of F^- (Cholak, 1959).

Barley roots discriminate markedly in favor of Cl^- and against F^- absorption, with a 100-fold difference in the concentrations of Cl^- and F^- when the roots are exposed to equal initial concentrations (Venkateswarlu et al., 1965). Whereas Cl^- absorption was dependent on metabolism, F^- uptake was a diffusion process unaffected even by anaerobic conditions. In view of the increasing pollution of the atmosphere in many areas, it is possible that F_2 and F^- toxicities may become much more prevalent in the future.

Aluminum. Sweet leaf *(Symlocos tinctoria)* and Princess pine *(Lycopodium flabelliforme)* are high-Al plants. Calculated as Al_2O_3, leaves of the former contained as high as 7.67% (dry wt basis), and the latter 1.6%. Alumina in the ash of hickory varied from 3 to 37.5%; pecan tended to be somewhat lower in Al^{3+} than hickory (Robinson and Edgington, 1945). Al^{3+} is accumulated in relatively high concentrations in *Lycopodium, Hicoria,* and *Symplocus* (Hutchinson, 1945).

Arsenic. In general, the quantities of As found in plants are low since for the most part plants die before they accumulate as much as 10 ppm (Robinson and Edgington, 1945).

Cobalt. Co is absorbed only very sparingly by most plants (Robinson and Edgington, 1945). However, samples of swamp black gum [*Nyssa sylvatica* Marsh. var. *biflora* (Walt) Sarg.] collected in the Atlantic coastal plain contained approximately 60 ppm (Beeson et al., 1955) to 845 ppm Co (Kubota, et al., 1960), whereas other species collected from the same sites and from the same soils contained generally less than 0.1 ppm.

Copper. Some plants accumulate Cu^{2+} even from relatively low external concentrations of the element. In soils having 0.25–0.35 ppm of Cu^{2+}, leaves of *Thymus serpyllum,* dandelion, and *Viola hirto* contained 223, 320, and 560 ppm, respectively (Robinson and Edgington, 1945).

Manganese. Samples of tobacco from Connecticut contained as much as 2262 ppm Mn^{2+} (Robinson and Edgington, 1945).

Molybdenum. In Wyoming, woody aster and devil's plant brush contained 221 and 333 ppm Mo, respectively (Robinson and Edgington, 1945). The concentration of Mo is quite low in most plants—usually not more than 10 ppm.

Rare Earths. Robinson and Edgington (1945) considered the following elements together in the rare earth group: Sc, Y, La, Ce, Pr, Nd, Sm, Eu, Gd, Tb, Dy, Ho, Er, Tm, Yb, and Lu. Where rare earths were in high concentration in the soil, hickory leaves accumulated this group of elements. Leaves of a hickory at Rockville, Maryland, had 2296 ppm (Robinson, 1943; Robinson and Edgington, 1945). In one hickory tree near Falls Church, Virginia, nut meats, shells, husks, and leaves had 5, 7, 17, and 981 ppm, respectively, of rare earths.

Hickory absorbs relatively large quantities of rare earths from soil, and therefore analysis of its leaves may serve as an indication of the concentration of rare earths in a given soil (Robinson, 1943).

Selenium. The occurrence of Se in plants and soils has been discussed in a review (Rosenfeld and Beath, 1964).

The variation in concentration of Se in plants seems to exceed that of any other element (Robinson and Edgington, 1945). The extreme variation was from less than 0.1 to 14,900 ppm. *Astragalus* species are notorious accumulators of Se, and concentrations varied from 4,000 to 14,900 ppm for different species.

Certain plants, such as woody aster *(Xylorhiza parryi* Gray) (Beath et al., 1934) and *Astragalus* (Beath et al., 1935, 1937; Knight and Beath, 1937; Byers, 1935, 1936; Byers et al., 1938), absorb Se from a soil from which crop plants do not absorb the element. Those plants capable of absorbing large amounts of Se from soils from which common crop plants and grasses absorb only small amounts have been called "converter" plants. The most important converter plants in South Dakota were reported to be *Astragalus racemosus, Stanleya bipinnata, Aster multiflorus, Gutierrezia sarothrae,* and *Grindelia squarrosa* (Moxon et al., 1939). When a converter plant dies, the Se it contains may be absorbed by a common crop plant which previously obtained little or no Se from the soil. Nonseleniferous crop plants appear to take up Se only after it has been converted to an organic form by a converter plant (Moxon et al., 1939).

Se is of considerable importance to the livestock industry, inasmuch as "alkali disease" has been shown to be caused by the grazing of seleniferous plants containing high concentrations of Se.

Silicon. Most plants do not contain high concentrations of Si, but scouring rushes may contain as much as 70% silica in the ash, and cereal straw frequently contains as much as 50% or more (Robinson and Edgington, 1945).

LITERATURE CITED

Ayers, A. D., C. H. Wadleigh, and O. C. Magistad. 1943. The interrelationships of salt concentration and soil moisture content with the growth of beans. J. Amer. Soc. Agron. 35:796–810.

Basslavskaya, S., and M. Syroeshkina. 1936. Influence of the chloride ion on the content of chlorophyll in the leaves of potatoes. Plant Physiol. 11:149–157.

Bearce, B. C., and H. C. Kohl, Jr. 1970. Measuring osmotic pressure of sap within live cells by means of a visual melting point apparatus. Plant Physiol. 46:515–519.

Beath, O. A., J. H. Draize, and C. S. Gilbert. 1934. Plants poisonous to livestock. Wyoming Agr. Exp. Sta. Bull. 200:1–84.

Beath, O. A., H. F. Eppson, and C. S. Gilbert. 1935. Selenium and other toxic minerals in soils and vegetation. Wyoming Agr. Exp. Sta. Bull. 206:1–55.

Beath, O. A., H. F. Eppson, and C. S. Gilbert. 1937. Selenium distribution in and seasonal variation of type vegetation occurring on seleniferous soils. J. Amer. Pharm. Ass. 26:394–405.

Beeson, K. C., V. A. Lazar, and S. G. Boyce. 1955. Some plant accumulators of the micronutrient elements. Ecology 36:155–156.

Bernstein, L. 1961. Osmotic adjustment of plants to saline media. I. Steady state.

Amer. J. Bot. 48:909–918.

Bernstein, L. 1963. Osmotic adjustment of plants to saline media. II. Dynamic phase. Amer. J. Bot. 50:360–370.

Bernstein, L., J. W. Brown, and H. E. Hayward. 1956. The influence of rootstock on growth and salt accumulation in stone-fruit trees and almonds. Proc. Amer. Soc. Hort. Sci. 68:86–95.

Boyer, J. S. 1967. Leaf water potentials measured with a pressure chamber. Plant Physiol. 42:133–137.

Boyer, J. S. 1968. Relationship of water potential to growth of leaves. Plant Physiol. 43:1056–1062.

Boyer, J. S. 1969. Measurement of the water status of plants. Annu. Rev. Plant Physiol. 20:351–364.

Byers, H. G. 1935. Selenium occurrence in certain soils in the United States, with a discussion of related topics. U. S. Dep. Agr. Tech. Bull. 482:1–48.

Byers, H. G. 1936. Selenium occurrence in certain soils in the United States with a discussion of related topics. Second rep. U. S. Dep. Agr. Tech. Bull. 530:1–79.

Byers, H. G., J. T. Miller, K. T. Williams, and W. H. Lakin. 1938. Selenium occurrence in certain soils of the United States with a discussion of related topics. 3rd Rep. U. S. Dep. Agr. Tech. Bull. 601:1–75.

Chatterton, N. J., C. M. McKell, F. T. Bingham, and W. J. Clawson. 1970. Absorption of Na, Cl, and B by desert saltbush in relation to composition of nutrient solution culture. Agron. J. 62:351–352.

Cholak, J. 1959. Fluorides: A critical review. I. The occurrence of fluoride in air, food, and water. J. Occupational Med. 1:501–511.

Clinton, G. P. 1915. Tobacco . . . . phyllodiniation or string leaves, pp. 27–29. *In* G. P. Clinton, Part I, Report of the Botanist for 1913. Connecticut Agr. Exp. Sta. 38th Annu. Rep. pp. 1–42.

Connor, D. J. 1969. Growth of grey mangrove (*Avicennia marina* var. *resinifera*) in nutrient culture. Biotropica 1:36–40.

Ehlig, C. F. 1962. Measurement of energy status of water in plants with a thermocouple psychrometer. Plant Physiol. 37:288–290.

Ehlig, C. F., W. R. Gardner, and M. Clark. 1968. Effect of soil salinity on water potentials and transpiration in pepper (*Capsicum frutescens*). Agron. J. 60:249–253.

Gale, J., and A. Poljakoff-Mayber. 1970. Interrelations between growth and photosynthesis of salt-bush (*Atriplex halimus* L.) grown in saline media. Australian J. Biol. Sci. 23:937–945.

Gale, J., R. Naaman, and A. Poljakoff-Mayber. 1970. Growth of *Atriplex halimus* L. in sodium chloride salinated culture solutions as affected by the relative humidity of the air. Australian J. Biol. Sci. 23:947–952.

Gardner, W. R., and C. F. Ehlig. 1965. Physical aspects of the internal water relations of plant leaves. Plant Physiol. 40:705–710.

Gardner, W. R., and R. H. Nieman. 1964. Lower limit of water availability to plants. Science 1943:1460–1462.

Gauch, H. G., and O. C. Magistad. 1943. Growth of strawberry clover varieties and of alfalfa and ladino clover as affected by salt. J. Amer. Soc. Agron., 35:871–880.

Gauch, H. G., and C. H. Wadleigh. 1942. The influence of saline substrates upon the absorption of nutrients by bean plants. Proc. Amer. Soc. Hort. Sci. 41:365–369.

Gauch, H. G., and C. H. Wadleigh. 1944. Effects of high salt concentrations on growth of bean plants. Bot. Gaz. 105:379–387.

Gauch, H. G., and C. H. Wadleigh. 1945. Effect of high concentrations of sodium, calcium, chloride, and sulfate on ionic absorption by bean plants. Soil Sci. 59:139–153.

Gauch, H. G., and C. H. Wadleigh. 1951. Salt tolerance and chemical composition of Rhodes and Dallis grasses grown in sand culture. Bot. Gaz. 112:259−271.

Greenshpan, H., and B. Kessler. 1970. Sodium fluxes in excised citrus roots under steady-state conditions, with particular reference to salinity resistance. J. Exp. Bot. 21:360−371.

Gwirtsman, J., R. Mavrodineau, and R. R. Coe. 1957. Determination of fluorides in plant tissue, air, and water. Anal. Chem. 29:887−892.

Haas, A. R. C., and J. N. Brusca. 1955. Fluorine toxicity in citrus. California Agr. 9(3):15−16.

Haise, H. R., H. J. Haas, and L. R. Jensen. 1955. Soil moisture studies of some Great Plains soils. II. Field capacity as related to 1/3 atmosphere percentage, and "minimum point" as related to 15- and 26-atmosphere percentages. Proc. Soil Sci. Soc. Amer. 19:20−25.

Hasson-Porath, E. and A. Poljakoff-Mayber. 1969. The effect of salinity on the malic dehydrogenase of pea roots. Plant Physiol. 44:1031−1034.

Hasson-Porath, E., and A. Poljakoff-Mayber. 1971. Content of adenosine phosphate compounds in pea roots grown in saline media. Plant Physiol. 47:109−113.

Hayward, H. E., and L. Bernstein. 1958. Plant-growth relationships on salt-affected soils. Bot. Rev. 24:584−635.

Hayward, H. E., and W. M. Blair. 1942. Some responses of Valencia orange seedlings to varying concentrations of chloride and hydrogen ions. Amer. J. Bot. 29:148−155.

Hayward, H. E., and E. M. Long. 1942. Vegetative responses of Elberta peach on Lovell and Shalil rootstocks to high chloride and sulfate solutions. Proc. Amer. Soc. Hort. Sci. 41:149−155.

Hayward, H. E., and O. C. Magistad. 1946. The salt problem in irrigation agriculture (Research at the United States Regional Salinity Laboratory). U.S. Dep. Agr. Misc. Pub. 607.

Hayward, H. E., and W. B. Spurr. 1944a. Effects of isosmotic concentrations of inorganic and organic substrates on entry of water into corn roots. Bot. Gaz. 106:131−139.

Hayward, H. E., and W. B. Spurr. 1944b. The tolerance of flax to saline conditions: Effect of sodium chloride, calcium chloride, and sodium sulfate. J. Amer. Soc. Agron. 36:287−300.

Hayward, H. E., and C. H. Wadleigh. 1949. Plant growth on saline and alkali soils. Advance. Agron. 1:1−38.

Hayward, H. E., W. M. Blair, and P. E. Skaling. 1942. Device for measuring entry of water into roots. Bot. Gaz. 104:152−160.

Hayward, H. E., E. M. Long, and R. Uhvits. 1946. Effect of chloride and sulfate salts on the growth and development of the Elberta peach on Shalil and Lovell rootstocks. U. S. Dep. Agr. Tech. Bull. 922.

Hsiao, E., and H. G. Gauch. 1971. Unpub. data.

Hutchinson, E. 1945. Aluminum in soils, plants, and animals. Soil Sci. 60:29−40.

Joshi, G., T. Dolan, R. Gee, and P. Saltman. 1962. Sodium chloride effect on dark fixation of CO_2 by marine and terrestrial plants. Plant Physiol. 37:446−449.

Kahane, I., and A. Poljakoff-Mayber. 1968. Effect of substrate salinity on the ability for protein synthesis in pea roots. Plant Physiol. 43:1115−1119.

Kaufmann, M. R., and A. N. Eckard. 1971. Evaluation of water stress control with polyethylene glycols by analysis of guttation. Plant Physiol. 47:453−456.

Kirkham, M. B., W. R. Gardner, and G. C. Gerloff. 1969. Leaf water potential of differentially salinized plants. Plant Physiol. 44:1378−1382.

Klepper, B., and H. D. Barrs. 1968. Effects of salt secretion on psychrometric determinations of water potential of cotton leaves. Plant Physiol. 43:1138−1140.

Knight, S. H., and O. A. Beath. 1937. The occurrence of selenium and seleniferous vegetation in Wyoming. Part I. The rocks and soils of Wyoming and their relations to the selenium problem. Part II. Seleniferous vegetation of Wyoming. Wyoming Agr. Exp. Sta. Bull. 221.

Kubota, J., V. A. Lazar, and K. C. Beeson. 1960. The study of cobalt status of soils in Arkansas and Louisiana using the black gum as the indicator plant. Proc. Soil Sci. Soc. Amer. 24:527—528.

Lagerwerff, J. V., and H. E. Eagle. 1961. Osmotic and specific effects of excess salts on beans. Plant Physiol. 36:472—477.

LaHayte, P. A., and E. Epstein. 1969. Salt toleration by plants: Enhancement with calcium. Science 166:395—396.

Lawler, D. W. 1970. Absorption of polyethylene glycols by plants and their effects on plant growth. New Phytol. 69:501—513.

Magistad, O. C., and R. F. Reitemeier. 1943. Soil solution concentrations at the wilting point and their correlation with plant growth. Soil Sci. 55:351—360.

Magness, J. R., L. O. Regeimbal, and E. S. Degman. 1932. Accumulation of carbohydrates in apple foliage, bark, and wood as influenced by moisture supply. Proc. Amer. Soc. Hort. Sci. 29:246—252.

Masaewa, M. 1936. Zur Frage der Chlorophobie der Pflanzen. Bodenk. Pflanzenernähr. 1:39—57.

Meiri, A., E. Mor, and A. Poljakoff-Mayber. 1970. Effect of time of exposure to salinity on growth, water status, and salt accumulation in bean plants. Ann. Bot. 34: 383—391.

Miller, G. W., and D. W. Thorne. 1956. Effect of bicarbonate ion on the respiration of excised roots. Plant Physiol. 31:151—155.

Milthorpe, F. L. 1956. The growth of leaves. Butterworths, London. 223 pp.

Moxon, A. L., O. E. Olson, and W. V. Searight. 1939. Selenium in rocks, soils, and plants. South Dakota Agr. Exp. Sta. Tech. Bull. 2.

Mozafar, A., and J. R. Goodin. 1970. Vesiculated hairs: A mechanism for salt tolerance in *Atriplex halimus* L. Plant Physiol. 45:62—65.

Nieman, R. H. 1962. Some effects of sodium chloride on growth, photosynthesis, and respiration of twelve crop plants. Bot. Gaz. 123:279—285.

Nieman, R. H. 1965. Expansion of bean leaves and its suppression by salinity. Plant Physiol. 40:156—161.

Noggle, J. C. 1966. Ionic balance and growth of sixteen plant species. Proc. Soil Sci. Soc. Amer. 30:783—766.

Pollak, B., and Y. Waisel. 1970. Salt secretion in *Aeluropus litoralis* (Willd.) Parl. Ann. Bot. 34:879—889.

Porath, E., and A. Poljakoff-Mayber. 1964. Effect of salinity on metabolic pathways in pea root tips. Israel J. Bot. 13:115—121.

Rémy, T., and H. Liesegang. 1926. Untersuchungen über die Rückwirkungen der Kaliversorgung auf Chlorophyllgehalt, Assimilationsleistung, Wachstum und Ertrag der Kartoffeln. Landwirt. Jahr. 64:213—240.

Richards, L. A., and L. R. Weaver. 1944. Moisture retention by some irrigated soils as related to soil-moisture tension. J. Agr. Res. 69:215—235.

Robinson, W. O. 1943. The occurrence of rare earths in plants and soils. Soil Sci. 56:1—6.

Robinson, W. O., and G. Edgington. 1945. Minor elements in plants, and some accumulator plants. Soil Sci. 60:15—28.

Robinson, W. O., H. W. Lakin, and L. E. Reichen. 1947. The zinc content of plants on the Friedensville zinc slime ponds in relation to biogeochemical prospecting. Econ. Geol. 42:572—582.

Rosenfeld, I., and O. A. Beath. 1964. Selenium. Academic, New York.

Schneider, G. W., and N. P. Childers. 1941. Influence of soil moisture on photosynthesis, respiration, and transpiration of apple leaves. Plant Physiol. 16:565–583.

Scholander, P. F., H. T. Hammel, E. Hemmingsen, and W. Garey. 1962. Salt balance in mangroves. Plant Physiol. 37:722–729.

Scholander, P. F., H. T. Hammel, E. Hemmingsen, and E. D. Bradstreet. 1964. Hydrostatic pressure and osmotic potential in leaves of mangroves and some other plants. Proc. Nat. Acad. Sci. 52:119–125.

Scholander, P. F., H. T. Hammel, E. D. Bradstreet, and E. A. Hemmingsen. 1965. Sap pressure in vascular plants. Science 148:339–346.

Scholander, P., E. D. Bradstreet, H. T. Hammel, and E. A. Hemmingsen. 1966. Sap concentrations in halophytes and some other plants. Plant Physiol. 41:529–532.

Shalhevet, J., and L. Bernstein. 1968. Effects of vertically heterogeneous soil salinity on plant growth and water uptake. Soil Sci. 106:85–93.

Shantz, H. L., and R. L. Piemeisel. 1940. Types of vegetation in Escalante Valley, Utah, as indicators of soil conditions. U. S. Dep. Agr. Tech. Bull. 713.

Shull, C. A. 1916. Measurement of the surface forces in soils. Bot. Gaz. 62:1–31.

Siegel, S. M. 1969. Microbiology of saturated salt solutions and other harsh environments. V. Relation of inosine-5 -phosphate and carbohydrate to growth of wildtype and mutant *Penicillium* in boric acid and potassium chloride selective media. Physiol. Plantarum 22:1152–1157.

Spencer, E. L. 1937. Frenching of tobacco and thallium toxocity. Amer. J. Bot. 24:16–24.

Spencer, E. L., and G. I. Lavin. 1939. Frenching of tobacco. Phytopathology 29:502–503.

Spoehr, H. A., and H. W. Milner. 1939. Starch dissolution and amylolytic activity of leaves. Proc. Amer. Phil. Soc. 81:37–78.

Stern, W. L., and G. K. Voigt. 1959. Effect of salt concentration on growth of red mangrove in culture. Bot. Gaz. 121:36–39.

Venkateswarlu, P., W. D. Armstrong, and L. Singer. 1965. Absorption of fluoride and chloride by barley roots. Plant Physiol. 40:255–261.

Wadleigh, C. H. 1946. The integrated soil moisture stress upon a root system in a large container of saline soil. Soil Sci. 61:225–238.

Wadleigh, C. H., and A. D. Ayers. 1945. Growth and biochemical composition of bean plants as conditioned by soil moisture tension and salt concentration. Plant Physiol. 20:106–132.

Wadleigh, C. H., and H. G. Gauch. 1942. Assimilation in bean plants of nitrogen from saline solutions. Proc. Amer. Soc. Hort. Sci. 41:360–364.

Wadleigh, C. H., and H. G. Gauch. 1944. The influence of high concentrations of sodium sulfate, sodium chloride, calcium chloride, and magnesium chloride on the growth of guayule in sand cultures. Soil Sci. 58:399–403.

Wadleigh, C. H., and H. G. Gauch. 1948. Rate of leaf elongation as affected by the intensity of the total soil moisture stress. Plant Physiol. 23:485–495.

Wadleigh, C. H., H. G. Gauch, and M. Kolisch. 1951. Mineral composition of orchard grass grown on Pachappa loam salinized with various salts. Soil Sci. 72:275–282.

Wadleigh, C. H., H. G. Gauch, and V. Davies. 1943. The trend of starch reserves in bean plants before and after irrigation of a saline soil. Proc. Amer. Soc. Hor. Sci. 43:201–209.

Wadleigh, C. H., H. G. Gauch, and O. C. Magistad. 1946. Growth and rubber accumulation in guayule as conditioned by soil salinity and irrigation regime. U. S. Dep. Agr. Tech. Bull. 925.

Wadleigh, C. H., H. G. Gauch, and D. G. Strong. 1947. Root penetration and moisture extraction in saline soil by crop plants. Soil Sci. 63:341–349.

Weigel, R. C., Jr. 1968. Maximal growth and yield of snapbeans (*Phaseolus vulgaris*). M. S. Thesis, Univ. Maryland, College Park.

Weigel, R. C. 1970. Mineral nutrition of soybeans (*Glycine max*). Ph.D. Thesis, Univ. Maryland, College Park.

Weimberg, R. 1970. Enzyme levels in pea seedlings grown on highly salinized media. Plant Physiol. 46:466–470.

Wilson, A. M., and G. A. Harris. 1968. Phosphorylation in crested wheatgrass seeds at low water potentials. Plant Physiol. 43:61–65.

Wilson, J. R. 1970. Response to salinity in *Glycine*. VI. Some effects of a range of short-term salt stresses on the growth, nodulation, and nitrogen fixation of *Glycine wightii* (formerly *javanica*). Australian J. Agr. Res. 21:571–583.

Wilson, J. R., K. P. Haydock, and M. F. Robins. 1970. The development in time of stress effects in two species of *Glycine* differing in sensitivity to salt. Australian J. Biol. Sci. 23:537–553.

CHAPTER 17

Theories of Limiting Factors

In the following discussion primary attention is given to the inorganic nutrition of plants, assuming that all other factors, such as temperature, light, organic nutrition, and so on, are optimal. Nightingale (1942a, b) is one of the few investigators who consistently evaluated the organic as well as the inorganic nutrition of the plant, and an exception is therefore made in presenting his concepts.

The principal question concerning limiting factors is whether *one* limiting factor controls growth at any given time or whether *two or more* factors may simultaneously control rate of growth. *The* controlling factor or factors could be either too low or too high. The fundamental question involved here can perhaps best be visualized by referring to Fig. 17.1.

Factors E and F are at two levels of excess, as far as the maximal theoretical yield is concerned. If E is K^+, it is possible that the excess, over and above that required for maximal yield, might have no deleterious effect, since it is generally believed that plants often have higher concentrations of K^+ than are required for maximal growth. However, if the plant contains this relative excess of an element that is required, but which is beneficial only within a fairly narrow range (e.g., B), the excess could be a limiting factor.

We are ordinarily more concerned, however, with deficiencies of one or more elements rather than excesses, and we therefore discuss in some detail the effect of a deficient factor (or factors) on growth. According to one school of thought, and according to a considerable proportion of the experimental data, growth appears to be controlled by *the* most limiting factor (i.e., C). According to this concept, additions of B and D, both of which are also limiting,

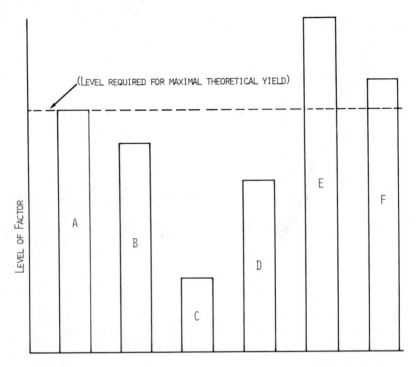

(LEVEL REQUIRED FOR MAXIMAL THEORETICAL YIELD)

LEVEL OF FACTOR

A

B

C

D

E

F

Fig. 17.1

Diagrammatic representation of six factors at various levels of insufficiency, sufficiency, or excess.

would be without effect; only upon the addition of C would there be an increase in growth. Other investigators argue that an increase in any limiting factor or factors increases growth—even though the factor is not *the* most limiting factor.

In addition, there is much controversy as to whether or not an increase in a limiting factor results in a linear response in growth (until the next-most-limiting factor becomes operative) or whether the response is nonlinear, that is, first increments have larger effects on growth than later increments.

There is even the possibility that response to certain nutrients may be linear, and to others nonlinear.

It would be of great economic value to know whether one or more than one limiting factor controls growth. Likewise, it would be helpful to know whether increments of a limiting factor produce a linear or nonlinear (i.e., ever-diminishing) response. If only *the most limiting* factor controls growth, then it would be important to determine, among various limiting factors that have been revealed by analysis, which one is *the* most limiting. Presumably, until *the* most limiting factor is eliminated, there is no reason to alleviating any other limiting factor(s). Next, the nature of the response, linear or nonlinear, would be important, inasmuch as it would have an influence on the increase in growth that might be expected from alleviating *the* most limiting factor by an increment of that factor.

The potentialities for growth are rarely appreciated, but scientists at the Plant Industry Station, Beltsville, Maryland, demonstrated how extraordinary these potentialities are. With improved cultural techniques and controlled environment growth chambers, phenomenal yields of 10 to 50 times above normal have been obtained (Cathey, 1967; Klueter et al., 1967, 1969; Krizek et al., 1967, 1968a, b, c, 1969; Bailey et al., 1968). For the most part these investigators used temperatures of 85°F day and 75°F night, and light was maintained as high as 4000 ft-c using cool-white fluorescent tubes supplemented by incandescent lighting for at least 15 hr/day. Relative humidity was held at 65% and CO_2 of the atmosphere was enriched to as much as 2000 ppm. Benefits of CO_2 enrichment of the atmosphere were much greater at 30°C (86°F) day temperature than at 24°C (75°F) (Krizek et al., 1968c).

Klueter et al. (1967) designed special plant growth chambers in which all operations are automatic, including the regulation of CO_2, light, and nutrition. Observations are automatically recorded by a data-recording system. By programming the system, the tape can be run through a computer and the data analyzed.

THEORIES

Liebig's Law of the Minimum

Liebig was one of the first plant physiologists to be interested in multiconditioned processes, such as growth, and factors that regulate them. In *Chemistry in its Application to Agriculture and Physiology,* Liebig (1840) discussed the "one indispensable factor," and stated that plant growth was directly proportional to the supply of the nutrient present in the minimum. His famous Law of the Minimum (Liebig, 1840, 1855) stated that, "the yield of any crop always depends on that nutritive constituent which is present in minimal amount." Note that Liebig's law concerned only nutrients, not water, O_2 supply, temperature, light intensity, and so on. In 1910 prizes were offered by the German Agricultural Society for the best symbolic illustration that would explain and dramatize the law of the minimum concept. The "minimum tub" was one of the more interesting portrayals of Liebig's Law (see Moulton, 1942, p. 78). In this illustration the various staves of a tub (or barrel) represent various essential nutrients and, in contrast with a regular tub, the staves are of various heights. The tub can be filled (the analogy to growth) only to the height of the shortest stave.

Liebig's idea was adopted by Blackman and broadened to include a multi-conditioned process—photosynthesis (Blackman, 1905). Incidentally, Blackman broadened Liebig's definition to include factors other than mineral nutrients: "When a process is conditioned as to its rapidity by a number of separate factors, the rate of the process is limited by the pace of the 'slowest' factor." Blackman was primarily interested in the effects of CO_2 supply and light intensity on the rate of photosynthesis (Blackman, 1905). The responses to increases in either of these factors were linear, and had sharp breaks where the next most limiting factor took over. Most investigators have found a linear response only in a restricted portion of the response curve, and a gradual tapering-off of the response rather than a sharp break as reported by Blackman (1905).

Mitscherlich's Law of Diminishing Returns

The law of diminishing returns was promulgated by Wollny (1891), who stated that the yield of crops, as regards quality and quantity, is determined by the factor that, under the existing conditions, functions in insufficient minimum or in oversufficient maximum amount or intensity. It was Mitscherlich (1909), however, who developed and popularized Wollny's law, and thus it has come to be associated with Mitscherlich rather than Wollny. In fact, Mitscherlich (1909) really proposed a different version of the law of the minimum when he stated that "the increase in yield per unit of limiting nutrient applied is directly proportional to the decrement from the maximal yield." This may be stated more simply by saying that a much larger increase in yield is obtained if the nutrient under consideration is far from adequate, as compared with the increase if that same nutrient is nearly adequate.

As may be seen in Fig. 17.2, some workers divide the nutrient–response curve into several segments. The minimum percentage segment is that portion of the curve over which there is very little growth response at the lowest of nutrient levels. The poverty adjustment segment is that portion of the curve over which there is a more-or-less linear increase in growth with increases in supply of the nutrient. The critical percentage or critical concentration (which has been called the sufficiency level or value by other workers) is the point on the nutrient–response curve above which luxury consumption occurs. Luxury consumption pertains to that part of the curve in which further increases in supply of a nutrient result in no further increase in growth.

Hanway and Dumenil (1965) define critical level as that level associated with maximal yield. For N and P in corn, the values were 3.2 and 0.34%, respectively. They also recognized an optimum level, the level associated with economic optimum (most profitable) yield.

Macy (1936) contended that the Mitscherlich law held only during the pov-

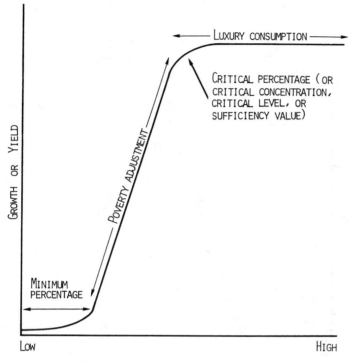

Fig. 17.2
General portrayal of the growth or yield response to increasing concentrations of a given nutrient—together with some of the terminology from the literature applied to various portions of the nutrient—yield response curve.

erty adjustment phase, while the Liebig law of the minimum held over the rest of the yield curve.

We see nothing "critical" in the critical percentage of critical concentration. It seems more pertinent to regard the poverty adjustment portion as a critical concentration range. In the poverty adjustment portion of the curve, the amount of nutrient supplied is critical. In Fig. 17.3 we suggest what we regard as appropriate terminology.

The Mitscherlich pot culture method for detecting soil deficiencies is rigidly prescribed with regard to container, amount of soil, variety of oats, and other details (Mitscherlich, 1930). For 100% fertility, the culture would contain 3.5 g of N, 0.7 g of P (calculated as P_2O_5), and 1.3 g of K (calculated as K_2O). For an acre of soil, these amounts would be 2250, 450, and 820 lb, respectively (Willcox, 1937). It followed, then, that a 5-1-2 fertilizer (or multiples thereof) was the best ratio for *all* kinds of plants.

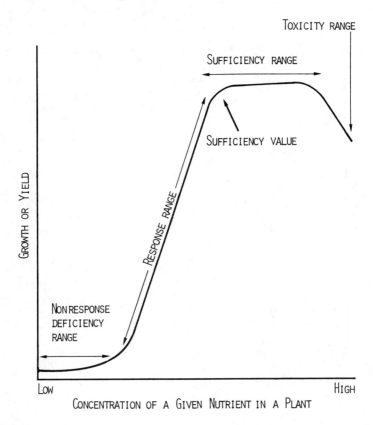

Fig. 17.3
Suggested terminology for the relationship between yield and the concentration of a given nutrient element.

Similarly to Liebig (1840), Wollny (1891) and Mitscherlich (1909) believed that a single factor was limiting at any one time, but Mitscherlich presented evidence that the response to additions of that factor was nonlinear (Mitscherlich, 1909). Mathematically, Mitscherlich's idea is

$$dy/dx = (A - y)C$$

where dy is the increase in growth or yield resulting from an increment of a growth factor dx, dx is an increment of a growth factor x, A is the maximal possible yield when all factors are at their optima, y is the yield after any given quantity of factor x has been applied, and C is a proportionality constant which varies with the nature of the growth factor. According to Mitscherlich, the values of C were: 0.122 for N, 0.60 for P_2O_5, and 0.40 for K_2O. In general, researchers have found that the value of C varies rather widely with different crops and with different environmental conditions.

Baule Product Law and Willcox's Fertility Index

Under conditions in which only one factor limits growth, Baule (see Will-cox, 1937) attempted to explain plant response to increments of this factor. Under such conditions he determined the amount of a nutrient required to pro-duce 50% of the maximal potential yield of the crop. He called this amount the "Baule unit." The next unit was assumed to give a further increase of 25%, the third unit 12½%, the fourth 6¼%, and so on, up to 10 units. In other words, each successive unit produces one-half the increase of the preceding unit. The Baule units for N, P_2O_5, and K_2O, in pounds per acre, are 223, 45, and 76, respectively.

Baule recognized that growth may be conditioned by two or more limiting factors. In such a case he reasoned that growth was paced or determined by the *product* of the degree of sufficiency of two or more deficient factors. He there-fore formulated the Baule product law which states that "if two or more limit-ing elements (or factors) are simultaneously increased, the combined effect will be the product of the individual effects." Inasmuch as 10 Baule units of any given factor would permit maximal potential growth, any deficient factor would be at some fraction of 10/10. To understand how the Baule product law operates, let us consider the case in which only three factors are limiting—N, K, and P. Let us suppose that K is at one-half sufficiency, N at three-fourths, and phosphorus at seven-eighths. The maximal growth that could be obtained in this case is then $K(1/2) \times N(3/4) \times P(7/8) = 32.81\%$. An increase in any one of the three deficient factors would produce an increase in growth, but the greatest effect would obviously come from an increase in the most deficient factor: $K(1) \times N(3/4) \times P(7/8) = 65.62\%$.

Willcox (1937) strongly championed Mitscherlich's idea of a nonlinear re-sponse to added increments of a deficient factor, but Willcox believed that two or more limiting factors could simultaneously determine growth. He agreed with Baule that an increase in any one of two or more limiting factors would increase growth. He therefore accepted the Baule unit concept and the Baule product law but chose to call the latter the "fertility index."

Willcox (1937) championed a concept which he called the "quantity of plant life." It was calculated as $\dfrac{318}{\% \text{ N in whole plant at maturity}}$. For ex-ample, in the case of a legume with 2.6% N, $\dfrac{318}{0.026} = 12{,}230$ lb plant ma-terial per acre; or sugarcane with 0.33% N, $\dfrac{318}{0.0033} = 93{,}366$ lb plant material per acre.

The formula agrees reasonably well with the maximal quantities of plant ma-terial per acre that have been attained on occasion under field conditions and appears to hold for many crops under ideal conditions for growth. The formula may therefore serve as an indication of the maximal amount of plant material that can be produced by a given crop. According to Willcox (1937), 318 represents the amount of N that can be absorbed in one season by an annual crop growing on 1 acre of land. He hypothesized that yield of a crop (or, rather, total dry weight of a crop) was inversely related to the percentage of N in its tissues.

The ideas of Mitscherlich, Baule, and Willcox are supported by the fact that growth, as a function of nutrient input, is logarithmic and follows a pattern of diminishing increases—as dictated by the Mitscherlich equation. Often plants that produce the larger yields of dry matter have lower percentages of N in their tissues than do those that produce smaller yields of dry matter.

Lagatu and Maume

These French scientists originated the concept that growth is a function of the "intensity" and "quality" of nutrition (Lagatu and Maume, 1934). Certainly most physiologists would agree that balance, quantitatively and qualitatively, is an important consideration in the nutrition of plants. Lagatu and Maume argued that, at any given intensity of nutrition, the quality of nutrition could make a great difference, that is, which salts contributed to this given intensity of nutrition. Conversely, at any given qualitative composition, the overall concentration or intensity of that type of a nutrient solution could make a great difference. The argument concerning acceptance or rejection of this general philosophy hinges partly on the question how large the differences in intensity and quality must be in order to affect growth appreciably. Balanced nutrition has clearly meant different things to different investigators.

Shear, Crane, and Myers

Following the lead of Lagatu and Maume (1934), the concept of intensity and quality was called "nutrient element balance" and was championed as such by Shear and Crane (1947) and Shear et al. (1946, 1948). The latter investigators presented considerable data emphasizing the significance of nutrient element balance. Working with tung trees in sand culture, Shear et al. (1953) observed and discussed the many interactions among elements—particularly those affecting the concentrations of various elements in leaves. For example, B and the heavy metals—Al^{3+}, Cu^{2+}, Fe^{3+}, Mn^{2+}, and Zn^{2+}—decreased in concentration in leaves as K^+ was increased. Total dry weights of leaves and whole plants were affected largely by the level of K^+ supply.

There may, however, be some question as to the significance of balance if all of the essential nutrients are at or above the levels required for maximal growth. Unless one or more of these should, in addition, be high enough to be toxic, a variation in concentration above adequacy may have little or no effect.

In this regard, Ulrich (1948) reported that "plants with widely different nutrient composition have similar yields as long as these nutrient concentrations are well above the critical level."

It is possible that the nutrient element balance concept may place too much emphasis on nutrient elements since, as Ulrich (1948) noted, "there is little value in stressing the proper mineral balance within the plant without simultaneously considering this balance in relation to the organic constituents required for growth." In this same regard, Nightingale (1942a) clearly showed for pineapple plants the importance of adjusting, through fertilization, the NO_3^- concentration within the plant in terms of the plant's carbohydrate reserves. Plants high in sugars could use extra N, whereas those low in sugars could not.

Numerous workers, among them Bear and Prince (1945), reported that there is a remarkably constant concentration of bases in a given species. In alfalfa, they noted that the sum of the equivalents of Ca^{2+}, Mg^{2+}, and K^+ per unit of dry matter tends to be a constant, approaching 170–187 meq/100 g of dry plant material. Their conclusion seems to devalue that of Shear et al. (1953) to the effect that balance (i.e., ratio) among ions is extremely important in plants. Bear and Prince (1945) concluded that each of these ions (Ca^{2+}, Mg^{2+}, K^+) has at least two functions in the plant—one specific and the other or others capable of being performed interchangeably by all three cations. They reasoned that once the supply of each cation is adequate to meet the specific needs for it, there could be a wide range in ratios in the remaining quantities absorbed by the plant to meet its total cation needs.

Nightingale

Nightingale (1942a, b) studied in detail the factors that limit growth of pineapples in Hawaii. The pineapple plant is ideally suited to this type of study inasmuch as: (1) the crop is vegetatively propagated and genetic variability is virtually nil; (2) the crop requires 2 years, and thus there is adequate time to study the plants and to manipulate nutrition to improve growth and yield; (3) irrigation is practiced and thus water need not be limiting; and (4) there is no varietal effect to be considered.

Nightingale (1942a) championed the idea of *a* limiting factor at any given time. He reasoned that it was important to know *the* limiting factor in growth at any given time even though it was determined that it was a factor over which man had no control, for example, inadequate sunlight. If light were *the* most limiting of all of the limiting factors observed, he believed that there was no reason to correct a lesser deficiency that *was* within his control. If the most limiting deficiency were one over which man had control, such as the matter of adding K^+, then the necessary step was taken to alleviate the deficiency. Therefore, Nightingale was able to effect savings for growers by recommending against the addition of fertilizer when, for example, light was *the* most limiting factor and fertilizer would have no beneficial effect on growth; conversely, he recommended the addition of a fertilizer element when it was known that its deficiency was *the* most limiting factor. In the latter case the use of fertilizer resulted in returns in excess of the cost of the fertilizer.

Before Nightingale could ascertain when a given nutrient was deficient, he had to determine the concentration coincident with maximum yield. Plotted data had to reveal that when the concentration of this nutrient was above some given level there was no further increase in yield. The concentration above which higher yields did not occur was regarded as a sufficiency for that nutrient. Sufficiency values for the various nutrients could only be obtained by amassing a large number of comparisons of nutrient concentration versus yield. Smith (1962) reported that, "it is a long and arduous task, encompassing studies made over a period of many years, to obtain even tentative [sufficiency] values." In at least *some* of these comparisons, no factor other than the one being studied and plotted is limiting. If a crop has a concentration of K^+, for example, as high as the value regarded as sufficient, but the yield is low, some factor other than K^+ is obviously responsible. Even the operation of nutrient element balance does not preclude determining the sufficiency concentration of a given nutrient. Nutrient element balance could be involved in the differing yields of crops that all have the same K^+ concentration, for example, but considerations of balance do not preclude determining the relationship between yield and the sufficiency level of a given nutrient. Several of these points of discussion are illustrated in Fig. 17.4. In this figure only the comparisons of yield versus nutrient concentration that fall near or on the curve are of importance in determining the relationship between yield and, in this hypothetical example, K^+. The yields for B and C theoretically *could have been* as high as those for A, since all three samples had the same concentration of K^+. The plants in sample B may have had sunlight as the limiting factor; the low yield in C might have been related to an unfavorable nutrient element balance. A represents a case in which no factor other than K^+ limited growth, hence the maximal growth with this concentration of K^+ became established. The K^+ concentration for point D is as favorable as that at point F, since as much growth or yield occurred at D as at F. Again, the yield at E, as compared with D, was the result of some factor other than K^+ and, again, it may have involved a matter of balance or a most limiting deficiency of some other nutrient or a factor such as sunlight. A high enough concentration of K^+ was present at G to result in some inhibition of growth. In view of the extreme massing of all of the plots below the curve, H and the other very scattered points above the curve must be assigned to an error in measurement of yield or of K^+ concentration.

With many crops the time of sampling could be standardized, and as early a

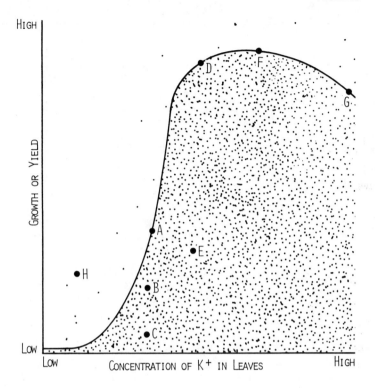

Fig. 17.4

Hypothetical relationship between yield and concentration of K^+ in leaves. (See text for explanation of letters A–H on graph.)

date as possible would probably be selected in order to have time for analyses and to supply any needed nutrient(s) to the crop. If the growth period of a crop were sufficiently long, as is the case with pineapple, for example, one would want to know the sufficiency values for K^+, Ca^{2+}, and other nutrients at different stages of growth. The same kind of observations would be plotted as just discussed, except this time determination would be made of the K^+ concentration, for any *given* stage of growth, above which there would be no reason to raise the K^+ concentration, inasmuch as higher K^+ concentrations *at that stage* never correlated with still higher ultimate yields. The relationship for NO_3^- concentration versus stage of growth is shown for pineapple in Fig. 17.5. (Nightingale, 1942a). Also, with regard to stage of growth of pineapple, leaf P was adequate at 0.02% during vegetative growth, but a concentration of at least 0.028% was required during flower bud differentiation. It was concluded that leaf K should probably not be permitted to drop much below 0.38% at any time (Nightingale, 1942b).

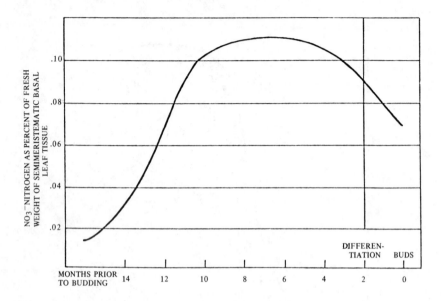

Fig. 17.5

Proximate maximal percentages of leaf NO_3^- attainable at different stages of growth of pine-apples. Maximal NO_3^- is desirable only if plants contain adequate carbohydrate reserves, as indicated in warm sunny location by not less than 15% and in cool or cloudy location by about 30% no. 1 yellow-green color of leaves. From G. T. Nightingale (1942a) Bot. Gaz., 103:409–456. The Univ. of Chicago Press and copyright by the University of Chicago—by permission.

One final question that could logically be asked concerns the matter of varie-ties of a given crop. Are the sufficiency values for a given nutrient essentially the same for different varieties? Although there is a possibility that there might be differences in the sufficiency values for different varieties, these differences have not been found to be very great. Even if varietal differences should prove worthy of consideration, a given variety of a crop is generally grown in sufficient acreage to justify determining its specific sufficiency values for the various essential elements.

In addition to determining the concentration of various inorganic nutrients in the meristematic zone at the base of leaf blades (such tissue is always physio-logically young), Nightingale also made certain other observations or determi-nations. Even when leaf NO_3 was found to be deficient, Nightingale would not consider adding NO_3^- when carbohydrate reserves in leaves were too low (that is, if carbohydrate reserves constituted *the* most limiting factor), since carbohydrate is required for reduction and utilization of NO_3^-. He found that leaf color could be correlated with carbohydrate reserve of leaves, and from then on leaf color was used as the index rather than time-consuming car-bohydrate analyses (Nightingale, 1942a). He also determined the percentage of root tips that were white, since the greater the percentage of white root tips the greater the capacity of the plant to absorb applied fertilizer. If roots were deemed to be dormant *primarily* because of insufficient soil moisture, soil moisture would be corrected and root growth thereby stimulated prior to the addition of a deficient fertilizer element. A "test pull" on representative plants indicated the extensiveness of the root system and in turn the overall capacity of the plant to absorb nutrients under consideration for addition to the crop.

In a nutrition study with algae, in which all other factors of the physical and chemical environment could be made optimal, it can be seen how readily the sufficiency value can be determined with considerable precision (Fig. 17.6) (Gerloff and Fishbeck, 1969).

Ulrich

Ulrich (1948) emphasized the complexities of plant growth, and the many factors that affect this multiconditioned process. He noted that: growth $=$ f (soil, climate, time, plant, management, and so on). Although it is admittedly difficult to solve such a complex equation or set of factors as they affect growth, it follows that an evaluation of only the soil factor, for example, cannot possibly be relied upon to produce maximal crop production. The same is true of course for each of the other factors entering into the equation for growth or yield.

Ulrich noted that two soils with the same concentration of a given nutrient may or may not meet the requirements of a crop depending on the concentration of other elements. For example, two soils each with adequate Fe^{3+} might differ radically in the concentration of Mn^{2+}. If one of the soils had an excessively high concentration of Mn^{2+}, that element might interfere with absorption of Fe.

Further, during favorable weather a crop utilizes more nutrients in growth, when they are available, than during unfavorable weather. Therefore, a given soil may meet the nutrient requirements of a crop when the growth rate and set of fruit are poor, but fail to do so when growth rate and set of fruit are high.

Clements

Clements (1964), working with sugarcane in Hawaii, regarded fitness of a crop to its energy regime (receipt and dispersal) as the dominant influence in crop productivity. He regarded the following indexes as the two most important with regard to the crop's fitness: (1) tissue moisture level (sheath moisture index) as the most important; and (2) carbohydrate level of this same tissue as the second most important index. He also determined and evaluated the N index, P index, Ca^{2+} index, Mg^{2+} index, and others.

Sheath moisture level of sugar cane dominated not only growth of the plants but also the levels of nutrients the plants maintained (Clements, 1951).

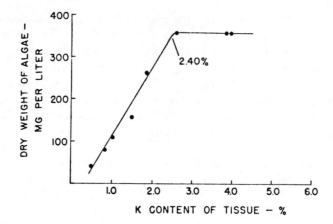

Fig. 17.6

Dry weight of a green alga, *Draparnaldia plumosa,* as related to the K^+ concentration of the tissue. A critical concentration (i.e., sufficiency value) of 2.40% (dry weight basis) is indicated. (Gerloff and Fishbeck, 1969.)

Tyner

Tyner (1946) studied the relationship between corn grain yields and the N, P, and K^+ concentrations of the sixth leaf from the bottom of the plant. He also attempted to establish the critical concentration for each element. Critical concentration was defined as "the optimum concentration of a nutrient above which response to further increments of this nutrient are doubtful or occur at rapidly diminishing rates." He considered this interpretation to be in accord with the one proposed by Macy (1936) which Macy called the "critical percentage."

Tyner (1946) obtained highly significant correlation and regression coefficients for the relationships of yields to percentages of N, P, and K^+. For each change of 0.1% N, P, or K^+ in the sixth leaf at the bloom stage, corn yields varied 4.43 ± 1.1, 25.3 ± 0.67, and 2.05 ± 0.93 bushels, respectively. The critical N, P, and K concentrations of the sixth leaf were tentatively set at 2.90% N, 0.295% P, and 1.30% K^+ (dry weight basis).

Bray

Bray suggested the nutrient mobility concept in which he claimed that crop yields obey the Mitscherlich nonlinear response curve for relatively immobile elements in the soil such as P and K^+, but Liebig's linear-response law of the minimum for mobile nutrients such as NO_3^-. An excellent discussion of the use of this concept appears in a paper by Vavra and Bray (1959). Confirmation of the concept has been reported by Bray (1963). For further discussion of Bray's and other concepts, the reader is referred to Tisdale and Nelson (1966).

Bray's concept may provide a reconciliation, then, for the conflicting evidence with regard to the response to increments of a deficient nutrient, that is, the response to some nutrients may be linear, and to others nonlinear. As is so often the case with strongly divergent points of view, the truth may ultimately involve, in part, both views; one view explains one portion of the evidence, and the other view the rest of the evidence.

LITERATURE CITED

Bailey, W. A., D. T. Krizek, and H. H. Klueter. 1968. Controlled environment and the genies of growth, pp. 2–12. *In* Science for better living (The Yearbook of Agr., 1968). U. S. Government Printing Office, Washington, D.C.

Bear, F. E., and A. L. Prince. 1945. Cation-equivalent constancy in alfalfa. J. Amer. Soc. Agron. 37:217–222.

Blackman, F. F. 1905. Optima and limiting factors. Ann. Bot. 19:281–295.

Bray, R. H. 1963. Confirmation of the nutrient mobility concept of soil-plant relationships. Soil Sci. 95:124–130.

Cathey, H. M. 1967. Photobiology in horticulture. Proc. 17th Int. Hort. Congr. 3: 385–391.

Clements, H. F. 1951. Environmental influences on the growth of sugar cane, pp. 451–469. *In* E. Truog (ed.), Mineral nutrition of plants. Univ. Wisconsin Press, Madison.

Clements, H. F. 1964. Foundations for objectivity in tissue diagnosis as a guide to crop control. Plant Anal. Fertilizer Problems 4:90–110.

Gerloff, G. C., and K. A. Fishbeck. 1969. Quantitative cation requirements of several green and blue-green algae. J. Phycol. 5:109–114.

Hanway, J. J., and L. Dumenil. 1965. Corn leafn analysis – The key is correct interpretation. Plant Food Rev. 11:5–8.

Klueter, H. H., W. A. Bailey, and D. T. Krizek. 1969. Plant growth chambers for engineered environments. Agr. Res. 17:5.

Klueter, H. H., W. A. Bailey, H. M. Cathey, and D. T. Krizek. 1967. Development of an experimental growth chamber system for studying the effects of major environmental factors on plant growth. Paper presented at winter meeting, Amer. Soc. Agr. Eng., 67–112.

Krizek, D. T., H. M. Cathey, H. H. Klueter, and W. A. Bailey. 1967. Escalating the growth and development of F_1 hybrid annuals by use of controlled environments. Abstr. Amer. Soc. Hort. Sci. Meeting.

Krizek, D. T., W. A. Bailey, and H. H. Klueter. 1968a. Accelerated growth of seedlings and precocious flowering under controlled environments at elevated CO_2, light, and temperature. Plant Physiol. 43(Suppl.):S-32.

Krizek, D. T., W. A. Bailey, and H. H. Klueter. 1968b. Benefits of carbon dioxide enrichment and direct seeding of F_1 hybrid annuals under controlled environments. Abstr. Amer. Soc. Hort. Meeting.

Krizek, D. T., W. A. Bailey, H. H. Klueter, and H. M. Cathey. 1968c. Controlled environments for seedling production. Paper presented at 18th annual meeting, Int.

Plant Propagators' Soc., Eastern Region, Toronto, Canada. Dec. 6, 1968.

Krizek, D. T., H. H. Klueter, and W. A. Bailey. 1969. From seed to flower, fast . . . . in controlled environments. Agr. Res. 17:3—4.

Lagatu, H., and L. Maume. 1934. Recherches sur le diagnostic foliare. Ann. Ecole Nat. Agr. Montpellier 22:257—306.

Liebig, J. von. 1840. Chemistry in its application to agriculture and physiology. Peterson, Philadelphia.

Liebig, J. von. 1855. Die Grundsätze der Agricultur-chemie mit Rücksicht auf die in England angestellten Untersuchungen. F. Vieweg und Sohn, Braunschweig, Germany.

Macy, P. 1936. The quantitative mineral nutrient requirements of plants. Plant Physiol. 11:749—764.

Mitscherlich, E. A. 1909. Das Gesetz des Minimums und das Gesetz des abnehmenden Bodenertrages. Landwirt. Jahrb. 38:537—552.

Mitscherlich, E. A. 1930. Die Bestimmung des Dungerbedurfnisses des Bodens. 3rd ed. Paul Parey, Berlin.

Moulton, F. R. 1942. Liebig and after Liebig — A century of progress in agricultural chemistry. Pub. 16. Amer. Soc. Advance. Sci., Washington, D. C. 111 pp.

Nightingale, G. T. 1942a. Nitrate and carbohydrate reserves in relation to nitrogen nutrition of pineapple. Bot. Gaz. 103:409—456.

Nightingale, G. T. 1942b. Potassium and phosphate nutrition of pineapple in relation to nitrate and carbohydrate reserves. Bot. Gaz. 104:191—223.

Shear, C. B., H. L. Crane, and A. T. Myers. 1946. Nutrient-element balance: A fundamental concept in plant nutrition. Proc. Amer. Soc. Hort. Sci. 47:239—248.

Shear, C. B., and H. L. Crane. 1947. Nutrient-element balance. U. S. Dep. Agr. Yearbook, 1943—1947:592—601.

Shear, C. B., H. L. Crane, and A. T. Myers. 1948. Nutrient-element balance: Application of the concept to the interpretation of foliar diagnosis. Proc. Amer. Soc. Hort. Sci. 51:319—326.

Shear, C. B., H. L. Crane, and A. T. Myers. 1953. Nutrient element balance: Response of tung trees grown in sand culture to potassium, magnesium, calcium, and their interactions. U. S. Dep. Agr. Tech. Bull. 1085.

Smith, P. F. 1962. Mineral analysis of plant tissues. Annu. Rev. Plant Physiol. 13: 81—108.

Tisdale, S. L., and W. L. Nelson. 1966. Soil fertility and fertilizers. Macmillan, New York. 694 p.

Tyner, E. H. 1946. The relation of corn yields to leaf nitrogen, phosphorus, and potassium content. Proc. Soil Sci. Soc. Amer. 11:317—323.

Ulrich, A. 1948. Plant analysis — Methods and interpretation of results, pp. 157—199. *In* H. B. Kitchen (ed.) Diagnostic techniques for soils and crops (Their value and use in estimating the fertility status of soils and nutritional requirements of crops). Amer. Potash Inst., Washington, D.C.

Vavra, J. P., and R. H. Bray. 1959. Yield and composition response of wheat to soluble phosphate drilled in the row. Agron. J. 5:326.

Willcox, O. W. 1937. ABC of agrobiology. Norton, New York. 323 pp.

Wollny, E. 1891. Die Kultur der Getreidarten mit Rücksicht auf Erfahrung und Wissenschaft. Winter, Heidelberg, 67 pp.

CHAPTER 18

Specific Limiting Factors

NUTRIENT DEFICIENCIES AS LIMITING FACTORS

When essentiality was established for a given element, exhaustive research concerning the concentration of that element in plants associated with deficiency generally followed. Deficiency values for various essential elements have been determined for certain crops, and these values have been most helpful. They are of importance, following an analysis of plant material, in order to determine whether the concentration lies within or above the deficiency range.

It has been argued that *a* definite minimal leaf content for each essential element does not exist, inasmuch as the deficiency concentration varies depending on the concentrations of other nutrients (Shear and Crane, 1947). It was concluded that the only infallible sign of a deficiency is stunting of growth. There appears to be no disagreement as to the universality of stunting at the onset of a deficiency of any essential element. With regard to a deficiency concentration for each essential element, it would probably be generally agreed that it is indeed preferable to speak of a deficiency range (of concentration). However, even a range is useful in the interpretation of a chemical analysis of plants suspected of being deficient in an essential nutrient. From the standpoint of maximal crop production, however, the sufficiency value for each essential element is *really* the important consideration. Whereas it is relatively easy to observe deficiency symptoms for a given element and then to determine the concentration of the element in plants when the deficiency appears, it is more difficult to determine sufficiency values. For a given element the sufficiency value can be determined only when *all* factors, other than the one being studied, are at their optima, hence not limiting.

Both plant and soil analyses are useful for determining the relationship between concentration of a nutrient and growth. In comparing plant and soil analyses with ultimate yield, data are inevitably obtained that pertain to the deficiency range and to the sufficiency value or sufficiently level. Analyses of the crop and the soil, then, provide this type of information.

In *Diagnostic Criteria for Plants and Soils,* Chapman (1966) assembled material of value in diagnosing the nutritional status of plants, the fertility status of soils, and methods of dealing with them under various circumstances. Somewhat earlier, Ulrich (1952) discussed the physiological basis for assessing nutritional requirements of plants. Reuther (1961) covered plant analysis and fertilizer problems for many crops, and Ulrich (1961) presented a detailed and very illuminating evaluation of plant analysis in the nutrition of sugar beet plants.

Plant Tests

Sampling. Before describing analytical tests that can be made on plants, it is important to discuss the problem of sampling in regard to position on the plant and time and frequency of sampling. A chemical method might fail to provide any information about a certain deficiency when one part of the plant is sampled but provide a clear-cut indication of the deficiency when another plant part is sampled.

Analyses of leaf petioles have proved far superior to analyses of other plant parts in assessing the K^+ nutrition of grapes and ladino clover, and the NO_3^- status of grapes and sugar beets.

In tomatoes leaf blades proved to be better indicators of the nutrient status of the plant than rachises of leaves or the stem. Juice from the midrib of leaves from the lower half of a tobacco plant gave the best indication of N, P, and K deficiencies (Shear, 1943). In lima beans the best indicator of P status was the petiole of the second leaf, a recently matured leaf, which should have approximately 500 ppm of acetic acid-soluble P on a dry weight basis (Ulrich and Berry, 1961). This was regarded as the critical concentration.

Inasmuch as Na^+ tends to stay primarily in the roots of nonhalophytic plants, such as beans and peaches, roots are far more indicative of Na^+ concentration in the soil than are the leaves which receive relatively little of the Na^+.

Nightingale (1942a, b) used the semimeristematic white basal tissues of pineapple leaves for NO_3^-, total-K, and total-P analyses.

In general, analyses of fruits have been very poor indicators of the nutritional status of a plant. Even in the presence of rather pronounced deficiency, the composition of fruits tends to remain remarkably constant. The constancy of floral and fruit characteristics, as compared with vegetative characteristics, has long been recognized by plant taxonomists, and taxonomic keys have been primarily based on floral and fruit characteristics. It is perhaps not too surprising, then, that the composition of fruits tends to be remarkably stable. The striking example in this regard pertains to the cotton plant. Although a given plant may form several flowers, if the plant has reserves for the development of only one boll, only one flower sets and that one boll is remarkably normal. If the plant can form 50 bolls, then each of the 50 bolls is fully formed.

The most recently fully matured leaf is often chosen for analysis (Hylton et al., 1967). This leaf is less variable in percent dry weight than young or old leaves. There is always a question as to what constitutes young and old, whereas the most recently fully matured leaf is fairly easy to identify. In general, it has proved a good indicator of the nutrient status of a plant.

For certain purposes it may be desirable to collect leaves from the top part of the plant, and for other purposes from the bottom. For example, Mg^{2+} is mobile within the plant and moves from older to younger leaves in a deficiency. For deficiencies of mobile elements, then, older leaves or stems may be better indicators than younger leaves (Nightingale, 1942b). However, Mn^{2+} does not move from older to younger leaves; younger leaves show Mn-deficiency symptoms when the element is deficient, and in this case the younger leaves are better indicators of the deficiency than older ones. When accumulation to the point of toxicity is suspected, as in B toxicity, older leaves, having been on the plant the longest, will most strikingly indicate accumulation of the toxic constituent. Inasmuch as B is nonmobile, if a deficiency is suspected, the younger parts may give a better indication of deficiency.

It is difficult to suggest any specifics regarding the time and frequency of sampling, since only experimentation can ultimately settle these questions. The only general dictum that can be stated safely is that early sampling is most desirable before secondary effects obscure the picture. For example, let us suppose that a crop is growing in a soil in which there is a deficiency of K^+ and of NO_3^-. Let us further assume that K^+ first becomes limiting, and that K-deficiency symptoms develop. If K^+ later becomes available more rapidly than NO_3^-, NO_3^- then becomes the limiting factor in growth and, because NO_3^- limits growth, K^+ could actually accumulate in the plants. An analysis would reveal a high K^+ concentration and yet, still visible on the plants, would be the K-deficiency symptoms. An early sampling, when K-deficiency symptoms first appeared, would have indicated a K^+ deficiency at least at that time, and the high-K^+ reading *and* evident K^+-deficiency symptoms at the later sampling would not be irreconcilable.

Types of Analyses. With the development of instrumental methods of analysis, Chapman (1966) concluded that there is little to be said in favor of quick tests and the various extraction methods. He reported that to date total values (expressed on a dry mater basis) correlate just as well with plant performance as the values determined by various extraction procedures.

There are various solutions or reagents for determining the concentrations of various nutrients in fresh plant tissue or juice squeezed from fresh tissue. The best known and most widely used reagent is probably the diphenylamine reagent which gives a blue color with NO_3^-. It may be applied to a cross section or longitudinal stem section or to juice from leaves, stems, or other plant parts. The intensity of the blue color is a measure of NO_3^- concentration. Tests for NO_3^- and other constituents have been described in detail (Emmert, 1934, 1942; Pettinger, 1931). In sap from corn plants, the following concentrations of nutrients were regarded as sufficient: NO_3^-, 300 ppm; total P, 0.2 mg/ml as P_2O_5; and, K^+, 2.0 mg/ml as K_2O (Pettinger, 1931). It was noted that K^+ concentration in sap is a good indicator of K^+ nutrition of corn plants. Low K^+ was associated with accumulation of Fe^{3+} in nodes of the plants.

"Tissue-test" kits contain paper strips impregnated with various reagents for testing for different nutrients. Similarly to the test solutions, these strips can be applied to the surface of a cut stem or to juice obtained from a plant part. Color reactions of the paper indicate the relative abundance of the nutrient.

In general, chemical analyses are far more specific and accurate than quick tests, but the analyses are more time-consuming. With the advent of more rapid analytical procedures involving flame emission or absorption spectrophotometers, autoanalyzers, and other instruments, quick tests are being supplanted by the more reliable measurements of analytical procedures. Chemical analyses can be made on expressed sap or on fresh or dried plant material.

Soil Tests

Soil tests have the unique advantage over plant tests in that they can be made prior to planting when it is easiest to supply fertilizers to correct shortages of essential nutrients. In 1963 nearly 3,000,000 soil samples were analyzed in the United States to help farmers and homeowners make proper soil treatments (Enfield, 1964).

For years soil-testing laboratories have primarily employed rapid spot tests to indicate the relative amounts of nutrients in soils. These have undoubtedly served a very useful purpose despite their recognized lack of accuracy. With the advent of more rapid procedures, chemical analyses are beginning to replace the less reliable quick tests.

There can be no question that plant and soil tests both have their place and value. Whenever possible, both the plant and the soil should be analyzed because of the different purposes these tests serve; one test aids in the interpretation of the other. There are certain advantages and disadvantages of soil and of plant tests. The primary advantage of soil tests is that they can be performed before the crop is planted, so that remedial steps can be taken prior to planting. The grower does not have to wait until a deficiency appears on the crop before attempting to correct it. Crop growth can never be as good following the correction of a deficiency as it would have been had the deficiency never been permitted to occur.

The principal disadvantage of the soil test is that it may not accurately predict availability to and uptake by the plant. There may be little correlation between fertilization or soil analysis and the composition of plants even for a given area of similar soils (Sims and Volk, 1947). Second, a consideration only of the amounts of nutrients in soil may prove misleading if there is an unassessed soil condition that interferes with uptake of one or more nutrients. A heavy soil may be well supplied with nutrients but so limit root extension that there is insufficient absorption of nutrients for normal growth (Lilleland and Uriu, 1960). We have already noted that "adequate" Fe^{3+} can prove inadequate, as far as the plants are concerned, in the presence of excessively high concentrations of Mn^{2+}. Likewise, a high pH may render "adequate" levels of Zn^{2+}, Cu^{2+}, Mn^{2+}, and B inadequate for a crop.

Stout and Overstreet (1950) concluded that "since the mineral content of the growing plant itself is the result of integration of the factors of its soil and climatic environment and of its own genetic powers of absorption, chemical tests of plant tissue should be indicative of the plant's own degree of success in extracting an adequate mineral supply from the soil in which it is growing."

In some cases plant tests have proved more reliable for determining deficiencies of nutrients than have soil tests. Analyses of juice from the midrib of leaves from the lower half of tobacco plants formed a more accurate basis for making recommendations as to supplemental fertilizer treatment than an analysis of the soil for N, P, and K (Shear, 1943). For fruit trees in California, Lilleland and Uriu (1960) reported that leaf analyses proved more accurate than soil analyses, particularly for assessing elements such as Zn^{2+}, Fe^{3+}, Mn^{2+}, and B.

Tyner (1946), Hill and Cannon (1948), and Kenworthy and Gilligan (1948) obtained significant relationships between tissue analyses and growth responses of various crops. Plant tissue analyses can serve as guides for developing efficient fertilizer programs for cotton (Tucker, 1965). N shortages, for example, can be detected before deficiencies appear, and fertilizer can be applied to prevent any loss in yield. Plant analyses are much less costly than well-conducted field fertilizer trials (Lorenz, 1965).

With the possible exception of N, some workers concluded that there is not a satisfactory relationship between tissue tests or plant analyses and crop yields (Wolf and Ichisaka, 1947; Atkinson et al., 1948; Chubb and Atkinson, 1948).

Vandecaveye (1948), and others, studied various responses of microorganisms in regard to their reliability in determining the concentrations of Cu^{2+}, Mg^{2+}, K^+, and P in soils. Analyses for these ions by biological assays called for very careful purification of salts and rigorous attention to procedural details. A few hundredths of a part per million in the soil solution or a nutrient solution constitute sufficiency for most of these elements. The biological assays were quite reliable in the hands of skilled researchers, but the assays were nevertheless time-consuming; it was not easy to run large numbers of samples.

Flame emission and absorption spectrophotometry have largely supplanted biological assays. In atomic absorption spectrophotometry, for example, 1 ng of Zn^{2+} can be detected.

Field trials have the advantage of allowing the soil and the plant to be considered simultaneously. It is reasoned that over a period of years field fertilizer trials will indicate the best treatment on the average. The fact that one treatment is best one year and another treatment another year indicates that crop requirements vary according to environmental conditions. There are few if any years that are average years.

The results of field fertilizer trials are generally regarded to be strictly applicable only to a given soil type and to a given locale. These limitations would not be serious if the results accurately predicted the best treatment year after year for a given soil and a given locale. Unfortunately, the best treatment on the average is not necessarily the best one for any given year.

PLANT FACTORS

Varietal Differences

Differences in yields of varieties are well documented and well known; they are too numerous to consider in this regard.

Varieties versus Nutrition and Efficiency of Utilization

Nutritional mutants in *Neurospora crassa* and *N. sitophila* (Beadle and Tatum, 1945) and other microorganisms (Tatum, 1945; Davis, 1950) have been studied in detail, but there has been little interest in the possibility of selecting higher plants according to variations in nutrient uptake and assimilation (Vose, 1963). Varieties grown under the same conditions but having different yields must clearly differ in their efficiencies in dry matter production, described by Vose (1963) as "differential yield response." Vose also noted that such varieties may differ markedly in "differential nutrient uptake," as shown by the concentrations and total amounts of elements in the shoots and roots.

At one time there was considerable interest in the water requirement differences of crops and/or varieties. It appears equally desirable to explore varieties for possible differences in their capacities to absorb and/or to utilize nutrients under different fertility levels.

There are various possible mechanisms that might be involved in the nutritional variations among varieties (Vose, 1963). The number and type of roots may have an effect on the nutrition. Smith (1934) reported that P efficiency of corn varieties was directly related to the ratio of secondary to primary roots; the higher the ratio, the more efficient the variety. Cation exchange capacity of roots and the factors that affect carriers may also be important.

Organic Acids

Noggle (1966) reported a positive correlation between yield and concentration of organic acids in cells of plants. Nutrient treatments that increased organic acids also increased yields, and those that decreased organic acids decreased yields.

SOIL FACTORS

Aeration

The percentage of O_2 in soil atmosphere may vary depending on the type of soil. Thus Leonard (1945) reported the following percentages of O_2 in Houston clay, Sarpy fine sandy loam, and Ruston sandy loam: 0–21, 10–21, and 18–21%, respectively.

There are numerous reports to the effect that low O_2 partial pressures around roots retard their growth (Table 18.1) (Livingston and Free, 1917; Cannon, 1925; Loehwing, 1931; Boynton, 1939; Vlamis and Davis, 1944; Erickson, 1946; Leonard and Pinckard, 1946; Hopkins et al., 1950). Plants often have few if any root hairs under conditions of low-O_2 supply (Loehwing, 1931, 1934).

Hopkins et al. (1950) noted that root growth of tomato, soybean, and tobacco stopped at 0.5% O_2 in the gas around the roots; top growth continued. In general, growth of tops, roots, a selected leaf, and stems increased as the O_2 concentration was increased from 0.5 to 6.4% and was often best of all with forced aeration (21% O_2). In pea roots there was an almost linear correlation between aeration and root elongation, but branching increased as rate of aeration decreased (Geisler, 1965).

The degree of soil aeration may affect not only absorption of salts but also the loss of organic substances from roots, particularly the tips. Loss of organic substances from cotton radicles was enhanced by chilling, low pH, or anaerobic conditions (Christiansen et al., 1970). Ca^{2+} or Mg^{2+} reversed or prevented this loss when induced by chilling or anaerobiosis but not when induced by low pH.

Especially when aeration around roots is at a low level, the form in which N

Table 18.1

Fresh Weights and K Concentration of Expressed Sap of Plants Grown 6 Wk in Drained and Submerged Clay[a, b]

		Shoots		Roots	
Plant	Soil treatment	Fresh weight (g)	K concentration (meq/liter)	Fresh weight (g)	K concentration (meq/liter)
Barley	Drained	18.0	167	2.2	57
	Submerged	2.0	—	0.2	—
Tomato	Drained	12.2	—	1.2	—
	Submerged	2.7	—	1.4	—
Lowland	Drained	20.0	196	9.2	55
rice	Submerged	34.5	209	38.6	32

[a] Vlamis and Davis (1944).

[b] Submerged plant roots were gradually exposed to increasing soil moisture until the water level was 1 in. above the soil surface. Plants of both sets were frozen and sap expressed and analyzed for K concentration.

is supplied can be important with regard to root growth and salt absorption. Barley grew as well under anaerobic as aerobic conditions provided the NO_3^- concentration was increased severalfold under the anaerobic condition (Woodford and Gregory, 1948). There is good evidence that under such conditions roots obtain appreciable quantities of O_2 from NO_3^- when N is supplied in this form rather than as the NH_4^+ ion (Arnon, 1937; Gilbert and Shive, 1945).

The requirement for O_2 by microorganisms actively decomposing organic matter is comparatively high (Page and Bodman, 1951). This may explain the commonly observed depression in plant growth that occurs immediately after the addition of large amounts of organic matter, since plant growth may be impeded by a deficiency of O_2 when plants are in competition with microorganisms for available O_2.

Soil Moisture

In crested wheatgrass seeds some phosphorylation occurred with a water potential of -880 atm, but the seeds did not incorporate ^{32}P into NAD, ATP, and uridine diphosphate until a water potential of -130 atm was reached (Wilson and Harris, 1968).

When soil moisture is limiting, root growth is generally reduced and, along with this reduction, absorption of nutrients is restricted. Low soil moisture limits absorption of some nutrients more than others, for example, B. However, when soil spaces are filled with water, soil O_2 is drastically reduced and so is nutrient uptake. For an excellent, detailed coverage of the effects of soil moisture on the mineral nutrition of plants, the review by Wadleigh and Richards

(1951) should be consulted. Kozlowski (1968a,b) provides thorough coverage of the relationship between soil water and plant growth.

C/N Ratio of Organic Matter

The C/N ratio of organic matter added to a soil is of considerable importance in crop growth and yield. If the applied organic matter contains too little N, with respect to C, microorganisms will tie up available N to the detriment of the plants. Plant residues with an initial C/N ratio of approximately 35:1 contain adequate N for decomposition (Norman, 1951).

Excesses of Salts (See Chapters 15 and 16)

Organic Toxins and Root Excretions

According to Reed (1908), plant physiologists have been interested in the question of root excretions since the time of De Candolle around 1850. De Candolle's Theory of Root Excretions proposed that excreted substances have a deleterious effect when absorbed by other plants belonging to the same order as the plants from which the excretions came, but that these substances are harmless or even beneficial to plants belonging to a different order. Reed (1908), while believing in the occurrence of root excretions, was of the opinion that in most soils kept in "good tilth" root excretions do not accumulate to a harmful extent. He was of the opinion that oxidation and the action of soil microorganisms destroys these deleterious substances.

Schreiner and Lathrop (1911c) examined soils from 18 states, which showed varying fertility and infertility. When the soils were separated into good and poor soils, the relationship of dihydroxystearic acid with poor soils was striking. The acid was found in every infertile soil. They concluded that it was not possible to state that this acid was the only factor contributing to infertility or unproductivity, but that it was a readily recognized symptom of poor soil conditions.

The subject of root toxins is still discussed, and Mojé (1966) has described organic compounds actually isolated and identified in soils, root excretions, and plant residues that have been shown to be toxic to plant growth. He also discusses the history of the subject, including the early work by Livingston and also by Schreiner and his colleagues. Livingston et al. (1905) suggested that "so-called exhausted soils are poisoned soils, and crop rotation is beneficial in agriculture because it prevents the accumulation in the siol of the injurious excreta of any one form of plant life." Shortly following this pronouncement, Schreiner and his colleagues were very active champions of toxic compounds in soils as frequent causes of poor plant growth and yield (Schreiner and Reed, 1907, 1909; Schreiner et al., 1907; Schreiner and Sullivan, 1908, 1909; Schreiner and Shorey, 1908a, b, 1909; Schreiner and Lathrop, 1911a, b, c; and Schreiner, 1923). Dihydroxystearic acid, in particular, received much attention, from the early 1900s, as a phytotoxic compound in soils (Mojé, 1966).

Subsoil Compaction, Acidity, and Nutrient Deficiencies

Drought damage to crops often occurs in the southeastern United States even in areas with a rainfall of 50 in. per year as reported by Pearson et al. (1963). It was suggested that roots did not grow deeply enough into the subsoil. These investigators concluded that poor root growth into subsoil might be caused by any one or more of the following factors: (1) subsoil acidity (often too high in Al^{3+} or Mn^{2+}); (2) subsoil nutrient deficiencies (e.g., cotton roots do not grow into a nonacid subsoil containing all essential nutrients but lacking in Ca^{2+}); and (3) inability of roots to penetrate a compacted layer of soil just below plow depth.

Bases on Exchange Complex

High-yielding capacity of soils, for alfalfa, appears to be associated with a high concentration of Ca^{2+} in the soil and a high proportion of Ca^{2+} on the base exchange complex of colloids (Bear and Prince, 1945).

ABOVE-GROUND FACTORS

Night Temperature

Went (1944a, b, 1945) demonstrated that for tomatoes stem elongation is optimal and fruit set abundant only at night temperatures between 15 and 20°C. Above 18°C there is too little translocation of sugar; below 18°C rates of growth processes become limiting (Went, 1944b). The effect of night temperature on stem elongation was later confirmed in a comparison between field- and greenhouse-grown (air-conditioned) tomatoes (Went and Cosper, 1945).

The optimal night temperature may be a function of the stage of growth and variety, since Went (1945) reported a shift in the optimal night temperature from 30°C for small plants to 18°C for San Jose Canner tomato plants in the early fruiting stage, and from 30 to 13°C for Illinois T-19 tomatoes.

Photoperiod

Length of the photoperiod has been shown to affect the nutrition of higher plants (Hamner, 1938; Neidle, 1939). Photoperiod also affected the growth and reproduction of *Marchantia polymorpha* in a study of its nutritional requirements and responses (Voth and Hamner, 1940). Irrespective of nutrient supply, plants on short photoperiod produced a greater total number of gemma cups, but on long photoperiods a greater number of gametangiophores, than comparable plants on short photoperiod.

Carbon Dioxide

Three factors have been listed as controlling the rate of photosynthesis: (1) intensity of illumination; (2) temperature of leaf; and (3) concentration of CO_2 in the air (Blackman and Matthaei, 1905). It was concluded that in nature the limiting CO_2 concentration of the atmosphere prevents the high rates of photosynthesis obtained experimentally in the presence of additional CO_2. Recent work, particularly in the greenhouse, also indicates that greater plant growth and yield often occur when additional CO_2 is supplied.

LITERATURE CITED

Arnon, D. I. 1937. Ammonium and nitrate nitrogen nutrition of barley at different seasons in relation to hydrogen ion concentration, manganese, copper, and oxygen supply. Soil Sci. 44:91–121.

Atkinson, H. J., L. M. Patry, and R. Levick. 1948. Plant tissue testing. III. Effect of fertilizer applications. Sci. Agr. 28:223–228.

Beadle, G. W., and E. L. Tatum. 1945. *Neurospora*. II. Methods of producing and detecting mutations concerned with nutritional requirements. Amer. J. Bot. 32: 678–686.

Bear, F. E., and A. L. Prince. 1945. Cation-equivalent constancy in alfalfa. J. Amer. Soc. Agron. 37:217–222.

Blackman, F. F., and G. L. C. Matthaei. 1905. Experimental researches in vegetable assimilation and respiration. IV. A quantitative study of carbon-dioxide assimilation and leaf-temperature in natural illumination. Proc. Roy. Soc. B76:402–460.

Boynton, D. 1939. Soil atmosphere and the production of new rootlets by the apple tree root systems. Proc. Amer. Soc. Hort. Sci. 37:19–26.

Cannon, W. A. 1925. Physiological features of roots, with especial reference to the relations of roots to aeration of the soil. Carnegie Inst. Wash. Pub. 368.

Chapman, H. D. (ed.). 1966. Diagnostic criteria for plants and soils. Div. Agr. Sci., Univ. California, Berkeley, Calif. 793 pp.

Christiansen, M. N., H. R. Carns, and D. J. Slyter. 1970. Stimulation of solute loss from radicles of *Gossypium hirsutum* L. by chilling, anaerobiosis, and low pH. Plant Physiol. 46:53–56.

Chubb, W. O., and H. J. Atkinson. 1948. Plant tissue testing. II. A study of the method of foliar diagnosis. Sci. Agr. 28:49–60.

Davis, B. D. 1950. Studies on nutritionally deficient bacterial mutants isolated by means of penicillin. Experientia 6:41–50.

Emmert, E. M. 1934. Tests for phosphate, nitrate and soluble nitrogen in conducting tissue of tomato and lettuce plants, as indicators of availability and yield. Kentucky Agr. Exp. Sta. Circ. 43.

Emmert, E. M. 1942. Plant-tissue tests as a guide to fertilizer treatment of tomatoes. Kentucky Agr. Exp. Sta. Bull. 430.

Enfield, G. H. 1964. Soil testing gains ground. Plant Food Rev. 10:2—3, 9.

Erickson, L. C. 1946. Growth of tomato roots as influenced by oxygen in the nutrient solution. Amer. J. Bot. 33:551—561.

Geisler, G. 1965. The morphogenetic effect of oxygen on roots. Plant Physiol. 40: 85—88.

Gilbert, S. G., and J. W. Shive. 1945. The importance of oxygen in the nutrient substrate for plants — Relation of the nitrate ion to respiration. Soil Sci. 59:453—461.

Hamner, K. C. 1938. Correlative effects of environmental factors on photoperiodism. Bot. Gaz. 99:615—629.

Hill, H., and H. B. Cannon. 1948. Nutritional studies by means of tissue tests with potatoes grown on a muck soil. Sci. Agr. 28:185—199.

Hopkins, H. T., A. W. Specht, and S. B. Hendricks. 1950. Growth and nutrient accumulation as controlled by oxygen supply to plant roots. Plant Physiol. 25:193—209.

Hylton, L. O., A. Ulrich, and D. R. Cornelius. 1967. Potassium and sodium interrelations in growth and mineral content of Italian ryegrass. Agron. J. 59:311—314.

Kenworthy, A. L., and G. M. Gilligan. 1948. Interrelationships between the nutrient content of soil, leaves, and trunk circumference of peach trees. Proc. Amer. Soc. Hort. Sci. 51:209—215.

Kozlowski, T. T. 1968a. Water deficits and plant growth. Vol. I, Development, control and measurement. Academic, New York.

Kozlowski, T. T. 1968b. Water deficits and plant growth. Vol. II, Plant water consumption and response. Academic, New York.

Leonard, O. A. 1945. Cotton root development in relation to natural aeration of some Mississippi blackbelt and delta soils. J. Amer. Soc. Agron. 37:55—71.

Leonard, O. A., and J. A. Pinckard. 1946. Effect of various oxygen and carbon dioxide concentrations on cotton root development. Plant Physiol. 21:18—36.

Lilleland, O., and K. Uriu. 1960. Nutritional needs of fruit trees indicated by leaf analysis. California Agr. 14:12.

Livingston, B. E., and E. E. Free. 1917. The effect of deficient soil oxygen on the roots of higher plants. Johns Hopkins Univ. Circ. 293:183—185.

Livingston, B. E., J. C. Britton, and F. R. Reid. 1905. Studies on the properties of an unproductive soil. U. S. Dep. Agr. Bur. Soils Bull. 28:1—39.

Loehwing, W. F. 1931. Effects of soil aeration on plant growth and root development. Proc. Iowa Acad. Sci. 38:71—72.

Loehwing, W. F. 1934. Physiological aspects of the effect of continuous soil aeration on plant growth. Plant Physiol. 9:567—583.

Lorenz, O. A. 1965. Better vegetable yields and quality with plant analysis. Plant Food Rev. 11:2—4.

Mojé, W. 1966. Organic soil toxins, pp. 533—569. In H. D. Chapman (ed.) Diagnostic criteria for plants and soils. Div. Agr. Sci., Univ. California, Berkeley.

Neidle, E. K. 1939. Nitrogen nutrition in relation to photoperiodism in Xanthium pennsylvanicum. Bot. Gaz. 100:607—618.

Nightingale, G. T. 1942a. Nitrate and carbohydrate reserves in relation to nitrogen nutrition of pineapple. Bot. Gaz. 103:409—456.

Nightingale, G. T. 1942b. Potassium and phosphate nutrition of pineapple in relation to nitrate and carbohydrate reserves. Bot. Gaz. 104:191—223.

Noggle, J. C. 1966. Ionic balance and growth of sixteen plant species. Proc. Soil Sci. Soc. Amer. 30:763—766.

Norman, A. G. 1951. Role of soil microorganisms in nutrient availability, pp. 167—183. In E. Truog (ed.) Mineral nutrition of plants. Univ. Wisconsin Press, Madison.

Page, J. B., and G. B. Bodman. 1951. The effect of soil physical properties on nutrient availability, pp. 133–166. *In* E. Truog (ed.) Mineral nutrition of plants. Univ. Wisconsin Press, Madison.

Pearson, R. W., Z. F. Lund, C. R. Camp, Jr., and F. Adams. 1963. What restricts root growth? Agr. Res. 12:6–7.

Pettinger, N. A. 1931. The expressed sap of corn plants as an indicator of nutrient needs. J. Agr. Res. 43:95–119.

Reed, H. S. 1908. Modern and early work upon the question of root excretions. Popular Sci. Monthly 73:257–266.

Reuther, W. 1961. Plant analysis and fertilizer problems. Pub. 8. Amer. Inst. Biol. Sci., Washington, D. C. 454 pp.

Schreiner, O. 1923. Toxic organic soil constituents and the influence of oxidation. J. Amer. Soc. Agron. 15:270–276.

Schreiner, O., and E. C. Lathrop. 1911a. Dihydroxystearic acid in good and poor soils. J. Amer. Chem. Soc. 33:1412–1417.

Schreiner, O., and E. C. Lathrop. 1911b. Examination of soil for organic constituents, especially dihydroxystearic acid. U. S. Dep. Agr. Bur. Soils Bull. 80:1–33.

Schreiner, O., and E. C. Lathrop. 1911c. Examination of soils for organic constituents, especially dihydroxystearic acid. U. S. Dep. Agr. Bur. Soils Bull. 80.

Schreiner, O., and H. S. Reed. 1907. Some factors influencing soil fertility. U. S. Dep. Agr. Bur. Soils Bull. 40:5–40.

Schreiner, O., and H. S. Reed. 1909. The role of oxidation in soil fertility. U. S. Dep. Agr. Bur. Soils Bull. 56:7–52.

Schreiner, O., and E. C. Shorey. 1908a. The isolation of picoline carboxylic acid from soils, and its relation to soil fertility. J. Amer. Chem. Soc. 30:1295–1307.

Schreiner, O., and E. C. Shorey. 1908b. The isolation of dihydroxystearic acid from soils. J. Amer. Chem. Soc. 30:1599–1607.

Schreiner, O., and E. C. Shorey. 1909. The isolation of harmful organic substances from soils. U. S. Dep. Agr. Bur. Soils Bull. 53:5–53.

Schreiner, O., and M. X. Sullivan. 1908. Toxic substances arising during plant metabolism. Proc. Amer. Soc. Biol. Chem. 1907–1908(4):26–27.

Schreiner, O., and M. X. Sullivan. 1909. Soil fatigue caused by organic compounds. J. Biol. Chem. 6:39–50.

Schreiner, O., H. S. Reed, and J. J. Skinner. 1907. Certain organic constituents of soil in relation to soil fertility. U. S. Dep. Agr. Bur. Soils Bull. 47:7–52.

Shear, C. B., and H. L. Crane. 1947. Nutrient-element balance, U. S. Dep. Agr. Yearbook 1943–1947:592–601.

Shear, G. M. 1943. Plant tissue tests versus soil tests for determining the availability of nutrients for tobacco. Virginia Agr. Exp. Sta. Tech. Bull. 84.

Sims, G. T., and G. M. Volk. 1947. Composition of Florida-grown vegetables. I. Mineral composition of commercially-grown vegetables in Florida as affected by treatment, soil type and locality. Florida Agr. Exp. Sta. Tech. Bull. 438.

Smith, S. N. 1934. Response of inbred lines and crosses in maize to variations of nitrogen and phosphorus supplied as nutrients. J. Amer. Soc. Agron. 26:785–804.

Stout, P. R., and R. Overstreet. 1950. Soil chemistry in relation to inorganic nutrition of plants. Annu. Rev. Plant Physiol. 1:305–342.

Tatum, E. L. 1945. X-Ray induced mutant strains of *Escherichia coli*. Proc. Nat. Acad. Sci. 31:215–219.

Tucker, T. C. 1965. The cotton petiole – Guide to better fertilization. Plant Food Rev. 11:9–11.

Tyner, E. H. 1946. The relation of corn yields to leaf nitrogen, phosphorus, and potassium content. Proc. Soil Sci. Soc. Amer. 11:317–323.

Ulrich, A. 1952. Physiological basis for assessing the nutritional requirements of plants. Annu. Rev. Plant Physiol. 3:207–228.

Ulrich, A. 1961. Plant analysis in sugar beet nutrition, pp. 190–211. *In* W. Reuther (ed.) Plant analysis and fertilizer problems. Pub. 8. Amer. Inst. Biol. Sci., Washington, D.C. 454 p.

Ulrich, A., and W. L. Berry. 1961. Critical phosphorus levels for lima bean growth. Plant Physiol. 36:626–632.

Vandecaveye, S. C. 1948. Biological methods of determining nutrients in soil, pp. 199–230. *In* H. B. Kitchen (ed.) Diagnostic techniques for soils and crops (Their value and use in estimating the fertility status of soils and nutritional requirements of crops). Amer. Potash Inst., Washington, D. C. 308 pp.

Vlamis, J., and A. R. Davis. 1944. Effects of oxygen tension on certain physiological responses of rice, barley, and tomato. Plant Physiol. 19:33–51.

Vose, P. B. 1963. Varietal differences in plant nutrition. Herbage Abstr. 33:1–13.

Voth, P. D., and K. C. Hamner. 1940. Responses of *Marchantia polymorpha* to nutrient supply and photoperiod. Bot. Gaz. 102:169–205.

Wadleigh, C. H., and L. A. Richards. 1951. Soil moisture and the mineral nutrition of plants, pp. 411–450. *In* E. Truog (ed.) Mineral nutrition of plants. Univ. Wisconsin Press, Madison. 469 pp.

Went, F. W. 1944a. Plant growth under controlled conditions. II. Thermoperiodicity in growth and fruiting of the tomato. Amer. J. Bot. 31:135–150.

Went, F. W. 1944b. Plant growth under controlled conditions. III. Correlation between various physiological processes and growth in the tomato plant. Amer. J. Bot. 31: 597–618.

Went, F. W. 1945. Plant growth under controlled conditions. V. The relation between age, light, variety and thermoperiodicity of tomatoes. Amer. J. Bot. 32:469–479.

Went, F. W., and L. Cosper. 1945. Plant growth under controlled conditions. VI. Comparison between field and air-conditioned greenhouse culture of tomatoes. Amer. J. Bot. 32:643–654.

Wilson, A. M., and G. A. Harris. 1968. Phosphorylation in crested wheatgrass seeds at low water potentials. Plant Physiol. 43:61–65.

Wolf, B., and V. Ichisaka. 1947. Rapid chemical soil and plant tests. Soil Sci. 64:227–244.

Woodford, E. K., and F. G. Gregory. 1948. Preliminary results obtained with an apparatus for the study of salt uptake and root respiration of whole plants. Ann. Bot. 12: 335–370.

Author Index *

* Numbers in brackets indicate the page on which the complete reference appears. Other numbers indicate the page on which it is cited.

Subject Index